LEAVING CERTIFICATE

CHEMISTRY LIVE!

DECLAN KENNEDY ***MSc, MEd, HDE, FICI***

FOLENS

Editor
Aidan Culhane
Design and Layout
Gary Dermody, Liz Murphy, Paula Byrne
Illustrations
Michael Phillips
Picture Research
Sinéad Kelleher

ISBN 1 84131 392 0

Folens Publishers, Hibernian Industrial Estate, Greenhills Road, Tallaght, Dublin 24.

The publishers would like to acknowledge the following for permission to reproduce material and for allowing photographs to be taken on their premises:

Seán Downey, Aughinish Alumina, Colaiste Muire Cobh, Irish Fertiliser Industries, Premier Periclase Ltd, Bob Gillen, Ballincollig Sewage Treatment Plant, Cork County Council, Seán Finn, Irish Ispat Ltd, Pat Murphy, Inniscarra Water Treatment Plant, Forensic Science Laboratory Dublin, Dr. John Fox, Brian Kennedy, Alan Kennedy, Trustee Savings Bank Cobh, Donncha O'Connell, Therese O'Donovan, Chemistry Department UCC, Physics Department UCC, KB Scientific, Kevin Lawless, Lennox Laboratories, Murty O'Dowd, Water Technology Ltd., Whitegate Oil Refinery, Ann Woods, Wellcome Trust Photo Library, Science Photo Library, Science and Society Picture Library, Corbis Images Ltd, Slidefile, UKAEA, Allsport Photographic Ltd, Irish Farmers' Journal, The Telegraph Colour Library, Johnson Matthey Ltd, I.C.I., Newbridge Cutlery Co. Ltd, Holt Studios International, Banana Rite Ltd, Shout Picture Library, Golden Vale Plc, Pfizer Pharmaceuticals, The Anthony Blake Photo Library, Topham Picturepoint, Finbarr O'Connell Photographer, Oxford Scientific Film.

Contents

PREFACE

This textbook is specifically written for the Leaving Certificate Chemistry syllabus introduced in September 2000. All of the core material for Higher and Ordinary level students is included in this textbook. All Higher Level material is indicated by a vertical line drawn down the left-hand side of the margin.

The two Options are included in the accompanying Workbook. Placing the Options in the Workbook has enabled me to include many worked examples throughout the textbook. Only one of the two Options need be covered, i.e. either CHAPTER 25 and CHAPTER 26 (OPTION 1) or CHAPTER 27 and CHAPTER 28 (OPTION 2).

All of the Mandatory Student Experiments on the syllabus are included in this textbook. Full step-by-step instructions are given for each experiment. A full list of Mandatory Experiments is given on the following page. The accompanying Teachers Manual gives full details of the preparation of all reagents, lists of apparatus and chemicals required, worked solutions to numerical problems, background information, etc.

In keeping with the recommendations of the syllabus committee, I have not assumed any previous knowledge of chemistry among the students for whom this textbook is written. For this reason, the opening chapters of the book introduce students to the fundamental concepts before moving on to some of the major areas of the syllabus.

In line with modern trends in chemical education, throughout the textbook the symbol L is used to represent litres rather than the small letter l which may be confused with the number 1.

The use of the letters [LCH] and [LCO] after a question indicate that this question is taken either from a past Leaving Certificate Higher or Ordinary level paper, respectively. It is strongly recommended that students use the Workbook in conjunction with the textbook. Questions in the Workbook are prefixed by the letter W. These questions play an important role in testing the student and in preparing him/her for the Leaving Certificate examination. The permission of the Department of Education and Science to reproduce questions from past Leaving Certificate examination papers is acknowledged.

Writing this textbook has been a very challenging task and I am extremely grateful to the many people who have given me so much support and encouragement. I wish to express my sincere thanks and deep gratitude to Oliver Broderick and Janet Cotter who proof-read the entire manuscript and made very useful suggestions and comments on the material. I also wish to acknowledge the help of those who have supported me in so many ways in my teaching and those who proof-read selected chapters. The comments of the proof-readers have helped me greatly in the production of the final draft of this textbook.

Niall Brett, Diarmuid Cashman, Dr Peter Childs, Tim Cronin, John Daly, Kieran Dennehy, Dr Fiona Desmond, Sr Mercedes Desmond, Peter Desmond, Dr Francis Douglas, Seán Finn, Eileen Fitzpatrick, Pat Hanratty, Prof. Brian Hathaway, Jim Hayes, Reg Hayes, Goretti Healy, Prof. Aine Hyland, Dr Matthias Jauch, Kathleen Kelly, Michael King, Ann Lawrence, Sr Ann Lenihan, John Levis, Chris Lordan, Pat McCleery, Aidan McNally, Bro. Bede Minihane, Brian Murphy, Pat Murphy, Fergus O'Brien, Brid O'Connor, Ger O'Driscoll, Ted O'Keeffe, Celia O'Riordan, Don O'Shea, Gerry O'Shea, Joanne O'Shea, Michael Redmond, Terry Spacey, John Toomey, Sr Rita Twomey, Prof. David Waddington, Pat Walsh.

I wish to thank Seán Downey and Pat O'Connell, Department of Chemistry UCC, for their assistance in setting up and taking photographs of many of the experiments shown in this textbook. I also wish to express my thanks to Aidan Culhane, Gary Dermody, Paula Byrne, Liz Murphy, Sinead Kelleher and, of course, John O'Connor, Managing Director of Folens Publishers, whose persistence persuaded me to write this textbook. I am delighted with the high standard of artwork provided by Michael Phillips and I thank him for his excellent work.

Last but not least, a special word of thanks to Brian, Esther, Alan and Karen for all their support and encouragement.

I would like to dedicate this textbook to my late parents, Larry and Noreen Kennedy.

Declan Kennedy
August 2000

MANDATORY STUDENT EXPERIMENTS

The following is a list of Mandatory Student Experiments in the order in which they appear in the textbook. Experiments in bold are Higher Level only.

1. Flame tests (Li, Na, K, Ba, Sr, Cu only). EXPT. 3.2 PAGE 13
2. Recrystallisation of benzoic acid and determination of its melting point. EXPT. 5.1 PAGE 54
3. Tests for anions in aqueous solutions: chloride, carbonate, nitrate, sulfate, **phosphate, sulfite, hydrogencarbonate**. EXPT. 6.1 PAGE 71
4. Determination of the relative molecular mass of a volatile liquid (conical flask or gas syringe may be used). EXPT. 10.1 PAGE 117
5. Preparation of a standard solution of sodium carbonate. EXPT. 13.1 PAGE 153
6. Standardisation of a hydrochloric acid solution using a standard solution of sodium carbonate. EXPT 13.2 PAGE 158

6A. (Ordinary Level only) A hydrochloric acid/sodium hydroxide titration and its use to prepare the salt sodium chloride. EXPT 13.3 PAGE 165

7. **Determination of the concentration of ethanoic acid in vinegar**. EXPT. 13.4 PAGE 166
8. **Determination of the amount of water of crystallisation in hydrated sodium carbonate**. EXPT. 13.5 PAGE 169
9. Redox reactions of Group VII elements – halogens as oxidising agents (reactions with bromides, iodides, Fe^{2+} and sulfites). Displacement reactions of metals (Zn with Cu^{2+}, Mg with Cu^{2+}). EXPT. 14.1 PAGE 183
10. **A potassium manganate(VII) / ammonium iron(II) sulfate titration.** EXPT. 15.1 PAGE 193
11. **Determination of the amount of iron in an iron tablet.** EXPT. 15.2 PAGE 197
12. **An iodine/thiosulfate titration**. EXPT 15.3 PAGE 202
13. **Determination of the percentage (w/v) of hypochlorite in bleach.** EXPT 15.4 PAGE 204
14. Monitoring the rate of production of oxygen from hydrogen peroxide using manganese dioxide as a catalyst. EXPT. 16.1 PAGE 211
15. Studying the effects on the reaction rate of (i) concentration, (ii) temperature, using sodium thiosulfate solution and hydrochloric acid. EXPT. 16.2 and EXPT. 16.3 PAGE 217 AND PAGE 220
16. Simple experiments to illustrate Le Chatelier's Principle:
    (i) $CoCl_4^{2-} + 6H_2O \rightleftharpoons Co(H_2O)_6^{2+} + 4Cl^-$ (to demonstrate the effects of both temperature changes and concentration changes on an equilibrium mixture).
    (ii) $Cr_2O_7^{2-} + H_2O \rightleftharpoons 2CrO_4^{2-} + 2H^+$ (iii) $Fe^{3+} + CNS^- \rightleftharpoons Fe(CNS)^{2+}$
    [(ii) and (iii) to demonstrate the effects of concentration changes on an equilibrium mixture.] EXPT. 17.1 PAGE 236
17. **Estimation of total hardness using ethylenediaminetetraacetic acid (edta). (Balanced ionic equation required.)** EXPT. 19.1 PAGE 267
18. Determination of total suspended and dissolved solids (expressed as p.p.m.) by filtration and evaporation respectively. Determination of pH. EXPT.19.2 PAGE 271
19. **Estimation of dissolved oxygen by redox titration.** EXPT. 19.3 PAGE 277
20. Colorimetric experiment to estimate the concentration of free chlorine in swimming-pool water or bleach (using a colorimeter or a comparator). EXPT. 19.4 PAGE 284
21. Preparation and properties of ethene [combustion, tests for unsaturation using bromine water and acidified potassium manganate(VII) solution]. EXPT. 21.1 PAGE 310
22. Preparation and properties of ethyne [combustion, tests for unsaturation using bromine water and acidified potassium manganate(VII) solution.] EXPT. 21.2 PAGE 313
23. Determination of the heat of reaction of hydrochloric acid with sodium hydroxide. EXPT. 21.4 PAGE 329
24. Extraction of clove oil from cloves (or similar alternative) by steam distillation. EXPT. 22.2 PAGE 357
25. Preparation of soap. EXPT. 23.1 PAGE 365
26. Preparation and properties of ethanal [properties limited to reactions with (i) acidified potassium manganate(VII), (ii) Fehling's reagent and (iii) ammoniacal silver nitrate]. EXPT. 23.2 PAGE 372
27. Preparation and properties of ethanoic acid (properties limited to reactions with sodium carbonate, magnesium **and ethanol**). EXPT. 23.3 and **EXPT. 22.1** PAGE 376 AND PAGE 354
28. Separation of a mixture of indicators or coloured substances using paper chromatography or thin-layer chromatography or column chromatography. EXPT. 23.4 PAGE 384

Chemistry – The Creative Science

1

1.1 INTRODUCING CHEMISTRY

You are about to begin your study of chemistry at senior level. If you were stopped in the street and asked 'what is chemistry?', you might have to think a little before you answered. If you were to look up a definition of chemistry in a dictionary, you would find that **chemistry is a branch of science concerned with the structure and composition of materials and the changing of one substance into another**. Many careers involve a knowledge of chemistry – medicine, food science, materials science, engineering (civil, electrical, chemical, electronic, etc.), agricultural science, pharmacy, veterinary science, biotechnology, etc. (FIG 1.1).

Fig 1.1
Chemistry is used in medicine, engineering, electronics, forensic science and many more areas.

Everything around us is made from chemicals, from the food that we eat to the clothes that we wear and the air that we breathe. From the dawn of civilisation people have been changing certain substances into new substances. The production of metals, glass, dyes, etc. date back thousands of years. Chemistry as we know it, however, is a relatively new branch of science established about 400 years ago when it broke its links with the alchemists of old.

The most obvious impact of chemistry in our lives is the huge variety of new materials that we see all around us. Can you imagine life without modern medicines, synthetic clothes, plastics, petrol, detergents, soap, etc? There would be no modern medicines to relieve pain or cure infections, no unbreakable bottles, no hi-fi equipment, no CDs or cassettes, no colourful paints, no easy-clean worktops, etc.

Chemistry has made enormous improvements to the quality of our lives and has increased our life expectancy. We can now expect to live a lot longer than our ancestors. Chemistry influences our lives more than any other single technology.

1.2 LEAVING CERTIFICATE CHEMISTRY

The Leaving Certificate chemistry course covers many areas of the subject. It is designed to provide a relevant and modern course for those of you who will end your study of chemistry at Leaving Certificate level. It is also designed to provide a foundation course in chemistry for those of you who will continue your studies in chemistry or in related subjects at third level. The student practical work specified on the syllabus will go far to ensure that you develop important skills in laboratory procedures and techniques (FIG 1.2).

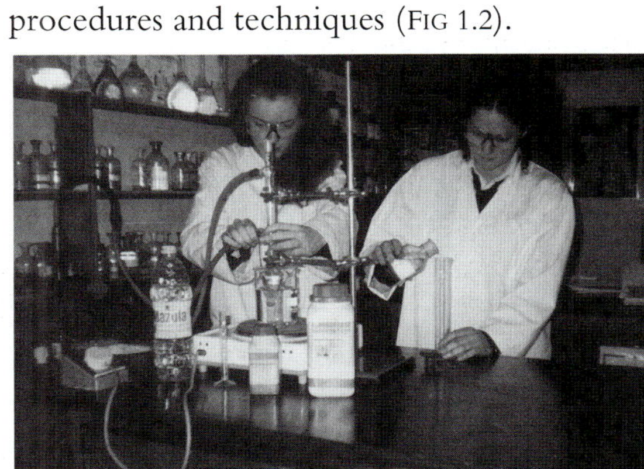

Fig 1.2
These students are preparing soap in the school laboratory.

In writing this textbook, great care has been taken to present the material as clearly as possible and in a logical sequence. It is hoped that you, the student, will find the study of chemistry both interesting and challenging. However, you may find some parts of the course more interesting than others! The first few chapters of this textbook are rather theoretical in nature. It is necessary to have this solid theoretical foundation in chemistry before proceeding further. The forensic scientist, fascinated with the study of the clues from the scene of a crime, at some stage had to study topics like chemical bonding! Once the theoretical foundation of chemistry has been laid, we will then move on to study topics like water analysis, oil refining, environmental chemistry, industrial chemistry, etc.

It is important that you work hard at your studies and written exercises. Study consistently and try to understand the various ideas presented as you work through the course. Don't leave all your study to the last few weeks before the examination! Get plenty of practice for your examination using the written exercises at the end of each chapter and in the accompanying workbook. You are about to study one of the most fascinating subjects of all - I hope you will enjoy it!

The Atom

2

2.1 EVIDENCE FOR THE EXISTENCE OF SMALL PARTICLES

Many observations in chemistry suggest that the materials around us are made up of small particles. The scientific name for materials is **matter**. Thus, the idea that materials are made up of small particles is often referred to as the **particulate nature of matter**. Consider the experiment shown in FIG 2.1.

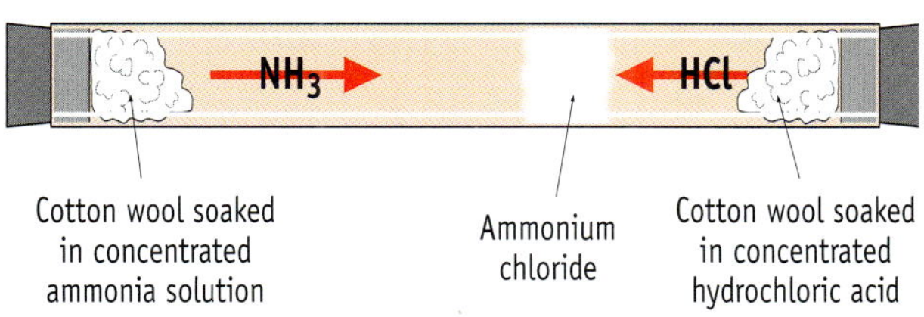

Fig 2.1
When the particles of ammonia meet the particles of hydrochloric acid (hydrogen chloride), a white ring of a substance called ammonium chloride is formed.

Using a tongs, a ball of cotton wool soaked in concentrated hydrochloric acid (hydrogen chloride solution) is placed in one end of a glass tube. At the same time, another ball of cotton wool, soaked in concentrated ammonia solution, is placed in the other end. After a short while, a white ring is formed as shown in FIG 2.1. Chemists explain this observation by saying that tiny particles of the ammonia gas moving or spreading from left to right meet up with particles of hydrogen chloride gas moving from right to left. A white cloudy substance is formed where the two types of particles meet. The spreading of gases in this way is often referred to as **diffusion**.

Similarly, consider what happens if a big crystal of copper sulfate is placed at the bottom of a large graduated cylinder filled with water. After a few hours the blue colour has spread throughout the water. We say that the small particles of copper sulfate have diffused throughout the water. Diffusion may also be demonstrated by placing a small amount of ink in water. After a short while the colour of the ink will have spread throughout the water.

The concept of the particulate nature of matter is not new in chemistry. You may be surprised to know that it was suggested as far back as 400 BC!

2.2 HISTORY OF THE ATOM *

It was the Greek philosophers (about 400 BC) who first proposed that matter was composed of minute particles. The Greek philosophers did not carry out any experiments – they thought manual work was demeaning. Their ideas arose from a certain amount of intuition and speculation arising from their logical way of reasoning. They believed that the tiny particles of which all matter was composed were so small that nothing smaller was possible. In other words, they were **indivisible**. The Greek word for 'indivisible' is 'atomos' from where we get the word **atom** to describe these indivisible particles.

Fig 2.2
John Dalton proposed a theory about atoms in 1808.

Because of the lack of experimental evidence, the idea of the existence of atoms was not developed for many years. In 1808, an English chemist called **John Dalton** (1766–1844) wrote a book in which he put forward an Atomic Theory to explain the results of a number of experiments he had performed with gases (FIG 2.2).

* It is not necessary for students to remember any of the dates quoted in this and subsequent chapters. They are given merely to help put various discoveries in their proper historical context.

Dalton's Atomic Theory may be summarised as follows:

1. All matter is made up of very small particles called atoms.
2. All atoms are indivisible. They cannot be broken down into simpler particles.
3. Atoms cannot be created or destroyed.

Dalton's Atomic Theory was readily accepted by nineteenth-century chemists. Even though Dalton was a very humble man and shunned any form of glory, he became very famous and many foreign chemists travelled to Manchester to meet him. A painting of Dalton collecting marsh gas with his students hangs in the town hall in Manchester.

However, Dalton's picture of the atom began to be questioned as a result of experiments carried out towards the end of the nineteenth century.

2.3 DISCOVERY OF THE ELECTRON

In the 1860s, there was a lot of scientific research taking place into what happens when an electric current is passed through a glass tube containing a gas at a pressure low enough for the electric current to flow. Because of the low pressure, such tubes are often called vacuum tubes since most of the air inside the tube had been pumped out of it. It was observed that when the gas conducted the electric current, a glow or luminescence was observed inside the tube (Fig 2.3).

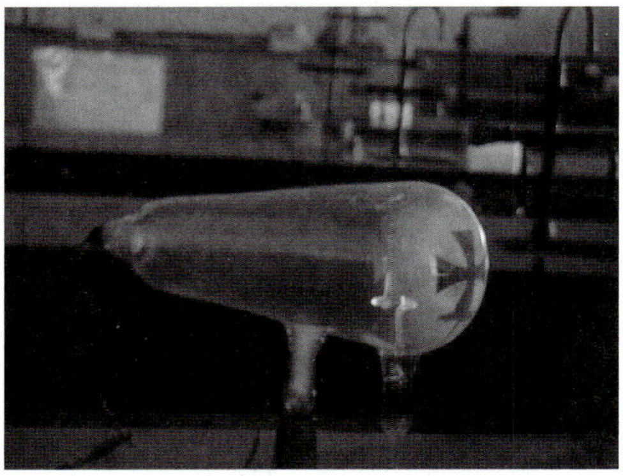

Fig 2.3
An electric current passing through air causes the air to glow.

In 1875, an English chemist called **William Crookes** (1832-1919) set about trying to discover what was happening inside the glass tube (Fig 2.4).

Crookes demonstrated that there was some sort of radiation coming from the end of the tube connected to the negative end of the battery. This negative metal plate inside the tube is called the **cathode**. He showed the existence of this radiation by placing a Maltese cross inside the tube. He found that a shadow was formed at the other end of the tube (Fig 2.5).

Fig 2.4
William Crookes investigated what happens when electricity is passed through gases.

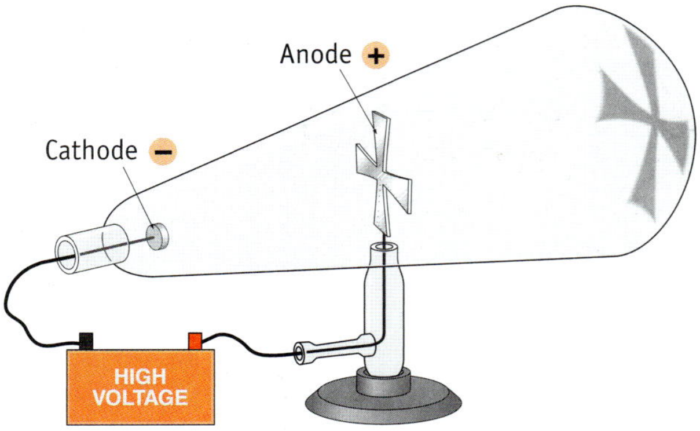

Fig 2.5
Crookes demonstrated that the rays coming from the negative electrode (cathode) cast a shadow at the far end of the tube.

Because the rays came from the cathode, they were called **cathode rays**. The cathode rays themselves are invisible but when they strike the glass they cause it to glow or fluoresce.

Crookes performed another experiment in which he placed a light paddle wheel, mounted on rails, in front of the cathode (Fig 2.6).

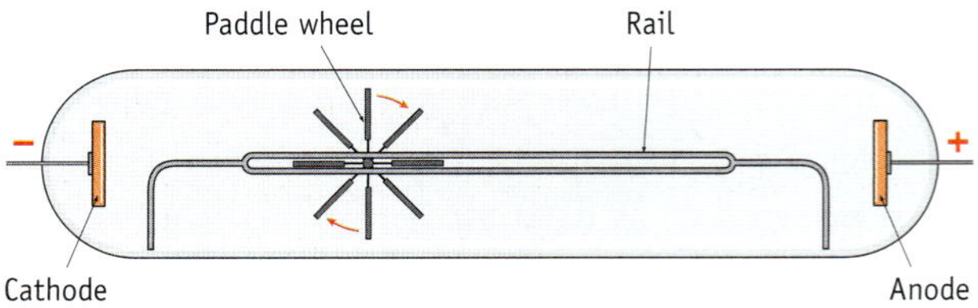

Fig 2.6
Crookes concluded that small particles which were coming from the cathode caused the paddle wheel to turn.

When the current was switched on, the paddle wheel rotated and travelled down the tube towards the positive electrode or **anode**. (To help you remember these terms think of the mnemonic CNAP – cathode negative, anode positive.) Crookes observed that the vanes always turned away from the cathode and concluded that these vanes were being struck by particles coming from the cathode.

In 1897, another English scientist called J.J. Thomson (FIG 2.7), working in the Cavendish Laboratory in Cambridge University, devised an experiment to investigate if these cathode rays consisted of charged particles. He obtained a narrow beam of cathode rays by passing the beam through a small hole in the positive electrode (FIG 2.8).

Fig 2.7
J.J. Thomson is credited with the discovery of the electron. He received the Nobel Prize in physics in 1906 for his work on the electron.

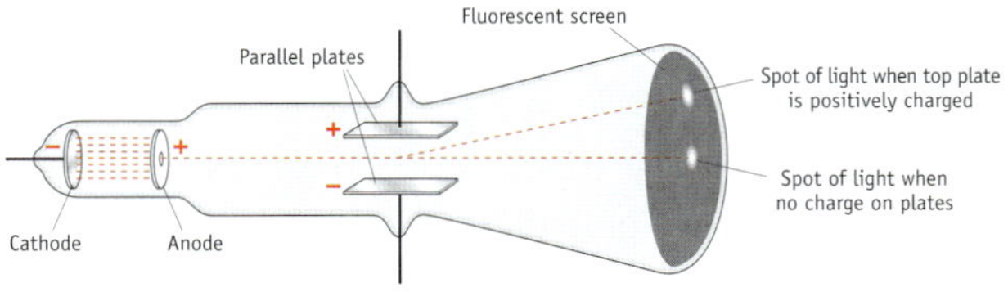

Fig 2.8
J.J. Thomson's apparatus which showed that the cathode rays consisted of negatively charged particles.

This narrow beam passed between two parallel metal plates about half way along the tube and then struck a fluorescent screen at the far end of the tube. Thomson found that, if there was no charge on the parallel plates, the beam of cathode rays passed straight down the tube and struck the fluorescent screen at the position shown in FIG 2.8. However, on placing a positive charge on the top plate, the position of the spot moved upwards as shown in FIG 2.8. **Thomson concluded that since the cathode rays were attracted up towards the positive plate, they consisted of negatively charged particles.** (Unlike charges always attract each other.) These negatively charged particles were called **electrons**. This name had been proposed as far back as 1891 by **George Stoney**, professor of physics in University College Galway. Stoney proposed that the smallest amount of electric charge be called an electron. When J.J. Thomson discovered that cathode rays consisted of negatively charged particles, the name electron was given to the particles rather than to the quantity of charge carried by them.

A modern version of J.J. Thomson's experiment is shown in FIG 2.9.

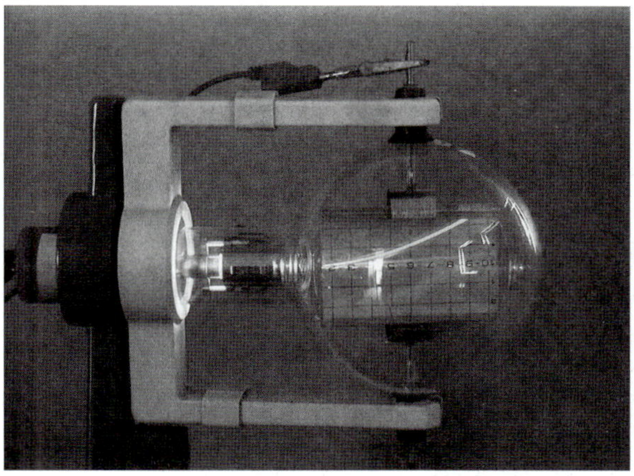

Fig 2.9
The negative beam of electrons is attracted up towards the positive plate. In this apparatus, fluorescent paper is used to show the path of the electron beam.

> Cathode rays are streams of negatively charged particles called electrons.

Thus, what is actually happening in the above experiment is that the high voltage is 'dragging' electrons out of the atoms of the cathode. These electrons then travel through the air to the positive electrode.

J.J. Thomson then went one step further with his experiment. He placed a large electromagnet outside the glass tube and adjusted the strength of the magnetic field. He found that the electrons were also deflected in the magnetic field. Using the magnetic field, he was able to bring the beam of electrons back down to its original position. From the results of his experiment, he was able to calculate the ratio of the charge of the electron to its mass. This ratio of charge to mass is often simply called the $\frac{e}{m}$ of the electron. Its value is given below but you do not have to remember the value.

$$\frac{\text{charge of electron}}{\text{mass of electron}} = 1.76 \times 10^{11} \text{ coulombs* per kg.}$$

From the results of his experiment, Thomson was not able to measure separately either the quantity of negative charge on the electron or its mass. The results of his experiment only gave him the ratio of charge to mass of the electron. Had he been able to measure one of these quantities, he could have worked out the second one mathematically from the value of the $\frac{e}{m}$ of the electron.

It took 12 years before an experiment was designed to measure the quantity of charge on the electron.

2.4 MILLIKAN'S OIL DROP EXPERIMENT

In 1909, an American physicist called **Robert Millikan** (FIG 2.10) set up an experiment to determine the size of the charge on the electron.

Fig 2.10
Robert Millikan measured the size of the charge on the electron. He received the Nobel Prize in physics in 1923 for this experiment.

Millikan sprayed tiny droplets of oil between two charged metal plates (FIG 2.11).

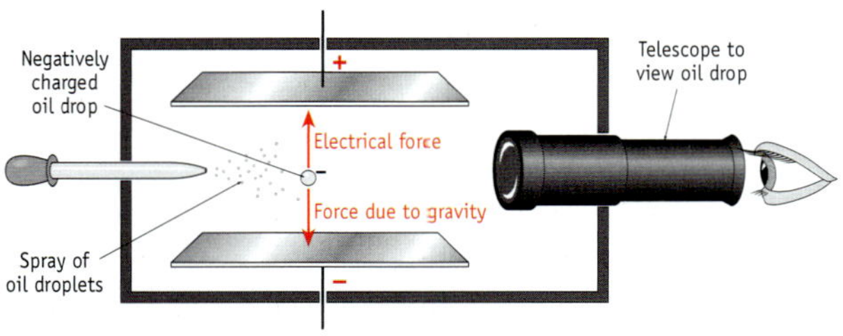

Fig 2.11
Millikan's Oil Drop experiment to measure the size of the charge on the electron.

* Coulombs are the units in which quantity of charge is measured.

He used X-rays to ionise the air between the plates, i.e. the molecules in the air lost electrons and formed ions. As the oil droplets fell through the air, they picked up these electrons. Thus, the oil droplets became negatively charged. Using a telescope, Millikan focused on a particular oil droplet as it fell through the air. He then placed a positive charge on the top plate and observed that the negatively charged oil droplet was attracted up towards the positive plate. He carefully adjusted the charge on the plate until the oil droplet was stationary, i.e. suspended in mid air! When the oil droplet was stationary, the weight of the oil droplet due to the force of gravity must have been equal to the electrical force pulling the droplet upwards. From various measurements that Millikan took, he was able to calculate the size of the charge on the electron. The value of the charge on the electron was measured as 1.6×10^{-19} coulomb.

Millikan substituted this value into the $\frac{e}{m}$ value obtained by Thomson:

$$\frac{\text{charge of electron}}{\text{mass of electron}} = 1.76 \times 10^{11}$$

$$\frac{1.6 \times 10^{-19}}{\text{mass of electron}} = 1.76 \times 10^{11}$$

From this equation, Millikan calculated the mass of the electron as 9.1×10^{-31} kg. The electron certainly has a very small mass! Thus, by 1909 the electron has been identified as a negatively charged particle of very small mass.

2.5 THOMSON'S 'PLUM PUDDING' MODEL OF THE ATOM

In 1898, J.J. Thomson proposed a very simple model of the structure of the atom. He proposed that, since atoms are neutral, there must be positive charges also in the atom to neutralise the negative electrons. He visualised the atom as a sphere with positive charges and electrons embedded in the sphere at random, i.e. something like raisins in a pudding. For this reason, Thomson's model of the atom became known as the 'Plum Pudding' model (FIG 2.12).

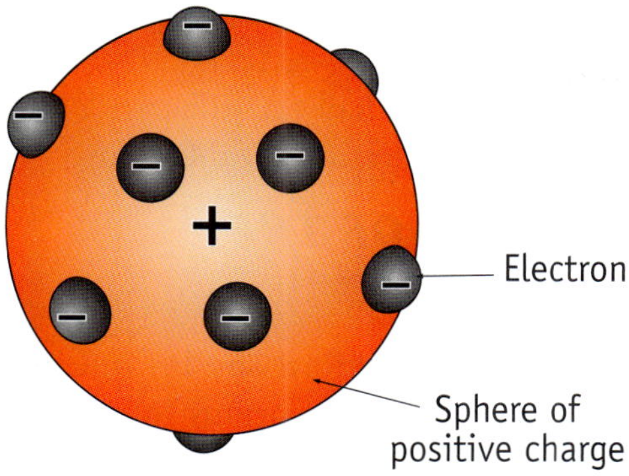

Fig 2.12
Thomson imagined an atom as a sphere of positive charge with negative electrons embedded in it at random. This model became known as the 'plum pudding' model of the atom.

The results of an interesting experiment performed in 1909 made scientists question Thomson's simple picture of the atom. This experiment is described in the next section.

2.6 DISCOVERY OF THE NUCLEUS

In 1909, two research students Hans Geiger and Ernest Marsden at the University of Manchester were studying the scattering of particles called alpha particles by very thin sheets of gold foil. Alpha particles are positively charged particles produced by certain radioactive substances. (Alpha particles are covered in more detail in CHAPTER 8.) The research work was being carried out under the supervision of **Ernest Rutherford**, a scientist from New Zealand who was particularly interested in radioactivity (FIG 2.13).

Fig 2.13
Ernest Rutherford (1871 – 1937) received the Nobel Prize in chemistry in 1908 for his work on radioactivity. In 1910, he and his research team performed a very important experiment which lead to the discovery of the nucleus.

The arrangement for the experiment is shown in FIG 2.14.

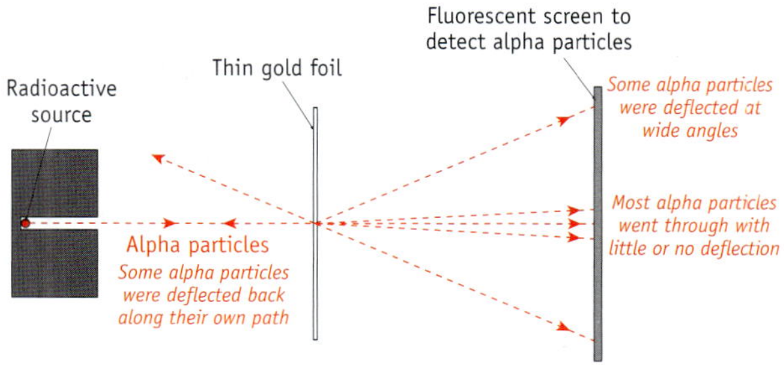

Fig 2.14
The experiment carried out by Rutherford and his research team to investigate how alpha particles were scattered by a piece of gold foil.

Rutherford and his fellow workers expected to find that the alpha particles would be deflected by small amounts as they passed through the thin gold foil (FIG 2.15). This small deflection would be what Thomson's 'Plum Pudding' model of the atom would predict.

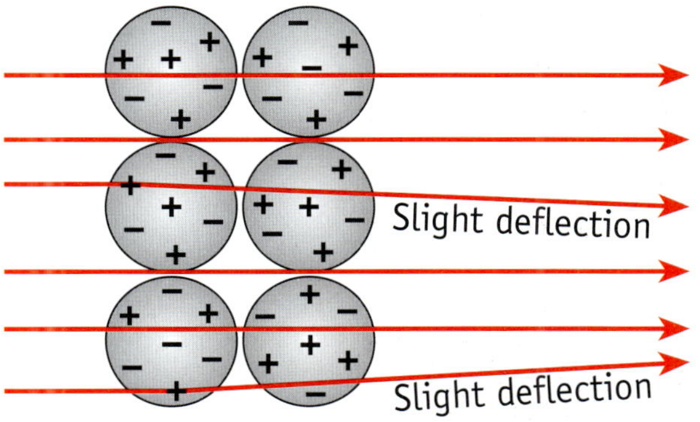

Fig 2.15
The expected result of the experiment if Thomson's model is correct. Most of the alpha particles should pass straight through the thin foil. Some alpha particles should suffer slight deflections.

Note in FIG 2.15 that, if the positive charge is distributed throughout the atom, the alpha particles will suffer only slight deflections when they meet one of these positive charges.

Rutherford's own description of what actually happened is worth quoting:

> *'Then I remember two or three days later Geiger coming to me in great excitement and saying "We have been able to get some of the alpha particles coming backwards." It was quite the most incredible event that has ever happened to me in my life. It was almost as incredible as if you fired a 15-inch shell at a piece of tissue paper and it came back and hit you.' (FIG 2.16)*

Fig 2.16
Rutherford was very surprised by the results of his experiment!

In short, Rutherford's research team found that most of the alpha particles went straight through the metal foil as expected. However, some of the alpha particles were deflected at large angles and a few were actually reflected back along their own path! This was totally unexpected. The only way to explain these results was to assume that the positive charge and the mass of the atoms of the metal foil were concentrated in a small dense core. This small dense core was called the **nucleus** of the atom (FIG 2.17).

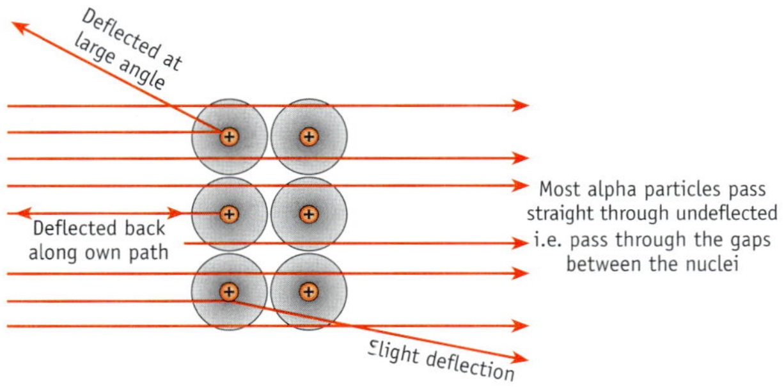

Fig 2.17
The large deflections can be explained if the mass and positive charge of the atom are concentrated in a small core called the nucleus.

Note in FIG 2.17 that most of the alpha particles pass straight through the foil because they never encounter the nucleus. However, occasionally one comes very close to the high concentration of positive charge in the nucleus. This high concentration of positive charge is immovable because of its mass. The positively charged alpha particles and the positively charged nucleus repel each other. This repulsion is so great as to cause the alpha particles to be deflected at large angles. In some cases, they bounce back along their original paths when they scored a direct hit with the nucleus.

From careful study of the scattering angles, Rutherford and his team deduced that the nucleus of an atom had a diameter of approximately $\frac{1}{100,000}$ of the size of the atom.

2.7 DISCOVERY OF THE PROTON

From 1911 to 1919, Rutherford and his research team continued bombarding various substances with alpha particles. They found that in the case of light atoms such as oxygen and nitrogen, small positively charged particles were given off as a result of the bombardment. This did not occur with the atoms of the heavier metals like gold. They explained their results by saying that the alpha particles were breaking up the nuclei of the lighter atoms to release these positively charged particles. The reason that this did not occur in the case of the heavier gold atoms was because the large, positive nuclear charge repelled the alpha particles before they had a chance to break up the nucleus. From about 1920 onwards, Rutherford began to refer to the small positive particles as **protons**. This term is still used to refer to the positive charges in the nucleus of the atom. By 1924, Rutherford and his co-workers had managed to knock protons out of the nuclei of most of the lighter elements.

2.8 DISCOVERY OF THE NEUTRON

From the results of his experiments on the scattering of alpha particles, Rutherford suspected that there was another particle in the nucleus of the atom. If only protons existed in the nucleus, the nucleus would fall apart because of the repulsive forces of the protons for each other. He reasoned that there must be some particles in the nucleus to help 'cement' the positively charged protons together. This other particle in

Fig 2.18
James Chadwick who received the Nobel Prize in physics in 1935 for discovering the neutron.

the nucleus was discovered by a scientist called **James Chadwick**, a former student of Rutherford (FIG 2.18).

In 1932, James Chadwick bombarded a sample of beryllium with alpha particles (FIG 2.19).

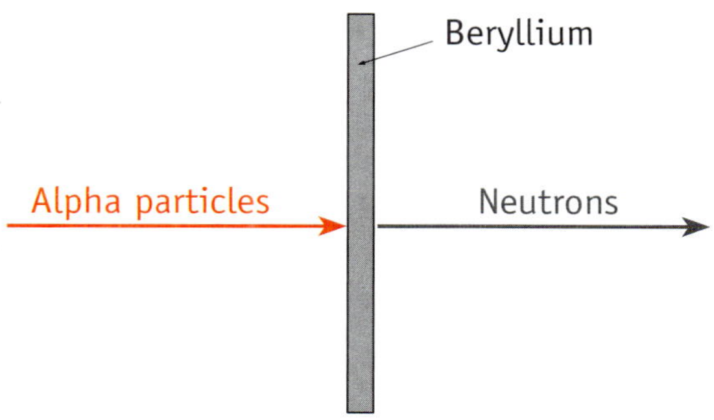

Fig 2.19
Alpha particles knocked neutrons out of the nuclei of beryllium atoms.

It was discovered that some type of radiation consisting of particles with no charge came from the beryllium. On further investigation, Chadwick concluded that the alpha particles were knocking these neutral particles out of the nuclei of the beryllium atoms. He found that these particles had about the same mass as the proton and, since they were neutral, he called them neutrons. The neutron proved to be one of the most important particles ever discovered. Neutrons were subsequently used to bring about the splitting of atoms of uranium to release nuclear energy in the atomic bomb and nuclear reactors.

2.9 PROPERTIES OF THE SUB-ATOMIC PARTICLES

With the discovery of the neutron in 1932, scientists were now aware of the existence of three particles inside in the atom. These three particles are often referred to as **sub-atomic** ('within the atom') **particles**. Protons and neutrons were located inside the nucleus and the electrons were located somewhere in the space outside the nucleus. (The location of the electrons is covered in more detail in CHAPTER 3.)

The charge on the electron was found to be -1.6×10^{-19} coulomb, that on the proton was $+1.6 \times 10^{-19}$ coulomb and, as we have seen, there is no charge on the neutron. The mass of the proton was found to be the same as the mass of the neutron, i.e. 1.67×10^{-27} kg. The mass of the electron was found to be much smaller at 9.1×10^{-31} kg. You are not required to remember any of these actual values but you should know the relative values as given in TABLE 2.1.

	Relative Charge	Relative Mass	Location
Proton	+1	1	Nucleus
Neutron	0	1	Nucleus
Electron	-1	$\frac{1}{1838}$	Outside Nucleus

Table 2.1
Summary of the properties of sub-atomic particles.

CHAPTER SUMMARY

- All matter consists of small particles.
- The spreading of particles is called diffusion.
- The idea of the atom originated with the ancient Greek philosophers.
- John Dalton said that atoms are indivisible and cannot be created or destroyed.
- William Crookes investigated cathode rays in vacuum tubes.
- Cathode rays consist of streams of electrons.
- J.J. Thomson showed that electrons are negatively charged and measured the $\frac{e}{m}$ of the electron.
- Robert Millikan measured the charge on the electron using his Oil Drop experiment.
- J.J. Thomson proposed a simple 'plum pudding' model of the atom.
- Rutherford and his co-workers discovered the nucleus of the atom and the existence of protons in the nucleus.
- James Chadwick discovered the neutron.

Exercises

2.1. (a) The scientific name for materials is ....

(b) The idea of atoms came originally from .... philosophers.

(c) An English chemist called .... proposed a theory about atoms in 1808.

(d) In 1875, an English chemist called .... studied the discharge of electricity through gases.

(e) The electrode connected to the negative end of the battery is called the ....

(f) The rays coming from the negative electrode are called..

(g) J.J. Thomson set up an experiment to investigate if cathode rays consisted of .... particles.

(h) J.J. Thomson showed that the cathode rays were attracted toward the .... charged plate.

(i) From the results of his experimental work, J.J. Thomson calculated the $\frac{e}{m}$ of the electron. What does $\frac{e}{m}$ represent? ....

(j) An American physicist called .... devised an experiment to measure the charge on the electron.

(k) Rutherford and his co-workers studied the scattering of .... particles by metal foils.

(l) Rutherford and his co-workers knocked protons out of the nuclei of some of the lighter elements using ....

(m) Name the scientist who discovered the neutron ....

(n) The neutron is an important particle in the area of nuclear energy. What is the reason for this?

....

(o) The mass of the proton is approximately .... times the mass of the electron.

2.2. What is meant by the term 'particulate nature of matter'?
Describe a laboratory experiment to demonstrate the diffusion of gases.
Explain how this experiment is evidence for the particulate nature of matter.

2.3. With regard to the development of atomic theory, write a brief note on the contribution of (a) John Dalton, (b) William Crookes and (c) James Chadwick.

2.4. Outline briefly the experiment performed by Rutherford and his co-workers that led to the discovery of the nucleus. Explain why their results were not consistent with Thomson's model of the atom.

Arrangement of Electrons in the Atom

3

3.1 INTRODUCTION

In the last chapter we saw how the proton, neutron and electron were discovered. We studied how scientists discovered that the protons and neutrons were concentrated in the nucleus of the atom. The next question to be answered concerned the location of the electrons. Ernest Rutherford proposed that the electrons were revolving around the nucleus just like the planets revolve around the sun. However, this simple picture defied the laws of physics. Electrons revolving around the nucleus would lose energy and would spiral into the nucleus, i.e. the atom would collapse. Clearly this does not happen. Therefore, some other model of the arrangement of electrons had to be devised. It was a young Danish physicist called **Niels Bohr** of the University of Copenhagen who provided an insight into the arrangement of electrons inside in the atom (FIG 3.1).

Fig 3.1
Niels Bohr won the Nobel Prize in physics in 1922 for his study of the arrangement of electrons in the atom.

3.2 BOHR'S STUDY OF SPECTRA

In order to understand how Bohr developed his theory about the arrangement of electrons in atoms, we first need to learn a little about spectra.

You may be aware that when white light (e.g. sunlight or light from an ordinary electric light bulb) is passed through a prism, the white light is broken up into an array of colours. Such an array of colours is called a spectrum (FIG 3.2).

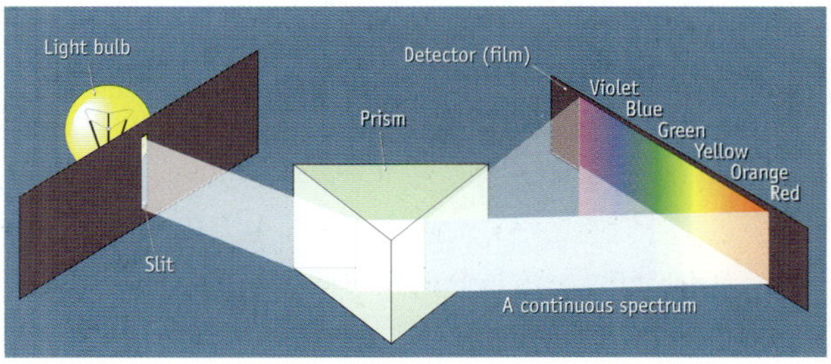

Fig 3.2
A continuous spectrum of colours from red to violet is formed when white light is passed through a prism.

This spread of colours is referred to as a **continuous** spectrum or rainbow. A rainbow is formed when the water droplets in the sky act as prisms and separate sunlight into its colours.

Consider what happens if the above experiment is repeated using the light coming from a hydrogen discharge tube instead of white light. A hydrogen discharge tube is simply a glass tube filled with hydrogen at low pressure through which an electric current is passed. Niels Bohr carried out this experiment and found that instead of seeing a continuous spectrum, he saw a series of narrow **lines** (FIG 3.3).

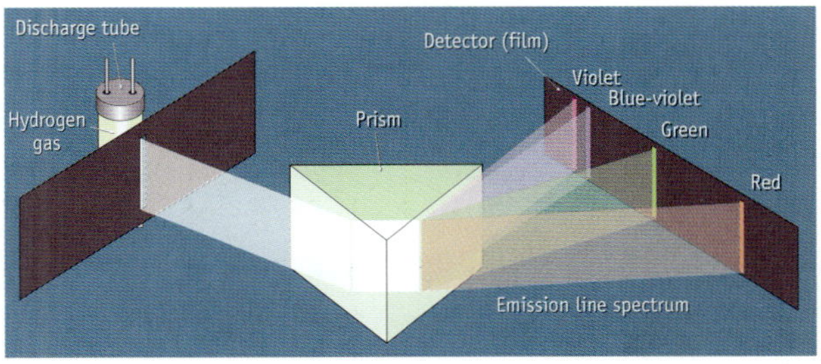

Fig 3.3
A line spectrum is formed when the light from a hydrogen discharge tube is passed through a prism.

Since the spectrum consists of lines, it is called a **line spectrum** to distinguish it from a continuous spectrum. Since the light is emitted when an electric current is passed through hydrogen, the spectrum is more accurately called an **emission line spectrum**. When the hydrogen in the discharge tube was replaced by other elements like sodium and mercury, it was found that these elements also produced line spectra. Some examples of line spectra are shown in FIG 3.4.

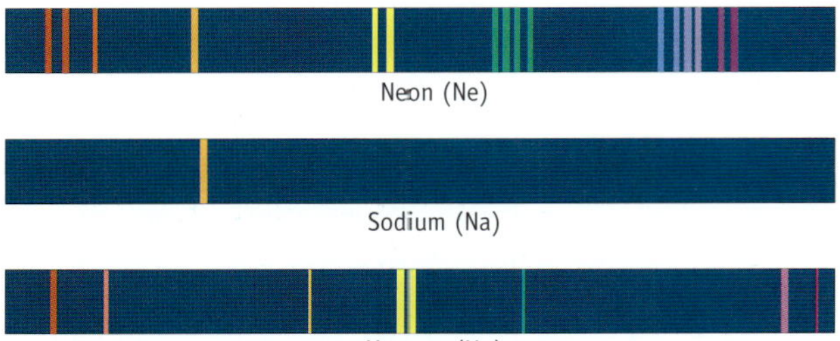

Fig 3.4
The emission line spectra of neon, sodium and mercury. These show how each element has its own unique emission line spectrum – a sort of 'fingerprint' spectrum.

Note that each element has its own unique emission line spectrum. This spectrum is like a 'fingerprint' for the element. In fact, the emission line spectrum of an element is often used by chemists to analyse materials for the presence of certain elements, e.g. in the analysis of steel an instrument called an emission spectrometer is used to analyse the sample of steel for different metals (FIG 3.5).

Fig 3.5
The instrument on the left of this photograph is called an optical emission spectrometer and is used by Irish Ispat Ltd. to identify the various metals in a sample of steel. Each metal in the steel can be identified because each has its own unique emission line spectrum.

You may already be aware that different elements emit different colours when an electric current is passed through them, e.g. sodium street lamps are yellow, mercury vapour lamps are blue and neon advertising signs are red.

EXPERIMENT 3.1

To study emission spectra using a spectrometer or a hand-held spectroscope.

Introduction

In this experiment, we study the emission line spectra of some elements. The most satisfactory way of doing this is to use a spectrometer and discharge tubes, if available (Method A). If this equipment is not available, then a hand-held spectroscope and salts of different metals may be used (Method B). (FIG 3.6)

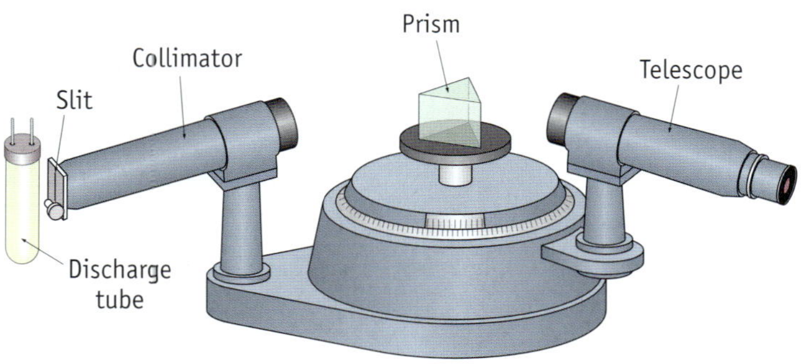

Fig 3.6
Examining an emission line spectrum using a spectrometer and a discharge tube.

Method A

A spectrometer and discharge tube is set up as shown in FIG 3.6. The spectrum of an ordinary electric light bulb is first viewed with the telescope of the spectrometer and is observed to be continuous. The electric light bulb is then replaced with the discharge tube of an element (e.g. mercury). An emission line spectrum is observed.

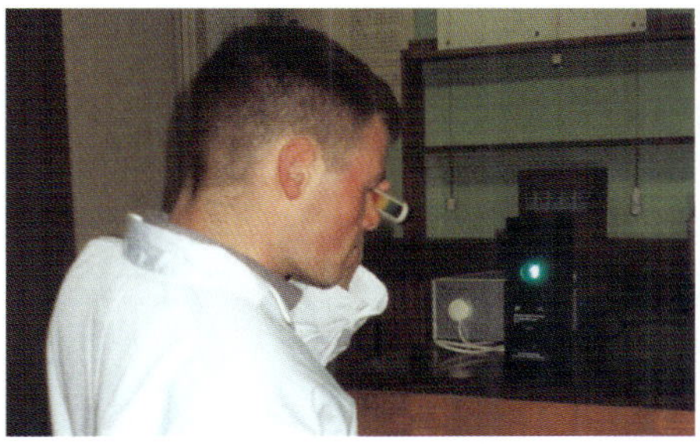

Fig 3.7
Studying an emission spectrum using a hand-held spectroscope.

Method B

This method involves heating a salt of the metal in the flame of a Bunsen burner. This method is more difficult to carry out as the flame colour lasts only for a few moments. It requires a little practice and concentration. A flame test rod (platinum wire held in a glass rod) is cleaned by dipping it into a test-tube of concentrated hydrochloric acid. A sample of a salt (e.g. potassium chloride or strontium chloride) is then held in the blue flame of the Bunsen burner. The coloured light given off is examined by looking at it through the direct vision spectroscope (FIG 3.7).

Compare the spectrum obtained with that obtained from an ordinary electric light bulb. Note that the ordinary electric light bulb produces a continuous spectrum but the light of the salt produces a line spectrum.

EXPERIMENT 3.2 MANDATORY STUDENT EXPERIMENT NO. I

To carry out flame tests with salts of lithium, sodium, potassium, barium, strontium and copper.

Introduction

In this experiment, we shall study the colours given off when salts of certain metals are heated in the flame of a Bunsen burner. The colours obtained can then be used to identify the metals in unknown compounds.

Apparatus required

Platinum wire held in glass rod, Bunsen burner, six small beakers, test-tube, pestle and mortar.

As platinum wire is expensive, many schools may not possess it in sufficient quantities for this experiment. Wooden splints soaked overnight in water work very well in place of the platinum. You will need a wooden splint for each sample of salt analysed to avoid cross contamination.

Chemicals required

Lithium chloride, sodium chloride, potassium chloride, barium chloride, strontium chloride, copper(II) chloride, concentrated hydrochloric acid (if using platinum wire method).

Procedure

1. Light the Bunsen burner and adjust it to get a blue flame.
2. Using the fume cupboard, place some concentrated hydrochloric acid in a test tube. Place the test-tube in a test-tube rack. Dip the platinum wire into the concentrated hydrochloric acid. This helps to clean the wire.
3. Hold the platinum wire at the tip of the blue cone of the Bunsen flame. If the wire is clean there should be no colour from it when it is placed in the Bunsen flame. If there is a colour, repeat the cleaning process.

4. Crush a little of the salt to be tested with a pestle and mortar. Place the salt in a small beaker and label the beaker.
5. Again dip the platinum wire in concentrated hydrochloric acid and then in the salt to be tested. Some of the salt should stick to the wire.
6. Hold the platinum wire in the flame of the Bunsen burner and note the colour given off as the salt burns.
7. Repeat the experiment for each of the other salts given to you by your teacher. (To avoid cross contamination, it is best to have a separate platinum wire for each sample of salt. If this is not possible, leave the sodium salt until last as its colour is very persistent. Make sure that the platinum wire is thoroughly cleaned using hydrochloric acid before each test.) Again, note the colour in each case. Summarise your results in the form of a table as follows:

Metal present	Colour
Lithium	
Potassium	
Barium	
Strontium	
Copper	
Sodium	

If you are given an unknown salt, you should be able to identify the metal in the salt from the results of your experiment.

Note: Potassium salts are often contaminated with sodium. For this reason the yellow colour of sodium may interfere with the lilac colour of the potassium.

Conclusions

Each metal gives a characteristic colour to the Bunsen flame. These colours are lithium (crimson), sodium (yellow), potassium (lilac), barium (green), strontium (red) and copper (blue-green). The flame tests for salts of some of these metals are shown in Fig 3.8. The various colours in fireworks displays are obtained by adding salts of various metals to the explosive mixture in the fireworks.

The experiment should be written up under the headings: Title, Date, Background Theory, Procedure, Results, Conclusions. Always include a labelled diagram when writing a report of an experiment.

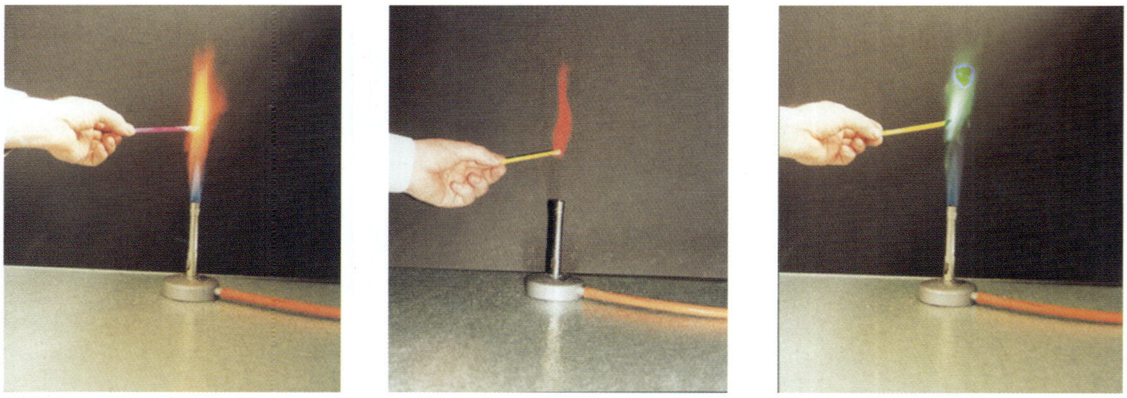

Fig 3.8

Flame tests may be carried out using a wooden splint soaked in water and dipped into a sample of the salt. These photos show the flame tests for sodium (left), strontium (centre) and copper (right).

But what has this study of spectra to do with the structure of the atom? It was the study of emission spectra that gave scientists a clue to the arrangement of electrons in an atom.

Niels Bohr realised that any model of the atom must explain (i) why the emission spectra of the elements are line spectra rather than continuous spectra and (ii) why the emission spectrum of each element is unique to that element. The answer to these two questions came in Bohr's Theory of Atomic Structure in 1913.

3.3 EXPLAINING THE EVIDENCE: THE BOHR THEORY

Bohr suggested a way to relate the emission line spectrum of hydrogen to the structure of an atom of hydrogen. Bohr's Theory may be summarised as follows:

1. Electrons revolve around the nucleus in fixed paths called orbits.
2. Electrons in any one orbit have a fixed amount of energy. For this reason orbits are also called **energy levels**. We say that the energy of the electron in a particular orbit or energy level is **quantised** or fixed at a definite value.

> An energy level is defined as the fixed energy value that an electron in an atom may have.

3. As long as an electron is in any one particular energy level, it neither gains nor loses energy.
4. When an atom absorbs energy, electrons jump from a lower energy level to a higher energy level. Electrons are less stable in the higher energy levels and do not remain there for long.
5. Energy is lost when an electron falls from a higher energy level to a lower energy level. Since an electron can only fall back to certain definite energy levels, only fixed amounts of light energy can be given off.

Some of the above ideas are summarised in Fig 3.9.

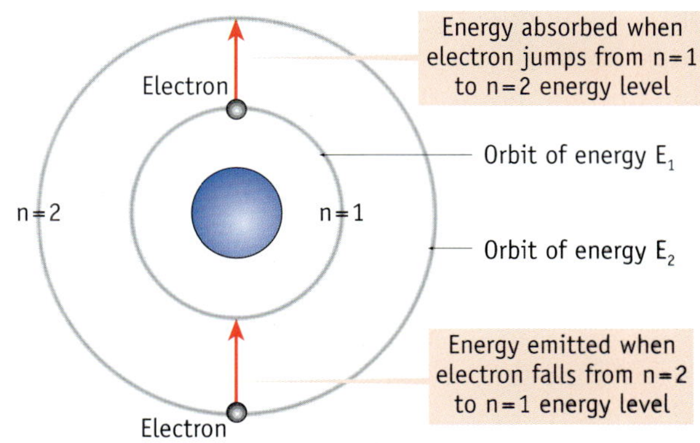

Fig 3.9
Bohr's Theory stated that the energy of the electron in any one energy level is fixed or quantised.

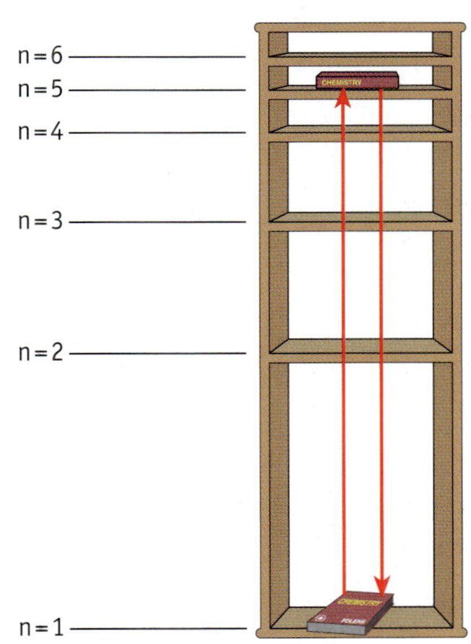

Fig 3.10
Electrons can only occupy certain energy levels. Each energy level has a fixed amount of energy associated with it. The book can only occupy certain shelves. Each shelf is a fixed distance from the ground and hence the book has a certain amount of potential energy on each shelf.

Up to the time that Bohr proposed his theory, scientists thought that electrons in an atom could have **any** energy value. Bohr's Theory introduced the idea of quantisation of energy, i.e. that electrons in atoms could only possess a certain quantum or fixed amount of energy Quantisation means that the energy of an electron in a particular energy level cannot have just any value but may only have a certain specific value. In other words, the energy of an electron in an atom is restricted to having only certain values. Bohr represented the energy levels by the letter **n**. The lowest energy level was the n = 1 level, the next highest was n = 2, etc. Bohr's mathematical calculations showed that the energy levels became closer to one another as the value of n increased (FIG 3.10).

By proposing that as long as an electron stays in one particular energy level it does not gain or lose energy, Bohr overcame the problem of the electron losing energy and spiralling into the nucleus.

But is Bohr's theory correct? What spectroscopic evidence is there for the existence of energy levels? Bohr's reasoning went as follows:

(a) Consider an atom of hydrogen. The one electron of hydrogen is normally found in its lowest energy level (n = 1). The electron is said to be in its **ground state**.

(b) When energy is supplied to an atom (by electricity or heating), the electron absorbs a certain amount of energy. This moves the electron into a higher energy level. The electron is now said to be in an **excited state**.

(c) The electron remains in the higher energy level for a short period of time and then falls back down to a lower energy level. When this occurs, a **definite amount of energy is emitted**. This definite amount of energy is equal to the difference in energy between the two energy levels.

(d) The definite amount of energy emitted appears as a line of a particular colour in the emission spectrum. Each colour corresponds to a particular wavelength or frequency. There is a mathematical equation relating the energy of light and its frequency:

$$E = hf$$

definite amount of energy emitted from atom → E; light of definite frequency or wavelength in emission spectrum → f

E is the energy of light emitted from the atom, h is just a number called Planck's constant and f is the frequency of the light emitted. Thus, each coloured line formed in the emission line spectrum has its own particular frequency.

(e) Each definite amount of energy emitted gives rise to a line in the emission spectrum (FIG 3.11).

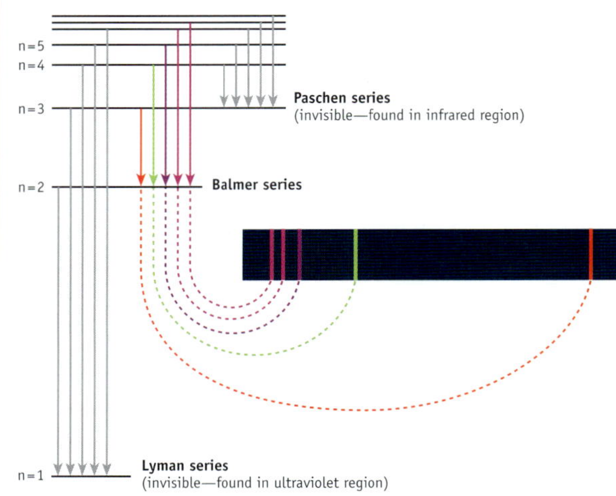

Fig 3.11
Each electron transition results in a line of particular wavelength in the emission spectrum of hydrogen.

If E_1 is the energy of the electron in the ground state and E_2 is the energy of the electron in the excited state (FIG 3.12), then the energy given out when the electron falls back down to the ground state may be calculated from the equation:

$$E_2 - E_1 = hf.$$

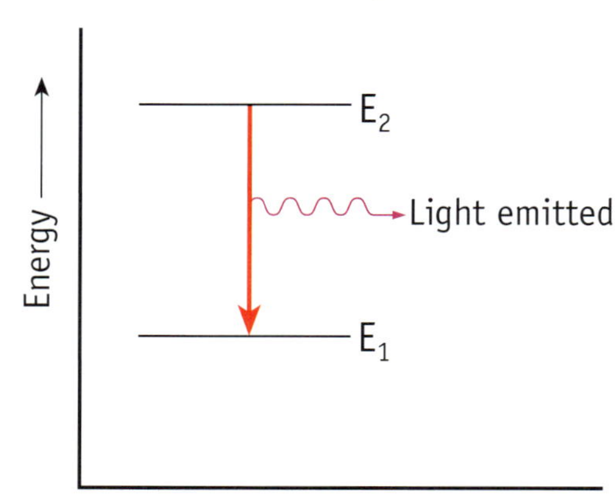

Fig 3.12
Energy is given out in the form of light when an electron falls back from the excited state to the ground state.

(f) Since only definite amounts of energy are emitted, this implies that electrons can occupy only definite energy levels. Therefore, energy levels must exist in the atom.

Bohr's work was of two kinds: theoretical and experimental. He calculated mathematically the energy corresponding to each of the energy levels n = 1, n = 2, n = 3, etc. Then he calculated the amounts of energy an electron would lose when it falls down from higher energy levels to lower energy levels. Finally, using the above equation, he calculated the frequency and wavelength of the coloured lines corresponding to these electron transitions. (The above equation gave him the frequency.

The relationship:

velocity of light = frequency × wavelength

enabled him to calculate the wavelength.)

His experimental work involved measuring the wavelengths of each line in the emission line spectrum of hydrogen using a spectrometer. Bohr found that the wavelengths of light in the emission spectrum of the hydrogen atom, calculated using his theory, were exactly the same values as those measured experimentally.

Note: In FIG 3.11 you will observe that it is a set of transitions called the 'Balmer' series which gives rise to the lines in the visible spectrum. (They are called Balmer after the man who first discovered them.) The other series of lines (Lyman, Paschen, etc.) appear in the ultraviolet and infra-red region of the spectrum and hence are invisible. However, Bohr predicted their existence and their discovery added further credibility to Bohr's theory.

Students often ask the question: 'If the hydrogen atom has only one electron, why is it that so many electron transitions (as shown in FIG 3.11) are possible?' The reason for this is that even in a small sample of hydrogen, there are millions of atoms of hydrogen. When energy is supplied to the sample of hydrogen, not all atoms may receive exactly the same amount of energy. In some atoms, the electron may be promoted from the n = 1 to the n = 2 energy level, in others it may be promoted from the n = 1 to the n = 3, etc. Thus, when these electrons fall back down, they are falling from different energy levels, so different lines will be formed in the emission spectrum.

Bohr's theory also explained why each element has its own unique emission line spectrum. Since each element has its own particular number of electrons, then there will be different numbers and types of transitions for each element, thus giving rise to a different emission line spectrum in each case.

Fluorescent light strips and lasers make use of electron transitions (FIG 3.13). A note on these is given in your Workbook.

Fig 3.13
This student is carrying out research into lasers in the Physics Department, University College, Cork. The laser light is emitted when electrons fall from high energy levels to lower energy levels.

Example 3.1

QUESTIONS

(i) What do you understand by energy levels?
(ii) Explain, in terms of energy levels, why the spectra of elements consist of a series of lines of definite frequency.
(iii) Different metals impart different colours to the Bunsen flame. In the case of any two named metals state the colour imparted to the flame.
(iv) Explain briefly why different metals produce different flame colours. *[LCH]*

Answers

(i) Energy levels are the different fixed energy values that electrons in an atom can have.

(ii) Electrons in an atom occupy definite energy positions (E_1). When an electron absorbs energy it moves up to a higher level (E_2). When it falls back to a lower level, it radiates the difference in energy in the form of light. The difference in energy is calculated from the formula $E_2 - E_1 = hf$ where h is Planck's constant and f is the frequency of light. Lines of definite frequencies must mean that the electron movements were between definite energy levels.

(iii) The colours are yellow (sodium), lilac (potassium), green (barium), red (strontium), blue-green (copper) and crimson (lithium).

(iv) Different metals produce different flame colours because the atoms of various elements have different numbers of electrons. Thus, electrons will be moving between different energy levels in each metal and this will cause the emission of light of different frequencies or energies, i.e. different colours will be produced in the visible region.

3.4 ATOMIC ABSORPTION SPECTROMETRY

In the previous section we saw how atoms can emit light when electrical or heat energy is supplied to them. However, scientists studying spectra have also found that atoms can **absorb** light. If white light is passed through a gaseous sample of an element, it is found that the light which comes out has certain wavelengths missing, i.e. dark lines are observed in the spectrum. Such a spectrum is called an **atomic absorption spectrum**. For example, when light coming from the sun was analysed by spectroscopists it was observed that dark lines appeared in it (Fig 3.14).

Fig 3.14
The absorption spectrum of the sun. The dark lines represent missing wavelengths and tell us what elements are present in the sun.

These dark lines correspond to missing wavelengths. From examining these lines, scientists were able to detect what elements are present in the sun since all elements absorb light of certain wavelengths. When light from the sun passes through these elements, each element absorbs certain wavelengths. Therefore, these missing wavelengths appear as dark lines in the spectrum of white light from the sun.

An absorption spectrum could be produced using the apparatus shown in Fig 3.15.

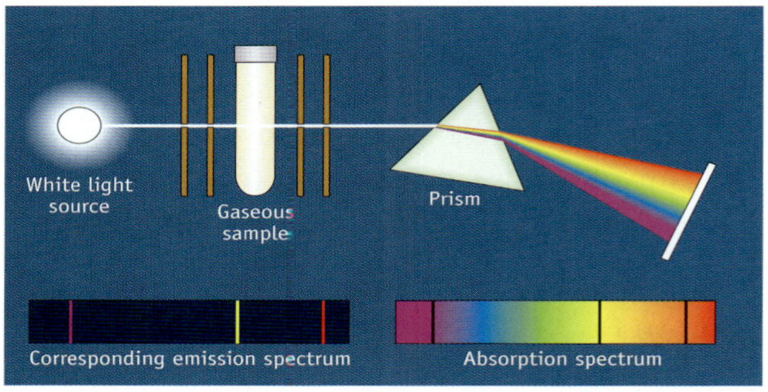

Fig 3.15
How an absorption spectrum is formed. Note that an absorption spectrum is like a photographic negative of an emission spectrum.

White light from a lamp is passed through a gaseous sample of an element. The light that comes out is passed through a prism and is found to have certain wavelengths missing. These wavelengths are absorbed by the sample and correspond exactly to the wavelengths that would be detected if an emission spectrum were recorded. The absorption spectrum is like a photographic negative of an emission spectrum. The reason for this is that atoms in the ground state absorb the same radiation as they emit in the excited state. An atom absorbs an amount of energy equal to the energy difference between the energy level of the electron in the ground state and the energy level that the electron occupies in the excited state.

An emission spectrum consists of coloured lines against a dark background. An absorption spectrum consists of dark lines against a coloured background. Atomic absorption spectrometry is a very useful analytical tool used by chemists to detect the presence of certain elements and to measure the concentrations of these elements.

In practice, chemists do not use white light as the source of radiation in atomic absorption spectrometry. Instead of white light, lamps which emit the characteristic radiation of the element being analysed (i.e. emit the line spectrum of the element) are used, e.g. a sodium lamp is used to measure the amount of sodium present in a particular sample being analysed. The sample containing the sodium is held in a flame in the instrument. (The instrument is called an atomic absorption spectrometer.) The intensity of the light transmitted through the flame before the sample is introduced and after the sample is introduced is then measured. The instrument indicates the amount of light absorbed and from this it is possible to measure the concentration of sodium in the sample. An example of the use of atomic absorption spectrometry is the analysis of water for metals like lead, mercury and cadmium. (These metals are often referred to as 'heavy metals' due to their relatively large atomic masses.) An atomic absorption spectrometer is shown in Fig 19.19.

3.5 ENERGY SUBLEVELS

As the study of spectra developed, more sophisticated types of spectrometers were developed. With these spectrometers, scientists were able to look more closely at lines formed in emission spectra. They found that in many cases, what appeared originally to be a single line, actually consisted of a number of lines very close together. For example, in studying the emission spectrum of sodium, what appeared to be a single yellow line was in fact made up of two yellow lines very close together. The two lines could not be as a result of electrons dropping to two different energy levels, as this would give rise to lines much farther apart in the spectrum.

Therefore, in order to explain this observation, scientists proposed that each main energy level, except the first, was made up of a number of sublevels all of which were close in energy. From their analysis of spectra, spectroscopists deduced that the number of sublevels was the same as the value of n for the main energy level. In other words, the n = 2 main energy level had two sublevels, the n = 3 main energy level had three sublevels and the n = 4 had four sublevels (Fig 3.16). (In this course we do not have to consider any energy levels beyond n = 4).

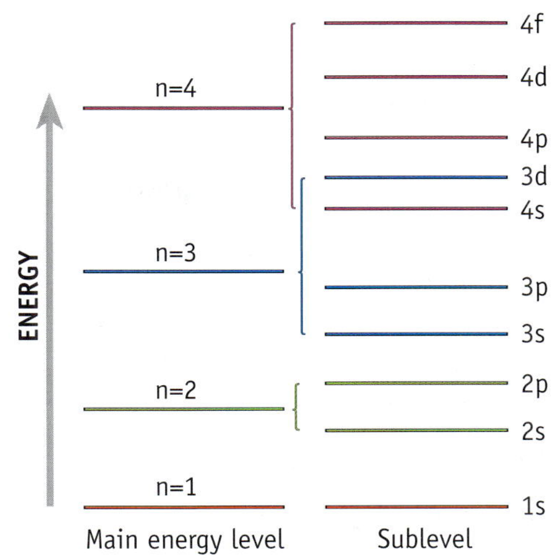

Fig 3.16
Sublevels associated with the main energy levels. Note that the 4s sublevel has been found to be lower in energy than the 3d sublevel.

Spectroscopists labelled the sublevels using the letters s, p, d and f. The sublevel of lowest energy was called the **s** sublevel, the one of next highest energy was called the **p** sublevel, the one above that the **d** sublevel and the one of highest energy was called the **f** sublevel. You will observe in Fig 3.16 that the 4s sublevel is in fact lower in energy than the 3d sublevel. This interesting observation will be discussed at a later stage (Chapter 4).

You might wonder why these particular letters were chosen to represent sublevels. Originally they were chosen from words used by spectroscopists to describe lines in line spectra: s for 'sharp', p for 'principal', d for 'diffuse' and f for 'fundamental'.

3.6 WAVE NATURE OF THE ELECTRON

For many purposes in chemistry, the Bohr picture of electrons travelling in fixed paths around the nucleus is quite an adequate model. However, in some cases where we move more deeply into the subject, it is necessary to look at some modifications or changes to the Bohr theory. This does not mean that the Bohr model is 'wrong'. The Bohr model works perfectly to explain some observations in chemistry. In other cases, we need a more modern model to explain our observations.

As already discussed, Bohr's theory worked perfectly to explain the emission spectrum of hydrogen. However, when his theory was applied to atoms with more than one electron, it failed to account for many of the lines in the emission spectra of these atoms.

In 1924, a French scientist called Louis De Broglie suggested that all moving particles had a wave motion associated with them. For fairly large objects like a tennis ball, this wave motion would be impossible to detect. However, for a small object like an electron the wave motion should be possible to detect. De Broglie was proved correct in 1927 when it was discovered that moving electrons had a wave motion associated with them. This idea that a particle can behave like a wave is often referred to as 'wave-particle duality'. What relevance has this to the Bohr theory you may ask? Consider FIG 3.17.

Fig 3.17
If the electron has a wave motion, then it cannot be travelling in a precise path at a fixed distance from the nucleus.

If the electron has a wave motion, then Bohr's picture of it travelling in a precise path at a precise distance from the nucleus cannot be true. This problem was treated mathematically by a German physicist called **Werner Heisenberg** (FIG 3.18). He put forward a very famous principle called **Heisenberg's Uncertainty Principle**. This principle states that **it is impossible to measure at the same time both the velocity and the position of an electron**.*

Fig 3.18
Werner Heisenberg who put forward his famous Uncertainty Principle.

Heisenberg reasoned as follows: consider a beam of light being used to detect the presence of an electron. The beam of light acts like a radar signal that bounces off an object. When the beam strikes the electron, the position of the electron is determined but since the electron has such a small mass, its velocity (speed) is immediately changed by the beam of light. In other words, Heisenberg said that the Bohr picture of the atom was not completely satisfactory. Bohr's model visualised electrons moving with a certain speed in orbits of fixed distances from the nucleus. Heisenberg stated that you cannot say this about the electron since you cannot measure at the same time both its speed and its distance from the nucleus (location).

It was clear therefore that the idea of an electron moving in a fixed orbit around the nucleus had to be modified. If the electron is not travelling in a fixed path, then we can only talk about the **probability** of finding it at a particular position inside the atom. This idea of probability led to a new picture being developed about the position of electrons in atoms.

---

* Heisenberg actually used the word momentum (= mass × velocity) instead of velocity in stating his principle. However, since the mass of the electron is a constant value and since we are not concerned with the mathematics of his principle, the simpler term velocity will be used here.

3.7 ATOMIC ORBITALS

Imagine that you are looking at an electron in an atom of hydrogen as it revolves around the nucleus. If you could take a large number of photographs of the electron and then put all of these photographs into one photograph, you would get a picture like that shown in FIG 3.19. Note from the diagram that the shading is heaviest near the nucleus and gets less as you move away from the nucleus. This means that there is a high probability of finding the electron near the nucleus of the atom and the probability gets less as the distance from the nucleus increases.

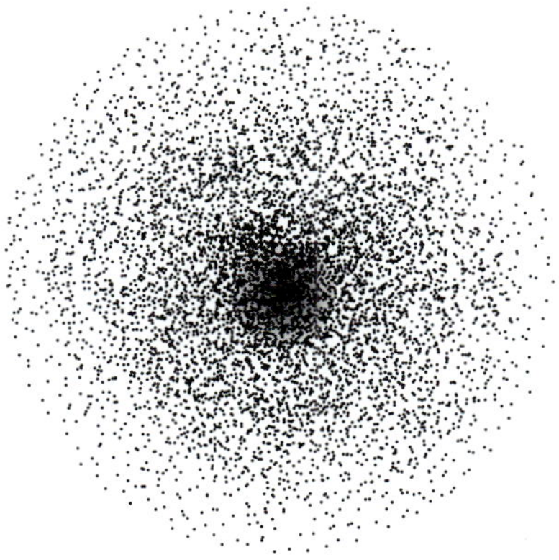

Fig 3.19
Diagrammatic representation of the position of an electron in an atom of hydrogen.

Although diagrams like FIG 3.19 give us a picture of the probability of finding an electron, they are quite tedious to draw. Instead, chemists draw a boundary within which there is a very high probability of finding an electron (approximately 95%). The region within this boundary is called an **orbital**.

> An orbital is a region in space within which there is a high probability of finding an electron.

The term orbital must not be confused with the term orbit. The term 'orbit' was used by Bohr to describe the path of an electron travelling with a certain velocity at a precise distance from the nucleus. As indicated above, the term orbital was introduced to describe the more realistic probability picture.

The orbital for the electron in a hydrogen atom is called a 1s atomic orbital and is indicated in FIG 3.20.

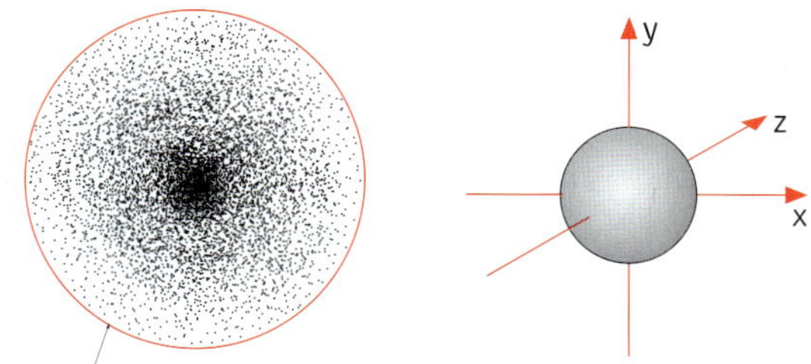

Within this boundary there is a 95% chance of finding an electron

Fig 3.20
An s orbital is spherical.

Erwin Schrödinger, an Austrian physicist, in 1926 devised mathematical equations to work out the probability of finding an electron in any particular sublevel in an atom (FIG 3.21).

Schrödinger worked at the Dublin Institute for Advanced Studies for a number of years and became an Irish citizen. Schrödinger's equations were solved to give us the shapes of the four types of orbitals, s, p, d, and f. Thus, an orbital could be said to be a pictorial representation of the solution of the Schrödinger wave equation describing a particular electron.

Fig 3.21
Erwin Schrödinger worked out the probability of finding a particular electron in an atom. It was his mathematical equations that gave us the shapes of orbitals.

It is found that **all s orbitals are spherical** and differ only in size (FIG 3.22). In other words, the probability of finding an electron in an s sublevel is identical in all directions from the nucleus.

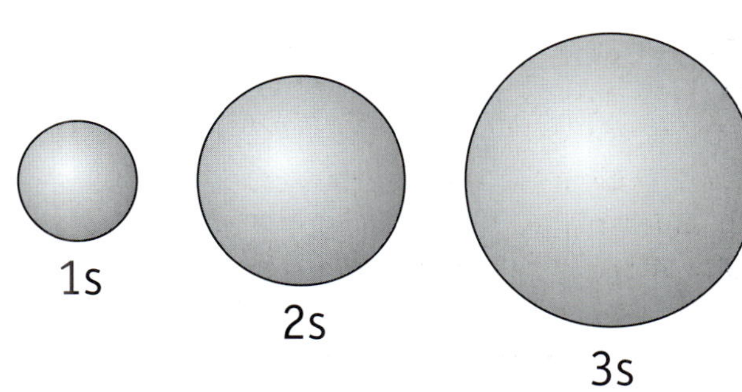

Fig 3.22
All s orbitals are spherical in shape. Even though the diagrams may look like circles, it must be remembered that they are spheres since atoms are three dimensional.

When Schrödinger's equations were solved to tell us the probability of finding an electron in a p sublevel, it was found that the resulting shapes resembled a dumb-bell, i.e. **p orbitals are dumb-bell shaped** (FIG 3.23). It was discovered from the splitting of lines in emission spectra that each p sublevel consisted of three parts p_x, p_y and p_z. The three corresponding p orbitals only differ in their orientation in space as shown in FIG 3.23. The letters x, y and z stand for the three main axes to represent direction. The three p orbitals are therefore at right angles to each other. Thus, the probability of finding an electron in a p sublevel is represented by these dumb-bell shapes. As one moves out from the nucleus along either the x, y or z axis, the probability of finding an electron is shown by the shapes of the p orbitals. Within the boundary of the region traced out by the dumb-bell, there is a high probability of finding an electron. Outside the dumb-bell, there is hardly any probability of finding an electron.

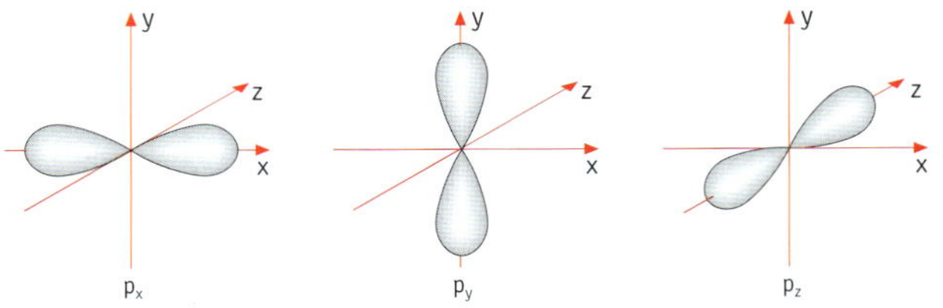

Fig 3.23
p orbitals are dumb-bell shaped. Each p orbital points along one of the three main axes.

The shapes of d and f orbitals are more complicated and need not be covered for the Leaving Certificate chemistry course.

When we have studied the Periodic Table (CHAPTER 4) we shall see how the electrons in atoms are distributed among the various orbitals.

CHAPTER SUMMARY

- The formation of emission line spectra indicates the presence of energy levels in atoms.
- An emission line spectrum consists of coloured lines against a dark background.
- The instrument used to examine spectra is called a spectrometer or a spectroscope.
- When salts are heated, different colours are emitted depending on the metal present in the salt.
- Bohr proposed that electrons revolve around the nucleus in fixed paths called orbits or energy levels.
- Bohr proposed that the energy of an electron in an energy level is quantised, i.e. fixed at a definite amount.
- When an atom absorbs energy, electrons jump from a lower energy level to a higher energy level.
- Energy is lost when an electron falls from a higher energy level to a lower energy level.
- Since only definite amounts of energy are emitted, this implies that electrons can occupy only definite energy levels.
- An absorption spectrum is produced when white light is passed through a gaseous sample of an element.
- An absorption spectrum consists of dark lines against a coloured background.
- Heisenberg's Uncertainty Principle states that it is impossible to measure at the same time both the velocity and the position of an electron.
- An orbital is a region in space within which there is a high probability of finding an electron.
- s orbitals are spherical. p orbitals are dumb-bell shaped.

EXERCISES

3.1. (a) A Danish scientist called .... provided much information about the arrangement of electrons in atoms.

(b) A rainbow of colours is obtained when light is passed through a ....

(c) Each element has its own .... emission line spectrum.

(d) To study spectra we use either a .... or a ....

(e) A sample of a salt of sodium when burned gives a .... colour.

(f) An unknown salt when burned gave a crimson colour. What metal is present in the salt? ....

(g) An energy level is defined as ....

(h) Bohr stated that the energy of the electron is quantised. What does this mean? ....

(i) Energy is .... when an electron falls from a higher energy level to a lower energy level.

(j) In the equation $E = hf$ what does h stand for? ....

(k) Distinguish between an emission spectrum and an absorption spectrum. ....

(l) Give one piece of evidence which suggested to scientists that sublevels existed ....

(m) State Heisenberg's Uncertainty Principle ....

(n) An Austrian physicist called .... used mathematical equations to work out the probability of finding a particular electron in an atom.

(o) Why are p orbitals labelled p_x, p_y and p_z? ....

3.2. Indicate briefly how the study of spectra provided experimental evidence for the existence of energy levels in atoms.
What is meant by the ground state of an atom?
Explain the terms (i) energy level and (ii) orbital. *[LCH]*

3.3. Describe an experiment you would carry out to investigate the flame colours of a series of salts given to you by your teacher.
Comment on the precautions that you would take to avoid cross contamination of the salts.
One of the salts when burned gave a lilac colour. What metal is present in this salt?
One of the salts has a very persistent colour that is difficult to eliminate from the other colours. Name this salt.

3.4. Distinguish between the terms 'orbit' and 'orbital'. Give a brief account of why it was necessary for scientists to introduce the concept of the orbital. Your account should refer to the work of (a) De Broglie, (b) Heisenberg and (c) Schrödinger.
Indicate using a diagram the difference between a 1s orbital and a 2s orbital.
The models of three orbitals are shown in FIG 3.24. Assign labels to the axes and name each orbital.

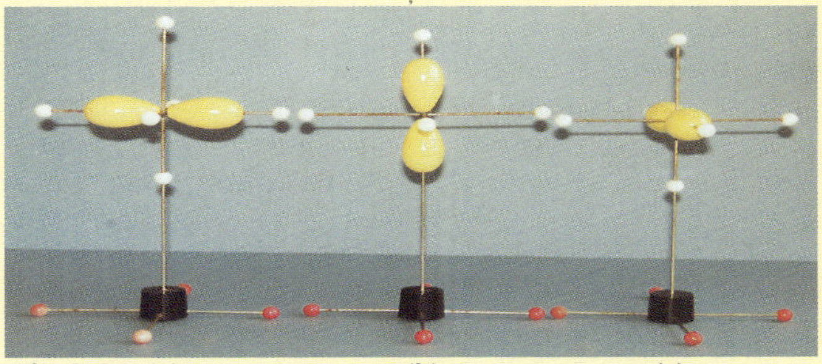

Fig 3.24 **(a) (b) (c)**

4 The Periodic Table

4.1 INTRODUCTION

In this chapter we shall study one of the most important foundation stones in chemistry – the Periodic Table of the Elements. We shall begin by discussing what an element is and consider some of the early ideas about elements. We shall then look at the work of some chemists who tried to put these elements into some sort of order. The Periodic Table itself will then be studied and we will learn how this table is of enormous benefit to us in deducing lots of information – thus making the study of chemistry much easier. Finally, we shall learn how to use the Periodic Table to deduce how electrons are arranged in atoms.

4.2 THE ELEMENTS

The Greek philosophers decided that the universe was made up of a few basic substances or elements. They said there were four elements – earth, air, water and fire. They stated that all substances were made up of various arrangements of these. Believe it or not, this idea remained the basis of chemical theory for almost 2,000 years! It has even passed into the English language – the term 'the raging of the elements' means that air and water are turning into a storm.

The first reasonably accurate definition of an element was given by an Irish chemist called Robert Boyle who was born in Lismore Castle, Co. Waterford (FIG 4.1). In his book, *The Sceptical Chemist*, published in 1661, he described elements as 'primitive and simple' substances. Boyle's description of an element is similar to the modern definition of an element.

> An element is a substance that cannot be split up into simpler substances by chemical means.

Fig 4.1
Robert Boyle was the first chemist to properly define the meaning of the term 'element'. He was born in Lismore Castle, Co. Waterford.

Copper is an example of an element. The above definition means that whatever chemical reactions you make copper undergo, you cannot break the copper up into any simpler chemicals. On the other hand, water is **not** an element because it can be broken up into hydrogen and oxygen. (In fact, Boyle also predicted that elements combine together in a chemical reaction to form what are now called **compounds**.)

It was an English chemist called **Humphry Davy** who made some of the most important discoveries in the history of the elements. After it was discovered that the compound water could be split up by electricity into the elements hydrogen and oxygen, Davy studied the effects of electricity on other compounds. For example, in 1807 he discovered potassium by passing electricity through a compound of potassium. When he saw the tiny globules of potassium catch fire, he is reported to have been as excited as a child and to have danced around the room in delight (FIG 4.2).

Fig 4.2
Humphry Davy discovered many elements by passing electricity through their compounds. In 1807, he discovered the element potassium by passing an electric current though melted potassium hydroxide.

Davy discovered sodium in a similar fashion and he passed an electric current through lime and isolated an element that he called calcium. Davy also discovered barium, strontium and magnesium. The discovery of the fact that an electric current could be used to obtain elements from their compounds was a turning point in the history of the discovery of the elements (FIG 4.3). After 1807, many new elements were discovered and approximately 100 elements have been discovered to date.

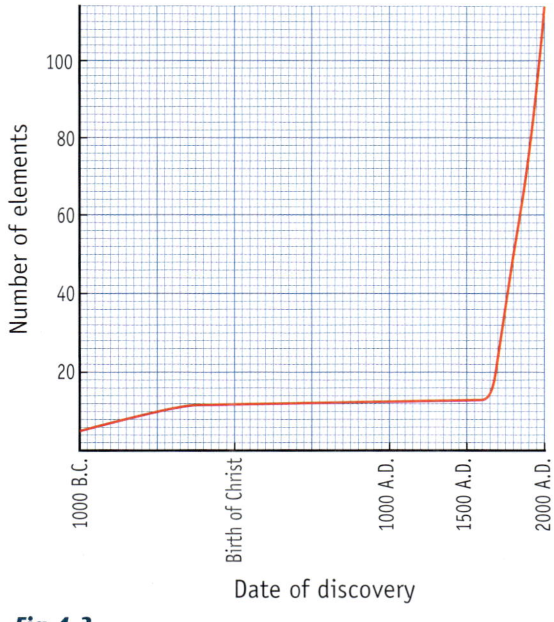

Fig 4.3
After Davy's discovery in 1807 that elements could be extracted from their compounds using electricity, there was a marked increase in the number of elements discovered.

Before concluding this section, it is worth mentioning an important discovery about elements made by an English scientist called Henry Moseley. In 1914, Moseley, using X-rays, discovered a method of determining the number of protons in the nucleus of an atom. He found that the main difference between one element and another was the number of protons in the nucleus of the atom. The number of protons is unique to that element and characterises the element, i.e. no two elements have the same number of protons. We shall return to Moseley's work in the next section.

4.3 LOOKING FOR A PATTERN: HISTORY OF THE PERIODIC TABLE

By 1800, approximately 50 elements had been discovered. As more and more elements were discovered, chemists were finding it increasingly difficult to understand and memorise the long lists of their chemical and physical properties. It was clear some kind of pattern was needed to help simplify matters. The work of four chemists eventually led to this pattern being developed. These four men were Johann Dobereiner, John Newlands, Dmitri Mendeleev and Henry Moseley.

1. Dobereiner's Triads

In 1810, Johann Dobereiner (FIG 4.4), was appointed professor of chemistry and physics at the University of Jena in Germany. As stated above, by this time approximately 50 elements had been discovered. Dobereiner looked at the properties of various elements in an attempt to find some order among them.

Fig 4.4
Dobereiner thought that elements could be categorised in groups of threes. He called such a group a triad.

In 1829, Dobereiner noticed that the recently discovered element bromine seemed half-way in its properties between chlorine and iodine*. He also

* i.e. there was a smooth graduation of properties going from chlorine to bromine to iodine.

looked at the atomic weights of these elements. (For the moment, we shall consider atomic weight as simply referring to the weight of an atom. The modern name for atomic weight is 'relative atomic mass'. It will be defined more fully in SECTION 4.7.) He observed that the atomic weight of bromine (80) was about half way between that of chlorine (35.5) and iodine (127) (FIG 4.5). He observed a similar pattern with calcium, strontium and barium and also with sulfur, selenium and tellurium. He called each of these groups of three elements a **triad**.

A triad is a group of three elements with similar chemical properties in which the atomic weight of the middle element is approximately equal to the average of the other two.

Chlorine (35.5)	**Calcium** (40)	**Lithium** (7)	**Sulfur** (32)
Bromine (80)	**Strontium** (88)	**Sodium** (23)	**Selenium** (79)
Iodine (127)	**Barium** (137)	**Potassium** (39)	**Tellurium** (128)

Fig 4.5
Four examples of Dobereiner's Triads. The elements in each triad have similar chemical properties and the atomic weight (relative atomic mass) of the middle element is approximately equal to the average of the other two.

Unfortunately for Dobereiner, few other cases of triads could be found and his triads were looked on as coincidences that were of little value. However, his work is important as it was the earliest attempt to trace a relationship between the properties of the elements and their atomic weights. His work foreshadowed the development of the Periodic Table that Dobereiner did not live to see.

2. Newlands' Octaves

John Newlands (FIG 4.6) was an analytical chemist at a sugar refinery in London. He grew interested in studying the properties of elements. In 1864 he arranged the known elements in order of their atomic weights. He noticed that the properties of the elements seemed to be repeated with every eighth element. He stated this fact in his **Law of Octaves.** (The word 'octave' comes from a musical scale where an octave is the eighth note above or below a given note.) Newlands drew up a table containing all the known elements in order of increasing atomic weight. A portion of Newlands' table is shown in TABLE 4.1.

H	Li	Be	B	C	N	O
F	Na	Mg	Al	Si	P	S
Cl	K	Ca	Cr	Ti	Mn	Fe

Table 4.1
Part of Newlands' table. After calcium, Newlands' Law of Octaves starts to break down.

Fig 4.6
John Newlands arranged elements in groups of eight. He said that the first and eighth element of each group has similar properties. He summarised his ideas in his 'Law of Octaves'.

Newlands' Octaves are groups of elements arranged in order of increasing atomic weight, in which the first and the eighth element of each group have similar properties.

If you examine TABLE 4.1, some problems emerge. Oxygen and sulfur were placed in the same group as iron. However, oxygen and sulfur do not resemble iron in chemical properties. Similarly, nitrogen and phosphorus do not resemble manganese in their properties. (Note that the noble gases are missing from the table as they had not been discovered at the time.)

The discrepancies in the table occurred because Newlands did not realise that some elements had not yet been discovered. Newlands should have left gaps in this table to take account of these elements. Unfortunately, he tried to force all the known elements to fit into his table. Newlands' Law of Octaves worked for only about the first 16 elements. After that, some of the elements were forced into the wrong groups. This meant that in some cases his Law of Octaves did not work, i.e. the eighth element was not similar to the first one. Because some elements did not fit the pattern, Newlands was sharply criticised when he read a paper to the Chemical Society in 1866. Newlands was unkindly asked if he might not get better results if he listed the elements in alphabetical order!

The rather hostile reception that the Law of Octaves received and the refusal of the Chemical Society to publish his paper seem to have discouraged him from following up his ideas. However, Newlands' ideas about putting the elements in the form of a table were basically correct. Twenty-one years after his address to the Chemical Society he was awarded the Davy Medal in recognition of his work. Newlands accepted the medal gracefully – he had not grown bitter over the treatment he received.

Newlands' Law of Octaves would not work for the modern Periodic Table as the noble gases have been discovered since he put forward his law. For example, the properties of the elements in the first two periods instead of repeating after every eighth element now repeat after every ninth element.

3. Mendeleev's Periodic Table

Dmitri Mendeleev (Fig 4.7) was professor of chemistry at the University of St. Petersburg, Russia. In 1869, he was writing a chemistry textbook and he wrote the properties of each known element (63 of them) on a separate card. He began arranging the cards in various different orders. He found that if the elements were arranged in order of increasing atomic weight (relative atomic mass), then every eighth element had similar physical and chemical properties. He summarised his findings in what he called his 'Periodic Law'.

Mendeleev's Periodic Law: When elements are arranged in order of increasing atomic weight (relative atomic mass), the properties of the elements vary periodically.

In order to demonstrate this regular repetition of properties, he placed all elements with similar properties in the same vertical column as part of a table. Because the properties of the elements repeated at regular intervals or periodically, the table was called the **Periodic Table of the Elements**. Mendeleev's table was different to Newlands' table in two respects.

(a) **He left gaps in his table to make the elements fit into the proper column**, i.e. so that all elements with similar properties would be in the same column or group of the Periodic Table, (Table 4.2).

Mendeleev then announced that the gaps represented elements that had not yet been discovered. Mendeleev, knowing the properties of the other elements in a particular group, even predicted the properties that the undiscovered element should have. For example, he predicted that there would be a new element discovered to fill the gap below silicon in the table. He called this 'eka-silicon'. The element was discovered 17 years later and called germanium (because it was discovered in Germany). Mendeleev also predicted the discovery of an element to fill

Fig 4.7
Dmitri Mendeleev arranged the elements in order of increasing atomic weight (relative atomic mass). His Periodic Table was quite similar to the modern Periodic Table.

	Group I	Group II	Group III	Group IV	Group V	Group VI	Group VII	Group VIII
Period 1	H							
Period 2	Li	Be	B	C	N	O	F	
Period 3	Na	Mg	Al	Si	P	S	Cl	
Period 4	K	Ca	*	*	As	Se	Br	

Table 4.2
A section of Mendeleev's Periodic Table. The asterisks represent the gaps that Mendeleev left for undiscovered elements.

the gap below aluminium in the table. This element which he called 'eka-aluminium' was discovered in 1875 and named gallium (because it was discovered in France). What was really remarkable about these elements was that the properties predicted by Mendeleev and the observed properties were almost identical. These discoveries led to Mendeleev's Periodic Law being accepted.

(b) **In a few cases he had to reverse the order of elements in order to make them fall into groups of elements with similar properties**. If you look at the modern form of the Periodic Table (inside back cover), you will observe that tellurium (element number 52) is placed before iodine (element number 53). If you look at the atomic weight of tellurium, you will observe that it is 127.6 whereas the atomic weight of iodine is 126.9. Since Mendeleev arranged the elements in order of atomic weight, he originally put iodine before tellurium. However, Mendeleev realised that the properties were more important than the atomic weight and he had to reverse the order so that the elements having similar properties would be in the same group. By reversing the order, iodine is put in the same group as fluorine, chlorine and bromine; all of these elements have similar properties. Similarly, tellurium is put in the same group as oxygen, sulfur and selenium; again all of these elements have similar properties.

At last, order had been brought into the list of elements. In 1955, a newly discovered element (number 101) was named mendelevium in recognition of Mendeleev's enormous contribution to the study of the elements.

4. Moseley – Atomic Number

We have seen above that Mendeleev had to reverse the order of some elements. He did not understand the reason for this. It was an English scientist called Henry Moseley who provided the answer. Moseley (Fig 4.8) was studying the frequencies of the X-rays emitted by atoms of the different elements. In 1913, he found that the frequencies varied with the quantity of positive charge in the nucleus. In other words, he discovered a method of determining the number of protons in the nucleus of an atom. This number is called the atomic number.

> The atomic number of an atom is the number of protons in the nucleus of that atom.

Fig 4.8
Moseley developed a method of determining the atomic number of an element using X-rays.

Moseley's discovery led to a major improvement in Mendeleev's table. Moseley showed that if the elements were arranged in order of increasing atomic number it was not necessary to change the order of some elements as Mendeleev had to do. Moseley's X-ray studies showed that Mendeleev had been right in placing tellurium (atomic number 52) before iodine (atomic number 53) in spite of the higher atomic weight of tellurium. Thus, by arranging the elements in order of increasing atomic number, the elements fall naturally into the correct groups. Moseley was the man who made the final step from Mendeleev's Periodic Table to the modern Periodic Table.

> The modern Periodic Table is an arrangement of elements in order of increasing atomic number.

Thus, the modern version of the Periodic Law may now be stated as follows:

> When elements are arranged in order of increasing atomic number, the properties of the elements vary periodically.

Although Moseley's work did not alter Mendeleev's Periodic Table to a great extent, it certainly confirmed the position of the elements. His X-ray technique was also used to confirm the discovery of new elements. Unfortunately, Moseley was transferred to Turkey at the outbreak of World War I and was killed at Gallipoli in 1915. He was only 27 years old and probably missed out on a Nobel Prize through his early death.

4.4 The Modern Periodic Table

There are a number of differences between Mendeleev's Periodic Table and the modern Periodic Table (shown on the inside back cover). The main differences may be summarised as follows:

1. In the modern Periodic Table, the elements are arranged in order of increasing **atomic number**. In Mendeleev's Periodic Table, the elements are arranged in order of increasing **atomic weight** (relative atomic mass).
2. There are **more elements** in the modern Periodic Table. When Mendeleev constructed his table some elements had not been discovered, e.g. the noble gases (group 0). There were 63 elements in Mendeleev's Periodic Table. There are 109 elements (to date) in the modern Periodic Table.
3. There are **no gaps** in the modern Periodic Table. Mendeleev's Periodic Table contained gaps for undiscovered elements.
4. In the modern Periodic Table, **the transition elements are listed in a separate block** in the middle of the table. Mendeleev did not put these elements in a separate block. He included them with the other elements. In Table 4.2, the transition elements have been left out for clarity but, in fact, Mendeleev included copper in the same box as potassium, zinc in the same box as calcium, etc.

4.5 Making use of the Periodic Table

The Periodic Table has been described as a 'chemical masterpiece'. It summarises an enormous amount of information and has brought order into chemistry. It greatly simplifies the study of chemistry and, since the Periodic Table can be used in the Leaving Certificate examination, it is worth learning how to use it to its maximum! With a copy of the Periodic Table and a knowledge of how it is constructed, we have hundreds of facts at our fingertips!

A complete modern form of the Periodic Table is given on the inside back cover of this textbook. For the sake of completion, the Lanthanide and Actinide elements are included. However, a lot of these elements are radioactive (Chapter 8) and little is known about their chemistry. Thus, we will not be studying them in this course.

In this chapter we shall consider three uses of the Periodic Table:

(i) Obtaining atomic numbers and mass numbers.

(ii) Obtaining relative atomic masses.

(iii) Writing electronic configurations.

Let us now take a close look at each of these three uses.

4.6 Atomic Numbers and Mass Numbers

We have already seen that the modern Periodic Table is an arrangement of elements in order of increasing atomic number. Thus, the number of the element in the Periodic Table tells us the number of protons in the nucleus, e.g. sodium is the eleventh element in the Periodic Table and therefore it has 11 protons in its nucleus. Since the electron has such a small mass, its mass is ignored when calculating the mass of an atom. Since almost the entire mass of an atom is due to the protons and neutrons in the nucleus of the atom, we define the mass number of an element as follows:

> The mass number of an element is the sum of the number of protons and neutrons in the nucleus of an atom of that element.

For example, sodium has 11 protons and 12 neutrons in its nucleus. Therefore, its mass number is 23 or we say that the mass of the atom is 23 atomic mass units. Because the masses of atoms are so small, they are not normally measured in grams. A special unit is used called the **atomic mass unit**. The recognised symbol for atomic mass unit is u but in this text, for clarity, we shall use a.m.u. where 1 a.m.u. = 1.66×10^{-24} gram.

The letter Z is usually used to indicate the atomic number. The letter A is used to indicate the mass number. You may recall learning about the **nuclear formula** of an element at Junior Certificate level. This is simply a method of summarising information about an element as shown in FIG 4.9.

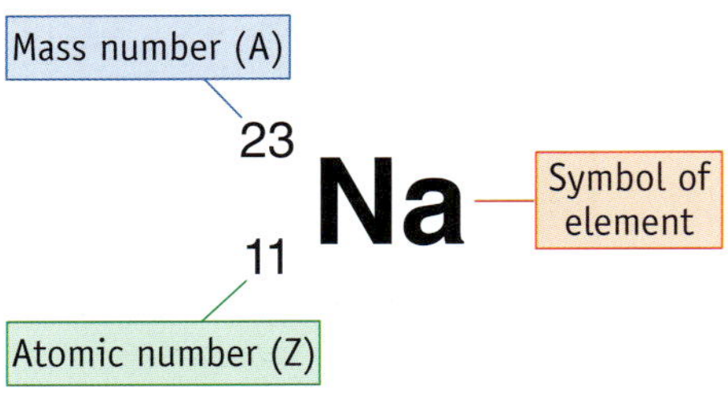

Number of neutrons = A − Z = 23 − 11 = 12

Fig 4.9
The nuclear formula for an element. Note that the number of neutrons is calculated by subtracting the atomic number from the mass number.

Note that the nuclear formula also contains the symbol for the element. For your examination you must know the symbols for the first 36 elements. To help you learn these, the name of each element is given under the symbol in the Periodic Table on the inside back cover of this book. However, the names of the elements do **not** appear on the Periodic Table in the Mathematics Tables that you are allowed to use in examinations. So learn the names and symbols of the first 36 elements! Note also that, in the nuclear formula, the atomic number is given at the bottom left-hand corner near the symbol. However, in each box in the Periodic Table the atomic number is given above the symbol.

Since the mass number is simply the number of protons plus neutrons, mass numbers must always be whole numbers. Though mass numbers are not given on the Periodic Table, the mass number of an element can usually be got by rounding off the number under the symbol to the nearest whole number, e.g. for sodium round off the 22.997 to 23. In fact, for most elements in the Periodic Table, the number under each symbol is not a whole number. This number is called the **relative atomic mass (atomic weight)** of the element. In the next section, we consider what this number actually means.

Example 4.1

How many (i) neutrons, (ii) electrons are there in the ion $^{45}_{21}Sc^{3+}$?

[LCH]

SOLUTION

(i) Number of neutrons = 45 - 21 = 24

(ii) Number of electrons in neutral atom = number of protons = 21

Since the atom has lost three electrons (ion is 3+)

Number of electrons = 21 - 3 = 18

Answer: (i) 24 neutrons, (ii) 18 electrons.

4.7 RELATIVE ATOMIC MASS ISOTOPES. MASS SPECTROMETER

In 1919, an English chemist called Francis William Aston built an instrument called a mass spectrometer to measure the masses of atoms (FIG 4.10). (The details of how this instrument works will be studied at a later stage.) He started working with a sample of neon gas and discovered something very unusual. He found that the neon gas consisted of two varieties of neon atom. One type of neon atom had a mass number of 20 and the other type had a mass number of 22. He concluded that the neon gas consisted of atoms of neon that differed in the number of neutrons in the nucleus $^{20}_{10}Ne$ and $^{22}_{10}Ne$. These two varieties of neon atom are said to be **isotopes** of neon. The first isotope, $^{20}_{10}Ne$, has 10 neutrons in its nucleus.

The second isotope, ${}^{22}_{10}Ne$, has 12 neutrons in its nucleus.

Isotopes are atoms of the same element (i.e. they have the same atomic number) that have different mass numbers due to the different number of neutrons in the nucleus.

Fig 4.10
Aston's original mass spectrometer

Having identified two isotopes of neon, Aston also identified two isotopes of chlorine: ${}^{35}_{17}Cl$ and ${}^{37}_{17}Cl$, commonly called chlorine-35 and chlorine-37.

There are many other examples of isotopes, e.g. carbon has three isotopes, ${}^{12}_{6}C$, ${}^{13}_{6}C$ and ${}^{14}_{6}C$, i.e. carbon-12, carbon-13 and carbon-14, respectively. Three isotopes of hydrogen have been found and these are shown in FIG 4.11.

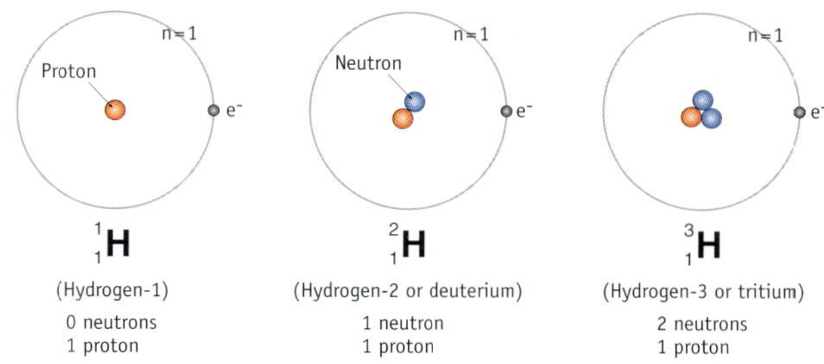

Fig 4.11
Hydrogen has three isotopes called hydrogen-1 (also called protium), hydrogen-2 (also called deuterium) and hydrogen-3 (also called tritium).

For his work on the discovery of isotopes Aston received the 1922 Nobel Prize in chemistry. Not only did Aston detect the presence of isotopes, but he also determined the percentages of each of the isotopes present. For example, in his study of chlorine gas he found that there were approximately three times as many chlorine-35 atoms as there were chlorine-37 atoms. He was then able to work out the average mass of an atom of chlorine. His method of calculation is shown in the following example:

Example 4.2

A sample of chlorine is found to consist of 75% ${}^{35}_{17}Cl$ and 25% ${}^{37}_{17}Cl$. Calculate the average mass of an atom of chlorine.

SOLUTION

In 100 atoms of chlorine there are:

75 atoms of mass 35 = 75 × 35 = 2625

25 atoms of mass 37 = 25 × 37 = 925

Total mass of 100 atoms = 3550

Average mass of 1 atom = 35.5

If you locate chlorine in the Periodic Table you will find that 35.5 is the number given under the symbol for chlorine. (To be more precise it is 35.453 because the percentages have been rounded off in the above problem.) This 'average mass' of an atom is measured relative to the mass of the carbon-12 isotope. For this reason it is called its **relative atomic mass.** (The old name for relative atomic mass was **atomic weight**.) The symbol for relative atomic mass is $\mathbf{A_r}$.

> Relative atomic mass is defined as:
> - The average of the mass numbers of the isotopes of the element,
> - As they occur naturally,
> - Taking their abundances into account and,
> - Expressed on a scale in which the atoms of the carbon-12 isotope have a mass of exactly 12 units.

At one stage, the scale for relative atomic masses was based on hydrogen having a relative atomic mass of 1 and at another stage on oxygen having a relative atomic mass of 16. However, with the discovery that both of these elements have a number of isotopes, it was decided to base the scale on one single isotope, i.e. the carbon-12 isotope. An atom of $^{12}_{6}C$ has a relative atomic mass of exactly 12.

Note: Since relative atomic mass is a **ratio** of two masses, it has no units (i.e. the a.m.u. units above and below the line cancel out). Thus, we say that the average mass of an atom of chlorine is 35.5 a.m.u. but its relative atomic mass is 35.5.

$$\text{Relative atomic Mass} = \frac{\text{Mass of one atom of an element}}{\frac{1}{12}\text{ mass of one atom of carbon–12}}$$

Example 4.3

A mass spectrometer is used to analyse a sample of boron. It is found that the boron sample contains 18.7% $^{10}_{5}B$ and 81.3% $^{11}_{5}B$. Calculate the relative atomic mass of boron.

Solution

In 100 atoms of boron there are

18.7 atoms of mass 10 = 18.7 × 10 = 187

81.3 atoms of mass 11 = 81.3 × 11 = 894.3

Total mass of 100 atoms = 1081.3

Average mass of 1 atom = 10.813

i.e. relative atomic mass of boron = 10.813

So far we have seen that a mass spectrometer may be used to identify the presence of isotopes, measure their relative abundances and measure the relative atomic mass of an element. We now consider the principle behind the operation of this very important instrument.

The Mass Spectrometer

The main parts of a mass spectrometer are shown in Fig 4.12. In order to understand how a mass spectrometer works, we now consider the various parts of the instrument.

1. Sample inlet. A small amount of the gas or liquid is injected into the instrument. As there is a very good vacuum inside it, liquids turn to gases in the instrument. This first stage is called the **vaporisation** stage. If the material being studied is a solid, it may need to be heated in a special attachment to vaporise it.

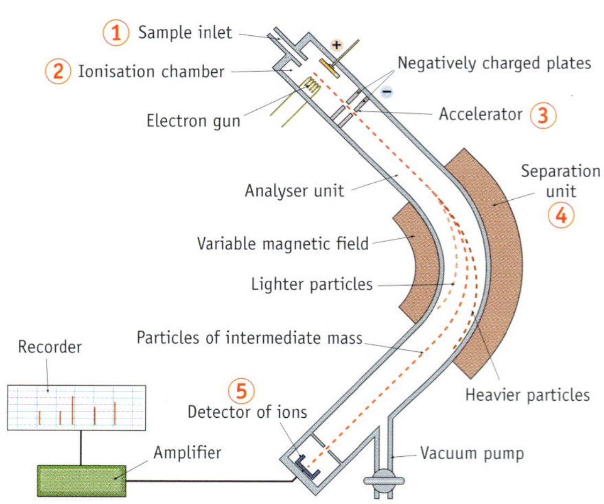

Fig 4.12
The main parts of a mass spectrometer.

2. Ionisation chamber. The purpose of this part of the apparatus is to turn the atoms or molecules into ions (i.e. charged atoms or molecules). This is done using an 'electron gun'. This is simply a heated filament that gives off electrons. The

electron gun fires high-energy electrons at the atoms or molecules of the sample gas. Electrons are knocked off these particles of the sample gas and **positive ions are produced**. (Since electrons are negatively charged, a particle with an overall positive charge is left behind.)

$$X + e^- \rightarrow X^+ + 2e^-$$

bombarding electron ion produced

3. Accelerator. The positive ions produced pass between a series of negatively charged plates. These plates attract the positive ions from the mixture in the ionisation chamber and **accelerate these positive ions** to high speeds. A fine beam of ions then passes into the analyser. (The reason why there must be a good vacuum inside the instrument is so that the positive ions can travel through it without colliding with any air molecules.)
4. Separation unit (magnetic field). In the analyser, the positive ions are made to move in curved paths using a magnetic field produced by a powerful electromagnet. The lighter particles are easier to deflect. The heavier particles are more difficult to deflect. Thus, the lighter ions are pushed further off course than the heavier ions. Therefore, the single beam of positive ions is split into a number of beams as shown in Fig 4.12. Thus, **the ions are separated according to their masses**.
5. Detector. The **detector** is a very sensitive instrument that responds to the number of ions hitting it. By changing the strength of the magnetic field, the operator of the mass spectrometer brings ions of different masses to a focus on the detector. The signal is then amplified into a sizeable electric current and viewed on a computer screen. Finally, the signal is sent to a recorder. A pen attached to a recorder traces out a series of peaks on the recorder paper. This series of peaks is called a **mass spectrum**. An example of a mass spectrum is shown in Fig 4.13. Mass spectra are interpreted by the chemist using his/her own experience and assisted by a library of different mass spectra (usually stored in a computer). The mass spectrum library in University College Cork holds over 90,000 entries! By comparing the mass spectrum of the sample being studied with the spectra in the library, it is possible for the chemist to readily identify compounds.

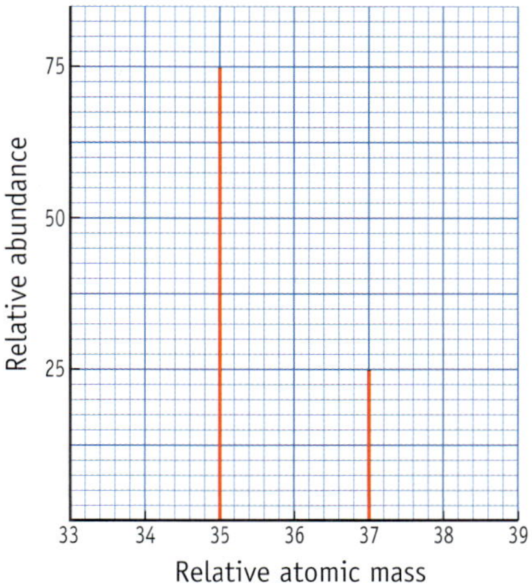

Fig 4.13
The mass spectrum of chlorine. Since the two isotopes of chlorine have different masses, they show up as different peaks on the mass spectrum. The percentage of each isotope present is worked out from the heights of the peaks.

In short, a mass spectrometer is used to (i) identify the presence of isotopes, (ii) measure the relative abundances of isotopes, (iii) measure relative atomic masses and relative molecular masses and (iv) identify unknown compounds.

In any description of how a mass spectrometer works, it is essential that you mention the five stages (vaporisation, ionisation, acceleration, separation and detection) as these are the five stages mentioned on the syllabus.

Note: The original instrument invented by Aston was actually called a 'mass spectrograph' as it used a photographic plate to detect ions. Modern mass spectrometers are much more sophisticated and use electrical detectors. These modern instruments are capable of detecting tiny amounts of substances – even less than one millionth of a gram!

Mass spectrometers are used in conjunction with a process called gas chromatography (Chapter 23) to detect banned drugs taken by athletes. The spectrum obtained is compared with a computer database of spectra of known drugs. Mass spectrometers are also used to test for drugs taken by drug addicts. Drugs leave traces in the hair of the addict. As hair grows about 1 cm per week, this test can reveal when the drug was last used. Hence, the success or failure of a rehabilitation programme can be monitored. A mass spectrometer is used in the Forensic Science Laboratory, Dublin, to analyse samples of suspected drugs confiscated by the Gardai. An example of a modern mass spectrometer is shown in FIG 4.14.

Fig 4.14
This chemist is injecting a sample of a compound to be analysed by the mass spectrometer in the Chemistry Department, University College, Cork.

In recent times, extremely small mass spectrometers have been developed for use in space travel. These mass spectrometers are about the size of a box of chocolates!

When the Viking Lancer space probe landed on Mars, it scooped up material from the planet's surface for analysis by an on-board mass spectrometer. Similarly, the Giotto spacecraft used a mass spectrometer to analyse the gases in the tail of Halley's comet in 1986.

4.8 WRITING ELECTRONIC CONFIGURATIONS OF ATOMS

We have already learned that Niels Bohr proposed the idea of electrons revolving around the nucleus in fixed paths called orbits or energy levels. The arrangement of electrons in an atom (i.e. the number of electrons in each main energy level) is called the **electronic configuration**. There are two methods of writing electronic configurations: either in terms of the simple Bohr model or in terms of energy sublevels. Using the Periodic Table, it is a very straightforward task to write the electronic configuration of an element. We now consider how this is done for both methods of writing electronic configurations.

The electronic configurations of the first 20 elements are given in TABLE 4.3. The Periodic Table is used in writing these electronic configurations.

Element	Atomic Number	Electrons in each shell				Electron configuration
		n=1	n=2	n=3	n=4	
Hydrogen	1	1				1
Helium	2	2				2
Lithium	3	2	1			2,1
Beryllium	4	2	2			2,2
Boron	5	2	3			2,3
Carbon	6	2	4			2,4
Nitrogen	7	2	5			2,5
Oxygen	8	2	6			2,6
Fluorine	9	2	7			2,7
Neon	10	2	8			2,8
Sodium	11	2	8	1		2,8,1
Magnesium	12	2	8	2		2,8,2
Aluminium	13	2	8	3		2,8,3
Silicon	14	2	8	4		2,8,4
Phosphorus	15	2	8	5		2,8,5
Sulfur	16	2	8	6		2,8,6
Chlorine	17	2	8	7		2,8,7
Argon	18	2	8	8		2,8,8
Potassium	19	2	8	8	1	2,8,8,1
Calcium	20	2	8	8	2	2,8,8,2

Table 4.3
The electronic configurations of the first 20 elements given in terms of the simple Bohr picture.

To illustrate the method, let us take potassium as an example.

Example 4.4

Write the electronic configuration for potassium showing the number of electrons in each main energy level.

SOLUTION

Step 1

Find potassium in the Periodic Table. Potassium is element number 19, i.e. there are 19 protons in the nucleus and, since all atoms are neutral, there must be 19 electrons in the energy levels around the nucleus in a neutral potassium atom. We must now write down how these electrons are arranged around the nucleus.

Step 2

Go to the beginning of the Periodic Table. Note that in the n = 1 period, there are only two elements. This means that the n = 1 main energy level can only hold two electrons. Of the nineteen electrons of potassium, **two electrons are in the n = 1 main energy level.**

Step 3

Start at the second period and count along from lithium. Note that there are eight elements in the second period, i.e. there are eight electrons in the n = 2 main energy level. Thus, we can say that potassium has **eight electrons in the n = 2 main energy level.**

Step 4

Start at the third period and count along from sodium. Note that there are eight elements in the third period, i.e. there are eight electrons in the n = 3 main energy level. Thus, we can say that potassium has **eight electrons in the n = 3 main energy level.**

Step 5

Start at the fourth period. Potassium is the first element you meet, i.e. potassium has **one electron in the n = 4 period.**

From the above information, we can write the electronic configuration of potassium as:

K = 2, 8, 8, 1

For your examination you must be able to write the electronic configurations of the first 20 elements in terms of the number of electrons in each main energy level. Study TABLE 4.3 carefully. Note that the number of electrons in the outer shell is the same as the group number, e.g. lithium (group I) has one outer electron, beryllium (group II) has two outer electrons, boron (group III) has three outer electrons, etc.

We now consider how to write the electronic configurations of the elements in terms of sublevels.

We have already learned that spectroscopists deduced that each main energy level, excluding the first, consisted of a number of sublevels. The n = 2 main energy level has two sublevels, the n = 3 main energy level has three sublevels, the n = 4 main energy level has four sublevels, etc. These sublevels are shown in FIG 4.15. From the study of spectra, chemists have learned that an s sublevel can hold two electrons – alternatively, we can say that an s **orbital** can hold two electrons. It has also been discovered that a p sublevel can hold six electrons – two electrons in each of the three p orbitals, p_x, p_y and p_z. Finally, a d sublevel can hold ten electrons – two electrons in each of the five d orbitals. (The f sublevel need not concern us in this course.)

There is a very simple rule for assigning electrons to the various sublevels. It is called the **Aufbau Principle**. (The word **Aufbau** is a German word meaning 'building up'.) The Aufbau Principle helps us to get a picture of how the electronic structure of an atom is built up as electrons are added to the atom.

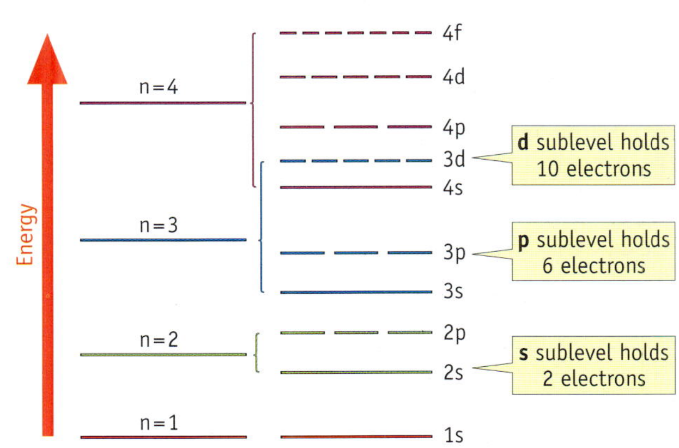

Fig 4.15
All of the main energy levels consist of sublevels. An s sublevel contains one s atomic orbital, a p sublevel contains three p atomic orbitals and a d sublevel contains five d atomic orbitals.

The Aufbau Principle states that when building up the electronic configuration of an atom in its ground state, the electrons occupy the lowest available energy level.

It is clear from Fig 4.15 that the order in which the sublevels are filled is 1s, 2s, 2p, 3s, 3p, 4s, 3d, 4p, etc. However, there is no need to learn off this order as it may easily be deduced from the Periodic Table. We consider the Periodic Table as being divided into four blocks – s block, p block, d block and f block. All elements in the s block have their outer (highest energy) electrons in an s sublevel, all those in the p block have their outer (highest energy) electrons in a p sublevel, all elements in the d block have their outer (highest energy) electrons in a d sublevel, etc. (Fig 4.16).

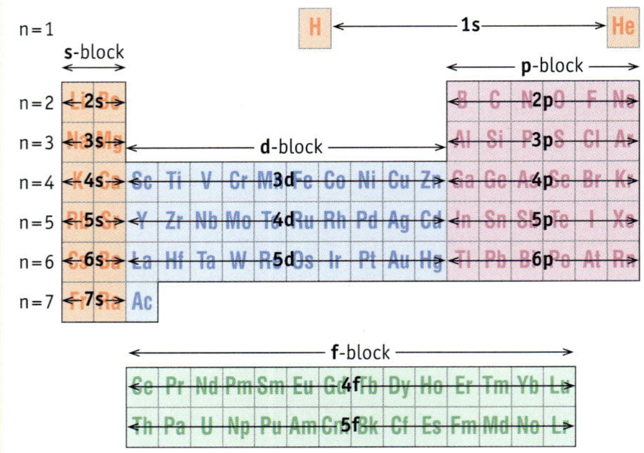

Fig 4.16
The electron configurations (s, p, d, etc.) may easily be deduced from the Periodic Table.

Study the following examples carefully.

Example 4.5

Write the electronic configuration (s, p) for an atom of oxygen.

Solution

Step 1

Locate the position of oxygen in the Periodic Table. Oxygen is element number 8, i.e. there are eight electrons to be distributed among the sublevels.

Step 2

Starting with hydrogen, we note that there are two electrons in the n = 1 main energy level. These two electrons are in the orbital of lowest energy, i.e. the 1s atomic orbital.
We indicate that there are two electrons in the 1s sublevel by writing $1s^2$.

Step 3

Starting at the second period with lithium, we note that there are two electrons in the 2s orbital ($2s^2$) and four electrons in the 2p orbital ($2p^4$).
Therefore, the electronic configuration of oxygen may be written as:
$O = 1s^2, 2s^2, 2p^4$
Check that the superscripts (2 + 2 + 4) add up to the atomic number of oxygen.

Note carefully the notation that is used in writing electron configurations! The number in front of the letter represents the main energy level, the letter itself represents the type of sublevel and the superscript represents the number of electrons in the sublevel.

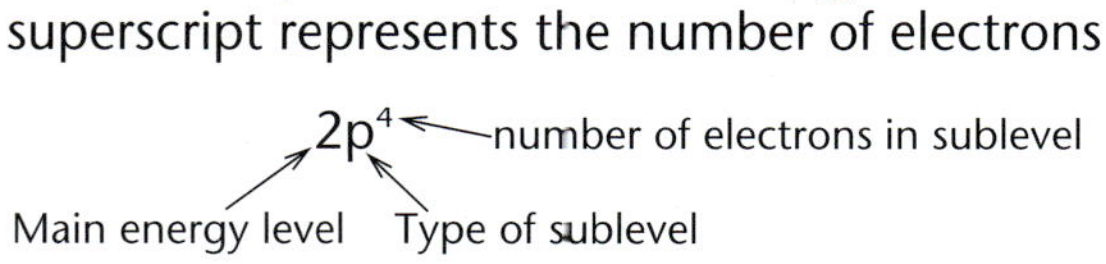

Example 4.6

Write the electronic configuration (s, p) for an atom of calcium.

SOLUTION

Step 1

Locate the position of calcium in the Periodic Table. Calcium is element number 20, i.e. there are 20 electrons to be distributed among the sublevels.

Step 2

Starting with hydrogen, we note that there are two electrons in the n = 1 main energy level. These two electrons are in the orbital of lowest energy, i.e. the 1s atomic orbital ($1s^2$).

Step 3

Starting at the second period with lithium, we note that there are two electrons in the 2s sublevel ($2s^2$) and six electrons in the 2p sublevel ($2p^6$).

Step 4

Starting at the third period with sodium, we note that there are two electrons in the 3s sublevel ($3s^2$) and six electrons in the 3p sublevel ($3p^6$).

Step 5

Starting at the fourth period with potassium, we note that there are two electrons in the 4s sublevel ($4s^2$). Therefore, the electron configuration of calcium may be written as:

$$Ca = 1s^2, 2s^2, 2p^6, 3s^2, 3p^6, 4s^2$$

ELECTRONIC CONFIGURATIONS OF d-BLOCK ELEMENTS

The electronic configurations of the first 36 elements are given in TABLE 4.4. Note that this table includes the electronic configurations of the first row of the d-block elements.

H =	$1s^1$
He =	$1s^2$
Li =	$1s^2, 2s^1$
Be =	$1s^2, 2s^2$
B =	$1s^2, 2s^2, 2p^1$
C =	$1s^2, 2s^2, 2p^2$
N =	$1s^2, 2s^2, 2p^3$
O =	$1s^2, 2s^2, 2p^4$
F =	$1s^2, 2s^2, 2p^5$
Ne =	$1s^2, 2s^2, 2p^6$
Na =	$1s^2, 2s^2, 2p^6, 3s^1$
Mg =	$1s^2, 2s^2, 2p^6, 3s^2$
Al =	$1s^2, 2s^2, 2p^6, 3s^2, 3p^1$
Si =	$1s^2, 2s^2, 2p^6, 3s^2, 3p^2$
P =	$1s^2, 2s^2, 2p^6, 3s^2, 3p^3$
S =	$1s^2, 2s^2, 2p^6, 3s^2, 3p^4$
Cl =	$1s^2, 2s^2, 2p^6, 3s^2, 3p^5$
Ar =	$1s^2, 2s^2, 2p^6, 3s^2, 3p^6$
K =	$1s^2, 2s^2, 2p^6, 3s^2, 3p^6, 4s^1$
Ca =	$1s^2, 2s^2, 2p^6, 3s^2, 3p^6, 4s^2$
Sc =	$1s^2, 2s^2, 2p^6, 3s^2, 3p^6, 4s^2, 3d^1$
Ti =	$1s^2, 2s^2, 2p^6, 3s^2, 3p^6, 4s^2, 3d^2$
V =	$1s^2, 2s^2, 2p^6, 3s^2, 3p^6, 4s^2, 3d^3$
Cr =	$1s^2, 2s^2, 2p^6, 3s^2, 3p^6,$ **$4s^1, 3d^5$**
Mn =	$1s^2, 2s^2, 2p^6, 3s^2, 3p^6, 4s^2, 3d^5$
Fe =	$1s^2, 2s^2, 2p^6, 3s^2, 3p^6, 4s^2, 3d^6$
Co =	$1s^2, 2s^2, 2p^6, 3s^2, 3p^6, 4s^2, 3d^7$
Ni =	$1s^2, 2s^2, 2p^6, 3s^2, 3p^6, 4s^2, 3d^8$
Cu =	$1s^2, 2s^2, 2p^6, 3s^2, 3p^6,$ **$4s^1, 3d^{10}$**
Zn =	$1s^2, 2s^2, 2p^6, 3s^2, 3p^6, 4s^2, 3d^{10}$
Ga =	$1s^2, 2s^2, 2p^6, 3s^2, 3p^6, 4s^2, 3d^{10}, 4p^1$
Ge =	$1s^2, 2s^2, 2p^6, 3s^2, 3p^6, 4s^2, 3d^{10}, 4p^2$
As =	$1s^2, 2s^2, 2p^6, 3s^2, 3p^6, 4s^2, 3d^{10}, 4p^3$
Se =	$1s^2, 2s^2, 2p^6, 3s^2, 3p^6, 4s^2, 3d^{10}, 4p^4$
Br =	$1s^2, 2s^2, 2p^6, 3s^2, 3p^6, 4s^2, 3d^{10}, 4p^5$
Kr =	$1s^2, 2s^2, 2p^6, 3s^2, 3p^6, 4s^2, 3d^{10}, 4p^6$

Table 4.4

The electronic configurations of the first 36 elements.

There are two important points to remember when writing the electronic configurations of these elements.

(i) Since the 4s sublevel is lower in energy than the 3d sublevel, the 4s sublevel is always filled before the 3d sublevel. Using the Periodic Table to help you write electronic configurations will help you in this regard. Simply write down the electronic configuration of calcium as shown above, count along the first row of the d-block (scandium, titanium, vanadium, etc.) and put the required number of electrons into the 3d sublevel. For example, since iron is the sixth element along the row, we put six electrons into the 3d sublevel, i.e. the electronic configuration of iron is $1s^2$, $2s^2$, $2p^6$, $3s^2$, $3p^6$, $4s^2$, $\mathbf{3d^6}$. Note that there are 10 elements in the first row of the d-block. This is because the d sublevel can hold ten electrons. Thus, the n = 3 main energy level can hold a total of 18 electrons: 2 electrons in the 3s sublevel, 6 electrons in the 3p sublevel and 10 electrons in the 3d sublevel. The number of electrons that any main energy level can hold may be calculated from the formula $2n^2$ where n = main energy level. Thus, the n = 1 main energy level can hold 2 electrons, the n = 2 main energy level can hold 8 electrons, the n = 3 main energy level can hold 18 electrons, etc.

(ii) If you study TABLE 4.4 you will see that the electronic configurations of chromium and copper are not what you might expect! You might think that the electronic configuration of chromium would be Cr = $1s^2$, $2s^2$, $2p^6$, $3s^2$, $3p^6$, $4s^2$, $3d^4$ rather than **Cr =** $\mathbf{1s^2, 2s^2, 2p^6, 3s^2, 3p^6, 4s^1, 3d^5}$ as written in TABLE 4.4. However, it has been found that p, d and f sublevels that are exactly half-filled or completely filled have extra stability. Therefore, one of the electrons in the $4s^2$ orbital 'flips over' to one of the d orbitals to give a d sublevel which is exactly half-filled. Hence, the electronic configuration of chromium is $1s^2$, $2s^2$, $2p^6$, $3s^2$, $3p^6$, $\mathbf{4s^1, 3d^5}$.

The situation with copper is rather similar to that of chromium. You might expect the electronic configuration of copper to be Cu = $1s^2$, $2s^2$, $2p^6$, $3s^2$, $3p^6$, $4s^2$, $3d^9$ rather than **Cu =** $\mathbf{1s^2, 2s^2, 2p^6, 3s^2, 3p^6, 4s^1, 3d^{10}}$ as shown in TABLE 4.4. However, it has been found that one of the electrons in the $4s^2$ orbital 'flips over' to one of the d orbitals. The completely filled 3d sublevel has extra stability associated with it. Hence, the electronic configuration of copper is $1s^2$, $2s^2$, $2p^6$, $3s^2$, $3p^6$, $\mathbf{4s^1, 3d^{10}}$.

For your examination you are expected to know how to write the electronic configurations of the first 36 elements in terms of their s, p, d configurations. Therefore, TABLE 4.4 should be studied carefully! Also, if you are asked to write the electronic configuration of either copper or chromium, alarm bells should start ringing in your head!

4.9 WRITING ELECTRONIC CONFIGURATIONS OF IONS

An atom that has lost or gained electrons is called an ion. Since electrons are negatively charged, an atom that gains electrons becomes negatively charged. On the other hand, an atom that loses electrons becomes positively charged. We make use of these facts when writing the electronic configuration of an ion.

Note: The electronic configurations of ions of the d-block elements are a little more complicated than those of the s and p-block elements. However, you need not be concerned about the ions of the d-block elements as they are not on the syllabus.

Example 4.7

Write the electronic configuration (s, p) of the Al^{3+} ion. What neutral atom has the same configuration? *[LCH]*

SOLUTION

Step 1

Locate the position of aluminium in the Periodic Table. Aluminium is element number 13, i.e. an atom of aluminium has 13 electrons.

Step 2

Work out the number of electrons in the ion. Since the aluminium ion has three positive charges, it must have lost three electrons. Thus, the Al^{3+} ion has 10 electrons, i.e. the same number of electrons as an atom of neon. Therefore, the electronic configuration of the Al^{3+} ion will be the same as an atom of neon.

$$Al^{3+} = 1s^2, 2s^2, 2p^6$$

Example 4.8

Write the electronic configuration (s, p) of the S^{2-} ion. What neutral atom has the same configuration?

SOLUTION

Step 1

Sulfur is element number 16, i.e. an atom of sulfur contains 16 electrons.

Step 2

Since the sulfide ion has two negative charges, it must have gained two electrons, i.e. the ion contains 18 electrons, which is the same number of electrons as an atom of argon. Thus, the electronic configuration of the S^{2-} ion will be the same as an atom of argon.

$$S^{2-} = 1s^2, 2s^2, 2p^6, 3s^2, 3p^6$$

Example 4.9

Identify the species represented by $[1s^2, 2s^2, 2p^6]^{2+}$

SOLUTION

This ion has 2 + 2 + 6 =10 electrons. The two positive charges on the ion indicate that it was formed from an atom by losing two electrons. Therefore, the atom from which the ion was formed must contain 12 electrons, i.e. the atom is magnesium. Thus, the species represented by $[1s^2, 2s^2, 2p^6]^{2+}$ is the Mg^{2+} ion.

4.10 ARRANGEMENT OF ELECTRONS IN ORBITALS OF EQUAL ENERGY

In writing electronic configurations we have not yet considered how electrons are distributed among orbitals of equal energy. For example, we have written the electronic configuration of nitrogen as $1s^2, 2s^2, 2p^3$. We now ask ourselves the question: how are the three electrons in the p sublevel distributed among the three p orbitals? In 1927, Friederich Hund, having studied the spectra of various elements, put forward an important rule to answer this question. This rule is called Hund's Rule of Maximum Multiplicity.

> Hund's Rule of Maximum Multiplicity states that when two or more orbitals of equal energy are available, the electrons occupy them singly before filling them in pairs.

If we apply this rule to the nitrogen atom it simply means that, of the three electrons added to the 2p sublevel, one electron will enter the p_x orbital, the second will enter the p_y and the third will enter the p_z. Thus, there is the maximum number of unpaired electrons. This is illustrated in FIG 4.17.

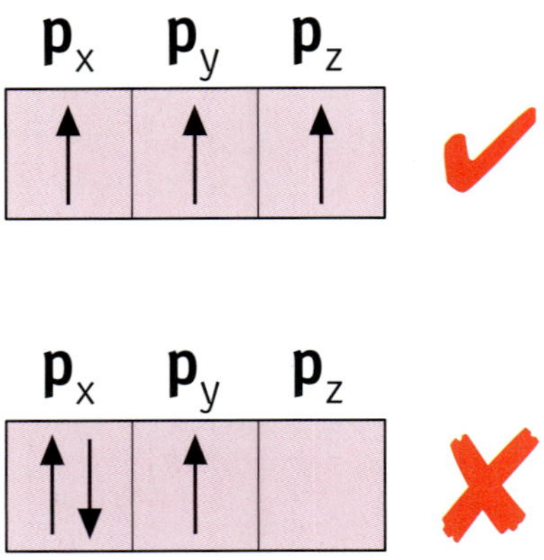

Fig 4.17
The three p electrons of nitrogen occupy the p orbitals singly to give the maximum number of unpaired electrons (Hund's Rule).

These 'arrows-in-boxes' diagrams are used to show how electrons are distributed in orbitals. The box represents the orbital and the arrow represents the electron. Arrows are used because it was discovered that, as electrons revolve around the nucleus, they spin on their own axis. (This is rather like the Earth spinning on its axis as it revolves around the sun.) However, the electrons can spin in either a clockwise direction or an anti-clockwise direction. In 1925, Wolfgang Pauli from his study of spectra made a very important deduction called the **Pauli Exclusion Principle**.

> The Pauli Exclusion Principle states that no more than two electrons may occupy an orbital and they must have opposite spin.

The arrows are used to indicate the direction of spin of the electrons. The ↑ represents an electron spinning clockwise and ↓ represents an electron spinning anti-clockwise. The 'arrows-in-boxes' method of representing the electron arrangements in atoms is shown for the atoms of the first ten elements in FIG 4.18.

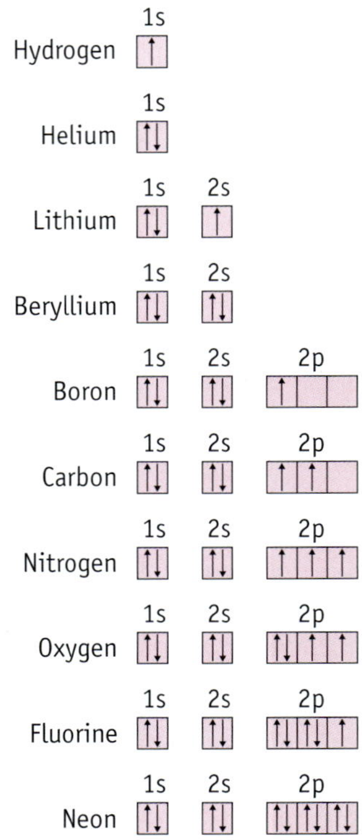

Fig 4.18
The 'arrows-in-boxes' method of representing the electronic configurations of the first ten elements.

The distribution of electrons among the p orbitals may also be shown using a more detailed form of the electronic configurations given in TABLE 4.4. For example, the electronic configurations of the elements from boron to neon may also be written as:

$$B = 1s^2, 2s^2, 2p_x^{\,1}, 2p_y^{\,0}, 2p_z^{\,0}$$
$$C = 1s^2, 2s^2, 2p_x^{\,1}, 2p_y^{\,1}, 2p_z^{\,0}$$
$$N = 1s^2, 2s^2, 2p_x^{\,1}, 2p_y^{\,1}, 2p_z^{\,1}$$
$$O = 1s^2, 2s^2, 2p_x^{\,2}, 2p_y^{\,1}, 2p_z^{\,1}$$
$$F = 1s^2, 2s^2, 2p_x^{\,2}, 2p_y^{\,2}, 2p_z^{\,1}$$
$$Ne = 1s^2, 2s^2, 2p_x^{\,2}, 2p_y^{\,2}, 2p_z^{\,2}$$

The 'arrows-in-boxes' diagrams or the extended forms of electronic configurations as written above need only be given if you are specifically asked about the distribution of electrons in the p sublevel. In all other cases, the electronic configurations as given in TABLE 4.4 are perfectly adequate.

CHAPTER SUMMARY

- An element is defined as a substance that cannot be split into simpler substances by chemical means.
- Johann Dobereiner classified elements in groups of three called triads.
- John Newlands classified elements in groups arranged in order of increasing atomic weight called octaves in which the first and eighth element of each group had similar properties.
- Dmitri Mendeleev arranged the elements in order of increasing atomic weight (relative atomic mass) showing a periodic repetition of chemical properties.
- Henry Moseley used X-rays to determine the atomic number of an element, i.e. the number of protons in the nucleus of an atom.
- The modern Periodic Table is an arrangement of elements in order of increasing atomic number.
- The mass number of an element is the sum of the number of protons and neutrons in the nucleus of an atom of that element.
- Isotopes are atoms of the same element (i.e. they have the same atomic number) which have different mass numbers due to the different numbers of neutrons in the nucleus.
- Relative atomic mass (atomic weight) is defined as the average of the mass numbers of the isotopes of the element, as they occur naturally, taking their abundances into account and expressed on a scale in which the atoms of the carbon-12 isotope have a mass of exactly 12 a.m.u.
- A mass spectrometer is used to (i) identify the presence of isotopes, (ii) measure their relative abundances, (iii) measure relative atomic masses and (iv) identify unknown compounds.
- The electronic configuration of an element shows the arrangement of electrons in an atom of that element.

EXERCISES

4.1. (a) The Greek philosophers suggested that all substances were made up of four elements. These were ....

(b) An English chemist called .... discovered sodium, potassium and a number of other elements.

(c) State briefly the contribution of Dobereiner to the classification of the elements ....

(d) In the history of the classification of the elements, name (i) the scientist who arranged elements in octaves, (ii) the scientist who predicted the discovery of new elements. (i).......... (ii) ....

(e) There were a number of discrepancies in the table of elements drawn up by Newlands. How did these discrepancies arise? ....

(f) Mendeleev's table of elements was different to that of Newlands in two main respects. Name any one of these differences ....

(g) Name the scientist who developed a method of determining the atomic number of an element..........

(h) State the Periodic Law ....

(i) Define relative atomic mass ....

(j) How many (i) neutrons, (ii) electrons are there in the ion $^{37}_{17}Cl^-$ ....

(k) Name the instrument used to identify isotopes of elements. ....

(l) Calculate the approximate relative atomic mass of Gallium given that it contains 60% of the isotope $^{69}_{31}Ga$ and 40% of the isotope $^{71}_{31}Ga$. ....

(m) Write the electronic configuration for (i) the fluorine atom and (ii) the calcium ion. ....

(n) Write down the s, p, d configuration for (i) chromium and (ii) manganese. ....

(o) What have the ions O^{2-}, F^-, Na^+ and Mg^{2+} got in common? ....

(p) What atoms are represented by the following electronic configurations: (i) $1s^2, 2s^2, 2p^3$ (ii) $1s^2, 2s^2, 2p^6, 3s^2, 3p^6, 4s^2, 3d^{10}$. (i)................... (ii) ....

(q) The electronic configuration for a metal ion M^{3+} is $1s^2, 2s^2, 2p^6$. Write the s, p configuration for the metal atom ....

(r) Show, using a diagram or otherwise, how the electrons in an atom of fluorine are distributed among the p orbitals ....

4.2. 'In Mendeleev's Periodic Table the elements were arranged in order of increasing relative atomic mass (atomic weight). In the modern Periodic Table the elements are arranged in order of increasing atomic number'.
Explain the meaning of the underlined terms.
What would be the effect on (a) argon and potassium and (b) tellurium and iodine in the modern Periodic Table if the elements were arranged in order of increasing relative atomic mass.
State the number of electrons, protons and neutrons in an atom of the isotope $^{26}_{12}Mg$.

4.3. With the aid of a labelled diagram, describe the mode of operation of a mass spectrometer. State two uses of a mass spectrometer. Why is it necessary to operate a mass spectrometer under high vacuum?

4.4. The electronic arrangement of the boron atom in its ground state can be shown as follows (where each arrow signifies an electron)

1s	2s	$2p_x$	$2p_y$	$2p_z$
↑↓	↑↓	↑		

Show in a similar manner the ground state electronic arrangement of an atom of (i) carbon, (ii) nitrogen, (iii) oxygen.
Name and state the rule illustrated by the electronic configurations you have shown.
Write the s, p, d ground state configuration of the atom with atomic number 21, putting the sublevels in the order in which they are filled. *[LCH]*

4.5. (a) State Pauli's Exclusion Principle.
Name the species represented by each of the following structures:
(i) $1s^2, 2s^2, 2p^6, 3s^2, 3p^6, 4s^2, 3d^8$ (ii) $1s^2, 2s^2, 2p^6, 3s^2, 3p^6]^{2-}$ *[LCH]*

(b) (i) Define the term isotope.

(ii) State the number of protons and the number of neutrons in the isotopes of copper $^{63}_{29}Cu$ and $^{65}_{29}Cu$.

(iii) It was found using a mass spectrometer that copper consists of 70% $^{63}_{29}Cu$ and 30% $^{65}_{29}Cu$. Calculate the relative atomic mass of copper.

(iv) What is the maximum number of electrons that can be accommodated in the third (n = 3) energy level?

(v) Why is the arrangement of electrons in energy levels for the ground state configuration of potassium 2, 8, 8, 1 and not 2, 8, 9? *[LCH]*

Chemical Bonding: Chemical Formulas

5

5.1 INTRODUCTION

If you look around you, you may not see many things that are pure elements. This is because most of the materials around us are made from combinations of elements, e.g. water is formed when hydrogen and oxygen combine together, table salt is formed by combining sodium and chlorine, carbon dioxide is formed when carbon and oxygen combine together, etc. Substances made by joining together two or more different elements are called **compounds**.

A compound is a substance that is made up of two or more different elements combined together chemically.

When elements combine to form compounds, there must be some sort of attractive force that holds the atoms together. These attractive forces are called **chemical bonds**. In this chapter we shall study the different types of chemical bonds that exist. We shall see how a knowledge of chemical bonding helps us to understand chemical formulas, e.g. why the formula for water is H_2O and not H_3O. We will learn how to write chemical formulas and how to name compounds.

It will become quite easy for us to understand chemical bonding if we first study an important rule called the Octet Rule.

5.2 THE OCTET RULE

In the early 1900s, chemists observed that the noble gases (helium, neon, argon, etc.) in Group 0 of the Periodic Table had very little reactivity, i.e. they formed practically no compounds. Hence, they are often called the 'inert' gases, (Fig 5.1). All of the noble gases consist of single atoms. It was reasoned that these elements do not react because they have stable electronic configurations. In 1916, two American chemists, Gilbert Lewis and Irving Langmuir, proposed a rule to predict how elements combine together to form compounds. Since all of the noble gases (except helium) have eight electrons in their outer main energy level, the rule was called the octet rule (octet = eight).

Fig 5.1
Helium and argon are two of the noble (inert) gases. Helium is a very light gas and is much safer than hydrogen for use in balloons. Argon is used to fill light bulbs. If the bulb contained oxygen, the filament would burn away.

The Octet Rule states that when bonding occurs, atoms tend to reach an electron arrangement with eight electrons in the outermost shell.

In other words, when elements react together to form compounds, their atoms tend to change their electron arrangement to try to end up with eight electrons in their outermost main energy level. It is important to realise that the octet rule is more a 'rule of thumb' rather than a strict chemical law. There are a number of exceptions to the octet rule. The main exceptions are:

(i) Transition metals do not usually obey the octet rule. In many of their compounds, transition metals can have more or less than eight electrons in their outermost energy level.

(ii) The elements near helium in the Periodic Table (e.g. hydrogen, lithium, etc.) tend to achieve the electronic arrangement of helium with **two** electrons in the outer shell rather than the eight electrons of the other noble gases.

We now study how the octet rule helps us to understand the two main types of chemical bonds – ionic bonds and covalent bonds.

5.3 IONIC BONDING – TRANSFER OF ELECTRONS

Atoms can obtain the electronic configurations of a noble gas by gaining or losing electrons. When an atom gains or loses electrons it becomes a charged atom, commonly referred to as an **ion**. As we shall see later, a group of atoms carrying a charge is also called an ion.

An ion is a charged atom or group of atoms.

For example, the elements in group I of the Periodic Table tend to lose one electron to form an ion with one positive charge. The elements in group II of the Periodic Table tend to lose two electrons to form an ion with two positive charges. In a positive ion, since the atom has lost electrons, there are more protons in the nucleus than there are electrons in orbit around it. The formation of some **positive ions** (also called **cations**) may be shown simply as follows:

$Na \rightarrow Na^+ + e^-$ (2,8,1 → 2,8) $Ca \rightarrow Ca^{2+} + 2e^-$ (2,8,8,2 → 2,8,8)

On the other hand, the elements in groups VI and VII of the Periodic Table tend to gain electrons to form **negative ions** (also called **anions**).

$Cl + e^- \rightarrow Cl^-$ (2,8,7 → 2,8,8) $O + 2e^- \rightarrow O^{2-}$ (2,6 → 2,8)

Note that in each case the ion formed has the electronic configuration of the nearest noble gas. It is the formation of this stable electronic configuration that is the 'driving force' behind the formation of the ion. Once the ions are formed, they are held together in the compound by the force of attraction of the oppositely charged ions. This force of attraction is called an ionic bond.

An ionic bond is the force of attraction between oppositely charged ions in a compound.

Common table salt (sodium chloride) is one of the best known examples of an ionic substance, i.e. a substance held together by ionic bonding. There are two ways of representing the formation of the ionic bonding in sodium chloride: either in terms of the Bohr-type circle diagram (FIG 5.2(a)) or in terms of the 'dot-and-cross diagram' (FIG 5.2(b)). In drawing electron dot-and-cross diagrams note that the outer electrons of one atom are represented by dots, with crosses used for the electrons of the other atom. Thus, the dots and crosses represent only the electrons in the outer energy level. The numbers under the chemical symbols show the arrangement of electrons in the energy levels.

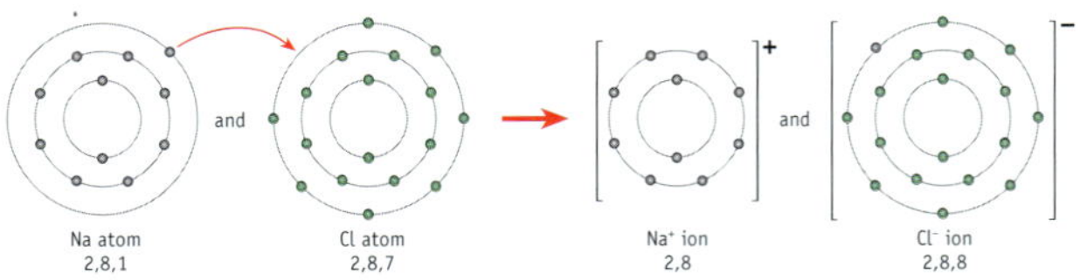

Fig 5.2 (a) The formation of the ionic bond in sodium chloride using Bohr-type diagrams.

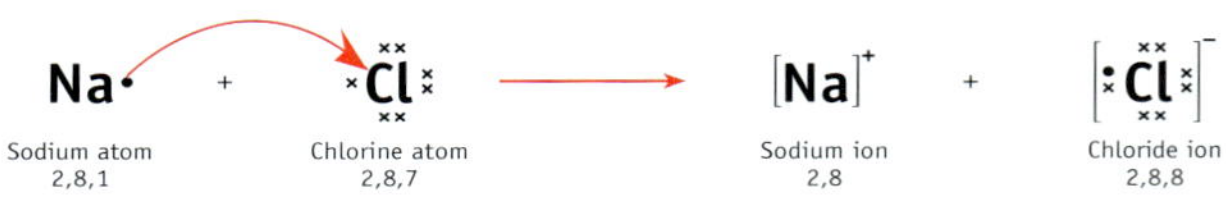

Fig 5.2 (b) The formation of the ionic bond in sodium chloride using dot-and-cross diagrams.

Ionic bonds are always formed by the complete transfer of electrons from one atom to another. This transfer of electrons is often shown by curved arrows (FIG 5.2 (a) and (b)).

Both methods of representing ionic bond formation are equally acceptable in an examination provided the electronic configuration is written as shown in FIG 5.2. Bohr-type circle diagrams are rather tedious to draw and give more information than we need since it is only the outer electrons that are involved in bonding. Therefore, we shall use dot-and-cross diagrams as much as possible in this textbook.

Example 5.1

Show the formation of the ionic bond in magnesium fluoride, MgF_2, by means of a suitable diagram.

SOLUTION

Each magnesium atom loses two electrons and each fluorine atom gains only one electron. Therefore, there must be two fluorine atoms for every one magnesium atom (FIG 5.3).

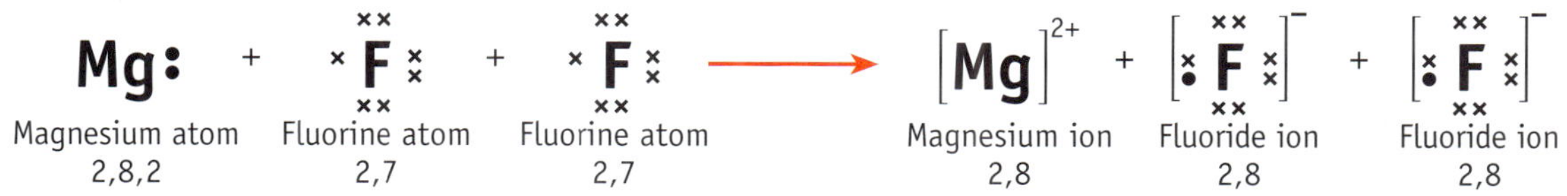

Fig 5.3
The formation of the ionic bond in magnesium fluoride, MgF_2.

X-ray studies enable chemists to work out the arrangement of ions in a crystal. For example, sodium chloride has a cubic structure (FIG 5.4).

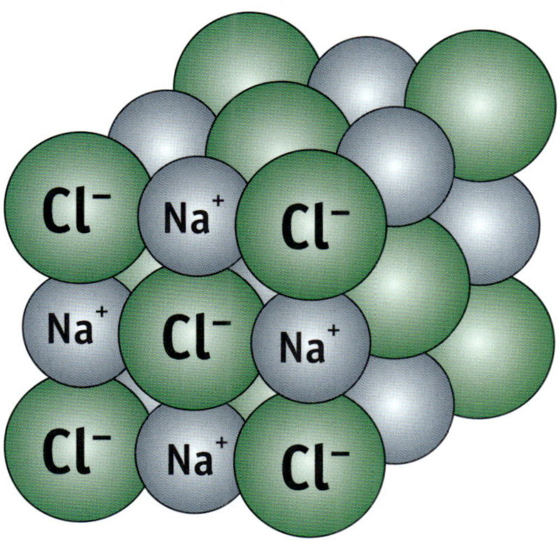

Fig 5.4
This diagram represents the crystal structure of sodium chloride, i.e. how the ions are packed together in the crystal.

Although FIG 5.4 shows only 27 ions, it must be remembered that the structure of sodium chloride consists of a regular three-dimensional arrangement of millions of sodium ions and chloride ions. This three-dimensional arrangement of ions is called a **crystal lattice.** FIG 5.4 shows the repeating unit in the crystal lattice. This repeating unit is called the **unit cell** of sodium chloride and it repeats itself in all directions to build up the crystal. (Under a powerful microscope the cubic shape of the sodium chloride crystals is very clear.) In the crystal lattice, each sodium ion is surrounded by six chloride ions and each chloride ion is surrounded by six sodium ions. In fact, ions are so small that the tiniest crystal of sodium chloride consists of millions of ions! It is a useful exercise to build a model of the unit cell of sodium chloride, (FIG 5.5).

Fig 5.5
A 'home-made' model of the unit cell of sodium chloride.

Sodium chloride (commonly called 'salt') is one of the most important ionic compounds in our lives. In Roman times, soldiers received part of their pay in salt – the Latin word for salt is 'sal', hence the word 'salary'. (If somebody tells you that you are 'not worth your salt', don't take this as a compliment!) Salt is an essential part of our diet. We need to take in about 0.5 g of salt every day to replace what we lose in sweat, tears and urine. However, most of us take in a lot more than this as much of the food we eat contains salt. Too little salt causes cramp in our muscles. Hence, many athletes take salt tablets to replace the salt they lose in sweating (FIG 5.6). Too much salt in your diet over a long period can cause high blood pressure – a major cause of heart disease and strokes. For centuries, salt has been used as a food preservative but it is also widely used in the chemical industry. It is used in the manufacture of soap, leather, sodium metal, chlorine gas, detergents, toothpaste, caustic soda, washing soda, etc. It is also spread on roads during the winter to help melt frost and snow. Unfortunately, salt spread on roads often gets splashed on to cars and this causes iron to rust more quickly.

Fig 5.6
Athletes take salt tablets to replace the salt they lose in sweating. Sweat is mainly salt solution.

5.4. FORMULAS OF IONIC COMPOUNDS

A **chemical formula** is a way of representing a compound using symbols for the atoms present and numbers to show how many atoms of each element are present. We have seen in EXAMPLE 5.1 that the chemical formula for magnesium fluoride is MgF_2. This tells us that the compound magnesium fluoride consists of magnesium and fluorine. It also tells us that there are twice as many fluoride ions as magnesium ions in a crystal of magnesium fluoride.

For your examination you must know how to write the formulas of ionic compounds of the first 36 elements. Before we begin, it is worth noting that ionic compounds are generally formed between the elements of groups I and II (metals) and those of groups VI and VII (non-metals) of the Periodic Table. Thus, it is true to say that ionic compounds are usually formed when metals react with non-metals. (The 'steps of stairs' going from boron to astatine divide the metals on the left from the non-metals on the right). As a general rule, metals have a tendency to lose electrons and non-metals have a tendency to gain electrons.

Some of the ions of the first 36 elements (excluding the d-block metals) are shown in FIG 5.7. In writing the formula of an ionic compound, it is important to remember that an ionic compound has no net charge, i.e. overall it is neutral. Therefore, there must be the same number of positive charges as negative charges present in the compound.

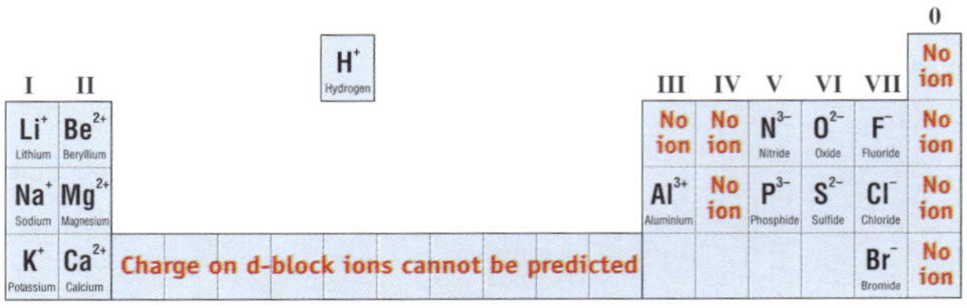

Fig 5.7
Some ions of the first 36 elements.

We now look at some examples that illustrate how we write the formulas of ionic compounds.

Example 5.2

Write the formula of potassium bromide.

SOLUTION

The potassium ion is K^+. The bromide ion is Br^-. The formula K^+Br^- is correct since there is the same number of positive and negative charges.

Answer: The formula of potassium bromide is KBr.

Example 5.3

Write the formula of calcium chloride.

SOLUTION

The calcium ion is Ca^{2+}. The chloride ion is Cl^-. The formula $Ca^{2+}Cl^-$ is not correct since there are two positive charges but only one negative charge. To get two negative charges we must have two Cl^- ions.

Answer: The formula of calcium chloride is $CaCl_2$.

Example 5.4

Write the formula of sodium sulfide.

SOLUTION

The sodium ion is Na^+. The sulfide ion is S^{2-}. The formula Na^+S^{2-} is not correct since there is one positive charge but two negative charges. To get two positive charges we must have two Na^+ ions.

Answer: The formula of sodium sulfide is Na_2S.

Example 5.5

Write the formula of aluminium oxide.

SOLUTION

The aluminium ion is Al^{3+}. The oxide ion is O^{2-}. The formula $Al^{3+}O^{2-}$ is not correct since there are three positive charges but only two negative charges. Bring all charges up to their lowest common denominator, i.e. 6. To get six positive charges we need two aluminium ions. To get six negative charges we need three oxide ions, i.e. $(Al^{3+})_2(O^{2-})_3$

Answer: The formula of aluminium oxide is Al_2O_3.

There are many ionic compounds which contain group (complex) ions, e.g. the sulfate ion, the carbonate ion, the nitrate ion, etc. For example, barium sulfate is used in X-ray examinations of internal organs, chalk is calcium carbonate, sodium hydrogencarbonate is used in baking soda, etc. The group ions commonly encountered are listed in TABLE 5.1. It is not possible to predict the formulas of group ions from the Periodic Table – so learn them off by heart!

NAME	FORMULA	
Hydroxide ion	OH^-	One negative charge
Nitrate ion	NO_3^-	
Hydrogencarbonate ion	HCO_3^-	
Permanganate ion	MnO_4^-	
Carbonate ion	CO_3^{2-}	Two negative charges
Chromate ion	CrO_4^{2-}	
Dichromate ion	$Cr_2O_7^{2-}$	
Sulfate ion	SO_4^{2-}	
Sulfite ion	SO_3^{2-}	
Thiosulfate ion	$S_2O_3^{2-}$	
Phosphate ion	PO_4^{3-}	Three negative charges
Ammonium ion	NH_4^+	One positive charge

Table 5.1

Some common group ions.

You will observe in FIG 5.7 that we have indicated that it is not possible to predict the charges of the ions of the transition metals. The transition metals are said to exhibit **variable valency**. The word valency means 'combining power'. (A more formal definition of valency will be given in SECTION 5.6.) Iron combines with chlorine to form either $FeCl_2$ or $FeCl_3$. The compound $FeCl_2$ is an ionic compound and the +2 charge on the iron atom is represented by putting the Roman numeral II in brackets. Thus, $FeCl_2$ is called iron(II) chloride. Similarly, the compound $FeCl_3$ is called iron(III) chloride. Copper is also said to exhibit variable valency because it can combine with oxygen to form either Cu_2O or CuO. Cu_2O is called copper(I) oxide and CuO is called copper(II) oxide.

Chromium is another example of a transition metal that exhibits variable valency. Cr^{3+} ions exist in chromium(III) chloride, $CrCl_3$. However, in $Na_2Cr_2O_7$, sodium dichromate, it appears that there are Cr^{6+} ions! (We shall clarify this in CHAPTER 14.) Hence, sodium dichromate is also written sodium dichromate(VI).

A final transition metal about whose variable valency you should be aware is manganese. MnO_2 is an ionic compound called manganese(IV) oxide - commonly called manganese dioxide. $MnSO_4$ is called manganese(II) sulfate and $KMnO_4$, commonly called potassium permanganate is also called potassium manganate(VII). You do not have to remember the actual valencies of any of the transition metals but you should be able to interpret the names as shown in the examples that follow shortly.

The reason why transition metals have variable valency is because there is such a small energy difference between the 4s and 3d sublevels. This means that different possible numbers of electrons can be lost from these sublevels giving metal ions with different charges on them.

Example 5.6

Write the formula of potassium hydroxide.

SOLUTION

The potassium ion is K^+. The hydroxide ion is OH^-. The formula K^+OH^- is correct since there is the same number of positive and negative charges.

Answer: The formula of potassium hydroxide is KOH.

Example 5.7

Write the formula of sodium sulfate.

SOLUTION

The sodium ion is Na^+. The sulfate ion is SO_4^{2-}. The formula $Na^+SO_4^{2-}$ is not correct since there is one positive charge but two negative charges. To get two positive charges we must have two Na^+ ions.

Answer: The formula of sodium sulfate is Na_2SO_4.

Example 5.8

Write the formula of calcium hydrogencarbonate.

SOLUTION

The calcium ion is Ca^{2+}. The hydrogencarbonate ion is HCO_3^-. The formula $Ca^{2+}HCO_3^-$ is not correct since there are two positive charges but only one negative charge. To get two negative charges we must have two HCO_3^- ions, i.e. $Ca^{2+}(HCO_3^-)_2$

Answer: The formula of calcium hydrogencarbonate is $Ca(HCO_3)_2$.

Example 5.9

Write the formula of iron(II) carbonate.

SOLUTION

The Roman number (II) indicates Fe^{2+}. The carbonate ion is CO_3^{2-}. The formula $Fe^{2+}CO_3^{2-}$ is correct since there are two positive charges and two negative charges.

Answer: The formula of iron(II) carbonate is $FeCO_3$.

Example 5.10

Name the compound $Cr_2(SO_4)_3$.

SOLUTION

The three sulfate ions together give a total negative charge of -6. Since the overall charge must be zero, the two chromium atoms must carry a total charge of +6. Therefore, each Cr atom must be Cr^{3+}. Therefore, the name of the compound is chromium(III) sulfate.

Answer: Chromium(III) sulfate

Note: Do not get confused between compounds whose names end in **-ide** and those whose names end in **-ate**. A compound which contains just two elements always ends in –ide, e.g. potassium chloride (KCl), magnesium oxide (MgO), magnesium sulfide (MgS). Compounds whose names end in **-ate** contain oxygen as well as the other two elements, e.g. magnesium sulfate ($MgSO_4$), potassium nitrate (KNO_3), potassium chlorate ($KClO_3$), etc.

5.5 d-BLOCK ELEMENTS AND TRANSITION ELEMENTS

In many cases there is no distinction made between the d-block elements and the transition elements. When referring to the s, p, d and f blocks of the Periodic Table, transition elements are regarded as those in the d-block of the table. However, if one considers the first row of the d-block from scandium to zinc, it is found that the properties of scandium and zinc are quite different to those of the other eight elements in this row. All of the elements from titanium to copper inclusive are referred to as transition elements. These show certain characteristics:

(a) Transition elements have variable valency. However scandium only forms Sc^{3+} ions and zinc only forms Zn^{2+} ions.

(b) Transition elements usually form coloured compounds. However, scandium and zinc only form white compounds.

(c) Transition elements are widely used as catalysts (see CHAPTER 16). However, scandium and zinc show little catalytic activity.

Since scandium and zinc do not have typical transition metal properties, chemists formulated a definition of transition metals that excludes these two elements but includes the elements from titanium to copper.

A transition metal is one that forms at least one ion with a partially filled d sublevel.

If you study TABLE 5.2, you will notice that neither Sc^{3+} nor Zn^{2+} has a partly filled d sublevel. In the case of Sc^{3+}, the sublevel is completely empty and in the case of Zn^{2+}, the sublevel is completely full. In other words, all scandium compounds have an empty 3d sublevel and all zinc compounds have a filled 3d sublevel. Since the properties of transition metals are associated with partly filled d sublevels, scandium and zinc are not considered transition metals. Thus, all of the ten elements from scandium to zinc are d-block elements but only those eight elements from titanium to copper (inclusive) are transition metals.

Note: You are **not** required to be able to write the electronic configuration of transition metal ions. TABLE 5.2 is simply given to help you understand the definition of a transition metal. You will observe from TABLE 5.2 that when transition metals form ions, electrons are first lost from the 4s sublevel before the 3d. This may seem strange as the 4s sublevel is filled before the 3d sublevel. A simple explanation for this is that the electrons in the 4s sublevel are farther away from the nucleus than those in the 3d and therefore are more easily removed. Hence, the 4s electrons are lost before the 3d electrons.

Sc^{3+}	$= 1s^2, 2s^2, 2p^6, 3s^2, 3p^6$
Ti^{3+}	$= 1s^2, 2s^2, 2p^6, 3s^2, 3p^6\ 3d^1$
V^{3+}	$= 1s^2, 2s^2, 2p^6, 3s^2, 3p^6\ 3d^2$
Cr^{3+}	$= 1s^2, 2s^2, 2p^6, 3s^2, 3p^6\ 3d^3$
Mn^{2+}	$= 1s^2, 2s^2, 2p^6, 3s^2, 3p^6\ 3d^5$
Fe^{2+}	$= 1s^2, 2s^2, 2p^6, 3s^2, 3p^6\ 3d^6$
Co^{2+}	$= 1s^2, 2s^2, 2p^6, 3s^2, 3p^6\ 3d^7$
Ni^{2+}	$= 1s^2, 2s^2, 2p^6, 3s^2, 3p^6\ 3d^8$
Cu^{2+}	$= 1s^2, 2s^2, 2p^6, 3s^2, 3p^6\ 3d^9$
Zn^{2+}	$= 1s^2, 2s^2, 2p^6, 3s^2, 3p^6\ 3d^{10}$

Table 5.2

The electronic configurations of some ions of the first row elements of the d-block. Note that neither scandium nor zinc has a partially filled d sublevel. Hence, they are not classed as transition metals.

5.6 COVALENT BONDING – SHARING OF ELECTRONS

When studying ionic bonding we learned that ionic compounds are usually formed when metals react with non-metals. However, there are many compounds that contain only non-metals, e.g. water, ammonia, carbon dioxide, etc. These compounds do not contain any ions. Instead, the atoms achieve noble gas configurations by sharing electrons. To help you understand this idea of electrons being shared, we now consider a number of examples.

Example 5.11

THE HYDROGEN MOLECULE, H_2.

A hydrogen atom has just one electron in its outer shell. It is found that two hydrogen atoms join together to form a hydrogen molecule, H_2. A cylinder of hydrogen gas consists of millions of these tiny independent units, each of which is called a **molecule** of hydrogen.

A molecule is a group of atoms joined together. It is the smallest particle of an element or compound that can exist independently.

A hydrogen molecule is formed when one hydrogen atom shares its single electron with another hydrogen atom. In turn, the second hydrogen atom shares its single electron with the first hydrogen atom. This means that two electrons are shared between the two hydrogen atoms. Thus, each hydrogen atom achieves a stable electron structure like the noble gas helium (Fig 5.8).

The chemical bond formed by sharing a pair of electrons is known as a **covalent bond**. The shared pair of electrons in a covalent bond is often represented simply by a dash. Thus, H-H represents a molecule of hydrogen.

Note: The term molecule can only be used for covalent compounds. Since ionic compounds consist of giant structures of ions, it is not correct to refer to a 'molecule' of sodium chloride. This is because no one sodium ion is associated with any one particular chloride ion.

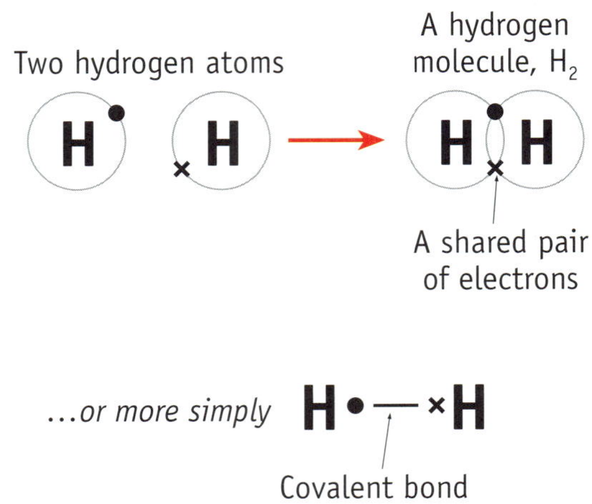

Fig 5.8
Covalent bond formation in a molecule of hydrogen.

Example 5.12

The chlorine molecule, Cl_2.

A chlorine atom has seven electrons in its outer shell. To achieve a stable octet of eight electrons, two chlorine atoms share a pair of electrons with each other, i.e. a covalent bond is formed between the two molecules. Thus, each atom has its own seven outer electrons and a share in one of the outer electrons of the other atom. The dot-and cross-diagram to illustrate the formation of the covalent bond in chlorine is shown in Fig 5.9.

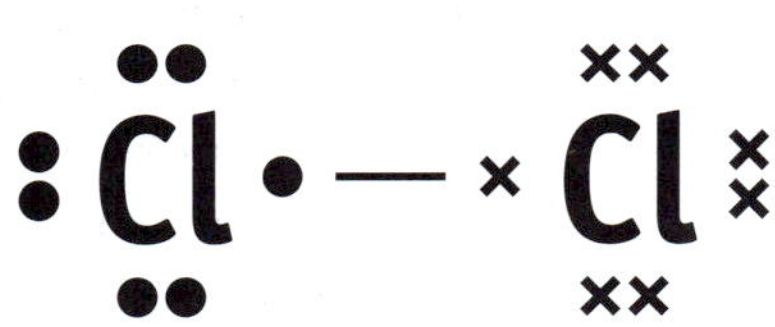

Fig 5.9
In the chlorine molecule, the two atoms are held together by a covalent bond.

Example 5.13

The water molecule, H_2O.

An oxygen atom has six electrons in its outer shell. This atom needs two more electrons to complete its octet. To achieve this, an oxygen atom shares two of its own electrons with two hydrogen atoms. Thus, two covalent bonds are formed as shown in Fig 5.10. The water molecule is a V-shaped molecule. We shall understand the reason for this at a later stage.

Note that there are two pairs of electrons on the oxygen atom that are not involved in bonding. These two pairs of electrons are called **'lone pairs'**. They are called 'lone pairs' because they are not bonded to another atom. Pairs of electrons that **are** involved in bonding (i.e. 'tied up' in covalent bonds) are called **'bond pairs'**. Thus, the oxygen atom in a water molecule has two lone pairs and two bond pairs.

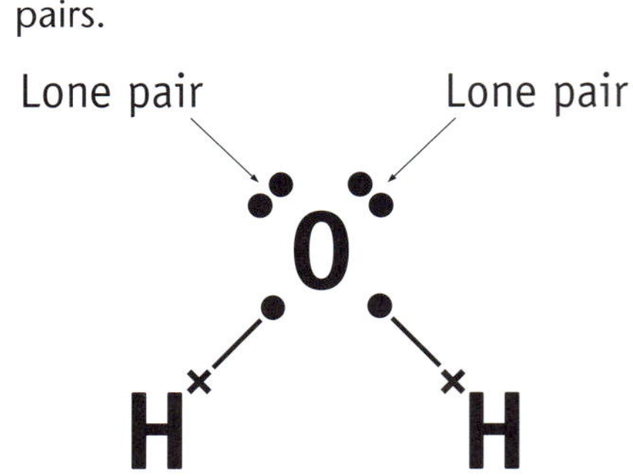

Fig 5.10
The water molecule contains two covalent bonds.

Example 5.14

The ammonia molecule, NH_3.

A nitrogen atom has five electrons in its outer shell. This atom needs three more electrons to complete its octet. To achieve this, a nitrogen atom shares three of its own electrons with three hydrogen atoms. Thus, three covalent bonds are formed as shown in FIG 5.11.

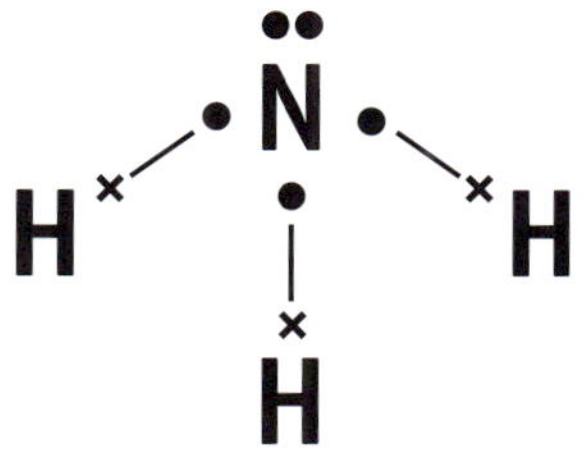

Fig 5.11
The ammonia molecule contains three covalent bonds.

Example 5.15

The methane molecule, CH_4.

A carbon atom has four electrons in its outer shell. This atom needs four more electrons to complete its octet. To achieve this, a carbon atom shares four of its own electrons with four hydrogen atoms. Thus, four covalent bonds are formed as shown in FIG 5.12.

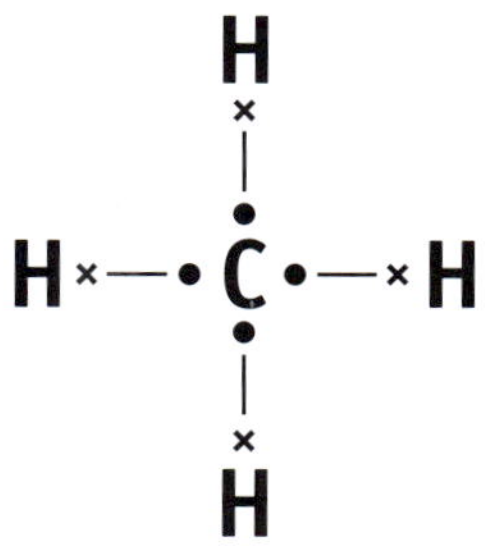

Fig 5.12
The methane molecule contains four covalent bonds.

From the above examples, it is clear that one can predict the number of covalent bonds formed around certain atoms: hydrogen atoms have one covalent bond, oxygen atoms have two covalent bonds, nitrogen atoms have three covalent bonds and carbon atoms have four covalent bonds. The number of covalent bonds formed by atoms of some selected elements is given in FIG 5.13.

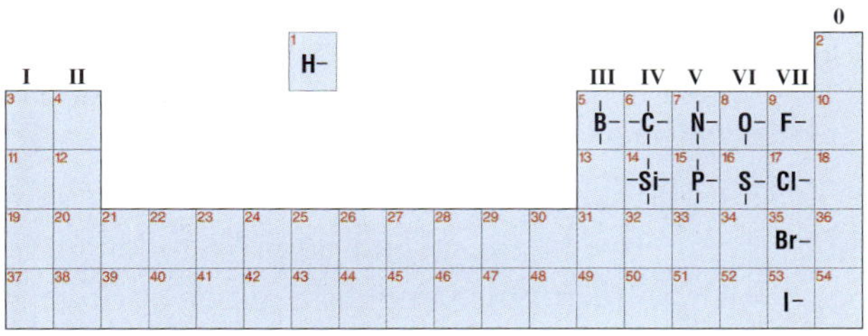

Fig 5.13
The number of covalent bonds formed by atoms of some selected elements.

Since a hydrogen atom cannot combine with more than one atom of any other element, we say its valency ('combining power') is one or it is monovalent. In fact, hydrogen is used as the standard by which valency is measured.

> The valency of an element is defined as the number of atoms of hydrogen or any other monovalent element with which each atom of the element combines.

Thus, since chlorine combines with one atom of hydrogen (forming HCl) we say it has a valency of one, oxygen has a valency of two (H_2O), nitrogen a valency of three (NH_3), carbon a valency of four (CH_4), etc.

DOUBLE AND TRIPLE BONDS

In EXAMPLES 5.11 – 5.15 we have seen how atoms share single pairs of electrons to form covalent bonds. Such bonds are more correctly referred to as **single** covalent bonds. Thus, a single bond is formed when **one** pair of electrons is shared. However, in some molecules the atoms share two or even three pairs of electrons. When **two** pairs of electrons are shared, we say that a **double bond** is formed. The oxygen molecule contains a double bond and the carbon dioxide molecule contains two double bonds (FIG 5.14). Note from this diagram how the formation of the double bonds enables each element to achieve the octet of electrons in the outer energy level.

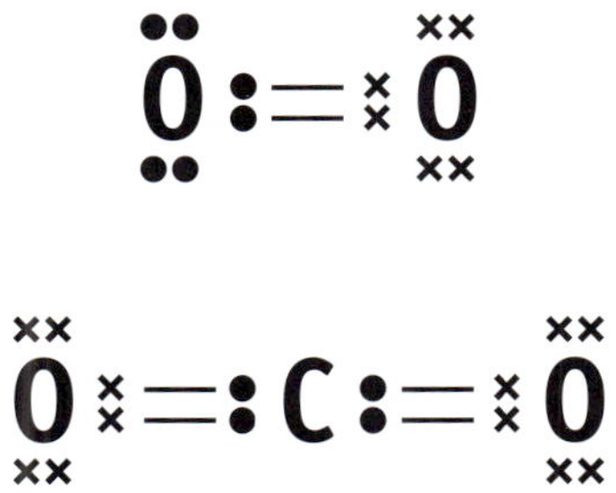

Fig 5.14
Both the oxygen molecule and the carbon dioxide molecule contain double bonds.

It is also possible for three pairs of electrons to be shared. When three pairs of electrons are shared we say that a **triple bond** is formed. The nitrogen molecule, N_2, is an example of a molecule that contains a triple bond (FIG 5.15).

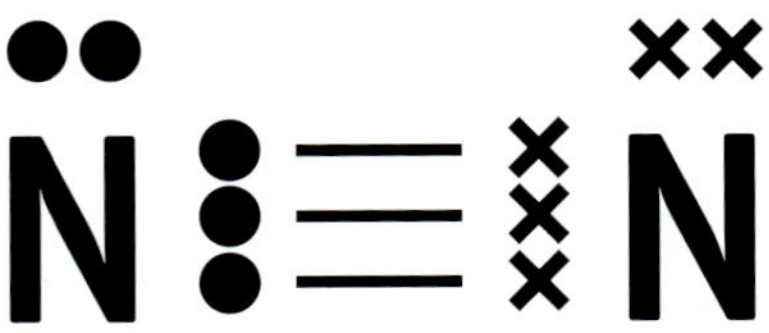

Fig 5.15
The nitrogen molecule contains a triple bond.

Example 5.16

Draw an 'electron dot' structure showing the bonding in HCHO

In showing the bonds in this molecule, remember that each hydrogen atom has just one bond around it, the carbon atom must have four bonds around it and the oxygen atom has two bonds around it. The valencies of these atoms are satisfied by the structure as shown in FIG 5.16.

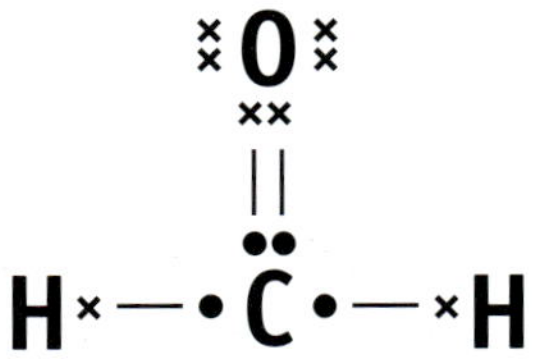

Fig 5.16

SIGMA AND PI BONDING

It is also possible to describe chemical bonding in terms of atomic orbitals. For example, one could describe the formation of the H_2 molecule by saying that the 1s orbitals of two hydrogen atoms overlap as shown in FIG 5.17.

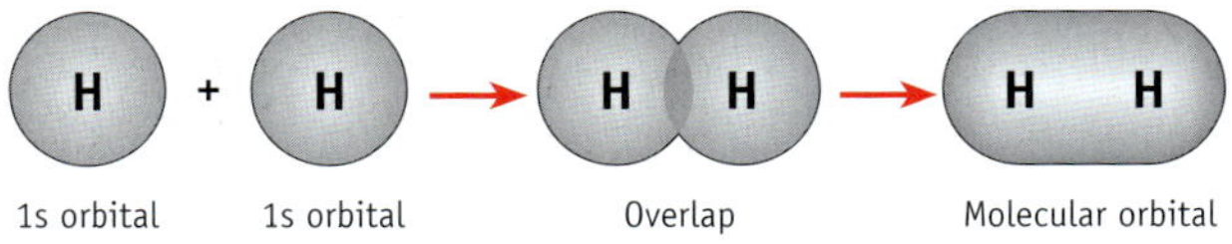

Fig 5.17
The covalent bond in H_2 being formed by the overlap of the 1s atomic orbitals of two hydrogen atoms.

When the two atomic orbitals overlap they form what is called a molecular orbital. This is rather like two soap bubbles coming together to form a large soap bubble. The covalent bond formed by the head-on overlap of the two atomic orbitals is called a **sigma bond**. The Greek letter σ (sigma) is used to indicate a sigma bond.

Similarly, the bonding in the chlorine molecule could be explained by overlap of orbitals. If we consider the electron configuration of chlorine we see that one of its p orbitals is half-filled.

$$Cl = 1s^2, 2s^2, 2p^6, 3s^2, 3p_x^{\,2}, 3p_y^{\,2}, 3p_z^{\,1}$$

The bonding in Cl_2 may be explained by saying that two half-filled p_z atomic orbitals overlap head-on to form a sigma bond (FIG 5.18).

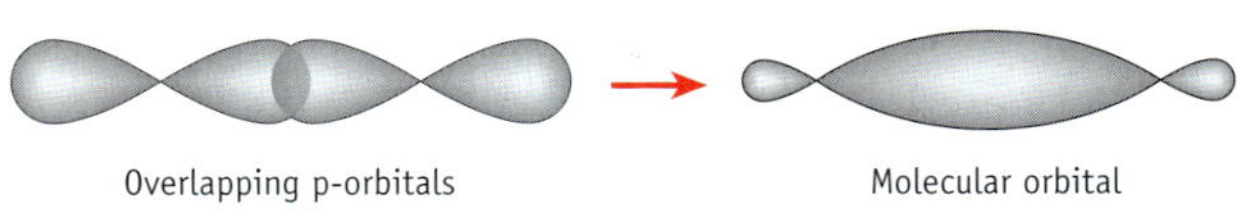

Fig 5.18
The covalent bond in Cl_2 being formed by the head-on overlap of two p orbitals.

You have probably noticed that there is very little difference between the orbital approach to single bond formation and the simple shared electron pair approach discussed in EXAMPLES 5.11 - 5.16 above. Both approaches describe the same occurrence, i.e. all single bonds can also be called sigma bonds. However, the two types of descriptions differ when we describe the formation of double and triple bonds. To illustrate how the orbital theory can give us a better picture, we now consider how the triple bond in the nitrogen molecule, N_2, is formed.

$$N = 1s^2, 2s^2, 2p_x^{\,1}, 2p_y^{\,1}, 2pz^1$$

We note that there are three half-filled p orbitals in the nitrogen atom. Imagine two nitrogen atoms approach each other so that the two $2p_x$ orbitals overlap head-on to form a sigma bond (FIG 5.19).

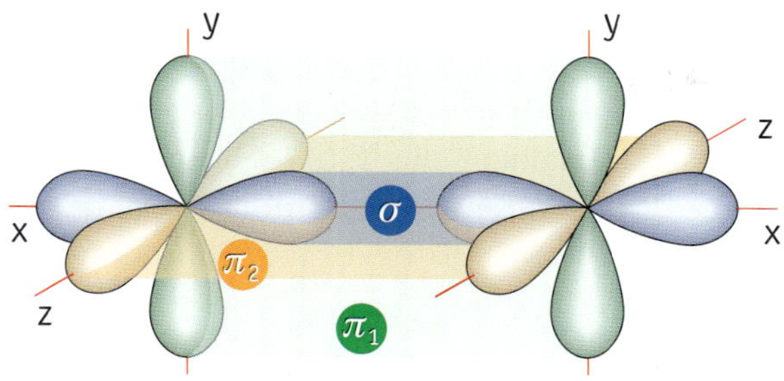

Fig 5.19
The triple bond in nitrogen consists of one sigma and two pi bonds.

The two $2p_y$ orbitals overlap sideways to form a type of covalent bond called a π bond (π is the Greek letter, pronounced 'pi'). Similarly, a second π bond is formed from the sideways overlap of the two $2p_z$ orbitals.

Thus, the triple bond in nitrogen consists of one sigma and two pi bonds. This is true of all triple bonds. In a double bond, one of the bonds is a sigma and one is a pi bond (FIG 5.20).

Fig 5.20
A double bond consists of one sigma and one pi bond. A triple bond consists of one sigma and two pi bonds.

Note: Since there is less overlapping of orbitals in forming pi bonds, these bonds are not as strong as sigma bonds. Thus, an O = O double bond is not twice as strong as an O - O single bond.

So, the orbital theory gives us a detailed picture of the formation of double and triple bonds. This is the only use we shall make of this theory in our study of chemistry for the Leaving Certificate. For all other cases of bond description the simpler 'shared-pair theory' is perfectly adequate.

5.7 CHARACTERISTICS OF IONIC AND COVALENT COMPOUNDS

We have seen that ionic compounds usually consist of a network of ions whereas covalent compounds generally consist of individual molecules. Many of the properties of ionic and covalent compounds can be understood if we keep these facts in mind.

(A) HARDNESS

Ionic compounds are usually difficult to cut. This is because each ion is held in the crystal lattice by strong attractive forces between it and the ions of opposite charge around it. On the other hand, covalent compounds consisting of molecules are usually quite soft. For example, iodine consists of dark grey crystals that are quite soft. A crystal of iodine consists of millions of iodine molecules, I_2. Hence, iodine is referred to as a molecular crystal. There are only very weak forces between one molecule and another. (The nature of these weak forces will be studied in SECTION 5.10.) Hence, iodine crystals are so soft that they could be crushed between your fingers!

(B) MELTING AND BOILING POINTS

In ionic crystals, the ions can only vibrate about fixed positions. When heat is supplied, the ions gain energy and vibrate more and more. Eventually, the ions vibrate so much that they break away from their fixed positions in the crystal. When this happens the crystal structure breaks down and we observe the substance melting. However, because the forces of attraction between the ions are so strong, most ionic substances have high melting points - and even higher boiling points! For example, the melting point of sodium chloride is 801 °C and its boiling point is 1467 °C. On the other hand, since there are weak forces between the molecules in crystals like iodine, the melting and boiling points are much lower, e.g. the melting point of iodine is only 114 °C and its boiling point is only 184 °C. Other covalent compounds have even lower melting and boiling points, e.g. water [MP (melting point) = O °C, BP (boiling point) = 100 °C], ammonia [MP = -77 °C, BP = -33 °C], methane [MP = -182 °C, BP = -161 °C]. Thus, at room temperature all ionic compounds are solids and most covalent compounds are usually either liquids or gases.

The melting point of a substance called benzoic acid will be measured in EXPERIMENT 5.2.

(C) CONDUCTION OF ELECTRICITY

Ionic compounds do not conduct electricity in the solid state. This is because the ions are in fixed positions and cannot carry the electric current through the crystal. However, if the ionic substance is melted or dissolved in water, the ions are now free to move. Therefore, molten ionic compounds or ionic compounds in solution conduct an electric current. The apparatus to test if there are ions in solution is shown in FIG 5.21.

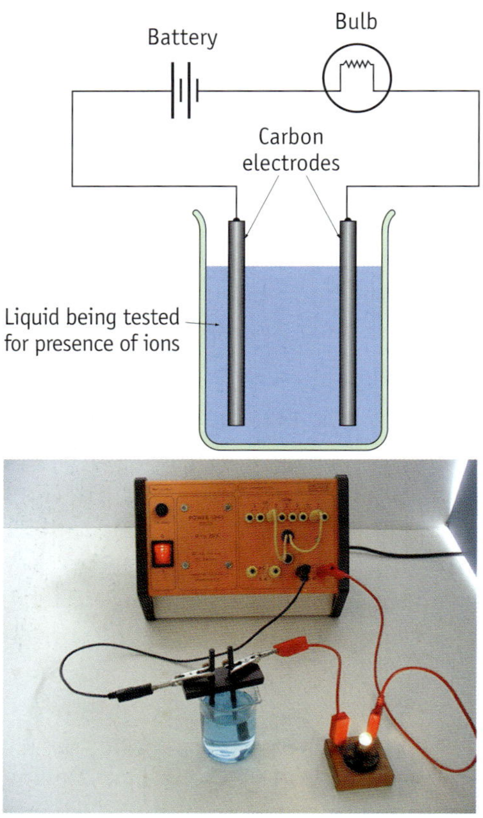

Fig 5.21
Apparatus to test if a solution will conduct electricity. The bulb lights if ions are present in the liquid being tested.

Since covalent substances do not contain ions, they do not usually conduct electricity as solids, liquids or when dissolved in water. Thus, if the beaker in FIG 5.21 contains substances like distilled water, alcohol, petrol or sugar solution, the bulb does not light.

We shall make a closer examination of the conduction of electricity through compounds at a later stage (CHAPTER 20).

Note: The fact that the bulb lights indicates that there are ions in solution. It does not necessarily mean that the substance that was originally dissolved in the water is ionic. For example, hydrogen chloride is a covalent substance which dissolves in water to give ions: $HCl \rightarrow H^+ + Cl^-$

The properties of ionic and covalent compounds are summarised in TABLE 5.3.

IONIC	COVALENT
1. Contain a network of ions in the crystal.	Contain individual molecules
2. Usually hard and brittle	Usually soft.
3. Have high melting points and boiling points.	Have low melting and boiling points.
4. Usually solid at room temperature.	Usually liquids, gases or soft solids at room temperature.
5. Conduct electricity in molten state or when dissolved in water.	Do not conduct electricity

Table 5.3
The properties of ionic and covalent substances

EXPERIMENT 5.1 MANDATORY STUDENT EXPERIMENT NO 2.

Part A: To obtain a pure sample of benzoic acid from an impure sample by the technique of recrystallisation.

Introduction

In this experiment, we use a very important technique called **recrystallisation** to purify a sample of impure benzoic acid. Benzoic acid is a crystalline substance that contains molecules of formula C_6H_5COOH. The process of recrystallisation is the most common method used to separate a solid material from solid impurities. The term 'recrystallisation' is used because we are starting off with a crystalline material, removing the impurities and making new crystals.

It is found by experiment that most substances dissolve more easily in hot liquids than in cold. For example, benzoic acid is very soluble in hot water but is only slightly soluble in cold water.

The word **solvent** is used to describe a liquid in which a substance is dissolved. The substance that is dissolved is usually referred to as the **solute**. The impure benzoic acid is dissolved in the minimum amount of hot water. This is done in order to get a solution which is almost saturated at the boiling point of the solvent, i.e. the water contains the maximum amount of dissolved benzoic acid. This hot, almost saturated solution is then filtered in order to remove any insoluble impurities. This filtration must be done quickly to prevent crystallisation during the filtration process. Large Buchner funnels, (FIG 5.22) are usually used for this. The funnels are usually heated to prevent crystallisation taking place while the hot solution is being filtered.

Insoluble impurities are left on the filter paper when the hot solution of benzoic acid is filtered. Soluble impurities (together with the dissolved benzoic acid) pass through the filter paper. On allowing the hot solution (called the **filtrate**) to cool, crystallisation takes place and crystals of pure benzoic acid are formed. (Any soluble impurities will not crystallise but will remain in solution due to their much lower concentration compared to the benzoic acid.)

The crystals formed are filtered off, washed with a little cold solvent and dried on filter paper or in a desiccator.

In the second part of the experiment we shall measure the melting point of the dried crystals.

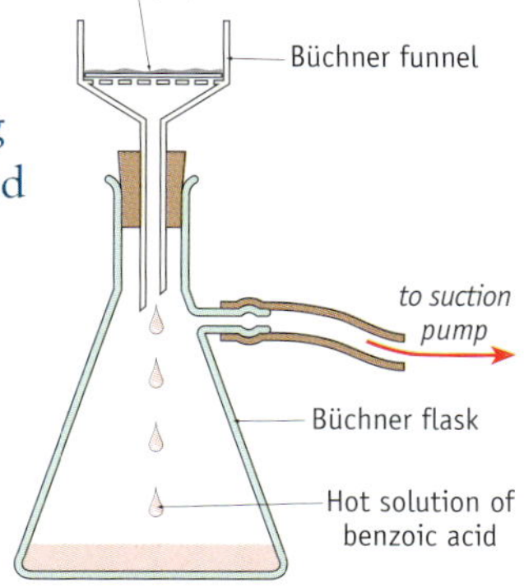

Fig 5.22
Filtering the hot solution of benzoic acid quickly using a Buchner funnel and suction pump.

Apparatus required

Beaker (Pyrex), stirring rod, graduated cylinder, spatula, clock glass, filter paper, conical flask, beaker tongs, Buchner flask, Buchner funnel, suction pump, retort stand.

Chemicals required

Sample of impure benzoic acid.

Procedure

1. Place one gram approx. of benzoic acid in a beaker. Using a graduated cylinder, add about 10 cm^3 of water to the beaker.
2. Place the beaker over a Bunsen burner or on a hot plate and boil the water.
3. Continue to add 10 cm^3 portions of water to the boiling water in the beaker until the benzoic acid has dissolved, i.e. until you can no longer see any of the white solid (only the specks of impurity, e.g. charcoal, will be visible in the water). While heating the water, keep the solution well stirred with the aid of a stirring rod.
4. While the solution is boiling, hold a Buchner funnel by its stem and heat it by immersing it in a beaker of hot water.
5. Set up the Buchner flask, funnel and suction pump as shown in FIG 5.22. It is helpful to clamp the flask using a retort stand.
6. Turn on the water supply to operate the suction pump.
7. Place a filter paper in the Buchner funnel. Moisten the filter paper slightly to help it stick to the Buchner funnel.
8. Using a beaker tongs or suitable gloves to hold the hot beaker, quickly filter the boiling solution by pouring it into the Buchner funnel. If crystals form on the filter paper, add a small amount of boiling water to dissolve them so that they pass through the filter paper.

9. Note the impurities that are left behind on the filter paper. Take out the filter paper and discard it in the rubbish bin.
10. Stop suction either by lifting the Buchner funnel and stopper out of the Buchner flask or by disconnecting the rubber tubing to the suction pump. If you turn off the tap you may get a 'suck-back' of water into the Buchner flask.
11. Cool the filtered solution (the filtrate) by running cold water from the tap over the sides of the Buchner flask. If ice is available, stand the flask in iced water for a few minutes.
12. Note that as the solution cools, white crystals begin to form. If these crystals are slow to form, scratching the inside of the flask with a glass rod speeds up the process.
13. Using a conical flask, funnel and filter paper, filter off the white crystals (FIG 5.23). If necessary, rinse out the Buchner flask with filtrate (not extra water!) to help you transfer all of the crystals.
14. Wash the crystals on the filter paper with a few drops of cold water.
15. Spread the damp filter paper containing the crystals on a clock glass and allow to air dry until the next practical class when the melting point of these crystals will be measured.

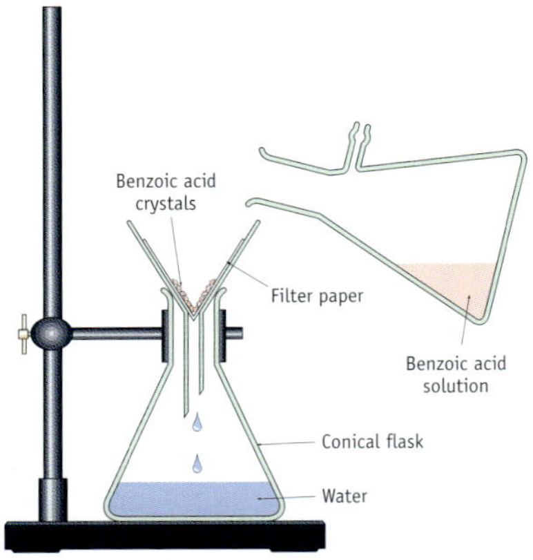

Fig 5.23
The crystals formed are filtered off using a filter funnel and filter paper.

EXPERIMENT 5.1

Part B: To measure the melting point of benzoic acid.

Introduction

In this experiment, we will measure the melting point of the benzoic acid crystals formed in Part A, using an aluminium melting point block. It is essential that the crystals are dry - otherwise an incorrect result will be obtained. Measuring the melting point of a substance is often used by chemists to test the purity of a substance and also to help identify the substance. Tables of melting points of most common substances are readily available to help in this task.

Apparatus required

Aluminium melting point block, retort stand and clamp, long glass tube or pipe, Bunsen burner or hotplate, capillary tube

Chemicals required

Benzoic acid.

Procedure

1. Seal the end of a capillary tube by rotating it for a few seconds in the blue flame of a Bunsen burner.
2. Rotate the open end of the capillary tube into the sample of benzoic acid crystals so that some of the material enters the capillary tube. If the crystals are not small enough to enter the capillary tube, they should be ground up using a pestle and mortar.
3. Drop the capillary tube through a long length of tubing, e.g. glass tubing standing vertically on the laboratory bench. When the bottom of the capillary tube hits the laboratory bench, the crystals will fall to the bottom. Repeat this procedure until the crystals are at a depth of about 1 cm at the bottom of the tube.

4. Place the capillary tube in the aluminium melting point block on a hotplate (FIG 5.24), and place a thermometer in the aluminium block. The thermometer you use must be capable of reading a temperature of at least 150 °C.
5. Turn on the hotplate and observe the crystals until they melt, i.e. until they change to a colourless liquid. Note the reading on the thermometer. This is the melting point of the benzoic acid.
6. Allow the aluminium block to cool and repeat the experiment using a second sample of benzoic acid. Take the average of the two readings as the melting point of the benzoic acid.

If the crystals of benzoic acid are pure and dry, their melting point should be the same as that quoted in the various chemistry data books. If the crystals are impure, they will melt over a range of temperatures. Your teacher will tell you if your result is good or bad!

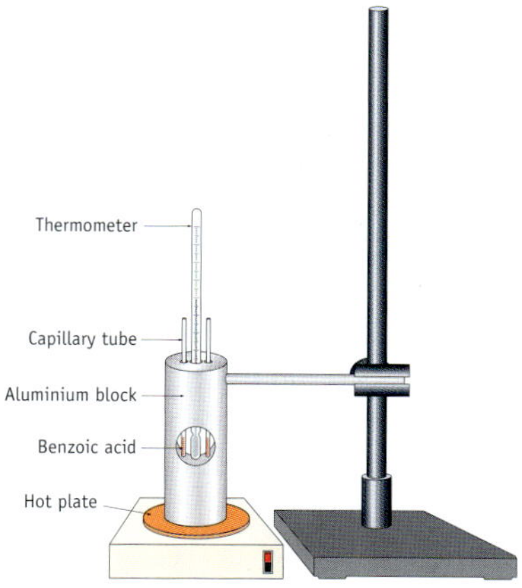

Fig 5.24
Measuring the melting point of benzoic acid using an aluminium melting point block.

Conclusion

Since the melting point of benzoic acid is reasonably low, we conclude that it must be a molecular crystal rather than an ionic crystal. If the substance were ionic, the melting point would be very high (greater than 700 °C at least).

5.8 SHAPES OF COVALENT MOLECULES

As discussed in the previous section, ionic compounds consist of giant crystal lattices but covalent compounds consist of separate molecules. These individual molecules have a particular shape. In 1940, Sidgwick and Powell proposed a theory to account for the shapes of molecules. Their theory is known as the **Valence Shell Electron Pair Repulsion Theory** or **VSEPR Theory** for short. (The word 'valence shell' simply means 'outer shell'.) This theory states that the shape of a molecule depends on the number of pairs of electrons in the valence shell of the central atom. Since electrons are negatively charged, the electron pairs repel each other and arrange themselves in space so as to be as far apart as possible.

In order to see how the VSEPR theory works, let us consider some examples.

Example 5.17

Beryllium chloride, $BeCl_2$.

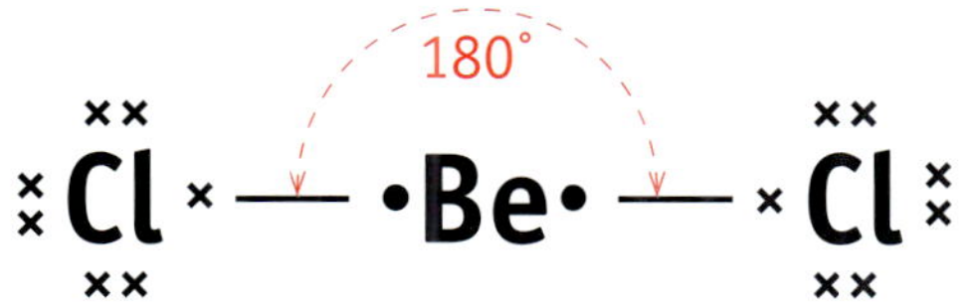

Fig 5.25
The beryllium chloride molecule consists of two pairs of electrons around the central atom. The bond angle is 180 °.

A molecule of beryllium chloride consists of **two pairs** of electrons around the central atom, (FIG 5.25). (The lone pairs of electrons on the chlorine atoms themselves are ignored by the VSEPR theory – we need only consider the electrons which are actually around the central atom.) By experiment it is found that $BeCl_2$ is a **linear** molecule, i.e. the bond angle is 180°. This bond angle is explained by saying that the two pairs of electrons, since they are negatively charged, repel each other and arrange themselves in space as far apart as possible from each other. When the bond angle is 180° the repulsion between the pairs is minimised. This shape is the shape adopted by any molecule that has two pairs of electrons around the central atom.

Example 5.18

Boron trichloride, BCl_3

A molecule of boron trichloride consists of **three pairs** of electrons around the central atom (FIG 5.26). When the bond angle is measured experimentally, it is found to be 120°, i.e. the three pairs of electrons are all equally far apart to minimise repulsion. The three pairs of electrons are all in the one plane. Hence, the shape of BCl_3 is said to be **triangular planar** or **trigonal planar**. This is the shape adopted by any molecule that has three pairs of electrons around the central atom.

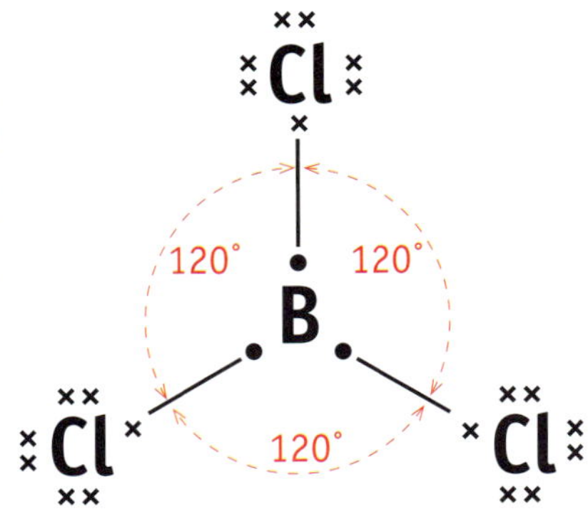

Fig 5.26
The boron trichloride molecule consists of three pairs of electrons around the central atom. The bond angle is 120°.

Example 5.19

Methane, CH_4

A molecule of methane consists of **four pairs** of electrons around the central atom (FIG 5.27(a)). When the bond angle is measured experimentally, it is found that it is 109.5°, i.e. the four pairs of electrons take up the shape of a **tetrahedron**. By adopting this shape, the repulsion between the four pairs of electrons is minimised.

It is difficult to represent the shape of a tetrahedron on paper. The wedge drawn in FIG 5.27(b) indicates that this particular C-H bond is coming out of the plane of the paper, the C-H dashed line indicates the bond is going behind the paper, and the final two C-H bonds are in the plane of the paper. Examining a model of a tetrahedron in your school laboratory will help you understand this.

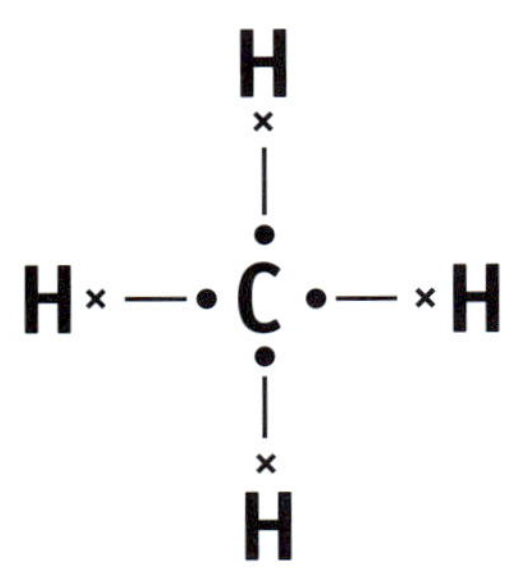

Fig 5.27 (a)
The methane molecule consists of four pairs of electrons around the central atom. The bond angle is 109°.

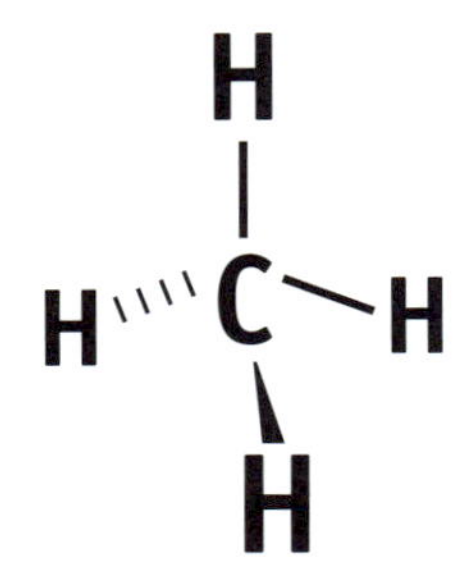

Fig 5.27 (b)
Representing a tetrahedron on paper.

The above three shapes can be demonstrated with molecular models or even with balloons knotted together (FIG 5.28).

You will observe that in the above three examples all pairs of electrons are bond pairs, i.e. there are no lone pairs in the molecules. A slight refinement of the VSEPR theory was necessary to take account of lone pairs on molecules. This modification states that the repulsion between two lone pairs in a molecule is greater than the repulsion between a lone pair and a bond pair which in turn is greater than the repulsion between two bond pairs:

lone pair/lone pair > lone pair/bond pair > bond pair/bond pair

Since lone pairs are not involved in bonding, they are closer to the nucleus than bond pairs – the electrons in bond pairs spend part of their time near the other atom in the bond. Because lone pairs are closer to the nucleus, they exert a greater force of repulsion on each other than do bond pairs on each other. This causes slight changes in the predicted bond angles. This is illustrated in EXAMPLES 5.20 and 5.21.

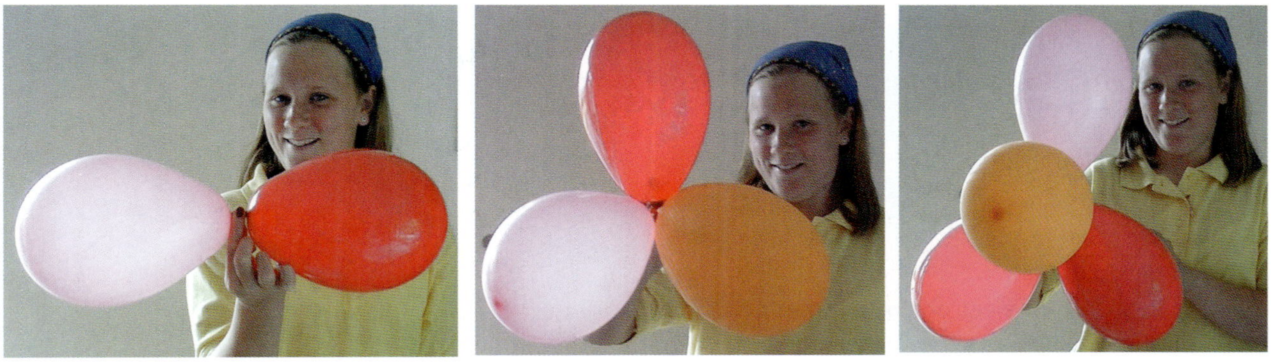

Fig 5.28
Two balloons take up a linear shape, three balloons take up a triangular planar shape and four balloons take up a tetrahedral shape.

Example 5.20

Use the VSEPR theory to deduce the shape of the ammonia molecule, NH_3.

SOLUTION

Step 1: Locate nitrogen in the Periodic Table. It is in group V, i.e. it has five electrons in its outer shell.

Step 2: Work out the number of electrons contributed by the three hydrogen atoms to the bonds around the central atom. Since each hydrogen atom contributes one electron, the three hydrogen atoms will contribute a total number of three electrons (FIG 5.29).

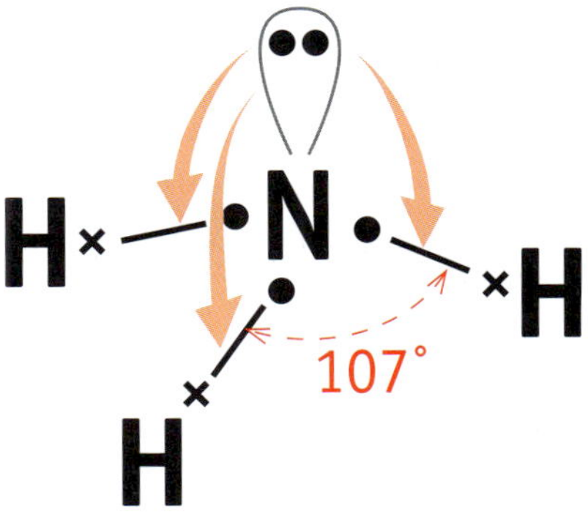

Fig 5.29
Ammonia has the shape of a distorted tetrahedron or pyramid.

Step 3: Work out the total number of electron pairs around the central atom

N	= 5 electrons
3H = 3 × 1	= 3 electrons
Total	= 8 electrons
	= four pairs of electrons
	⇒ tetrahedral shape.

We would therefore predict that ammonia has a basic tetrahedral shape. By experiment, it is found that the bond angle is 107°. It is less than the regular tetrahedral bond angle of 109.5° due to the presence of the lone pair of electrons. (The ammonia molecule has one lone pair and three bond pairs of electrons.) The increased repulsion caused by the presence of the lone pair causes the bond angle to decrease by about 2.5° to 107°, (FIG 5.29). Thus, the shape of the ammonia molecule may be described as distorted tetrahedral or **pyramidal** (FIG 5.30).

Fig 5.30
Model of ammonia molecule.

Example 5.21

Use the VSEPR theory to deduce the shape of the water molecule.

Solution

Proceeding as in Example 5.20.

O = 6 electrons
2H = 2 electrons
Total = 8 electrons
= four pairs of electrons ⇒ tetrahedral shape

We would predict that the water molecule has a basic tetrahedral shape. By experiment, it is found that the bond angle is 104.5°. It is less than the regular tetrahedral bond angle of 109.5° because of the presence of the two lone pairs of electrons. (The water molecule has two lone pairs and two bond pairs of electrons.) The increased repulsion caused by the presence of the two lone pairs causes the bond angle to decrease by about 5° from 109.5° to 104.5° (Fig 5.31). Thus, the shape of the water molecule may be described as distorted tetrahedral or **V-shaped**.

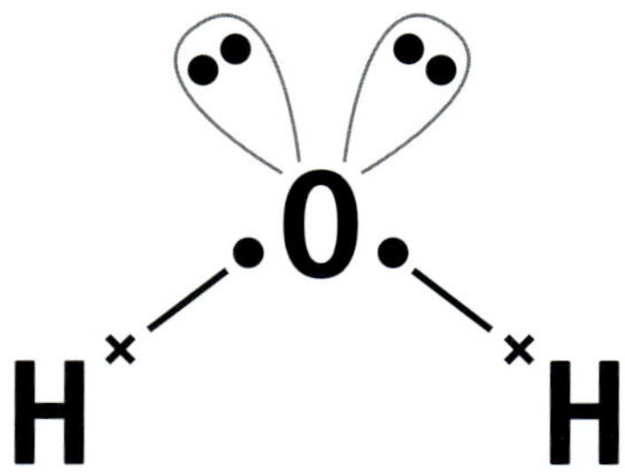

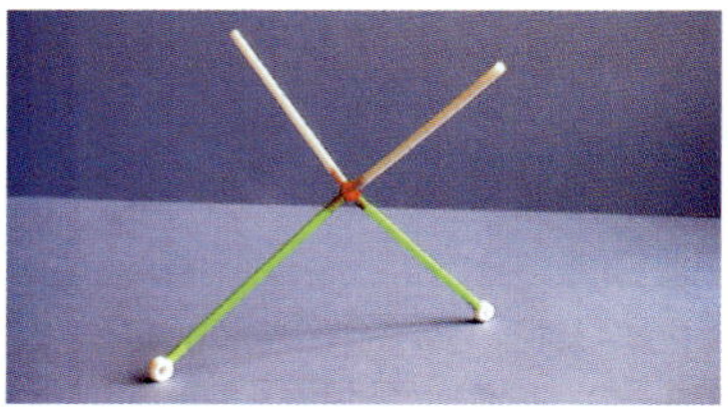

Fig 5.31
The increased repulsion due to the lone pairs causes the bond angle to decrease from 109.5° to 104.5°.

Example 5.22

Use the VSEPR theory to deduce the shape of the $SiCl_4$ molecule.

Solution

Si = 4 electrons
4 Cl = 4 electrons
Total = 8 electrons
= four pairs of electrons ⇒ tetrahedral shape

There are no lone pairs present, so $SiCl_4$ will have a regular tetrahedral shape.

Example 5.23

Use the VSEPR theory to deduce the shape of the PBr_3 molecule.

Solution

P = 5 electrons
3 Br = 3 electrons
Total = 8 electrons
= four pairs of electrons ⇒ tetrahedral shape

PBr_3 has one lone pair of electrons and three bond pairs of electrons. The lone pair will decrease bond angle to about 107°.

Therefore, the shape is distorted tetrahedral or pyramidal.

The shapes of molecules are summarised in Table 5.4

Total number of electron pairs	Number of bond pairs	Number of lone pairs	Shape of molecule	Examples
2	2	0	Linear	$BeCl_2$, BeH_2
3	3	0	Triangular planar	BCl_3, AlH_3
4	4	0	Tetrahedral	CH_4, $SiCl_4$
4	3	1	Pyramidal	NH_3, PH_3
4	2	2	V-shaped	H_2O, H_2S

Table 5.4
Summary of the shapes of molecules.

Note: The VSEPR Theory can also predict the shapes of molecules with five and six pairs of electrons around the central atom as well as the shapes of atoms containing double and triple bonds. However, this will not be discussed here as it is outside the scope of the syllabus.

5.9 ELECTRONEGATIVITY – 'TUG-OF-WAR' FOR ELECTRONS

In a covalent bond between identical atoms (e.g. in H_2, Cl_2, etc.), the pair of electrons is shared equally between the two atoms in the molecule. However, chemists have found that in bonds between different atoms, the pair of electrons is often more attracted to one of the atoms than the other. For example, in hydrogen chloride, HCl, it is found that the two electrons in the bond are more attracted to the chlorine than to the hydrogen. Since the two electrons spend more of their time nearer to the chlorine atom, this gives the chlorine atom a slight negative charge and leaves the hydrogen atom with a slight positive charge. This small amount of charge is indicated by the Greek letter δ (delta).

partial positive charge → $^{\delta+}$ H - Cl $^{\delta-}$ ← partial negative charge

The ability of an atom in a covalent bond to attract the electrons in the bond to itself is given a special name and is called the **electronegativity** of the atom.

> Electronegativity is the relative attraction that an atom in a molecule has for the shared pair of electrons in a covalent bond.

Thus, in the above example since the 'electron pulling power' of chlorine is greater than that of hydrogen, we say that chlorine is more electronegative than hydrogen. A very famous American chemist called Linus Pauling (Fig 5.32) studied the amounts of energy needed to break certain bonds. From his results he set up a scale of relative values of electronegativity. For example, he found that fluorine had four times the 'electron pulling power' of calcium. Therefore, he assigned an electronegativity value of 4.0 to fluorine and 1.0 to calcium. A table of electronegativity values is shown in Fig 5.33.

Fig 5.32
In 1939, the American chemist Linus Pauling made an enormous contribution to our understanding of chemical bonding in his famous book *The Nature of the Chemical Bond.* In 1954, he was awarded the Nobel Prize in chemistry and he was awarded the Nobel Peace prize in 1963.

I	II											III	IV	V	VI	VII	0
						H 2.1											He –
Li 1.0	Be 1.5											B 2.0	C 2.5	N 3.0	O 3.5	F 4.0	Ne –
Na 0.9	Mg 1.2											Al 1.5	Si 1.8	P 2.1	S 2.5	Cl 3.0	Ar –
K 0.8	Ca 1.0	Sc 1.3	Ti 1.5	V 1.6	Cr 1.6	Mn 1.5	Fe 1.8	Co 1.8	Ni 1.8	Cu 1.9	Zn 1.6	Ga 1.6	Ge 1.8	As 2.0	Se 2.4	Br 2.8	Kr –
Rb 0.8	Sr 1.0	Y 1.2	Zr 1.4	Nb 1.6	Mo 1.8	Tc 1.9	Ru 2.2	Rh 2.2	Pd 2.2	Ag 1.9	Cd 1.7	In 1.7	Sn 1.8	Sb 1.9	Te 2.1	I 2.5	Xe –
Cs 0.7	Ba 0.9	La 1.1	Hf 1.3	Ta 1.5	W 1.7	Re 1.9	Os 2.2	Ir 2.2	Pt 2.2	Au 2.4	Hg 1.9	Tl 1.8	Pb 1.8	Bi 1.9	Po 2.0	At 2.2	Rn –
Fr 0.7	Ra 0.9	Ac 1.1															

Fig 5.33
Values of electronegativity of the elements.

Note from the table of electronegativity values that the more electronegative elements are the very reactive non-metals like fluorine and chlorine. The least electronegative elements are the very reactive metals like potassium and sodium. Elements with low electronegativity values are said to be **electropositive**. We shall explain the trends in electronegativity values in Chapter 7.

Since chlorine is more electronegative than hydrogen, it is more correct to refer to the bond in H-Cl as a **polar covalent** bond. Thus we say that hydrogen chloride is a polar covalent molecule or simply a **polar molecule**. The word polar simply means that there is a positive and negative charge on the molecule and these two charges are separated by some distance.

Note: In writing chemical formulas, the more electronegative element is usually put last. For example, the compound formed between chlorine (E.N. = 3.0) and oxygen (E.N. = 3.5) is written as Cl_2O. Similarly, the compound formed between magnesium (E.N. = 1.2) and chlorine (E.N. = 3.0) is written as $MgCl_2$. However, there are many exceptions to this convention, e.g. NH_3, CH_4, etc.

USES OF ELECTRONEGATIVITY VALUES.

1. To predict polarity of covalent bonds

By referring to the table of electronegativity values, we can tell if the bonds in a molecule are polar. The greater the electronegativity difference, the more polar the bond.

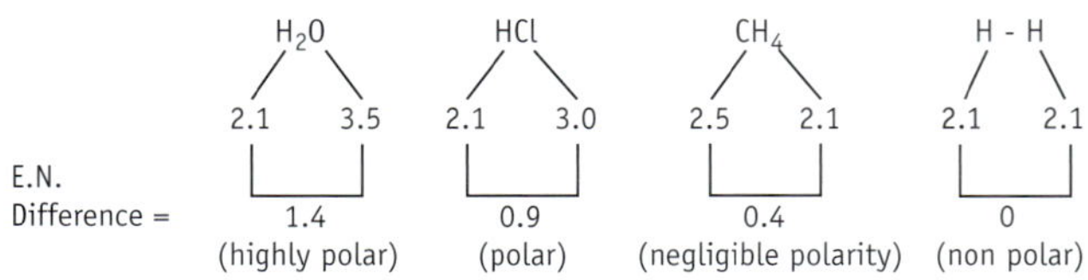

Most molecules that have polar bonds are polar molecules, i.e. in the molecule the partial positive and partial negative poles are separated by a distance (FIG 5.34).

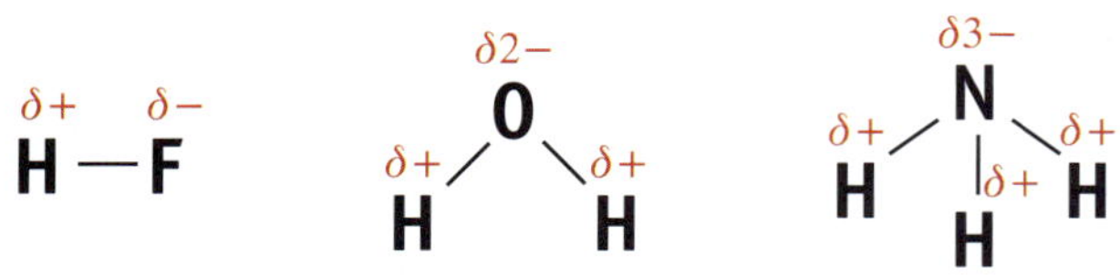

Fig 5.34
Hydrogen fluoride, water and ammonia are all polar molecules. Note that since each molecule is overall neutral, the sum of all the partial charges must add up to zero.

However, **there are some molecules which, even though they have polar bonds, are not polar molecules**. Studying FIG 5.35 will help you to understand this.

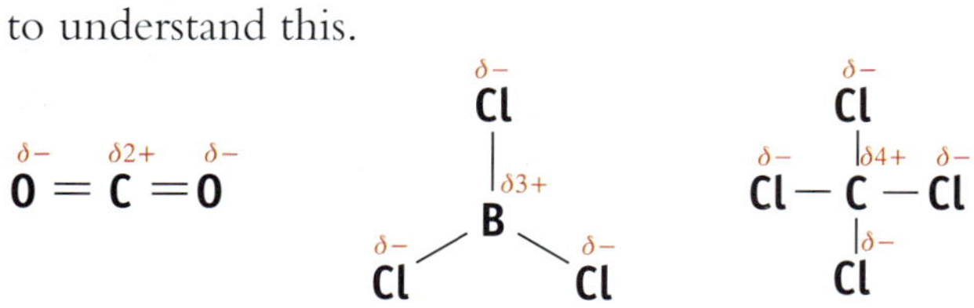

Fig 5.35
Since the 'centre of gravity' of the partial negative charges coincides with that of the partial positive charges, none of these molecules are polar.

None of the molecules in FIG 5.35 are polar molecules since the centres of partial positive charge and centres of partial negative charge coincide, i.e. cancel each other out. Indeed, the fact that water is a polar molecule gave chemists an indication of the shape of the molecule. If water were a linear molecule rather than V-shaped, then it would not be a polar molecule. (FIG 5.36)

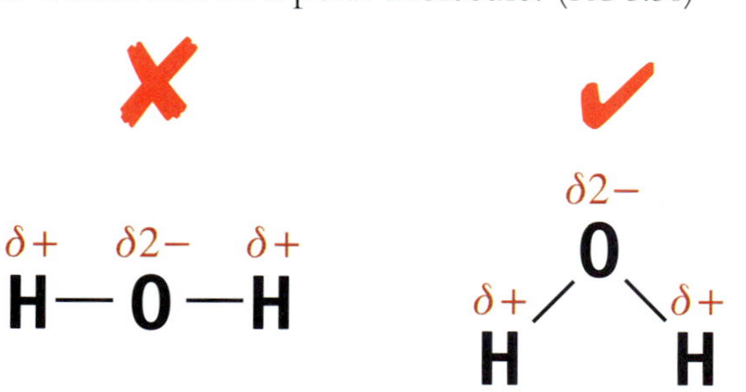

Fig 5.36
If the water molecule were linear it would have no dipole moment. The fact that water is V-shaped means that the partial positive and negative charges do not cancel out and hence, water is a polar molecule.

A simple way of demonstrating whether a liquid is polar or non-polar is to study the effect of a charged plastic rod on a stream of the liquid from a burette. A deflection of the stream indicates that the liquid consists of polar molecules .

In short, using electronegativity values and our knowledge about the shape of molecules, we can predict if a molecule is polar or non-polar.

So what difference does it make whether water is polar or non-polar? All the difference in the world! If water were not polar, it would be a gas at room temperature (SECTION 5.12) and life, as we know it, could not exist on Earth. Also, one of the most important properties of water is the fact that it is an excellent solvent. This property depends on the fact that water is a polar molecule. It is found that most ionic substances and most polar covalent substances dissolve in water. Consider, for example, an ionic substance such as sodium chloride dissolving in water (FIG 5.37).

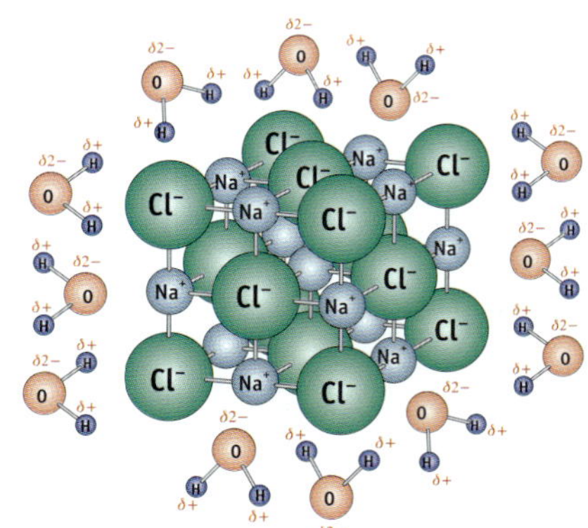

Fig 5.37
Sodium chloride dissolves readily in water. Note how the polar water molecules line up around the ions and draw them into solution.

The ionic bonding in NaCl is overcome by the strong attraction between the ions and the polar water molecules. These ions are dragged away from the crystal lattice and become surrounded by water molecules, i.e. the salt dissolves.

Polar covalent substances like HCl and NH_3 dissolve in water for the same reason. The general rule 'like dissolves like' summarises a lot about solubility. Thus, polar substances like HCl and NH_3 readily dissolve in water. However, non-polar substances like methane or oil or most plastics do not dissolve in water. This is because non-polar molecules have no charged ions or partially charged molecules. Therefore, the positive and negative ends of the water molecules are not attracted to them. Hence, oil and water do not mix. Therefore, dry cleaners use a non-polar solvent like perchloroethene (tetrachloroethene), C_2Cl_4, to dissolve out oily stains from clothing. Similarly, iodine crystals (non-polar) will not dissolve in water (FIG 5.38) but will dissolve in a non-polar solvent like hexane (C_6H_{14}). A gas called phosphine, PH_3, is virtually insoluble in water since phosphine is non-polar (E.N. value of both P and H = 2.1).

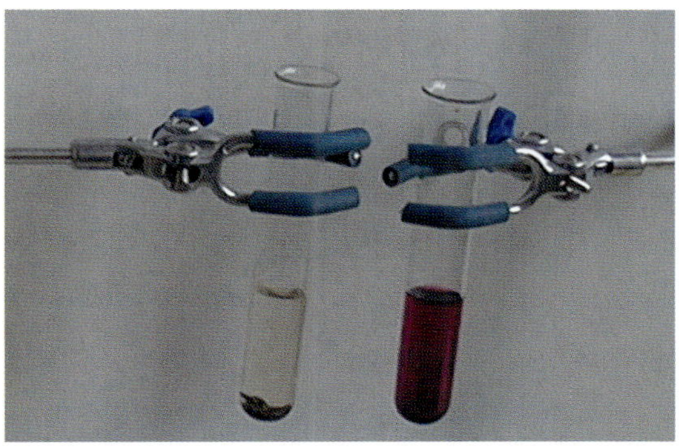

Fig 5.38
Like dissolves like. Iodine is non-polar. Therefore, it does not dissolve in water (polar) but will dissolve in hexane (non-polar).

2. To predict which compounds are ionic and which are covalent.

There is a useful 'rule-of thumb' which assists us in predicting whether a compound is ionic or covalent.

> An electronegativity difference greater than 1.7 indicates ionic bonding in a compound. An electronegativity difference of less than or equal to 1.7 indicates covalent bonding in a compound.

For example, consider the electronegativity difference of the elements in potassium fluoride and methane.

KF	CH_4
0.8 4.0	2.5 2.1
E.N. Difference = 3.2 i.e. > 1.7	E.N. Difference = 0.4 i.e. < 1.7
⇒ IONIC BONDING	⇒ COVALENT BONDING

Note: There are some exceptions to the above rule-of-thumb. The main ones are lithium hydride (LiH), sodium hydride (NaH), potassium hydride (KH) and calcium hydride (CaH_2). All of these compounds are ionic and contain the hydride ion, H^-. (The above rule-of-thumb would indicate incorrectly that the bonding is covalent.)

So far we have studied the two main types of bonding – ionic bonding and covalent bonding. Since these types of bonds hold atoms or ions together within the molecule or crystal lattice, this type of bonding is often referred to as **intramolecular** bonding (Latin, intra = within). However, there are other types of bonds that exist **between** one molecule and another. This type of bonding is referred to as **intermolecular** bonding (Latin, inter = between). There are three types of forces between molecules: van der Waals forces, Dipole-Dipole forces and Hydrogen Bonding.

5.10 VAN DER WAALS FORCES

A Dutch scientist called Johannes van der Waals (1837–1923) carried out research into the forces between molecules. He discovered that small attractive forces exist between molecules as a result of temporary internal shifts in the distribution of electrons within a molecule. To understand van der Waals forces, consider a molecule of hydrogen (FIG 5.39). As the two electrons move inside the molecule, it may happen that, at any one instant, both electrons may be closer to one end of the molecule than the other, i.e. one side of the molecule may be slightly negative leaving the other side of the molecule slightly positive. Thus, a temporary dipole is set up in the molecule. A dipole is a pair of separated opposite charges. (It is temporary because it only exists for a short time – in contrast to the permanent dipole in a molecule of HCl.) This temporary dipole could induce a similar dipole in a nearby molecule. There is then an attraction between the opposite charges. This attraction is called a van der Waals force.

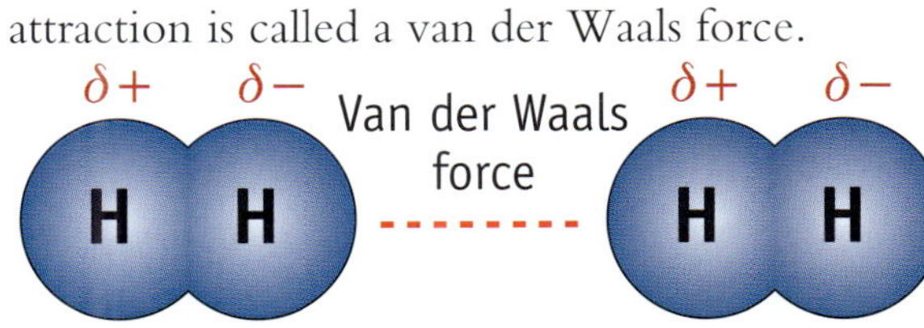

Fig 5.39
Van der Waals forces are weak attractions between temporary dipoles.

Van der Waals forces are weak attractive forces between molecules resulting from the formation of temporary dipoles.

Since the electrons are moving about at high speed, the attractions between any two given molecules only exist for an instant. Van der Waals forces are very weak. It has been estimated that this type of force is only about one thousandth as strong as the covalent bond holding the atoms together in the molecule.

Van der Waals forces are the only forces of attraction that exist between non-polar molecules, e.g. H_2, O_2, N_2, Cl_2, etc. Good evidence for the existence of these forces comes from the fact that gases that consist of non-polar molecules can be liquefied. Therefore, there must be some attractive forces between the molecules of non-polar substances.

It has been found that the strength of van der Waals forces increases as the molecules get bigger. This is because the greater number of electrons in the electron clouds allow the temporary dipoles to form more easily. For example, at room temperature Cl_2 is a gas, Br_2 is a liquid and I_2 is a solid due to the increasing strength of the van der Waals forces as the relative molecular mass increases. Similarly, H_2 has a boiling point of -252 °C but O_2 does not boil until -183 °C. The higher boiling point is due to the fact that more heat is required to break the stronger van der Waals forces formed between the bigger molecules of oxygen. Van der Waals forces also exist in the case of the noble gases, all of which consist of atoms. The boiling points of these gases increase with the size of the atoms due to the increasing strength of the van der Waals forces. Van der Waals forces are also used to explain the difference in the boiling points of molecules like propane, C_3H_8, and butane, C_4H_{10}. Butane has a higher boiling point because it has more electrons than propane. Therefore, temporary dipoles are more easily formed in butane giving greater numbers of van der Waals forces.*

5.11 DIPOLE-DIPOLE FORCES

In the previous section we saw how temporary dipoles can lead to an attraction between one molecule and another. However, in compounds whose molecules are polar, these dipole-dipole forces exist permanently.

Dipole-dipole forces are forces of attraction between the negative pole of one molecule and the positive pole of another.

For example, since hydrogen chloride is a polar molecule, it has a permanent electric dipole (Fig 5.40). The partially negative end of one molecule can attract the partially positive end of a neighbouring molecule. It must be remembered that because dipole-dipole forces are between **partial** electric charges (δ- and δ+), these forces are much weaker than the ionic bonds holding ionic compounds together.

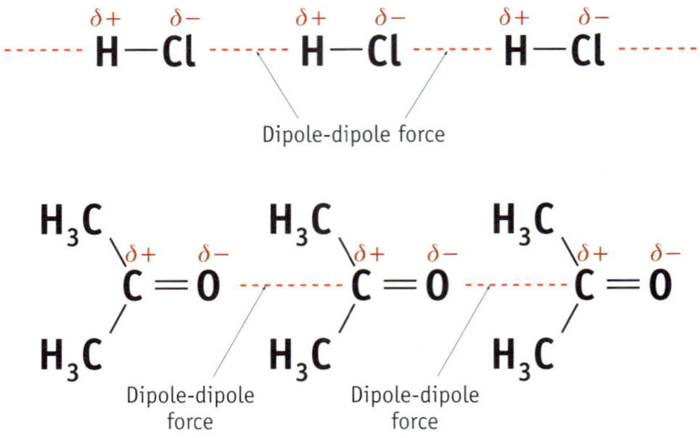

Fig 5.40
Dipole-dipole forces exist between polar molecules. Two examples are shown here – HCl and propanone (also called acetone), $(CH_3)_2CO$, used in nail varnish remover. For the sake of clarity, only three molecules are shown. It must be remembered that these forces act in all directions.

These attractive forces between the molecules give rise to higher boiling points than those of similar non-polar molecules. For example, the boiling point of HCl (-85 °C) is much higher than that of H_2 (-252 °C) as heat must be supplied to overcome the dipole-dipole forces in HCl. Similarly, the boiling point of a substance called methanal (formaldehyde), HCHO, is much higher (-21 °C) than that of ethene, C_2H_4, (-104 °C) for the same reason (Fig 5.41).

* Some chemists use the term 'van der Waals' forces to include not only the temporary dipole-dipole attractions (in non-polar molecules) but also the permanent dipole-dipole attractions (in polar molecules). Since the explanation of the forces in non-polar molecules was first given by Fritz London in 1930, the forces of attraction in non-polar molecules are also called 'London' forces. In keeping with the syllabus, this author has interpreted van der Waals forces to include London forces only.

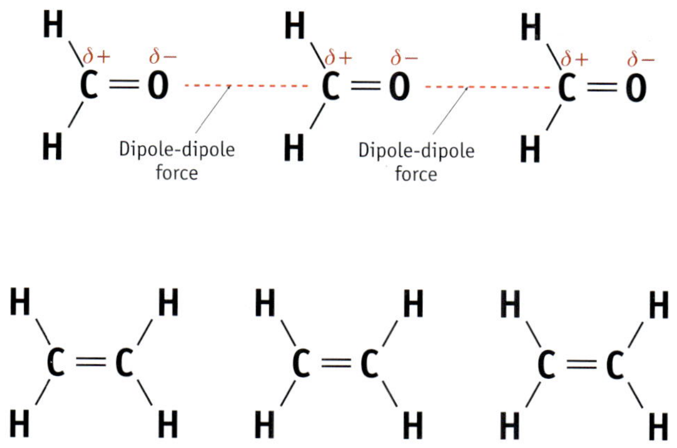

Fig 5.41
The dipole-dipole attractions between the molecules in methanal (formaldehyde) give it a much higher boiling point than that of ethene. There are no permanent dipole-dipole forces in ethene since it is a non-polar molecule.

5.12 HYDROGEN BONDING

When chemists studied the boiling points of water, ammonia and hydrogen fluoride they found that their boiling points were far higher than expected. For example, H_2O has a boiling point of 100 °C but H_2S has a much lower boiling point of only -61 °C. Both oxygen and sulfur are in the same group of the Periodic Table. Why the big difference? Similarly, NH_3 has a boiling point of -33 °C but PH_3 has a far lower boiling point of only -88 °C? Finally, HF has a boiling point of 20 °C but HCl has a boiling point of only -85 °C. The fact that the boiling points of water, ammonia and hydrogen fluoride are remarkably high is very obvious from Fig 5.42.

Clearly, the molecules of H_2O, NH_3 and HF must have fairly strong intermolecular forces pulling them together more strongly than if there were just van der Waals or dipole-dipole forces present. These intermolecular forces are called **hydrogen bonds**. Hydrogen bonding arises when an atom of hydrogen is bonded to a small electronegative atom like nitrogen, oxygen or fluorine (the three most electronegative elements in the Periodic Table). The hydrogen bonds in these three compounds are shown in Fig 5.43. Note that the hydrogen bonds are shown by dotted lines and that the hydrogen atoms form a sort of bridge between two molecules. The hydrogen atom is bonded to one atom by a covalent bond and to the other atom by a hydrogen bond.

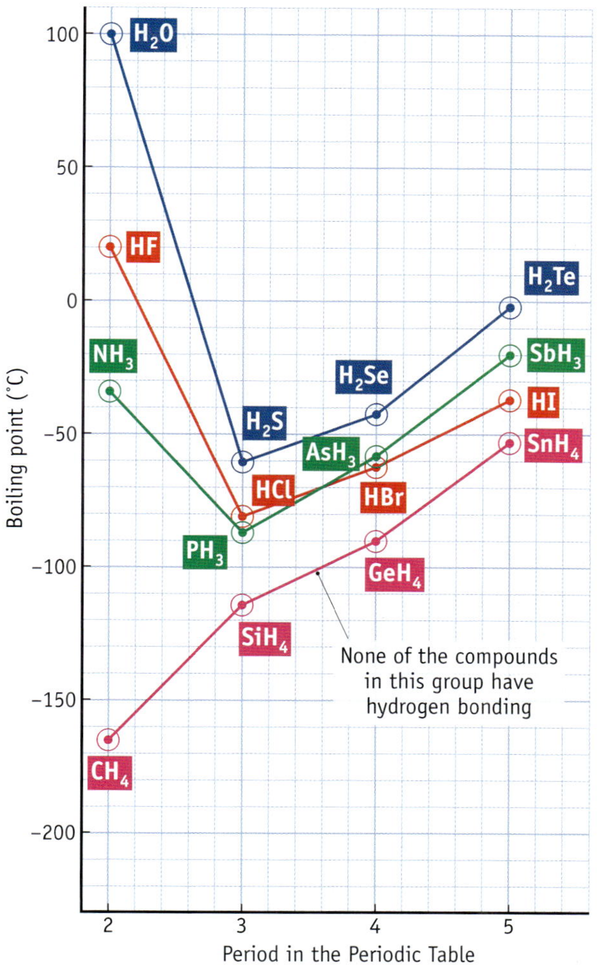

Fig 5.42
The high boiling points of H_2O, HF and NH_3 are evidence for the existence of hydrogen bonding. Note how much higher the boiling points of these compounds are compared with the boiling points of the hydrogen compounds of the other elements in the same group. The type of trend that would be expected if hydrogen bonding did not exist is shown by the bottom curve.

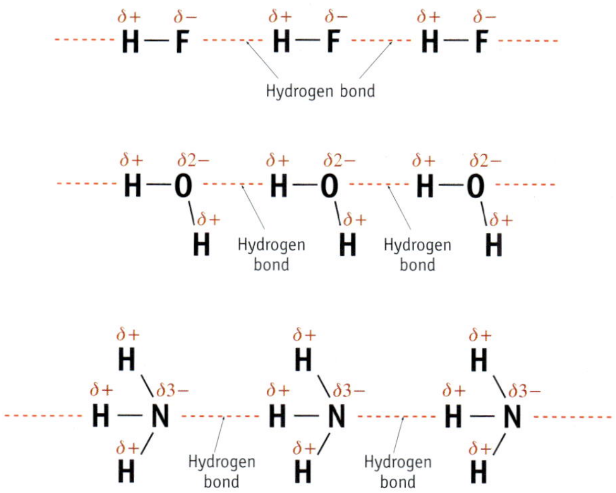

Fig 5.43
Hydrogen bonding occurs in hydrogen fluoride, water and ammonia.

The hydrogen atom in each of the three compounds has a fairly strong positive charge because in each case it is bonded to a very electronegative element. This causes a strong polarity in the molecules and they line up in molecular chains or groups that hold the molecules together quite strongly. Considerable energy must be supplied to break these molecular chains so that the molecules can escape from the liquid - hence the high boiling points. Hydrogen bonds are much stronger than either van der Waals forces or dipole-dipole forces. However, hydrogen bonds are much weaker than ordinary covalent bonds. It has been calculated that the strength of a hydrogen bond is about 10% of that of a covalent bond. However, because so many hydrogen bonds are present, their effect is quite appreciable. Note in FIG 5.42 that the boiling point of H_2O is higher than that of HF. This is because there are twice as many hydrogen bonds for each molecule of water as there are for every molecule of HF (even though the individual hydrogen bonds in HF are stronger due to the fact that fluorine is more electronegative than oxygen).

Hydrogen bonds are particular types of dipole-dipole attractions between molecules in which hydrogen atoms are bonded to nitrogen, oxygen or fluorine. The hydrogen atom carries a partial positive charge and is attracted to the electronegative atom in another molecule. Thus, it acts as a bridge between two electronegative atoms.

Hydrogen bonding is very important in synthetic clothing. Kevlar is a type of fibre developed by the Du Pont company (FIG 5.44). It is fire resistant, extremely strong and flexible. Kevlar is strong because long chains of molecules line up parallel to each other and are held together strongly by hydrogen bonding. This makes Kevlar an ideal material for sportswear, bullet-proof vests, jackets for fencers, etc.

Fig 5.44
Professional motorcyclists wear protective suits made of Kevlar. This material consists of chains of molecules held together by hydrogen bonding.

Hydrogen bonding also helps fibres absorb water. Wool contains chains of substances called proteins that are made up of many oxygen and nitrogen atoms. These atoms can form hydrogen bonds to water. This means that wool can absorb water and does not feel sticky to wear.

Ammonia is extremely soluble in water for two reasons. Firstly, like water it is a polar molecule and, as discussed already, 'like dissolves like' so we would expect it to dissolve in water, i.e. there are attractive forces between the partial charges in the polar molecules. Secondly, ammonia is very soluble in water due to the fact that it forms hydrogen bonds with the water. This helps the ammonia to form a perfect mixture with the water, i.e. it readily dissolves.

Hydrogen bonding in water also explains why water has a high surface tension. The attraction of the water molecules on the surface of water gives rise to a type of 'skin' on the water that enables insects to walk on it.

CHAPTER SUMMARY

- The octet rule states that, when bonding occurs, atoms tend to reach an electron arrangement with eight electrons in the outermost shell.
- Ionic bonding involves the transfer of electrons from one atom to another.
- Covalent bonding involves the sharing of electrons.
- A transition metal is one that forms at least one ion with a partially filled d sublevel.
- A sigma bond is formed by the 'head-on' overlap of atomic orbitals.
- A pi bond is formed by the sideways overlap of atomic orbitals.

CHAPTER SUMMARY

- Ionic compounds tend to be hard, have high melting and boiling points, and conduct electricity when molten or when dissolved in water.
- Covalent compounds tend to be soft, have low melting and boiling points, and do not conduct electricity.
- Recrystallisation is an important technique used to purify certain substances.
- The melting point of a substance can be used to help identify that substance and to check its purity.
- The VSEPR Theory can be used to predict the shapes of covalent molecules.
- Electronegativity is the relative attraction that an atom in a molecule has for the shared pair of electrons in a covalent bond.
- Electronegativity values may be used to predict the polarity of covalent bonds and also to predict which compounds are ionic and which are covalent.
- There are three types of intermolecular forces: Van der Waals forces, Dipole-dipole forces and Hydrogen bonds.

EXERCISES

5.1. (a) State the octet rule ....

(b) An ionic bond is the .... of attraction between oppositely charged ions in a compound.

(c) Give one use of sodium chloride in everyday life ....

(d) The formula of magnesium bromide is ....

(e) One of the characteristics of transition metals is that they exhibit variable ....

(f) Name the compound Cu_2O. ....

(g) A transition metal is defined as one which ....

(h) Why do most transition metal ions have variable valencies? ....

(i) In the first row of the d-block metals, name one metal which is not normally considered a transition metal ....

(j) Draw an electron dot structure showing the bonding in CH_3CHO.

(k) A pair of electrons which is not involved in bonding is called a ....

(l) A pi bond is formed by the .... overlap of atomic orbitals.

(m) Give two general characteristics of ionic compounds ....

(n) Give one item of information which may be deduced from the melting point of a solid ....

(o) A sample of benzoic acid prepared in the laboratory was found to melt over a range rather than at its expected melting point. Suggest a reason for this...........................

(p) VSEPR stands for ....

(q) How would you describe the shape of (a) the ammonia molecule....................... and (b) the $CHCl_3$ molecule? ....

(r) Predict the possible shapes of molecules of general formula XH_2 ....

(s) Which one of the following molecules is not polar:
HCl, CO_2, NH_3, CH_3Cl, H_2O ....

(t) An electronegativity difference greater than ....indicates ionic bonding.

(u) What term is used (i) for a bond in which the bonding electrons are not equally shared, (ii) for a bond in which there is a complete transfer of electrons?

....

(v) What type of bond exists (i) between the atoms in a molecule of hydrogen chloride, (ii) between molecules of hydrogen chloride? ....

(w) When liquid air is allowed to warm up, nitrogen boils off at -196 °C but oxygen does not boil off until -183 °C. Suggest a reason for this ....

(x) At room temperature, chloroform ($CHCl_3$), is a liquid but methane (CH_4), is a gas. Suggest a reason for this. ....

5.2. With reference to the Table of Electronegativity values, predict the type of bonding in (a) sodium bromide, (b) aluminium bromide, (c) nitrogen dioxide.
Indicate, showing outer electrons only, the type of bonding present in the compound that is formed when an atom of the element of atomic number 15 reacts with three atoms of chlorine. Use the VSEPR theory to deduce the shape of this chloride.
Write the formula of a chloride in which the type of bonding is different from the chloride just described. Using diagrams, show clearly how the binding forces arise.

5.3. (i) Suggest two possible shapes for molecules of general formula QX_3 where X is an atom of a monovalent element. Give a reason for the difference between these shapes.

(ii) Write the formula of the compound formed by the reaction between chlorine and the metallic element of atomic number 31.

(iii) Give two reasons why ammonia (NH_3) is very soluble in water. Why is the solubility of phosphine (PH_3) in water very low? *[LCH]*

5.4. Define (i) ionic bond, (ii) polar bond, (iii) hydrogen bond. Give one example of (i) and one example of (ii).
The table shows boiling points in kelvins (K) of the chlorides of three elements from the third period of the Periodic Table.

Formula	$NaCl$	$AlCl_3$	PCl_3
Boiling point (K)	1790	453	349

Explain why the boiling point of sodium chloride is so much higher than those of the other two chlorides. *[LCH]*

5.5. Answer the following questions by referring to the first 36 elements of the Periodic Table and to the table of electronegativities.

(i) Name one non-metal which exists as single atoms.

(ii) Give the formula of an ionic compound formed by an element in group II and show the charge on each ion.

(iii) Give the formula of a covalent compound formed by an element in group V and show clearly the electronic structure of the molecule.

(iv) What do you understand by electronegativity?

(v) Comment on the nature of the bonds you would expect in each of the compounds MgS and PH_3.

(vi) Which two of the first 36 elements would form the compound you would expect to have the greatest ionic character? *[LCH]*

Chemical Equations: Tests For Anions

6

6.1 INTRODUCTION

We have already seen that elements combine together to form a compound. We have seen how the atoms are held together in a compound by chemical bonds. When elements react together to form a compound, we say that a **chemical reaction** has taken place. A new substance is always formed when a chemical reaction occurs. This new substance may be formed by elements or compounds reacting together. There is usually some evidence to suggest that a chemical reaction has occurred, e.g. bubbles of gas may be formed, colour changes may be observed, a temperature change may be observed, light or sound may be given off, the solution may turn cloudy, etc. Some of these ideas are summarised in Fig 6.1.

Fig 6.1 Some evidence to indicate that a chemical reaction is taking place.

In this chapter we shall study how chemists represent chemical reactions. We shall also carry out some experiments to test for the presence of certain negative ions (also called **anions**) in solution. These tests will then be explained in terms of the chemical reactions that have taken place.

6.2 CHEMICAL EQUATIONS

When a chemical reaction takes place, the substances that react together are changed into new substances. The starting substances are called the **reactants** and the substances formed are called the **products**. The changing of reactants to products could be described by means of a sentence, e.g. *carbon reacts with oxygen to form carbon dioxide*. Alternatively we would write this sentence as a **word equation**:

$$\text{carbon} + \text{oxygen} \rightarrow \text{carbon dioxide}$$

However, chemists throughout the world have developed a 'shorthand' system for representing chemical reactions. For convenience, they represent a chemical reaction using the formulas for elements and compounds. This arrangement is known as a **chemical equation**. Thus the chemical equation for the above word equation is:

$$C + O_2 \rightarrow CO_2$$

Note that the reactants are written on the left-hand side of the arrow and the products on the right-hand side of the arrow. The '+' sign means 'and' and the '→' sign means 'react to give'. Thus, a chemical equation may be regarded as a sentence in which words are replaced by formulas.

6.3 BALANCING CHEMICAL EQUATIONS

A French chemist called Antoine Lavoisier (1743–1794) carried out extensive research into the masses of reactants and products in chemical reactions. He studied many different chemical reactions in which he carefully measured the masses of all the reactants before the chemical reaction took place. Then he collected all the products formed in the chemical reactions and carefully measured their masses.

For each chemical reaction that he studied, he found that the total mass of the products was equal to the total mass of the reactants. In 1789, he summarised his findings in the Law of Conservation of Mass.*

> The Law of Conservation of Mass states that the total mass of the products of a chemical reaction is the same as the total mass of the reactants.

For example, if 12 grams of carbon combines with 32 grams of oxygen it is found that 44 grams of carbon dioxide are formed. Thus, atoms cannot simply 'disappear' from the reactants. All atoms present in the reactants must also be present in the products. This idea is also stated in the Law of Conservation of Matter. (The term 'matter' is simply another word for 'materials' or 'substances'.)

* This Law is now more correctly called the Law of Conservation of Mass-Energy since Albert Einstein proposed in 1905 that mass and energy could be interconverted. However, for ordinary chemical reactions, the law of conservation of mass is perfectly adequate.

> The Law of Conservation of Matter states that in any chemical reaction, matter is neither created nor destroyed but merely changes from one form into another.

It is clear therefore that, if there is a particular number of atoms of an element present at the beginning of a reaction, the exact same number of atoms of this element must be present at the end of the reaction. We acknowledge this situation by saying that a chemical equation must be balanced. A balanced equation is one in which the total number of atoms of each element on the left-hand side of the equation is equal to the total number of atoms of the same element on the right-hand side of the equation.

Is the equation $C + O_2 \rightarrow CO_2$ balanced?

There is one carbon atom on the left and one carbon atom on the right. There are two oxygen atoms on the left and two oxygen atoms on the right. Thus, the equation is balanced.

To help you to see how to approach balancing an equation, consider the following example.

Example 6.1

Balance the equation $CH_4 + O_2 \rightarrow CO_2 + H_2O$

SOLUTION

Step 1: Count the number of atoms of each element.

Left-hand side	Right-hand side
1 C atom	1 C atom
4 H atoms	2 H atoms
2 O atoms	3 O atoms

Step 2: Adjust the numbers in front of the formulas.

To get four H atoms on the right-hand side of the equation, put a '2' before the H_2O:

$CH_4 + O_2 \rightarrow CO_2 + 2H_2O$

The '2' before the H_2O tells us that there are two water molecules, i.e. four hydrogen atoms and two oxygen atoms. Thus, there are now four H atoms on the right-hand side. Therefore, the H atoms are balanced. Now check for balancing of oxygen atoms.

Left-hand side	Right-hand side
1 C atom	1 C atom
4 H atoms	4 H atoms
2 O atoms	4 O atoms

Note that there are two oxygen atoms on the left-hand side of the equation. However, there are four oxygen atoms on the right-hand side (two in CO_2 and two in $2H_2O$). To get four oxygen atoms on the left-hand side, put a '2' before the O_2. This means we have two oxygen molecules, i.e. four oxygen atoms.

$CH_4 + 2O_2 \rightarrow CO_2 + 2H_2O$

Step 3: Re-check the balancing at the end. The number of atoms of each element on the left-hand side must be equal to the number of atoms of each element on the right-hand side.

Left-hand side	Right-hand side
1 C atom	1 C atom
4 H atoms	4 H atoms
4 O atoms	4 O atoms

The equation is now balanced.

Note one very important point about balancing equations. **When balancing an equation, the formulas cannot be changed in any way.** The equation may only be balanced by putting numbers in front of the formulas. Thus, you could not change H_2O to H_4O as the compound H_4O does not exist!

If you find it difficult to balance chemical equations do not worry! When you have finished working through the equations given in the questions at the end of this chapter and in the workbook, you will become quite an expert! Even if you don't become an expert, balancing equations is only a tiny part of your chemistry course.

The 'trial and error' method of balancing equations as outlined above is known as 'balancing by inspection'. In CHAPTER 14 we will study a more systematic method of balancing equations.

Finally, it is important to realise that a balanced equation does not tell us anything about the speed of a reaction or if a reaction will take place at all! For example, one may write an equation like:

$$2Ar + He \rightarrow HeAr_2$$

However, this reaction does not take place!

6.4 TESTS FOR ANIONS IN AQUEOUS SOLUTION

We have already seen in CHAPTER 3 that the presence of positive ions (cations) in a substance may be detected by means of flame tests. In this section we shall study tests to detect the presence of certain negative ions (anions) dissolved in water. This type of analysis where we detect the presence of certain substances is known as **qualitative analysis**. The ions we shall investigate are (i) the chloride ion (Cl^-), (ii) the sulfate ion (SO_4^{2-}), (iii) the sulfite ion (SO_3^{2-}), (iv) the carbonate ion (CO_3^{2-}), (v) the hydrogencarbonate (HCO_3^-), (vi) the nitrate ion (NO_3^-) and (vii) the phosphate ion (PO_4^{3-}).

You are not required to write fully balanced equations for all of the following tests. In most cases, simple equations involving ions (called **ionic equations**) are perfectly acceptable.

Note: The old name for the hydrogencarbonate ion was the bicarbonate ion.

EXPERIMENT 6.1 MANDATORY STUDENT EXPERIMENT NO. 3

EXPERIMENT 6.1A

To detect the presence of chloride ions in aqueous solution.

Apparatus required

Test-tubes, dropper, spatula, wash bottle.

Chemicals required

Any soluble chloride salt, e.g. NaCl, KCl, etc., silver nitrate solution (0.1 M), dilute ammonia solution (3 M), deionised water.

Procedure

1. Using a wash bottle of deionised water, fill a test-tube to about one quarter of its capacity.

2. Add a small amount of the chloride salt (a few crystals are sufficient) to the water. Shake the test-tube to help the salt dissolve. A clear solution should be obtained.
3. Using a clean dropper, add a few drops of silver nitrate solution ($AgNO_3$) to the clear solution. Note that the clear solution in the test-tube turns cloudy and a white solid material settles out of solution (FIG 6.2). The general name given to a material that settles out of solution is a **precipitate**. The insoluble material formed is called silver chloride and it is formed when the silver ions from the silver nitrate combine with the chloride ions dissolved in the water.

   $Ag^+ + Cl^- \rightarrow AgCl \downarrow$

   The arrow pointing down as shown indicates a precipitate is formed. Thus, the cloudiness is observed because insoluble silver chloride is formed.
4. To confirm that the white precipitate is silver chloride, add about a quarter of a test-tube of dilute ammonia solution and shake gently. Silver chloride is soluble in dilute ammonia solution. Thus, the cloudiness disappears.

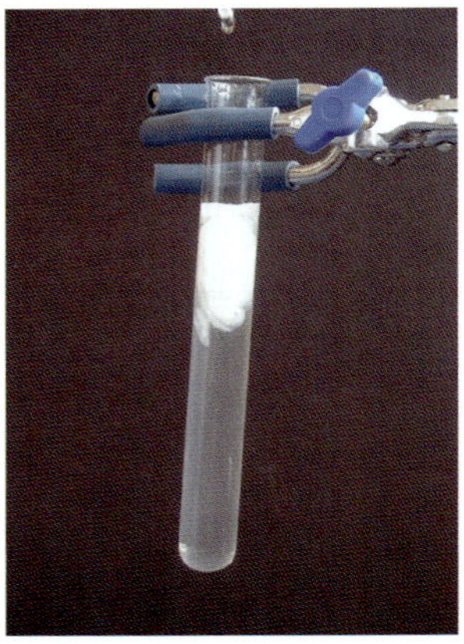

Fig 6.2
A white precipitate is formed when silver nitrate solution is added to a solution of chloride ions.

Conclusion

Chloride ions are present in solution.

Note: In chemistry the substances which cause a chemical reaction to take place are often referred to as **reagents.** For example, in the above reaction, silver nitrate is a reagent.

EXPERIMENT 6.1 B

To detect the presence of sulfate ions (SO_4^{2-}) and sulfite ions (SO_3^{2-}) in aqueous solution.

Apparatus required

Test-tubes, dropper, spatula, wash bottle.

Chemicals required

Any soluble sulfate salt, e.g. Na_2SO_4, K_2SO_4, etc., any soluble sulfite salt, e.g. Na_2SO_3, K_2SO_3, etc., barium chloride ($BaCl_2$) solution (0.25 M), dilute hydrochloric acid (3 M), deionised water.

Procedure

1. Using a wash bottle of deionised water, fill two test-tubes to about one quarter of their capacity.
2. Add a small amount of the sulfate salt (a few crystals are sufficient) to the water in one test-tube. Shake the test-tube to help the salt dissolve. A clear solution should be obtained. Make up a solution of the sulfite salt in a similar fashion. Label each test-tube.
3. Using a clean dropper, add a few drops of barium chloride solution ($BaCl_2$) to the clear solution in each test-tube. Note that the clear solution in each test-tube turns cloudy and a white solid material settles out of solution (FIG 6.3). The insoluble materials formed are barium sulfate and barium sulfite. These compounds are formed when the barium ions from the barium chloride solution combine with the sulfate ions or the sulfite ions dissolved in the water.

$$Ba^{2+} + SO_4^{2-} \rightarrow BaSO_4 \downarrow$$
barium sulfate

and

$$Ba^{2+} + SO_3^{2-} \rightarrow BaO_3 \downarrow$$
barium sulfite

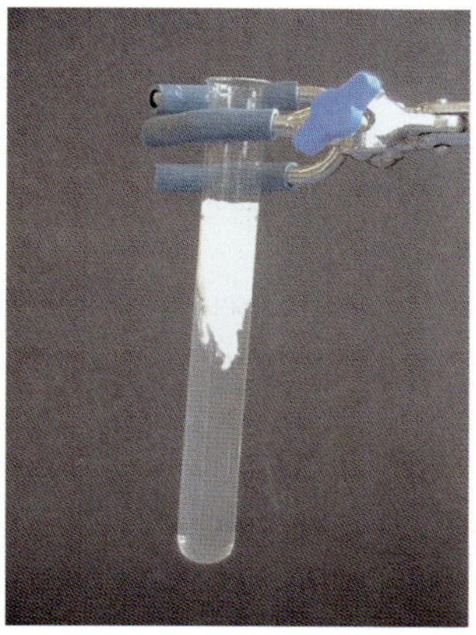

Fig 6.3
A white precipitate is formed when barium chloride solution is added to a solution of sulfate or sulfite ions.

To distinguish between sulfate and sulfite ions

To distinguish between the barium sulfate and the barium sulfite, add about a quarter of a test-tube of dilute hydrochloric acid to each of the above test-tubes and shake gently. Barium sulfate is insoluble in dilute hydrochloric acid. However, barium sulfite dissolves in dilute HCl. Thus, the cloudiness disappears in the case of the sulfite ion and remains in the case of the sulfate ion.

$$BaSO_4 + HCl \rightarrow \text{no reaction}$$

$$BaSO_3 + 2HCl \rightarrow BaCl_2 + SO_2 + H_2O$$

i.e. $SO_3^{2-} + 2H^+ \rightarrow SO_2 + H_2O$

Conclusion

Sulfate and sulfite ions are detected using barium chloride solution. They are distinguished from each other using dilute hydrochloric acid solution. Barium sulfate is insoluble in dilute hydrochloric acid but barium sulfite is soluble.

EXPERIMENT 6.1 C

To detect the presence of carbonate ions (CO_3^{2-}) and hydrogencarbonate ions (HCO_3^-) in aqueous solution.

Apparatus required

Boiling tube, test-tubes, dropper, wash bottle, spatula, one-holed rubber stopper to fit boiling tube, delivery tube.

Chemicals required

Any soluble carbonate salt, e.g. Na_2CO_3, K_2CO_3, etc., any soluble hydrogencarbonate salt, e.g. $NaHCO_3$, $KHCO_3$, etc., dilute hydrochloric acid (3 M), limewater, magnesium sulfate solution (0.25 M), deionised water.

Procedure

1. Place a small amount of the carbonate salt in a boiling tube to a depth of about 1 cm.
2. Insert a one-holed rubber stopper into the boiling tube and set up the apparatus as shown in FIG 6.4.
3. Remove the one-holed rubber stopper, add about a quarter of a test-tube of dilute hydrochloric acid to the carbonate salt in the boiling tube and quickly replace the rubber stopper.
4. A brisk effervescence (fizzing) is observed as the acid comes in contact with the carbonate. The carbonate ions react with the acid to liberate carbon dioxide:

$$CO_3^{2-} + 2H^+ \rightarrow CO_2 + H_2O$$

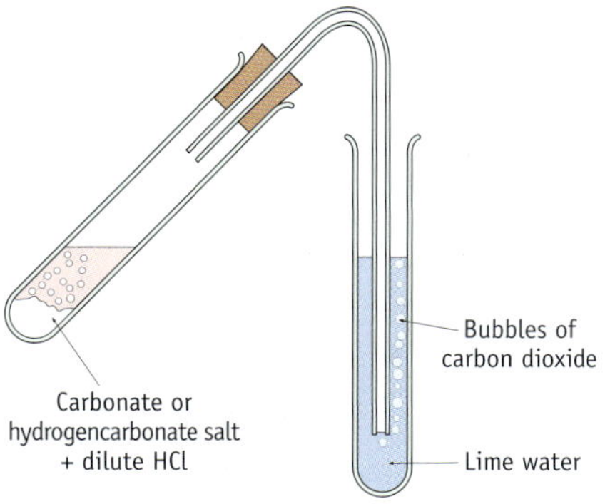

Fig 6.4
Carbon dioxide is given off when dilute hydrochloric acid is added to a carbonate or a hydrogencarbonate. This carbon dioxide turns limewater milky.

The limewater* turns milky due to the formation of calcium carbonate (chalk).

$Ca(OH)_2 + CO_2 \rightarrow CaCO_3 \downarrow + H_2O$

limewater — chalk (limestone, marble)

Note: If you continue to bubble carbon dioxide through the limewater, the milky colour disappears. This is because the insoluble calcium carbonate eventually reacts with the carbon dioxide to form soluble calcium hydrogencarbonate.

$CaCO_3 + CO_2 + H_2O \rightarrow Ca(HCO_3)_2$

insoluble — soluble

5. Repeat the experiment using the hydrogencarbonate salt instead of the carbonate salt. As in the case of the carbonate, the hydrogencarbonate reacts with the acid to liberate carbon dioxide.

   $HCO_3^- + H^+ \rightarrow CO_2 + H_2O$

   Thus, both a carbonate and a hydrogencarbonate react with dilute acids to give off carbon dioxide. This effervescence is also observed when you add substances like *Andrews Liver Salts* to water. *Andrews Liver Salts* contains sodium hydrogencarbonate and citric acid. When dissolved in water the citric acid reacts with the hydrogencarbonate to give off carbon dioxide. This causes the fizzing to occur.

To distinguish between carbonate and hydrogencarbonate ions

Since both carbonates and hydrogencarbonates react in a similar way with dilute acids, we must have some method of distinguishing between the two ions. This is done using a solution containing magnesium ions, e.g. magnesium sulfate ($MgSO_4$). Add a small amount of the carbonate salt to about a quarter of a test-tube of deionised water. Then add about a quarter of a test-tube of magnesium sulfate solution to the solution of the carbonate. Note the formation of a white precipitate of insoluble magnesium carbonate.

$Mg^{2+} + CO_3^{2-} \rightarrow MgCO_3 \downarrow$

insoluble

However, if this test is repeated with the hydrogencarbonate salt, no precipitate is formed as magnesium hydrogencarbonate is soluble in water

$Mg^{2+} + 2HCO_3^- \rightarrow Mg(HCO_3)_2$

soluble

Note: If the test-tube containing the magnesium hydrogencarbonate is placed in a beaker of boiling water for a few minutes, the clear solution turns cloudy. This is because the magnesium hydrogencarbonate is converted to magnesium carbonate.

$Mg(HCO_3)_2 \rightarrow MgCO_3 \downarrow + CO_2 + H_2O$

soluble — insoluble

* As limewater reacts with the carbon dioxide in the air, it is best to make up a fresh solution. This may easily be done by adding a small quantity of calcium hydroxide to a beaker of deionised water. Filter the cloudy suspension into a test-tube using a filter paper and funnel. The clear liquid that filters through is limewater.

Conclusion

Carbonate and hydrogencarbonate ions are detected using dilute acid. They are distinguished from each other using magnesium sulfate solution. Magnesium carbonate is insoluble in water but magnesium hydrogencarbonate is soluble.

EXPERIMENT 6.1 D

To detect the presence of nitrate ions (NO_3^-) in aqueous solution.

Preliminary note: Since all nitrate salts are soluble in water, it is not possible to detect the presence of the nitrate ion by a precipitation reaction. Instead, a special test called the **Brown Ring Test** is used to detect the presence of the nitrate ion.

Apparatus required

Test-tube, dropper, spatula, wash bottle.

Chemicals required

Any nitrate salt, e.g. $NaNO_3$, KNO_3, etc., iron(II) sulfate solution (0.5 M), concentrated sulfuric acid, deionised water.

Procedure

1. Using a wash bottle of deionised water, fill a test-tube to about one quarter of its capacity.
2. Add a small amount of the nitrate salt (a few crystals are sufficient) to the water. Shake the test-tube to help the salt dissolve. A clear solution should be obtained.
3. Add the same quantity of freshly prepared iron(II) sulfate solution to the test-tube.
4. This step involves the use of concentrated sulfuric acid. **Extreme care must be taken with this substance**. Under the direct supervision of your teacher, use a dropper to add about 3 cm^3 of concentrated sulfuric acid down the inside of the test-tube. The test-tube should be slanted slightly as shown in FIG 6.5. Since the concentrated sulfuric acid is so dense, it settles at the bottom of the test-tube and forms a layer beneath the mixture.
5. Note that a brown ring is formed at the junction of the two layers (FIG 6.6). This brown ring is formed as a result of nitrate ions being present and hence is the confirmatory test for the presence of the nitrate ion. The brown colour is due to the formation of the substance $FeSO_4.NO$. As the chemical reactions involved are rather complicated, we need not concern ourselves with the details of these reactions.

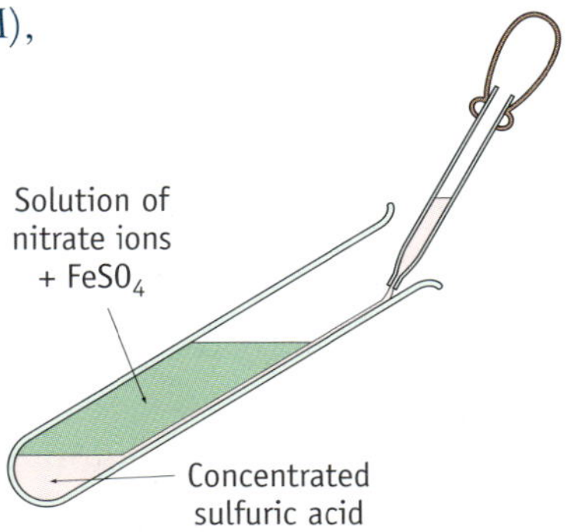

Fig 6.5
Adding concentrated sulfuric acid to the mixture of nitrate ion solution and iron(II) sulfate solution.

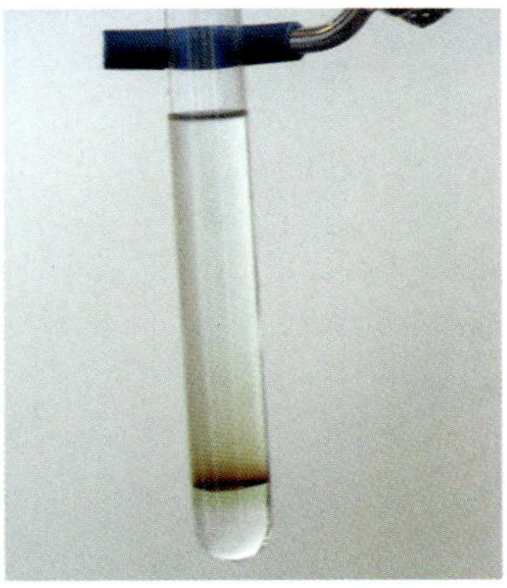

Fig 6.6
A brown ring is formed at the junction between the concentrated sulfuric acid at the bottom and the mixture of the nitrate ion solution and iron(II) sulfate solution on top.

Conclusion

Nitrate ions are present in a solution if a brown ring is formed when concentrated sulfuric acid is added to a mixture of the solution and a freshly prepared iron(II) sulfate solution.

EXPERIMENT 6.1 E

To detect the presence of phosphate ions in aqueous solution.

Apparatus required

Test-tube, dropper, spatula, wash bottle, hot plate or Bunsen burner.

Chemicals required

Any soluble phosphate salt, e.g. Na_2HPO_4, ammonium molybdate solution (0.5M), concentrated nitric acid, deionised water.

Procedure

1. Using a wash bottle of deionised water, fill a test-tube to about one tenth of its capacity.
2. Using a spatula, add a small amount of the phosphate salt (a few crystals are sufficient) to the water. Shake the test-tube to help the salt dissolve. A clear solution should be obtained.
3. Pour in ammonium molybdate solution until the test-tube is about half full. Ammonium molybdate is a compound of molybdenum (element number 42 in the Periodic Table).
4. This step involves the use of concentrated nitric acid. **Extreme care must be taken with this substance**. Under the direct supervision of your teacher and using a dropper, add five drops of concentrated nitric acid to the test-tube.
5. Note the formation of a yellow precipitate (FIG 6.7). If this does not form, place the test-tube in warm (about 60 °C) water. This yellow precipitate is a substance called ammonium phosphomolybdate. You are not required to know the equation for this reaction. Hence, this reaction will not be discussed in any further detail.

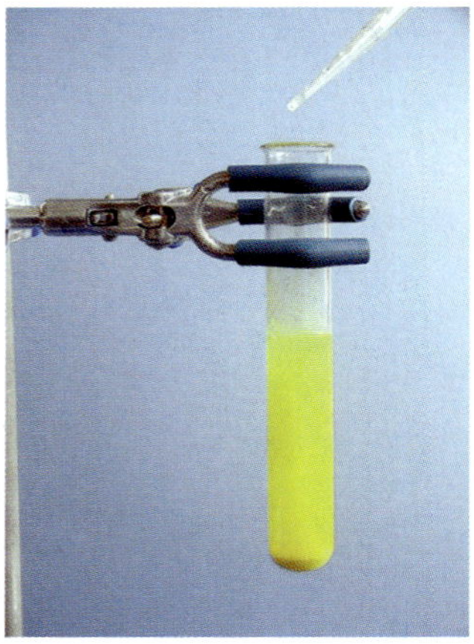

Fig 6.7
A yellow precipitate is formed when concentrated nitric acid is added to a solution containing phosphate ions and ammonium molybdate.

Conclusion

Phosphate ions are present in solution.

CHAPTER SUMMARY

- The Law of Conservation of Mass states that the total mass of the products of a chemical reaction is the same as the total mass of the reactants.
- When balancing an equation the formulas cannot be changed in any way.
- In a balanced equation, number of atoms of each element on left-hand side of equation = number of atoms of each element on right-hand side of equation.
- The tests for the various ions on the syllabus are summarised in TABLE 6.1.

CHAPTER SUMMARY

ANION	TEST	OBSERVATION	INFERENCE
Chloride	Add a few drops of $AgNO_3$ solution to a solution of the solid.	A white precipitate is formed which is soluble in dilute ammonia solution.	$Ag^+ + Cl^- \longrightarrow AgCl\downarrow$
Sulfate or **Sulfite**	1. Add $BaCl_2$ solution to a solution of the solid.	A white precipitate is formed.	$Ba^{2+} + SO_4^{2-} \longrightarrow BaSO_4\downarrow$ $Ba^{2+} + SO_3^{2-} \longrightarrow BaSO_3\downarrow$
	2. Distinguishing test. Add dilute HCl.	If precipitate remains ⇒ sulfate. If precipitate dissolves ⇒ sulfite.	$BaSO_4 + HCl \longrightarrow$ no reaction $SO_3^{2-} + 2H^+ \longrightarrow SO_2 + H_2O$
Carbonate or **Hydrogen-carbonate**	1. Add dilute HCl to the solid.	Fizzing. A gas is given off that turns limewater milky.	$CO_3^{2-} + 2H^+ \longrightarrow CO_2 + H_2O$ $HCO_3^- + H^+ \longrightarrow CO_2 + H_2O$ $Ca(OH)_2 + CO_2 \longrightarrow CaCO_3\downarrow + H_2O$
	2. Distinguishing test. Add $MgSO_4$ to a fresh solution of the solid.	White precipitate forms ⇒ carbonate. No precipitate forms ⇒ hydrogencarbonate. Boiling gives a precipitate.	$Mg^{2+} + CO_3^{2-} \longrightarrow MgCO_3\downarrow$ $Mg^{2+} + 2HCO_3^- \longrightarrow Mg(HCO_3)_2$ $Mg(HCO_3)_2 \longrightarrow MgCO_3\downarrow + CO_2 + H_2O$
Nitrate	Brown Ring test. To a solution of the solid add a freshly prepared $FeSO_4$ solution. Add concentrated sulfuric acid.	A brown ring is formed at the junction of the two liquids. ⇒ nitrate ion present	Brown ring is due to presence of nitrate ion.
Phosphate	Ammonium molybdate is added to a solution of the solid. Concentrated nitric acid is added. Solution is warmed.	Yellow precipitate is formed.	Phosphate ions are present.

Table 6.1
Summary of tests for anions.

EXERCISES

6.1. Balance each of the following equations.

(a) $NO + O_2 \rightarrow NO_2$
(b) $Fe + O_2 \rightarrow Fe_3O_4$
(c) $Al + Fe_2O_3 \rightarrow Al_2O_3 + Fe$
(d) $P + O_2 \rightarrow P_2O_5$
(e) $Al + Cl_2 \rightarrow AlCl_3$
(f) $Mg(OH)_2 + HCl \rightarrow MgCl_2 + H_2O$
(g) $HNO_3 + Ca(OH)_2 \rightarrow Ca(NO_3)_2 + H_2O$
(h) $Na_2CO_3 + HCl \rightarrow NaCl + H_2O + CO_2$
(i) $N_2 + O_2 \rightarrow NO$
(j) $K + H_2O \rightarrow KOH + H_2$
(k) $AgNO_3 + HCl \rightarrow AgCl + HNO_3$
(l) $BaCl_2 + Na_2SO_4 \rightarrow NaCl + BaSO_4$
(m) $AsCl_3 + 3H_2O \rightarrow H_3AsO_3 + HCl$
(n) $FeCl_3 + 3NaOH \rightarrow Fe(OH)_3 + 3NaCl$
(o) $C_8H_{18} + O_2 \rightarrow CO_2 + H_2O$

6.2. (a) What is an anion? ....

EXERCISES

(b) A .... substance is always formed when a chemical reaction occurs.

(c) State the Law of Conservation of Mass ....

(d) An equation may only be balanced by changing the .... in front of the ....

(e) The formula for the sulfate ion is .... and that of the sulfite ion is ....

(f) A material which settles out of solution is called a ....

(g) Silver chloride is .... in dilute ammonia solution.

(h) The formula for silver nitrate is ....

(i) Barium chloride forms a white precipitate with either .... or .... ions.

(j) Effervescence is observed when a dilute acid is added to a solution containing either a .... or a .... ion.

(k) Write down the equation for the reaction of carbon dioxide with limewater.
....

(l) Name two reagents used to detect the presence of the nitrate ion in aqueous solution.
.... *[LCH]*

(m) Ammonium molybdate is used to detect the presence of the .... ion.

(n) In testing for the phosphate ion, a .... coloured precipitate is obtained.

(o) What reagents would you use to test for the presence of chloride ions in aqueous solution? .... *[LCH]*

(p) Write an equation for the reaction between a dilute acid and a sulfite.
.... *[LCH]*

(q) How would you distinguish between sulfite and sulfate ions in aqueous solution?
.... *[LCH]*

(r) In the brown ring test, name the substance present in the bottom layer in the test-tube.
....

(s) The following reagents, in dilute aqueous solution, are used in qualitative analysis: HCl, NH_3, $BaCl_2$, $FeSO_4$, $AgNO_3$.
Which reagent, or combination of reagents, would best give a positive identification test for (i) the sulfate ion, (ii) the chloride ion ....

6.3. A school ordered five compounds, all of which were solids and soluble in water. When the order was delivered, the labels were missing. The delivery should have contained a nitrate, a carbonate, a sulfate, a sulfite and a chloride. All compounds were either the sodium or the potassium salt.
Describe the tests you would use to identify the five compounds. *[LCO]*

6.4. A student was given salt X for analysis in the laboratory. When dilute hydrochloric acid was added to the salt, there was a brisk effervescence of a colourless gas which gave a white precipitate when bubbled through limewater. Another white precipitate was observed when magnesium sulfate solution was added to an aqueous solution of the salt. Finally, it was found that salt X gave a lilac colour to the Bunsen flame.

(i) Identify X.

(ii) Give the names of the colourless gas and the two white precipitates.

(iii) Write a balanced equation for each reaction that took place. *[LCH]*

Trends In The Periodic Table

7

7.1 INTRODUCTION

In studying the Periodic Table in CHAPTER 4 we learned that, when the elements are arranged in order of increasing atomic number, the properties of the elements repeated periodically, i.e. at regular intervals. In this chapter we will first study how certain properties of the elements change within the Periodic Table. We will then see how these patterns of change may be used to explain the chemical properties of certain groups of elements.

The properties of elements whose pattern of change we will study are atomic radius, ionisation energy and electronegativity.

7.2 TRENDS IN ATOMIC RADII

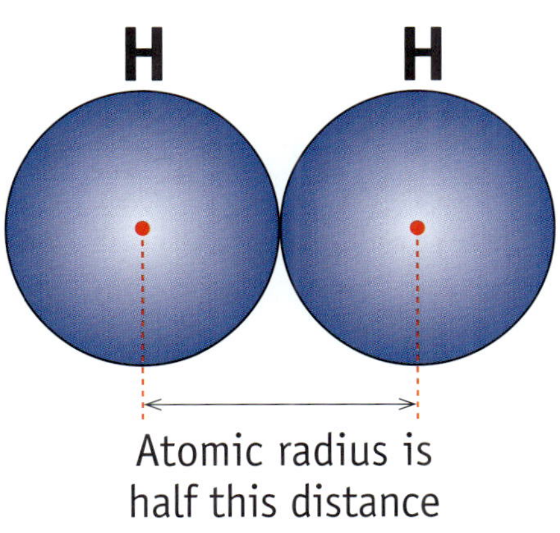

Fig 7.1
The atomic radius (covalent radius) is half the distance between the nuclei of two atoms of the same element that are joined together by a single covalent bond.

You might think that the atomic radius of an atom is the distance from the nucleus of the atom to the outermost electrons. However, since the position of an electron cannot be precisely located (Heisenberg Uncertainty Principle), it is not possible for chemists to measure exactly where the electron cloud of an atom ends. However, it is possible to measure experimentally the distance between the nuclei of atoms (FIG 7.1). This distance is known as the **bond length** and may be measured by methods known as 'X-ray diffraction' and 'electron diffraction'. In this course we shall be concerned with only one type of atomic radius* called the **covalent radius** which is defined as follows:

> The atomic radius (covalent radius) of an atom is defined as half the distance between the nuclei of two atoms of the same element that are joined together by a single covalent bond.

For example, in a molecule of hydrogen it is found that the distance between the two nuclei is 0.074 nanometres (1 nm = 1×10^{-9} m). Therefore the covalent radius of a hydrogen atom is $\frac{0.074}{2} = 0.037$ nm. Similarly, in a molecule of chlorine it is found experimentally that the distance between the two nuclei is 0.198 nm. Therefore the covalent radius of a chlorine atom is 0.099 nm. Some values of atomic radii (covalent radii) are given in FIG 7.2. Note that no values of covalent radii are quoted for the noble gases in FIG 7.2. This is because these compounds do not form covalent bonds with one another. (Even metals can form covalent molecules in the gas phase.)

* Van der Waals radii are measured for atoms which do not form chemical bonds (noble gases). Metallic radii values are obtained for metals by taking half the distance between two neighbouring metal ions in a crystal of the metal. Van der Waals radii and metallic radii are outside the scope of the syllabus. In this course the term atomic radius will always refer to covalent radius.

I	II											III	IV	V	VI	VII	0
H 0.037																	He –
Li 0.123	Be 0.089											B 0.080	C 0.077	N 0.074	O 0.074	F 0.072	Ne –
Na 0.157	Mg 0.136											Al 0.125	Si 0.117	P 0.110	S 0.104	Cl 0.099	Ar –
K 0.203	Ca 0.174	Sc 0.144	Ti 0.132	V 0.122	Cr 0.117	Mn 0.117	Fe 0.116	Co 0.116	Ni 0.115	Cu 0.117	Zn 0.125	Ga 0.125	Ge 0.122	As 0.121	Se 0.117	Br 0.114	Kr –

Fig 7.2
The atomic radii (covalent radii) in nanometres of some elements.

If we examine the values of atomic radii in FIG 7.2, a definite pattern emerges. To help you to see the trend, the values of atomic radii for the first 35 elements are plotted in FIG 7.3.

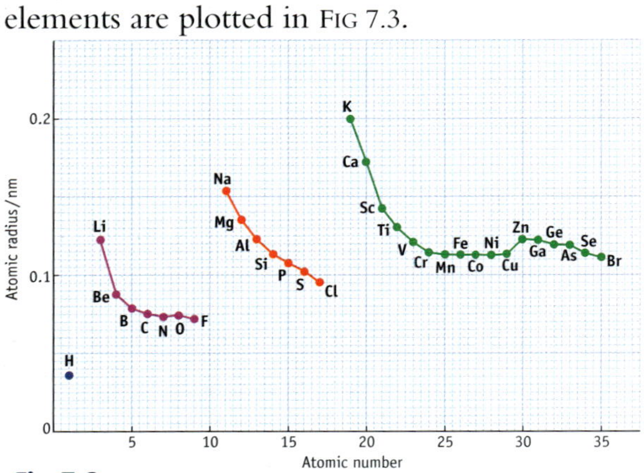

Fig 7.3
The values of atomic radii decrease along any one period (e.g. lithium to fluorine) and increase down any one group (e.g. lithium to potassium).

In general, the atomic radii values decrease across the period and increase down the group. This is shown in FIG 7.4.

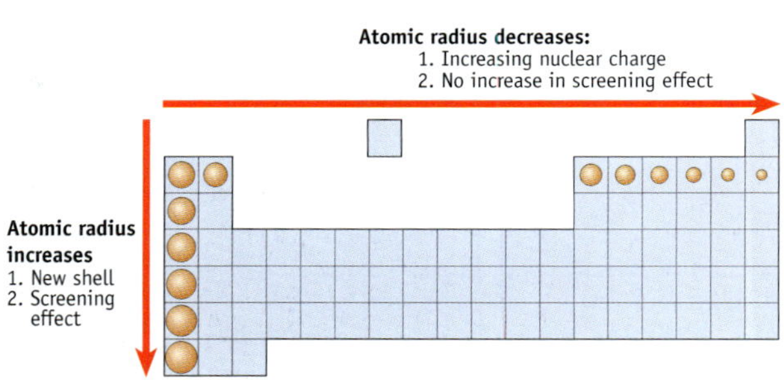

Fig 7.4
Atomic radii values decrease across the periods and increase down the groups.

To understand the trends in atomic radii values, it is important to remember that the size of an atom is governed by the electrostatic attraction between the positively charged protons and the negatively charged electrons. If this attraction is large, the positive protons will pull the outer electrons closer to the nucleus giving a smaller atomic radius. On the other hand, if the attraction is small, the electrons will be found further from the nucleus giving a larger atomic radius.

> The values of atomic radius increase down any one group in the Periodic Table for two reasons:

(a) The additional electrons are going into a **new shell** which is further from the nucleus, e.g. in lithium the electrons are being added to the n = 2 shell, in sodium they are being added to the n = 3 shell, etc. Since the outer electrons are becoming further away from the nucleus, the atomic radius increases.

(b) **Screening effect of inner electrons**. You might expect that, since the nuclear charge increases down a group, there would be an increased attractive force on the outer electrons causing the atomic radius to get smaller. However, even though the nuclear charge increases going from, say lithium to sodium to potassium, the increase in the nuclear charge is lessened because of what is called a 'screening effect'. This means that the inner shell or shells of electrons help to shield the outer electrons from the positive charge in the nucleus (FIG 7.5). Thus, some of the attractive force of the positive charges in the nucleus is lessened by the inner shells of electrons. Therefore, even though the nuclear charge increases, the increase is counteracted by the shielding effect of the inner shells of electrons. Thus, the atomic radius increases for the reason given in (a) above.

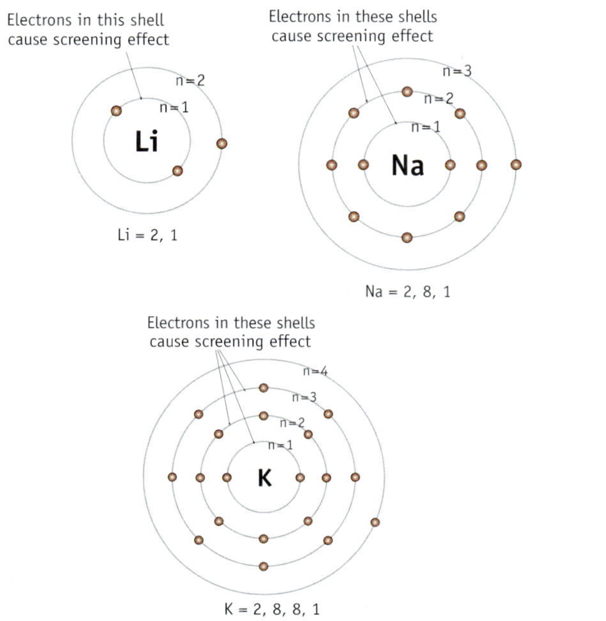

Fig 7.5
The increase in the nuclear charge is counteracted by the screening effect of the inner electrons.

> The values of atomic radius decrease from left to right across a period in the Periodic Table for two reasons:

(a) **Increasing nuclear charge**. The number of protons in the nucleus increases from left to right across any one period in the Periodic Table. This means that there is a greater attractive force on the outer electrons. The increased nuclear charge tends to draw the shells closer to the nucleus. As this attractive force increases from left to right across the Periodic Table, the nucleus pulls the outer electrons closer to it. Hence, the atoms become smaller, i.e. the atomic radius decreases.

(b) **No increase in the screening effect**. In moving from left to right across any one period in the Periodic Table, the extra electron which is being added is going into the **same** outer shell. Therefore, there is no increase in the screening effect. (For the n = 1 shell there is no screening effect at all. Beyond the n = 1 shell there is a screening effect but there is no increase in it along any one period.) Thus, there is no additional shell of electrons being added to counteract the increasing attractive force of the nucleus.

Hence, the size of the atom shrinks, i.e. the atomic radius decreases.

In some cases, the increase in nuclear charge is cancelled out by the shielding effect of inner electrons, e.g. in the case of aluminium and gallium which have exactly the same atomic radius.

7.3 TRENDS IN IONISATION ENERGY

Some elements lose electrons very easily, e.g. sodium and potassium lose their electrons so easily in a chemical reaction that they must be stored under oil to prevent them reacting. On the other hand, silver and gold have very little tendency to lose their electrons and hence are very unreactive. In order to measure the tendency of an atom to lose an electron, chemists define a term called the 'first ionisation energy'.

> The first ionisation energy of an atom is the energy required to completely remove the most loosely bound electron from a neutral gaseous atom.

We could represent the first ionisation energy of sodium by the equation:

$$Na \rightarrow Na^{+} + e^{-}$$

For example, the first ionisation energy of sodium is 494 units. (The actual units in which ionisation energies are measured are 'kilojoules per mole'. These units need not concern us at this stage.) A table of first ionisation energy values is given on page 45 of the Mathematics Tables and is reproduced in FIG 7.6.

I	II											III	IV	V	VI	VII	0
H 1310																	He 2370
Li 519	Be 900											B 799	C 1090	N 1400	O 1310	F 1680	Ne 2080
Na 494	Mg 736											Al 577	Si 787	P 1060	S 1000	Cl 1260	Ar 1520
K 418	Ca 590	Sc 632	Ti 661	V 649	Cr 653	Mn 715	Fe 761	Co 757	Ni 736	Cu 745	Zn 908	Ga 577	Ge 761	As 967	Se 941	Br 1140	Kr 1350
Rb 402	Sr 548	Y 636	Zr 669	Nb 653	Mo 695	Tc 699	Ru 724	Rh 745	Pd 803	Ag 732	Cd 866	In 556	Sn 707	Sb 833	Te 870	I 1010	Xe 1170
Cs 377	Ba 502	La 540	Hf 531	Ta 577	W 770	Re 761	Os 841	Ir 887	Pt 866	Au 891	Hg 1010	Tl 590	Pb 715	Bi 774	Po 812	At –	Rn 1040
Fr –	Ra 510	Ac 669															

Fig 7.6
The first ionisation energies of the elements.

The second ionisation energy of an element refers to the removal of a second electron from the positive ion formed when the first electron is removed. Thus, the second ionisation energy of sodium may be represented by the equation:

$$Na^+ \rightarrow Na^{2+} + e^-$$

Ionisation energy values are obtained from the emission spectra of atoms. If you examine the values of the ionisation energies in FIG 7.6, a definite pattern emerges. In general, ionisation energy values decrease down the groups and increase across the periods (FIG 7.7).

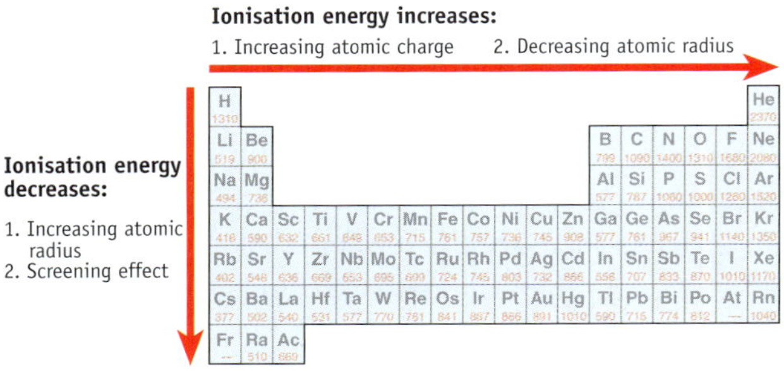

Fig 7.7
Ionisation energy values decrease down the groups and increase across the periods.

The values of ionisation energy decrease down the groups in the Periodic Table for two reasons:

(a) **Increasing atomic radius**. As already explained, the atomic radius increases as you move down any one group. This means that the outermost electrons are becoming further away from the attractive force of the nucleus. Therefore, it becomes easier to remove an electron from the outer shell, i.e. the ionisation energy values decrease.

(b) **Screening effect of inner electrons**. Even though the nuclear charge increases down a group, this increase in nuclear charge is essentially cancelled out by the screening effect of the intervening shells of electrons. This means that the outermost electrons are somewhat shielded from the attractive force of the positively charged nucleus, i.e. the ionisation energy values decrease.

The values of ionisation energy increase across the periods in the Periodic Table for two reasons:

(a) **Increasing nuclear charge**. Moving from left to right across the Periodic Table, the number of protons in the nucleus is increasing. This means that the attraction between the nucleus and the outer electrons is steadily increasing. Therefore, the electrons are being held more firmly and thus, it requires more energy to remove one of the electrons from the outermost shell, i.e. the ionisation energy values increase.

(b) **Decreasing atomic radius**. As already explained, within any period in the Periodic Table the atomic radius decreases from left to right. This means that an electron in the outermost level is becoming closer to the nucleus. Thus, it is more difficult to remove it due to the increased attraction between it and the nucleus, i.e. the ionisation energy values increase. Studying FIG 7.8 will help you to understand this.

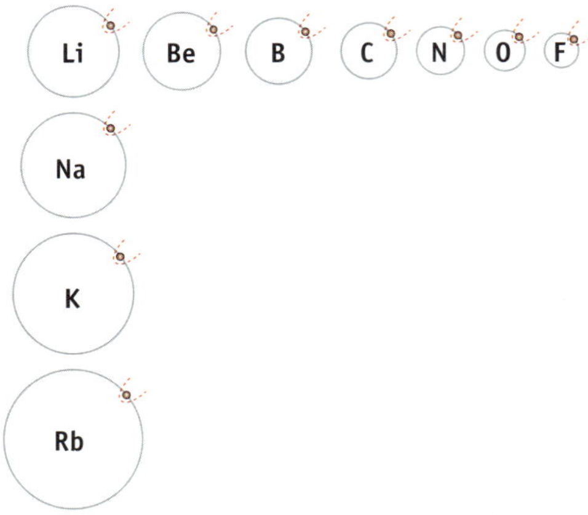

Fig 7.8
The increasing atomic radius down the groups makes it easier to remove an outer electron. The decreasing atomic radius across the periods makes it more difficult to remove an outer electron. The dotted line indicates that an electron is being removed from the outer shell.

It is clear from FIG 7.6 that the alkali metals are the elements with the lowest ionisation energies. Thus, these metals tend to lose electrons readily and this makes them very reactive. On the contrary, the elements with the highest ionisation energy values are the noble gases. Since these gases have very little tendency to lose electrons, they are very unreactive.

EXCEPTIONS TO THE GENERAL TREND ACROSS A PERIOD

If one plots a graph of ionisation energy versus atomic number for the first 20 elements (FIG 7.9), it is clear that, in any one period, some elements do not follow the smooth increase. For example, in the n = 2 period, beryllium and nitrogen have higher values than expected. The value for beryllium might be expected to lie on the dotted line between that of lithium and boron. Similarly, the value for nitrogen might be expected to lie between that of carbon and oxygen. It is clear from FIG 7.9 and from the values given in FIG 7.6 that the values for beryllium and nitrogen are much higher than expected.

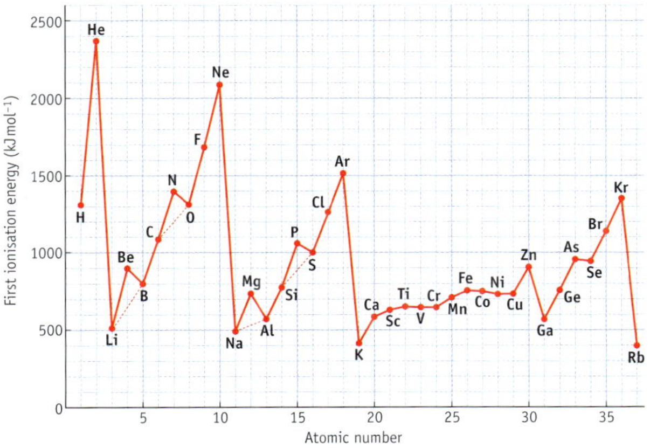

Fig 7.9
Plot of first ionisation energy vs. atomic number. Note that, in the second period, beryllium and nitrogen have higher values than expected. A similar situation occurs with magnesium and phosphorus in the third period.

These irregularities in ionisation energy trends are explained by the fact that any sublevel that is completely filled (as is the case with beryllium and magnesium) or exactly half-filled (as is the case with nitrogen and phosphorus) has extra stability. Because of the extra stability associated with these elements, their ionisation energy values are higher than expected since it requires extra energy to break up these stable arrangements by removing an electron.

Be = $1s^2, 2s^2$ N = $1s^2, 2s^2, 2p^3$

Mg = $1s^2, 2s^2, 2p^6, 3s^2$ P = $1s^2, 2s^2, 2p^6, 3s^2, 3p^3$

FURTHER EVIDENCE FOR EXISTENCE OF ENERGY LEVELS

In CHAPTER 2 we discussed the spectroscopic evidence for the existence of energy levels. However, evidence for the fact that electrons are arranged in shells of different energies is also provided by studying the values of a number of ionisation energies of any one particular element. For example, suppose we measure the first, second, third, etc. up to the nineteenth ionisation energy of potassium (FIG 7.10). Note that the ionisation energy values become so large that it is necessary to take the log of the values to condense the graph into a smaller space. To help explain the meaning of this graph, we have labelled the parts of the graph with the letters A-E. It will also help us to understand the significance of the graph if we keep the electron configuration of potassium in mind.

K = $1s^2, 2s^2, 2p^6, 3s^2, 3p^6, 4s^1$

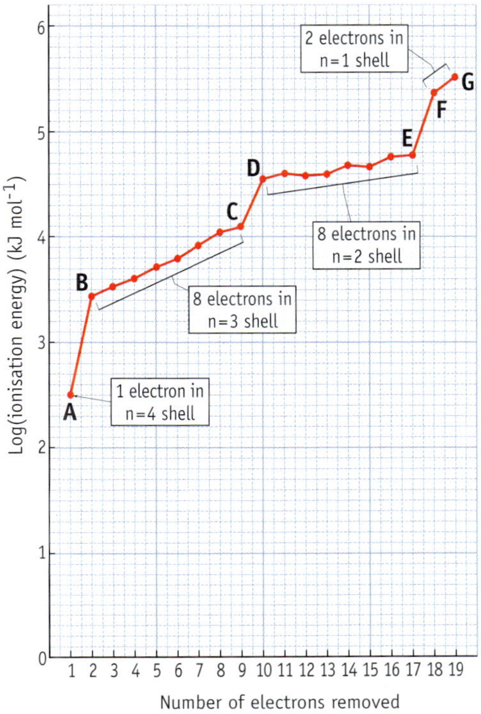

Fig 7.10
Plot of log of ionisation energy vs. number of electrons removed from an atom of potassium. This graph provides very good evidence for the existence of energy levels.

At A, one electron has been removed from potassium. The energy required for this is the first ionisation energy. This has the lowest ionisation energy value since the electron in the 4s sublevel is easiest to remove. The second electron is much more difficult to remove since this electron is being removed from the K^+ ion. This ion has eight electrons in the outer shell. The full outer sublevel ($3p^6$) has extra stability and therefore will require more energy to remove electrons from it. Also, the dramatic increase for the second ionisation energy suggests that these electrons must be closer to the nucleus and must experience less shielding. Therefore, we have very good evidence for the existence of an inner shell. This would explain why the second ionisation energy value (3070 units), point B on graph, is considerably higher than the first ionisation energy (418 units).

First ionisation energy $K \rightarrow K^+ + e^-$

Second ionisation energy $K^+ \rightarrow K^{2+} + e^-$

Third ionisation energy $K^{2+} \rightarrow K^{3+} + e^-$

From B to C we are removing eight more electrons. The values steadily increase as we remove these eight electrons from the n = 3 energy level. At this stage (point C on the graph), there is another sudden jump in the value of the ionisation energy. This implies that the electron at D is being removed from a shell which is closer to the nucleus, i.e. the n = 2 energy level. (This increase is actually much larger than it appears from the graph – due to the logarithmic scale, the jump is actually greater by a factor of ten.)

From D to E we are removing the eight electrons in this shell. At E there is again a sudden jump as we remove electrons from a shell which is closer to the nucleus, i.e. the n = 1 shell. At G we have removed all 19 electrons.

Thus, two general points emerge:

(a) There is a steady increase in ionisation energy values as electrons are removed from the same shell. All of the electrons in the same shell lie at approximately the same distance from the nucleus. Therefore, the ionisation energies will be approximately the same for all of these electrons. (For those of you who may lie awake at night wondering about this slight increase, the reason for it may be explained as follows: when an electron is removed from a neutral atom, the electrons remaining in the positive ion are more strongly attracted to the nucleus. This is because they are closer to the nucleus since it can be shown that a positive ion is smaller than the neutral atom from which it has been formed.)

(b) There is a large increase in ionisation energy whenever an electron is removed from a new shell. This is because the new shell is more strongly attracted to the nucleus because (a) it is closer to the nucleus and (b) the electron in this shell experiences less shielding. Also, because it is a filled shell, it has extra stability and requires more energy to remove electrons from it.

Example 7.1

The first four ionisation energies of the element of atomic number 5 are shown in TABLE 7.1.

electron removed	ionisation energy (kJ mol^{-1})
1st	799
2nd	2,421
3rd	3,650
4th	24,962

Table 7.1

Explain (a) why the second ionisation energy is greater than the first, (b) why the increase in ionisation energy is relatively small for the third electron but very large for the fourth electron. [LCH]

ANSWER (a):

The second ionisation energy is always greater than the first because the second electron is being removed from a positively charged ion. The electrons are held more firmly in a positively charged ion than in the corresponding neutral atom because a positive ion is smaller than the neutral atom from which it is formed. Since the electrons are closer to the nucleus, more energy is required to remove them, i.e. the second ionisation energy is greater than the first ionisation energy.

ANSWER (b):

The third electron, like the second electron, is being removed from the n = 2 shell. Hence, the increase in ionisation energy is fairly small. However, the fourth electron is being removed from the n = 1 shell. Since the n = 1 shell is nearer the nucleus than the n = 2 shell, it requires a lot more energy to remove the electron from the n = 1 shell. Also, the fourth electron is being removed from the stable $1s^2$ configuration.

7.4 TRENDS IN ELECTRONEGATIVITY

In SECTION 5.9 we discussed electronegativity and learned that electronegativity is the relative attraction that an atom in a molecule has for the shared pair of electrons in a covalent bond. From studying the table of electronegativity values (FIG. 5.33), we saw that elements like fluorine and chlorine on the right-hand side of the Periodic Table have high electronegativity values. On the contrary, elements like potassium and sodium on the left-hand side of the Periodic Table have low electronegativity values.

Let us now examine the trends in electronegativity values in the Periodic Table and explain the reasons for these trends. In general, electronegativity values decrease down the groups and increase across the periods (FIG 7.11). (Note that the trends in electronegativity follow the same pattern as those of ionisation energies.) Electronegativity values are not assigned to the noble gases, as these elements have no tendency to form bonds. Hence, the definition of electronegativity does not apply to them.

The values of electronegativity decrease down the groups in the Periodic Table for two reasons:

(a) **Increasing atomic radius**. As already discussed, the atomic radius increases as you move down any one group. This means that the outermost electrons are becoming further away from the attractive force of the nucleus. Therefore, there is a smaller attraction between the nucleus and the shared pair of electrons, i.e. the electronegativity decreases.

(b) **Screening effect of inner electrons**. Even though the nuclear charge increases down a group, this increase in nuclear charge is essentially cancelled out by the screening effect of the intervening shells of electrons. This means that the outermost electrons are somewhat shielded from the attractive force of the positively charged nucleus. Since it is these outermost electrons which are involved in bonding, the attraction of the nucleus for these electrons decreases going down the group, i.e. the electronegativity decreases.

The values of electronegativity increase across the periods in the Periodic Table for two reasons:

(a) **Increasing nuclear charge**. Moving from left to right across the Periodic Table, the number of protons in the nucleus is increasing. This means that the attraction between the nucleus and the outer electrons is steadily increasing. Therefore, the electrons involved in bonding are being more strongly attracted to the nucleus, i.e. the electronegativity increases.

(b) **Decreasing atomic radius**. As already explained, within any period in the Periodic Table, the atomic radius decreases from left to right. This means that the electrons in the outermost level are becoming closer to the nucleus. Thus, there is a greater attraction between the nucleus and these outer electrons, i.e. the electronegativity increases.

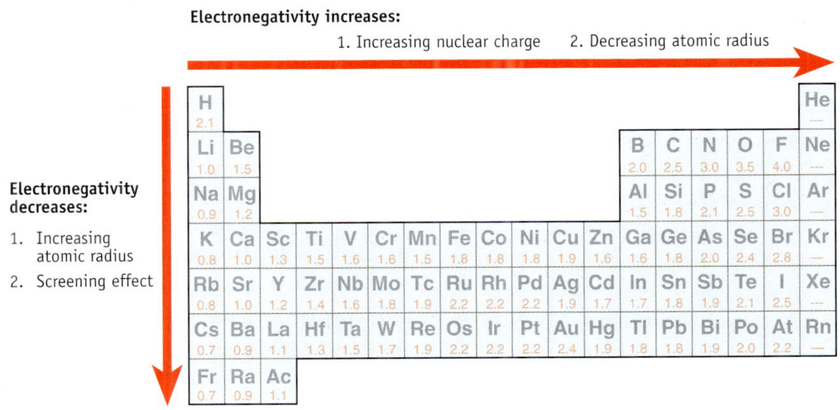

Fig 7.11
Electronegativity values decrease down the groups and increase across the periods.

The trends in electronegativity are clearly seen in a graph of electronegativity versus atomic number (FIG 7.12)

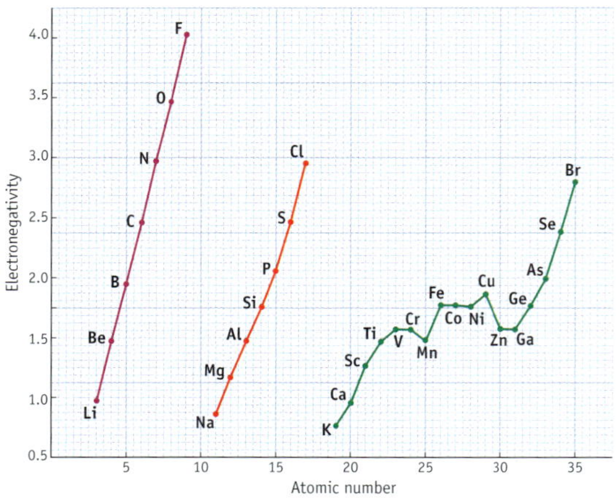

Fig 7.12
Plot of electronegativity values vs. atomic number for the first 35 elements. Note the increase across any one period (e.g. lithium to fluorine) and the decrease down a group (e.g. fluorine to chlorine to bromine).

So far we have seen that there is a definite repeating pattern or periodicity associated with atomic radii, ionisation energies and electronegativities of the elements. We now consider how the chemical properties of certain groups of elements are affected by these trends.

7.5 TRENDS WITHIN GROUPS

In our study of chemical bonding in CHAPTER 5, we have seen how bonds are formed when chemical reactions occur. In the case of ionic bond formation, the bonds are formed by the transfer of electrons from the outermost shell. In the case of covalent bond formation, the bonds are formed by the sharing of electrons in the outermost shell. In fact, **the chemical properties of elements are largely determined by the number of electrons in the outermost shell**. For example, all of the elements in group I (the alkali metals) have one electron in their outermost shell. As a result of this, all of these elements have similar chemical properties. Similarly, all of the elements in group VII (the halogens) have seven electrons in their outermost shell. Therefore, all of these elements also have similar chemical properties. To illustrate this similarity of elements in the one group, we now consider some of the properties of some groups of elements.

1. TRENDS IN CHEMICAL REACTIVITY OF ALKALI METALS

All of the alkali metals are very reactive elements. They are so reactive that none of them occur free in nature as the metals themselves. The alkali metals that you have in your school laboratory have been extracted from their compounds. (The method of extraction is covered in CHAPTER 28.) Since the alkali metals have low ionisation energies and low electronegativity values, they have a tendency to lose electrons to form ionic compounds. This tendency to lose electrons dominates all of their chemical reactions. In general, the reactivity of the alkali metals increases down the group as the ionisation energy decreases, i.e. the more easily the outer electron is lost, the more reactive the metal (FIG 7.13).

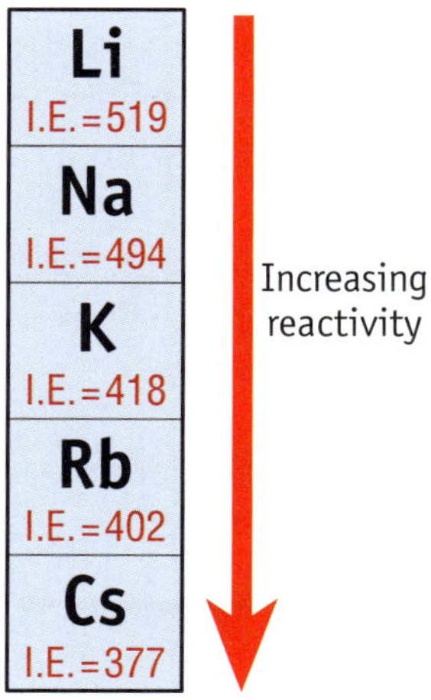

Fig 7.13
The alkali metals become more reactive as the ionisation energy decreases. The outer electron is more easily lost from the metal causing the metal to become more reactive.

(a) **Reaction with oxygen**: All of the alkali metals react with oxygen to form oxides, e.g.

potassium + oxygen → potassium oxide

$$2K + \tfrac{1}{2}O_2 \rightarrow K_2O$$

For this reason, the alkali metals must be stored under oil to prevent them reacting with oxygen in the air (FIG 7.14). A piece of lithium metal exposed to the air turns into a white powder within a few hours. It takes only a few minutes for a similar occurrence to take place with a piece of caesium! This is because the ionisation energy of caesium is much lower than that of lithium. Hence, the outermost electron is lost more easily to oxygen.

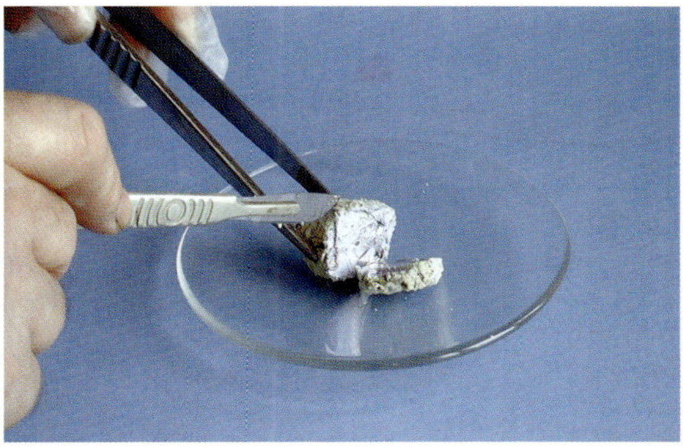

Fig 7.14
A piece of freshly cut potassium which had been stored in oil.

(b) **Reaction with water**: The reaction of the alkali metals with water illustrates very nicely the trend of increasing reactivity down the group. All of the alkali metals react with water to form the hydroxide of the metal and hydrogen gas is given off, e.g.

sodium + water → sodium hydroxide + hydrogen

$$Na + H_2O \rightarrow NaOH + \tfrac{1}{2} H_2$$

Lithium reacts quite vigorously with water, sodium more vigorously and potassium even more vigorously. In fact, the reaction is so vigorous in the case of potassium that the heat generated by the reaction causes the hydrogen to catch fire (Fig 7.15).

Fig 7.15
The reaction between potassium and water is so vigorous that the heat generated causes the hydrogen to catch fire.

Your teacher may demonstrate these reactions to you. Rubidium and caesium react so violently with water that their reactions are too dangerous to demonstrate. (Many chemistry CD ROMs of the Periodic Table have a video clip showing the reaction of rubidium or caesium with water.) Once again, the decreasing ionisation energy down the group means that the outer electron is lost more easily to water. Hence, moving down the group, the reaction with water becomes more vigorous.

Note: The reaction of the alkali metals with dilute acids is so dangerous that it must never be attempted. So much hydrogen is released that an explosion occurs.

sodium + hydrochloric acid → sodium chloride + hydrogen

$$Na + HCl \rightarrow NaCl + \tfrac{1}{2} H_2$$

2. TRENDS IN CHEMICAL REACTIVITY OF HALOGENS

When discussing electronegativity above, we have seen that the elements in group VII (the halogens) are the most electronegative elements in the Periodic Table. Since the electronegativity values decrease down the group, this means that fluorine is the most electronegative element of the halogens. In fact, fluorine is the most electronegative element in the Periodic Table. Since the halogens have such a great attraction for electrons, this makes them quite reactive. The halogens do not exist free in nature, e.g. chlorine gas is made from sodium chloride. In keeping with its high electronegativity, fluorine is the most reactive of all the halogens. (Fluorine is far too reactive to be kept in school laboratories.) The trend in reactivity of the halogens (Fig 7.16) is in the opposite direction to that of the alkali metals.

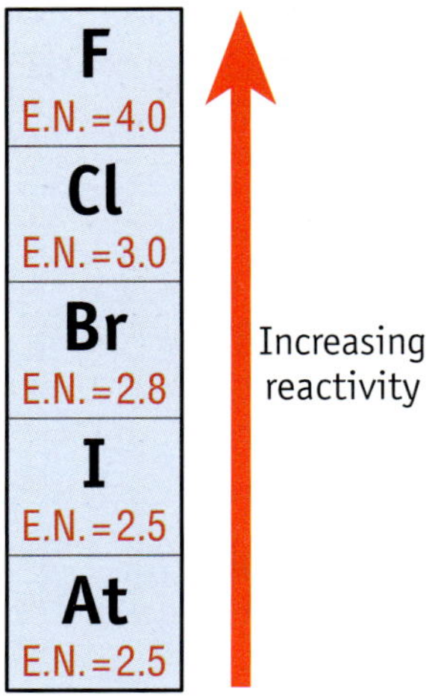

Fig 7.16
The reactivity of the halogens increases up the group. This is because as one moves up the group, the electronegativity values increase.

Because of their high electronegativities, halogens tend to remove electrons easily from other substances. A substance that removes electrons from other substances is called an 'oxidising agent'. (This term will be explained more fully in CHAPTER 14.) For example, if chlorine gas is bubbled through a solution containing bromide ions, the chlorine takes electrons from the bromide ions and converts them into bromine:

chlorine + bromide ion → chloride ion + bromine

$$\tfrac{1}{2}Cl_2 \quad + \quad Br^- \quad \rightarrow \quad Cl^- \quad + \tfrac{1}{2}Br_2$$

This shows that the electronegativity of the chlorine atom (E.N. = 3.0) is greater than that of the bromine atom (E.N. = 2.8) as the chlorine atom takes an electron from the bromide ion. This reaction is shown happening in FIG 7.17. The colourless solution of bromide ions turns red due to the formation of the red bromine.

Fig 7.17
Chlorine gas is bubbled into a solution containing Br^- ions. Since chlorine is more electronegative (more reactive) than bromine, it displaces Br_2 out of solution.

Similarly, if some bromine solution ('bromine water') is added to a solution containing iodide ions, iodine is displaced out of solution. The more electronegative bromine atoms (E.N. = 2.8) take electrons from the less electronegative iodine atoms (E.N. = 2.5) as shown in the following equation:

bromine + iodide ion → bromide ion + iodine

$$\tfrac{1}{2}Br_2 \quad + \quad I^- \quad \rightarrow \quad Br^- \quad + \tfrac{1}{2}I_2$$

Note that in each case the more reactive halogen displaces the less reactive halogen from solution.

You will study the reactivity of the halogens in an experiment you will be performing at a later stage (EXPERIMENT 14.1).

Note: Even though we have only considered the trends in chemical reactivity for groups I and VII, all groups of elements in the Periodic Table show trends in chemical reactivity. For example, in the group II elements (the alkaline earth metals), beryllium does not react with water, magnesium reacts very slowly with water and calcium undergoes a steady reaction with water. The reason for this is because ionisation energy decreases down the group: magnesium has a lower value than beryllium, calcium a lower value than magnesium, etc. Hence, calcium loses its outer electrons more readily than magnesium. Thus, a piece of calcium added to water reacts with the water to release hydrogen at a steady rate.

calcium + water → calcium hydroxide + hydrogen

$$Ca \quad + 2H_2O \rightarrow \quad Ca(OH)_2 \quad + H_2$$

Magnesium shows very little reactivity when added to water. It takes a few days for even a small amount of hydrogen to form. Strontium and barium react much more vigorously with water.

3. TRENDS IN PHYSICAL PROPERTIES

In our study of chemical bonding in CHAPTER 5, we learned that the noble gases are characterised by the fact that they form practically no compounds. Because they show such little reactivity, the noble gases are often called the **inert gases**. Even though it is not possible to consider any trends in chemical properties of the noble gases, it is possible to study the trends in their physical properties, e.g. boiling points. For example, if we consider the boiling points of the noble gases (TABLE 7.2), we see that there is a steady increase in boiling points down the group from helium to radon. This is related to the increasing atomic radius down the group. As the atoms increase in size, temporary dipoles are more easily created due to the larger electron clouds, i.e. van der Waals forces are stronger between bigger atoms. Therefore, it takes more energy to break these van der Waals forces giving rise to higher boiling points.

A similar trend is observed in the halogens. Fluorine and chlorine are gases at room temperature, bromine is a liquid, and iodine is a solid. The van der Waals forces between the individual molecules increase down the group due to the increasing size of the atoms in the molecules.

BOILING POINTS	
Noble Gases	**Halogens**
He = -269 °C	F_2 = -188 °C
Ne = -246 °C	Cl_2 = -34.7 °C
Ar = -186 °C	Br_2 = 58.8 °C
Kr = -152 °C	I_2 = 184 °C
Xe = -108 °C	
Rn = -61.8 °C	

Table 7.2

The boiling points of the noble gases and the halogens increase down the group.

CHAPTER SUMMARY

- The atomic radius of an atom is defined as half the distance between the nuclei of two atoms of the same element, which are joined together by a single covalent bond.
- Atomic radius values increase down any one group in the Periodic Table because the electrons are going into a new shell that is further from the nucleus and there is a screening effect of inner electrons.
- Atomic radius values decrease from left to right across a period because of increasing nuclear charge and no increase in the screening effect.
- The first ionisation energy of an atom is the energy required to completely remove the most loosely bound electron from a neutral gaseous atom.
- Ionisation energy values decrease down the groups because of increasing atomic radius and the screening effect of inner electrons.
- Ionisation energy values increase across the periods because of increasing nuclear charge and decreasing atomic radius.
- Further evidence for the existence of energy levels may be obtained from a study of the values of the successive ionisation energies of an element.
- Electronegativity values decrease down the groups because of increasing atomic radius and the screening effect of inner electrons.
- Electronegativity values increase across the periods because of increasing nuclear charge and decreasing atomic radius.
- The chemical properties of an element are determined by the number of electrons in the outermost shell.
- Within each group of the Periodic Table there are definite trends in the chemical and physical properties of the elements.

EXERCISES

7.1. (a) Define the term atomic radius ....

(b) Explain the meaning of the phrase 'screening effect of inner electrons'. ....

(c) Nuclear charge .... from left to right across the periods in the Periodic Table.

(d) Write down an equation to represent the first ionisation energy of potassium..........

(e) Explain why the first ionisation energy of element 33 is greater (i) than that of element 32, (ii) than that of element 34 .... *[LCH]*

(f) Beryllium and nitrogen are exceptions to the general increase in ionisation energies shown by the other elements in the period lithium to neon. Explain the reason for this .... *[LCH]*

(g) How do the values of successive ionisation energies of an element provide evidence for the existence of energy levels? ....

(h) What is meant by electronegativity? ....

(i) How are the electronegativities of elements related in general to their position in the Periodic Table? Give reasons in support of your answer .... *[LCH]*

(j) The reactivity of the alkali metals .... moving down the group.

(k) The reactivity of the halogens .... moving down the group.

(l) The boiling points of the noble gases .... down the group.

(m) Which one of the halogens is a solid at room temperature? ....

7.2. (a) The variation of first ionisation energy with atomic number for six successive elements is shown in FIG 7.18. Give reasons for the general increase indicated and explain why the increase is interrupted at one point.

(b) Discuss, with suitable explanations, how the following properties vary across a typical period (e.g. Na to Ar) of the Periodic Table: (i) atomic radius, (ii) first ionisation energy. *[LCH]*

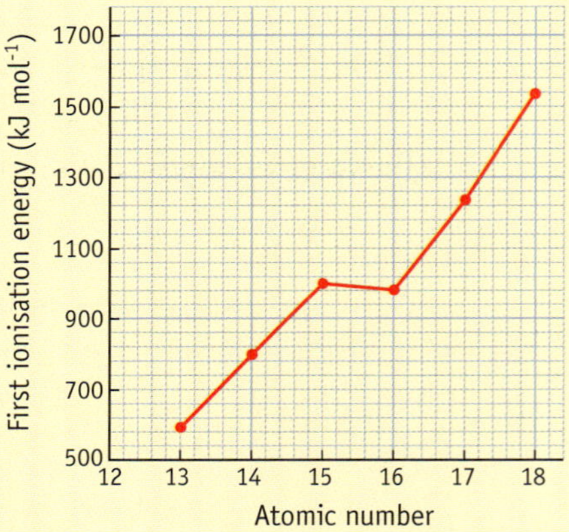

Fig 7.18

7.3. (a) What is meant by the first ionisation energy of an element? Explain (i) the decrease in first ionisation energies down a group of the Periodic Table, e.g. Be to Ba, (ii) the increase in electronegativity values across a row of the Periodic Table, e.g. Li to F. (Refer to Mathematics Tables p. 45 and p. 46) *[LCH]*

(b) The first and second ionisation energies of sodium and magnesium are shown in TABLE 7.3. Explain why sodium has a lower first ionisation energy but a much higher second ionisation energy than magnesium.

Ionisation energy (kJ mol^{-1})	Na	Mg
First	494	736
Second	4,550	1,450

Table 7.3

[LCH]

Radioactivity

8

8.1 INTRODUCTION

Fig 8.1
Henri Becquerel – the discoverer of radioactivity.

In 1896, a French physicist called Henri Becquerel (Fig 8.1) was studying the effect of sunlight on crystals of uranium salts. By chance, during a series of cloudy days, he left some of the salt resting on a photographic plate that was wrapped in black paper in the darkness of a drawer in his desk. To his amazement, he found that the photographic plate had become 'fogged' or darkened in the area near the uranium salt. The uranium salt was giving off some type of radiation. This radiation was penetrating the black paper and affecting the photographic plate. Becquerel investigated the radiation being given off and found it was being emitted continuously and did not depend on factors like light, heat, etc. In other words, the emission of radiation was **spontaneous**, i.e. it was a property of the uranium and occurred naturally. This process was called **radioactivity** ('giving off of rays') and the uranium salt was described as **radioactive**.

In 1898, two of Becquerel's colleagues, Pierre and Marie Curie, investigated the radioactivity of uranium salts in more detail (Fig 8.2). They obtained about eight tonnes of an ore of uranium called pitchblende. This ore was far more radioactive than could be accounted for simply by its uranium content. Over a four-year period, in a laboratory that consisted of an old wooden shed with a leaky roof, they carefully purified and re-purified the ore into smaller and smaller samples of highly radioactive material. Eventually, after thousands of re-crystallisations, they isolated two new elements. They named the first element **polonium** (after Madame Curie's native Poland) and the other **radium** (meaning 'giver of rays').

Fig 8.2
Pierre and Marie Curie isolated the radioactive elements radium and polonium.

Polonium is element number 84 and radium is element number 88 in the Periodic Table. The Curies shared the Nobel Prize for physics in 1903 with Becquerel for their study of radioactivity. In 1911, Marie Curie was awarded the Nobel Prize for chemistry for the work she and Pierre had carried out that led to the discovery of polonium and radium. (Sadly, Pierre was killed in a traffic accident in 1906 so Marie accepted the prize alone.)

8.2 THE NATURE OF RADIOACTIVE RADIATION

In Section 8.1, we have used the term 'radiation' without discussing what it actually means. After Becquerel's discovery of radioactivity, many scientists set about investigating the nature of the radiation. Within about ten years, it was discovered that the radiation coming from radioactive substances was of three types, alpha particles, beta particles and gamma radiation. These three names are taken from the first three letters in the Greek alphabet: alpha (α), beta (β) and gamma (γ). These three types of radiation were discovered by studying the behaviour of the radiation in electric and magnetic fields (Fig 8.3). (This diagram is really a composite diagram as it is not possible to study all three types of radiation in a single experiment.)

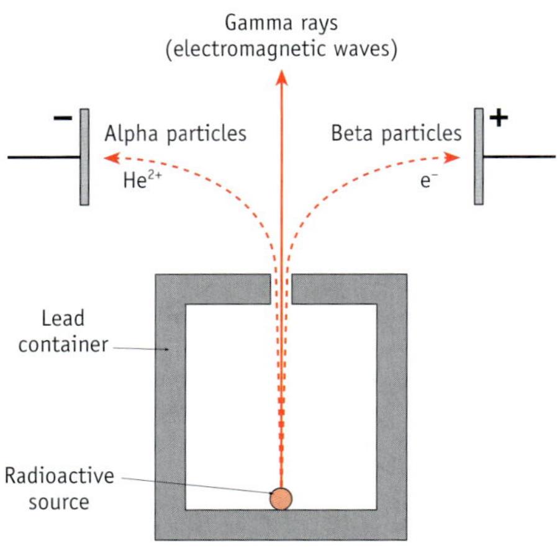

Fig 8.3
The electric field separates out the radiation into three types.

Alpha particles

Alpha particles are groups of two protons and two neutrons stuck together, i.e. the same as the nucleus of the helium atom. They may be represented as $^{4}_{2}He$. Alpha particles are emitted from the unstable nucleus of a radioactive element. These 'pieces' of the nucleus are thrown out of the unstable nucleus as it changes to a more stable nucleus (Fig 8.4).

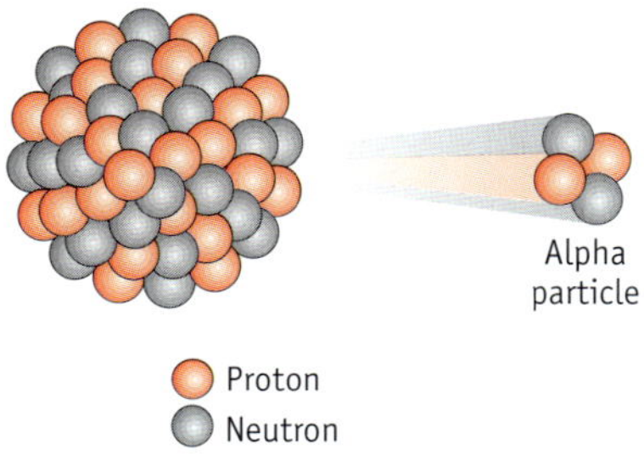

Fig 8.4
An alpha particle being emitted from an unstable nucleus.

Alpha particles have a low penetrating power and are stopped by a few centimetres of air or a sheet of paper (Fig 8.5).

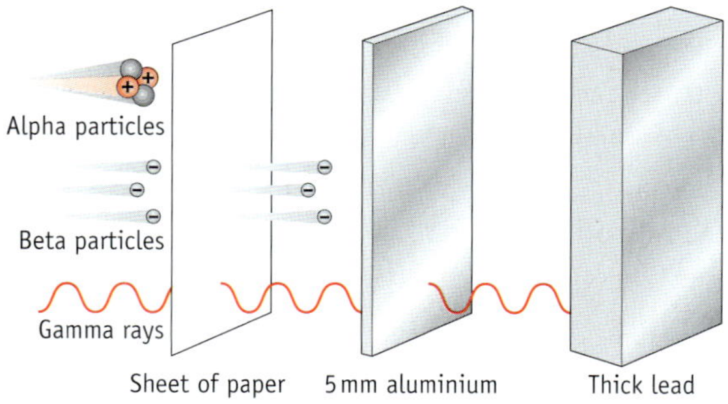

Fig 8.5
Alpha particles are stopped by a sheet of paper, beta particles are stopped by a few millimetres of aluminium, but gamma rays are only stopped by a thick piece of lead.

An example of a radioactive element that emits alpha particles is americium-241. Americium-241 is commonly used in smoke detectors (Fig 8.6). A number of the elements in the Lanthanide and Actinide series are radioactive (see Periodic Table inside back cover).

Fig 8.6
This smoke detector contains americium-241 which is a source of alpha particles.

Having identified alpha particles, scientists now had a better understanding of radioactivity. Since all radioactive substances have unstable nuclei, radioactivity is defined as follows:

> Radioactivity is the spontaneous breaking up of unstable nuclei with the emission of one or more types of radiation.

Beta particles

It was found that beta particles are simply electrons. A beta particle is formed when a neutron in an unstable nucleus is changed into a proton and an electron. This electron is then ejected from the nucleus (Fig 8.7).

$$\text{neutron} \rightarrow \text{proton} + \text{electron}$$

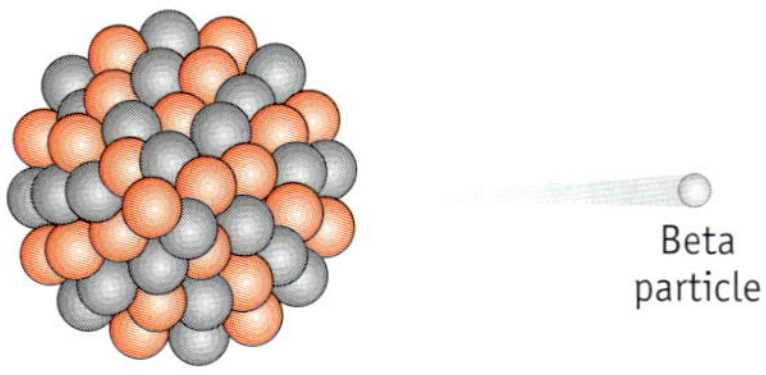

Fig 8.7
A neutron in the nucleus changes into a proton and an electron. The electron (beta particle) is then emitted from the nucleus.

Beta particles, since they are lighter than alpha particles, are more penetrating than alpha particles. As shown in Fig 8.5, beta particles can penetrate up to about 5 mm of aluminium.

An example of an element that emits beta particles is carbon-14. This substance is used to find out the age of objects in a process known as 'carbon dating'. The discovery of radioactivity in 1898 was a clear indication that the atom was not a solid sphere as Dalton had proposed but consisted of smaller particles. The only place that the emitted alpha and beta particles could be coming from was within the atoms of the radioactive material.

Gamma radiation

Gamma radiation is simply high-energy electromagnetic radiation, i.e. similar to X-rays. Since it does not consist of charged particles, it is not deflected in electric or magnetic fields. An unstable nucleus emits gamma radiation in order to lose surplus energy (Fig 8.8).

Gamma rays have a very high penetrating ability and can penetrate bricks, metal sheets, etc. In fact, they are only stopped by a thick slab of lead as shown in Fig 8.5. Because of their high penetrating ability, gamma rays are the most dangerous of the three types of radiation. If absorbed in large quantities into the body, they can change the structure of the chemicals in our bodies and cause cancer. However, gamma rays can also be used to kill cancerous cells by focusing the radiation on to the cancerous tissue. An example of an element that gives off gamma rays is cobalt-60. This is often used in hospitals in the treatment of cancer patients.

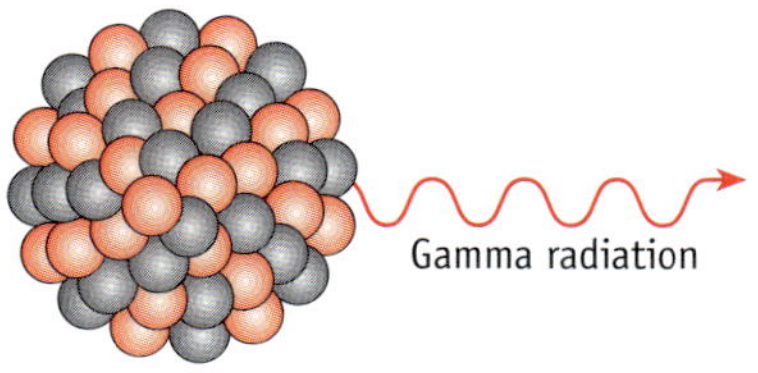

Fig 8.8
An unstable nucleus losing surplus energy by emitting gamma radiation.

Demonstration Experiment 8.1

To detect the presence of alpha particles, beta particles and gamma radiation, and to investigate the penetrating power of each of these three types of radiation.

Introduction

In this experiment, we shall use a radioactivity demonstration kit ('PANAX' demonstration kit). This kit contains samples of radioactive isotopes that emit each of the three types of radiation. It also contains an instrument called a **Geiger-Müller tube** (commonly referred to as a 'G-M tube') that is used to detect radioactivity.

The Geiger-Müller tube is normally connected to an instrument called a ratemeter. The ratemeter measures the number of radioactive particles entering the Geiger-Müller tube every second. Many ratemeters have a loudspeaker attached to them. Thus, the presence of radioactivity is indicated by a 'clicking' sound coming from the loudspeaker.

Apparatus required

Geiger-Müller tube; sources of alpha, beta and gamma radiation; tongs; ratemeter; paper, aluminium and lead absorbers; support for sources.

Procedure

1. Turn on the ratemeter and note that the Geiger-Müller tube registers the presence of radioactivity without any radioactive source being brought near it. This reading is known as the **background count**. It is caused by natural radioactivity in our surroundings due to rocks in the Earth and rays called cosmic rays from the Sun.
2. Using the tongs and holder supplied, bring the alpha particle source (americium-241) about 3 cm from the window of the Geiger-Müller tube (Fig 8.10). The ratemeter registers the presence of the alpha particles. A piece of paper is placed in front of the Geiger-Müller tube. Note that the count rate drops back to background count, i.e. the alpha particles have been stopped by the paper. If the alpha particle source is moved away from the Geiger-Müller tube, the reading drops to the background count when the source is only about 4 – 6 cm away. Thus, alpha particles have a range of only about 4 – 6 cm in air.

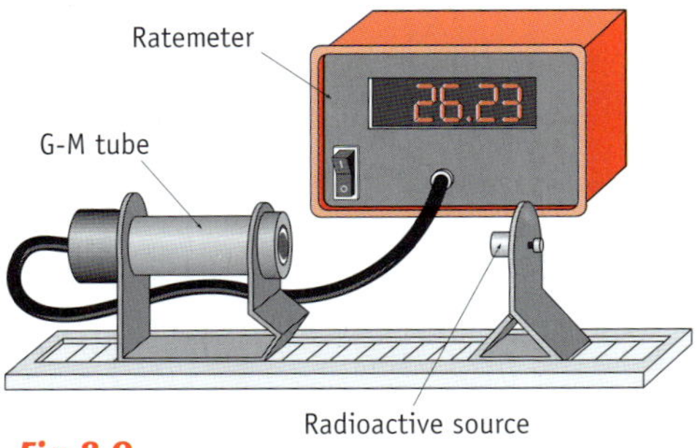

Fig 8.9
Apparatus to investigate the properties of radioactive radiation.

3. Repeat the experiment using the beta particle source (strontium-90). Note that the Geiger-Müller tube detects the presence of the beta particles also. Using paper and aluminium metal of various thicknesses, note that the beta particles are more penetrating than the alpha particles. A piece of aluminium metal of about 5 mm thickness is needed to absorb them fully. If the beta source is moved away from the Geiger-Müller tube, the reading drops to background count when the source is about 500 cm away. Thus, beta particles have a much higher range than alpha particles.
4. Repeat the experiment using the gamma source (cobalt-60). Note that the Geiger-Müller tube detects the presence of the gamma radiation. Using various sheets of lead of different thicknesses, note that the gamma radiation passes through a considerable thickness of lead before being absorbed.

Note: If this equipment is not available in your school, it may be possible for you to see the experiment on a videotape.

Conclusions

All three types of radioactive radiation may be detected using a Geiger-Müller tube. Alpha particles are the least penetrating and are stopped by a sheet of paper. Beta particles are more penetrating and can penetrate about 5 mm of aluminium. Gamma rays are the most penetrating and are only stopped by a thick block of lead. Some of the properties of alpha, beta and gamma radiation are summarised in Table 8.1.

	ALPHA PARTICLES	BETA PARTICLES	GAMMA RAYS
Nature	Helium nuclei 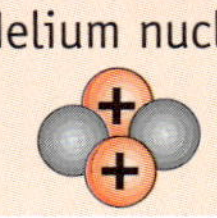	Electrons ⊖	Waves of very high energy 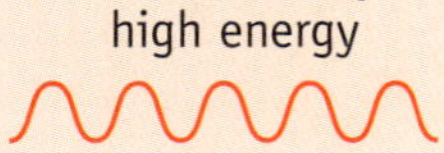
Affected by electric fields	Yes — attracted towards negative plate	Yes — attracted towards positive plate	No
Penetrating power	Stopped by paper	Stopped by 5 mm thickness of aluminium	Stopped by several cm of lead

Table 8.1
Some properties of alpha, beta and gamma radiation.

8.3 NUCLEAR REACTIONS

In describing the emission of radioactive radiation, we said that this radiation is emitted from the nuclei of unstable atoms. For this reason, the emission of radioactive radiation is called a **nuclear reaction**. Nuclear reactions are completely different from chemical reactions. In a chemical reaction, bonds are broken and new bonds are made. In other words, chemical reactions only involve electrons – the nuclei of the atoms stay the same. However, in a nuclear reaction, the change in the nucleus of an element forms a new element. For example, consider the emission of an alpha particle from an atom of radium-226. If the atom loses two protons and two neutrons from its nucleus, it must mean that the atomic number decreases by two and the mass number decreases by four. Since the atomic number decreases by two, this means that the element changes into a new element two places back along the Periodic Table. Thus, an atom of radium (atomic number 88) changes into an atom of radon (atomic number 86). This changing of one element into another may be represented by a nuclear equation.

radium-226 ⟶ radon-222 + alpha particle

mass number decreases by 4

$$^{226}_{88}\mathrm{Ra} \longrightarrow \, ^{222}_{86}\mathrm{Rn} + \, ^{4}_{2}\mathrm{He}$$

atomic number decreases by 2

Since an alpha particle has the same structure as the nucleus of an atom of helium, it is always represented as $^{4}_{2}\mathrm{He}$ in nuclear equations. As a double check that you have written a nuclear equation correctly, add up the atomic numbers of the products (86 + 2) and the mass numbers of the products (222 + 4) and these should be equal to those of the parent atom (88 and 226 respectively). The loss of an alpha particle from an element may be represented by a general equation as follows:

mass number — symbol — mass number decreases by 4 — new element two places before it in the Periodic Table

$$^{A}_{Z}\mathrm{X} \longrightarrow \, ^{A-4}_{Z-2}(\mathrm{X}-2) + \, ^{4}_{2}\mathrm{He}$$

atomic number — atomic number decreases by two — alpha particle

In other words:

When an atom loses an alpha particle it changes into an atom of the element two places before it in the Periodic Table. The mass number of the parent atom decreases by four.

Now consider what happens when an atom loses a beta particle. As explained above, a beta particle is ejected from the nucleus when a neutron is converted into a proton and an electron. This means that the new nucleus has one proton more and one neutron less. For example, consider what happens when an atom of carbon-14 loses a beta particle:

carbon-14 ⟶ nitrogen-14 + beta particle

mass number stays the same

$$^{14}_{6}\mathrm{C} \longrightarrow \, ^{14}_{7}\mathrm{N} + \, ^{0}_{-1}\mathrm{e}$$

atomic number increases by 1

Note how a beta particle is represented in a nuclear equation. (An electron cannot have a mass number or an atomic number. The values of 0 and -1 are given to the electron simply to balance the equation.) Note also that the carbon atom changes into a nitrogen atom with the same mass number. The mass number stays the same since the extra proton in the new nucleus is balanced out by the loss of the neutron. As a double check that you have written a nuclear equation correctly, add up the atomic numbers of the products (7 - 1) and the mass numbers of the products (14 + 0) and these should be equal to those of the parent atom (6 and 14 respectively).

The loss of a beta particle from an element may be represented by a general equation as follows:

mass number — symbol — mass number stays the same — next element in the Periodic Table

$$^{A}_{Z}\mathrm{X} \longrightarrow \, ^{A}_{Z+1}(\mathrm{X}+1) + \, ^{0}_{-1}\mathrm{e}$$

atomic number — atomic number increases by one

In other words:

When an atom loses a beta particle, it changes into an atom of the element one place after it in the Periodic Table. The mass number of the atom stays the same.

The loss of gamma radiation does not give rise to any new atoms – energy is simply lost from the nucleus. In the emission of alpha and beta particles from unstable nuclei, we have seen how one element may be changed into another. For most of the common elements around us, it is very difficult to change one element into another. The changing of one element into another is called **transmutation**.

8.4 HALF-LIFE

The emission of radiation is entirely a random process. When radioactive elements give out radiation, scientists do not understand fully why a particular atom breaks up at any particular moment. (You may have noticed this in the demonstration experiment to detect the presence of radioactive radiation. The ratemeter showed that the emission of radiation was occurring in a random rather than a regular manner.) If one studies the rate at which any particular radioactive isotope breaks up, it is found that each radioactive isotope decays at its own rate. In order to compare the rate of decay of radioactive isotopes, scientists define a term called the half-life.

The half-life of an element is the time taken for half of the nuclei in any given sample to decay.

A Geiger-Müller tube and ratemeter may be used to find out the half-life of a radioactive substance. FIG 8.10 shows a graph of the count rate plotted against time for one particular radioactive material. Note from the graph that it takes 30 minutes for the count rate to drop from 400 to 200 counts per minute. Thus, the half-life of the material is 30 minutes, i.e. after 30 minutes half of the nuclei which were originally present have decayed.

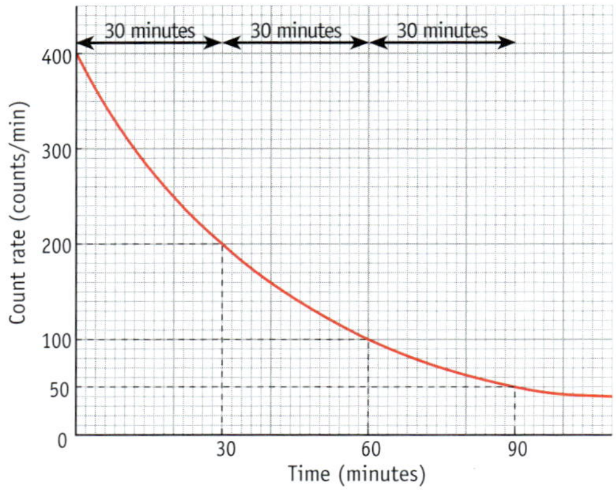

Fig 8.10
The count rate decreases by half every 30 minutes. Thus, the half-life of this particular radioactive substance is 30 minutes.

There is considerable variation in the half-lives of radioactive isotopes. For example, carbon-14 has a half-life of 5,700 years, cobalt-60 has a half-life of five years, and polonium-234 has a half-life of only 0.15 milliseconds.

8.5 USES OF RADIOISOTOPES

Radioactive isotopes, also called radioisotopes, have a wide variety of uses.

1. Medical uses

Gamma rays can be used to penetrate the body and kill cancerous cells. The dose of radiation must be carefully controlled to destroy the cancerous tissue but not healthy tissue. This treatment is often referred to as radiotherapy. Gamma radiation is also used to sterilise medical instruments, i.e. destroy germs.

2. Archaeological uses

Carbon consists of three isotopes, carbon-12, carbon-13 and carbon-14. The carbon-14 isotope is radioactive. Carbon-14 is present as a small percentage of the carbon dioxide in the air. Carbon dioxide is used by trees in the process of photosynthesis. When a tree dies, no more carbon-14 can be taken in and the carbon-14 already present decays slowly. By measuring the amount of carbon-14 in the remains of the plant, scientists can work out the age of the plant. This process is commonly referred to as carbon dating. Carbon dating was used to check the authenticity of the Dead Sea scrolls.

3. Agricultural research

Radioisotopes are used in agriculture to help scientists study things like the uptake of fertilisers by plants. A plant is fed a fertiliser containing a radioisotope like phosphorus-32. This isotope has a half-life of 14 days. By measuring the radioactivity of the leaves of the plant, scientists can find out how much of the fertiliser has reached them.

4. Food irradiation

The major cause of food spoilage is the bacteria, moulds and yeasts that grow on food. Irradiation of food with gamma rays kills disease-carrying organisms. For example, spices imported from tropical countries may contain micro-organisms. Irradiation kills these micro-organisms with no loss of flavour. Irradiation is not suitable for all foods, e.g. red meat turns brown and the taste changes.

5. Industrial uses

Radioisotopes are often used in industry. For example, leaks from underground water pipes may be located without digging up the entire pipe. A radioisotope is added at one end of the pipeline and the ground tested along the pipeline using a Geiger-Müller counter, (Fig 8.11).

Fig 8.11
Leaks from a pipeline can be traced by adding radioisotopes to the substance in the pipeline.

For your examination you must know three examples of the use of radioisotopes.

8.6 RADIOACTIVITY AROUND US

Before we conclude our study of radioactivity, it is important for us to realise that most of the radioactive radiation in our environment comes from natural sources. For example, the radioactive gas, radon, is produced by the decay of uranium and thorium found in rocks in the Earth. The gas is continually seeping out of the ground and into the air and buildings. Plants take up radioactive materials from the ground so that everything we eat or drink contains very small amounts of radioactivity. However, radioactivity is also spread throughout the world as a result of nuclear weapons tests, accidents at nuclear power plants (e.g. Chernobyl), discharges from nuclear fuel reprocessing plants, etc.

Many people are exposed to radiation because of the jobs they carry out, e.g. those who use radioactive isotopes in medicine, those working in the nuclear power industry, etc.

Great care should always be taken when working with radioactive sources, e.g. they should never be handled directly. Gamma sources must always be kept in a lead container and they must be stored securely.

CHAPTER SUMMARY

- Radioactivity was discovered by Henri Becquerel.
- Pierre and Marie Curie discovered polonium and radium.
- Radioactivity is the spontaneous breaking up of unstable nuclei with the emission of one or more types of radiation.
- Radioactive radiation is of three types – alpha, beta and gamma.
- Alpha particles are groups of two protons and two neutrons, i.e. helium nuclei, beta particles are electrons, gamma rays are high-energy electromagnetic radiation.
- Radioactive radiation is detected using a Geiger-Müller tube.
- Alpha particles are stopped by a sheet of paper, beta particles can penetrate about 5 mm of aluminium, and gamma rays are only stopped by a thick block of lead.
- The half-life of an element is the time taken for half of the nuclei in any given sample to decay.
- Radioisotopes are used in medicine, archaeology, agricultural research, food irradiation, smoke alarms and in industry.

EXERCISES

8.1. (a) Name the scientist who discovered radioactivity ....

(b) The three types of radioactive radiation are ....

(c) Of the three types of radioactive radiation, which one does not consist of particles? ....

(d) Name the least penetrating type of radioactive radiation ....

(e) Name an instrument commonly used to detect radioactivity ....

(f) Name the type of radiation which is stopped by about 5 mm thickness of aluminium metal

(g) Define the term 'half-life' ....

(h) Explain the meaning of the term 'radioisotope' ....

8.2. Write a brief note on radioactive radiation under the heading (a) types, (b) nature and (c) penetrating ability.

Describe an experiment to demonstrate the detection and penetrating ability of radioactive radiation.

8.3. Complete the following nuclear equations:

(a) $^{214}_{84}Po \longrightarrow$ .... $+\ ^{4}_{2}He$

(b) $^{198}_{79}Au \longrightarrow$ .... $+\ ^{0}_{-1}e$

(c) $^{24}_{11}Na \longrightarrow$ .... $+\ ^{0}_{-1}e$

(d) $^{238}_{92}U \longrightarrow\ ^{234}_{90}Th +$ ....

(e) $^{214}_{83}Bi \longrightarrow$ .... $+\ ^{0}_{-1}e$ ....

9 The Mole Concept

9.1 INTRODUCTION

We have already seen in CHAPTER 2 that the masses of atoms are very small. For this reason, chemists cannot count individual atoms or molecules directly. Instead, a chemist counts these particles by weighing a certain number of them. This is similar to the way that a bank cashier counts coins (FIG 9.1).

Fig 9.1
The bank cashier counts coins by weighing them. Similarly, a chemist counts atoms by weighing substances.

Knowing the weight of one coin, the bank cashier can work out from the weight of the bag of coins how many coins are in the bag. This means that it is not necessary to count out the individual coins. It was an Italian chemist called Amedeo Avogadro (1776 – 1856) who devised a similar method for chemists to count atoms and molecules. Following Avogadro's logic we can argue as follows:

1. We know that an atom of magnesium is twice as heavy as an atom of carbon (A_r of Mg = 24, A_r of C = 12).
2. Since one atom of magnesium is twice as heavy as one atom of carbon, then 100 magnesium atoms are twice as heavy as 100 carbon atoms. Similarly, one million magnesium atoms are twice as heavy as one million carbon atoms.
3. Therefore, if we have a piece of magnesium that is twice the weight of a piece of carbon, the two amounts must contain equal numbers of atoms. Think carefully about this.

For example, two grams of magnesium and one gram of carbon contain the same number of atoms. Similarly, 20 kilograms of magnesium and 10 kilograms of carbon contain the same number of atoms.

Thus, since an atom of carbon (A_r = 12) weighs twelve times as much as an atom of hydrogen (A_r = 1), one gram of hydrogen atoms and twelve grams of carbon atoms both contain the same number of atoms. Similarly, a silver atom (A_r = 108) has nine times the mass of a carbon atom (A_r = 12). If we have 12 g of carbon, this will contain a certain number of atoms. The mass of silver which will contain the same number of atoms of silver is equal to $9 \times 12 = 108$ g. The same argument applies to the other elements.

Therefore, if we measure out elements in the same proportions as their relative atomic masses, we will always be dealing with the same number of atoms of these elements. Thus, a chemist may be said to 'count the number of atoms by weighing'. Suppose we take the relative atomic mass in grams of a number of elements (TABLE 9.1).

12 g	24 g	27 g	40 g	56 g	63.5 g	108 g
carbon	magnesium	aluminium	calcium	iron	copper	silver

Table 9.1
All these masses contain the same number of atoms.

All of these amounts contain the same number of atoms. By experiment (e.g. the use of X-rays) it is found that this number of atoms is 6×10^{23}. (This number is more accurately 6.023×10^{23} but we do not need to use it to this degree of accuracy.) The number 6×10^{23} is called **Avogadro's number** or **Avogadro's constant** (in honour of Amedeo Avogadro) and it is given the symbol **L**. (This symbol is rarely used.) It is the number of atoms of carbon in 12 g of the carbon-12 isotope.

The amount of a substance which contains 6×10^{23} particles is called a **mole** of that substance.

One mole of a substance is the amount of that substance which contains 6×10^{23} particles of that substance.

The word 'mole' comes from the Latin word *moles* meaning a heap or pile. It draws attention to the particle nature of matter – a heap or pile of particles containing the Avogadro constant number of particles. In the above definition, the particles can be atoms, ions or molecules. The symbol for the mole is **mol.** Thus, when you weigh out 12 g of carbon, this amount of carbon contains 6×10^{23} atoms of carbon. We say that this **amount** of carbon is one mole (1 mol) of carbon atoms. Similarly, 48 g of magnesium is two moles (2 mol) of magnesium atoms. We say that the **amount** of magnesium is two moles (2 mol). The mole concept is a very useful one for chemists as it enables us to calculate the number of atoms (or molecules or ions) there are in a particular amount of substance which we have weighed out. Examples of moles of some elements are shown in FIG 9.2.

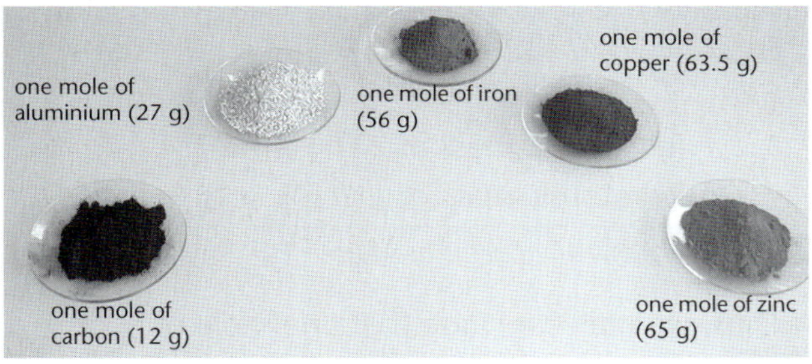

Fig 9.2
Each clock glass contains one mole of the substance.

Note:

1. The Avogadro number, 6×10^{23}, is an enormous number. If you had 6×10^{23} grains of pollen, they would cover the entire city of London to a depth of over 2 km!
2. When dealing with amounts of substances, you must be careful to specify what particle the amount refers to, e.g. the phrase 'one mole of hydrogen' should never be used, as it is confusing. It has two possible meanings: one mole of hydrogen atoms or one mole of hydrogen molecules. To avoid confusion, you must always specify the particle either by formula or words:
   i.e. 1 mole of hydrogen atoms or 1 mol of H (= 1 g of H atoms)
   1 mole of hydrogen molecules or 1 mol of H_2 (= 2 g of H_2 molecules).
   Similarly, 1 mole of O atoms = 16 g
   1 mole of O_2 molecules = 32 g
3. The mass of one mole of a substance is often referred to as the **molar mass**, e.g. the molar mass of carbon is 12 g/mol. Thus, the molar mass of an element is the relative atomic mass expressed in grams per mole.

9.2 CONVERTING MOLES TO GRAMS

Hopefully, it is clear from the above that the mass of one mole of atoms of any element is simply the relative atomic mass expressed in grams.

Mass of one mole of an element = Relative Atomic Mass in grams.

TABLE 9.2 lists the relative atomic masses of some common elements.

From TABLE 9.2, it is clear that one mole of Na = 23 grams, one mole of Cu = 63.5 grams, etc.

For all calculations involving the mole concept, you will need to refer to a list of relative atomic

Element	Symbol	Relative Atomic Mass
Hydrogen	H	1
Lithium	Li	7
Carbon	C	12
Nitrogen	N	14
Oxygen	O	16
Fluorine	F	19
Sodium	Na	23
Magnesium	Mg	24
Aluminium	Al	27
Phosphorus	P	31
Sulfur	S	32
Chlorine	Cl	35.5
Potassium	K	39
Calcium	Ca	40
Chromium	Cr	52
Iron	Fe	56
Copper	Cu	63.5
Zinc	Zn	65
Bromine	Br	80
Silver	Ag	108
Iodine	I	127
Lead	Pb	207

Table 9.2
The relative atomic masses of some common elements.

masses of elements. A list of relative atomic mass values is usually given at the top of the Leaving Certificate chemistry examination paper.

If we add up the relative atomic masses of all the atoms in a molecule of a compound, we obtain what is called the **relative molecular mass** (M_r) of the compound.

> The relative molecular mass of a compound is the sum of the relative atomic masses of all the atoms in a molecule of the compound.

> The relative molecular mass of a compound may also be defined as the mass of one molecule of that compound compared with one twelfth of the mass of the carbon-12 isotope.

Note: Relative molecular mass is sometimes referred to as **relative formula mass**. This is because some compounds (e.g. NaCl) contain ions rather than molecules. However, in this textbook we shall use the term relative molecular mass as this term is more widely used.

The method of calculating the relative molecular mass of a compound is shown in EXAMPLE 9.2.

From this example it is clear that:

> Mass of one mole of a compound = Relative Molecular Mass in grams.

In SECTION 4.7, we saw that the mass spectrometer may be used to measure relative atomic masses and abundances of isotopes. A second important use of the mass spectrometer is to measure relative molecular masses. The molecule is broken up into ions in the mass spectrometer and, from a study of the fragments formed, chemists can determine the relative molecular mass of the substance.

Example 9.1

What is the mass of three moles of aluminium atoms?

This problem is illustrated in FIG 9.3.

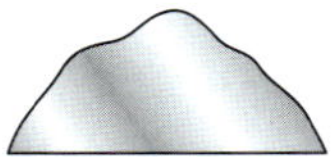

Fig 9.3
Converting moles to grams.

SOLUTION

1 mole of Al = 27 g
3 moles of Al = 27 × 3
= 81 g

Answer: 81 grams.

Example 9.2

What is the relative molecular mass of water, H_2O?

SOLUTION

Rel atomic mass of 2 H = 2 × 1	=	2
Rel atomic mass of 1 O = 1 × 16	=	16
Relative Molecular Mass	=	18

Thus, we may conclude that one mole of H_2O = 18 grams.

In this amount of water there are 6×10^{23} molecules of water.

Answer: 18

Example 9.3

What is the mass of one mole of sulfate ions, SO_4^{2-}?

SOLUTION

1 S = 1 × 32	= 32
4 O = 4 × 16	= 64
Rel Molecular Mass	= 96

i.e. one mole of sulfate ions = 96 grams

Note that the negative charges make no difference to the calculations (since electrons have negligible mass).

Answer: 96 grams

Example 9.4

What is the mass of one mole of aluminium sulfate, $Al_2(SO_4)_3$?

When working this out, remember that one molecule of aluminium sulfate contains two atoms of aluminium, three atoms of sulfur and twelve atoms of oxygen.

SOLUTION

$2\ Al = 2 \times 27 = 54$
$3\ S = 3 \times 32 = 96$
$12\ O = 12 \times 16 = 192$

Rel Molecular Mass = 342

i.e. one mole of $Al_2(SO_4)_3$ = 342 g
Answer: 342 g

Example 9.5

What is the mass of one mole of hydrated copper(II) sulfate, $CuSO_4.5H_2O$?

SOLUTION

$Cu = 63.5$
$S = 32$
$4\ O = 4 \times 16 = 64$
$5\ H_2O = 5 \times 18 = 90$

Rel Molecular Mass = 249.5

i.e. one mole of $CuSO_4.5H_2O$= 249.5 grams. This is illustrated in FIG 9.4.

Answer: 249.5 grams

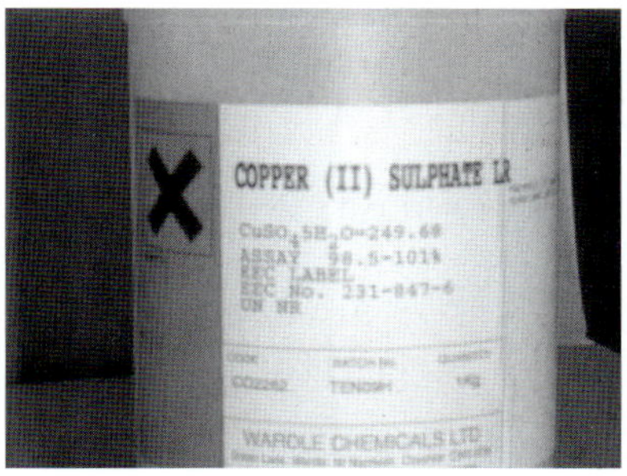

Fig 9.4
$CuSO_4.5H_2O$ has a relative molecular mass of 249.68.

Example 9.6

Calculate the mass of 0.125 moles of calcium chloride, $CaCl_2$.

SOLUTION

$Ca = 40$
$2\ Cl = 2 \times 35.5 = 71$

Rel Molecular Mass = 111

i.e. 1 mole $CaCl_2$ = 111 g
Therefore, 0.125 moles $CaCl_2$ = 111×0.125
= 13.875 g

Answer: 13.875 g

As a general rule:

Mass (in grams)
= Relative Molecular Mass × number of moles

Test Yourself:
Attempt questions 9.1 – 9.3 and W9.1 – W9.3.

9.3 CONVERTING GRAMS TO MOLES

The method of converting grams to moles is illustrated in the following examples:

Example 9.7

How many moles are there in 36 g of carbon atoms? (FIG 9.5)

SOLUTION

$$\begin{aligned} 12\text{ g carbon} &= 1\text{ mole} \\ \text{Therefore, } 1\text{ g carbon} &= \frac{1}{12}\text{ mole} \\ \text{Therefore, } 36\text{ g carbon} &= \frac{36}{12}\text{ moles} \\ &= 3\text{ moles.} \end{aligned}$$

Answer: 3 moles

Fig 9.5
Converting grams to moles.

From the above example, it is clear that:

$$\text{No. of moles of an element} = \frac{\text{mass of element}}{\text{relative atomic mass}}$$

Similarly, if one is dealing with a compound, the above relationship becomes:

$$\text{No. of moles of a compound} = \frac{\text{mass of compound}}{\text{relative molecular mass}}$$

This is illustrated in EXAMPLE 9.8.

Example 9.8

How many moles of water molecules are there in 117 g of water?

SOLUTION

First calculate the relative molecular mass of water

$$\begin{aligned} \text{O} &= 16 \\ 2\text{ H} = 2 \times 1 &= \ \ 2 \\ \hline \text{Rel Molecular Mass} &= 18 \end{aligned}$$

$$\text{Number of moles of water} = \frac{\text{Mass of water}}{\text{Rel molecular mass of water}} = \frac{117}{18} = 6.5\text{ moles}$$

Answer: 6.5 moles

We can summarise the above two relationships in the following general relationship:

$$\text{No. of moles of a substance} = \frac{\text{Mass of the substance}}{\text{Mass of one mole of the substance}}$$

This relationship is a very useful one in converting grams to number of moles. It should be learned off.

Example 9.9

How many moles of sulfuric acid, H_2SO_4, are there in 12.25 g of sulfuric acid?

SOLUTION

First calculate the relative molecular mass of sulfuric acid, H_2SO_4

$$\begin{array}{lll} 2\text{ H} & = 2 \times 1 & = 2 \\ 1\text{ S} & = 1 \times 32 & = 32 \\ 4\text{ O} & = 4 \times 16 & = 64 \\ \hline \multicolumn{2}{l}{\text{Rel Molecular Mass}} & = 98 \end{array}$$

$$\text{No of moles of } H_2SO_4 = \frac{\text{Mass of } H_2SO_4}{\text{Mass of one mole of } H_2SO_4} = \frac{12.25}{98} = 0.125 \text{ mole}$$

Answer: 0.125 mole

Test Yourself:
Attempt questions 9.4 – 9.5 and W9.4 – W9.5.

9.4 CALCULATIONS INVOLVING AVOGADRO'S NUMBER

The value of Avogadro's number, 6×10^{23}, is usually given at the top of your examination paper.

Calculations involving Avogadro's number may be divided into four types. All calculations are based on the following fact:

One mole of a substance contains 6×10^{23} particles of that substance.

The following examples indicate the type of calculations involved.

(A) CONVERTING MOLES TO NUMBER OF ATOMS OR MOLECULES

Example 9.10

How many atoms of sodium are present in 0.25 moles of this metal? This problem is illustrated in FIG 9.6.

SOLUTION

1 mole of sodium contains 6×10^{23} atoms.

$\therefore$ 0.25 moles of sodium contains $0.25 \times 6 \times 10^{23}$ atoms

$= 1.5 \times 10^{23}$ atoms

Answer: 1.5×10^{23} atoms

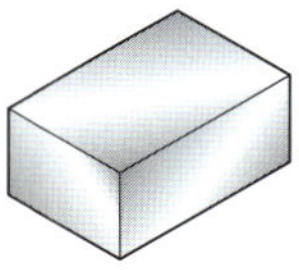

0.25 moles Na = ? atoms

Fig 9.6
Converting moles to numbers of atoms.

Example 9.11

How many (a) molecules and (b) atoms are present in 1.5 moles of water?

SOLUTION

1 mole of water contains 6×10^{23} molecules of water.

$\therefore$ 1.5 moles of water contains $1.5 \times 6 \times 10^{23}$ molecules of water.

$= 9 \times 10^{23}$ molecules of water ....(a)

Since each molecule of water contains three atoms

$\therefore$ No. of atoms in 1.5 moles of water $= 3 \times 9 \times 10^{23}$ atoms

$= 2.7 \times 10^{24}$ atoms............... (b)

Answer: (a) 9×10^{23} molecules, (b) 2.7×10^{24} atoms

(B) Converting grams to number of atoms or molecules

Example 9.12

How many atoms are there in 28.75 g of sodium?

Solution

1 mole of sodium contains 6×10^{23} atoms, i.e. 23 g sodium contains 6×10^{23} atoms

Therefore, 1 g sodium contains $\dfrac{6 \times 10^{23}}{23}$ atoms

28.75 g sodium contains $\dfrac{28.75 \times 6 \times 10^{23}}{23}$ atoms

$= 7.5 \times 10^{23}$ atoms

Answer: 7.5×10^{23} atoms.

Example 9.13

How many molecules of carbon dioxide are there in a fire extinguisher which contains 150 g of carbon dioxide, (Fig 9.7)?

Solution

First calculate the relative molecular mass of CO_2

1 C	$= 1 \times 12$	= 12
2 O	$= 2 \times 16$	= 32
Rel Molecular Mass		= 44

1 mole CO_2 contains 6×10^{23} molecules

i.e. 44 g CO_2 contains 6×10^{23} molecules

1 g CO_2 contains $\dfrac{6 \times 10^{23}}{44}$ molecules

150 g CO_2 contains $\dfrac{150 \times 6 \times 10^{23}}{44}$ molecules

$= 2.05 \times 10^{24}$ molecules

Answer: 2.05×10^{24} molecules

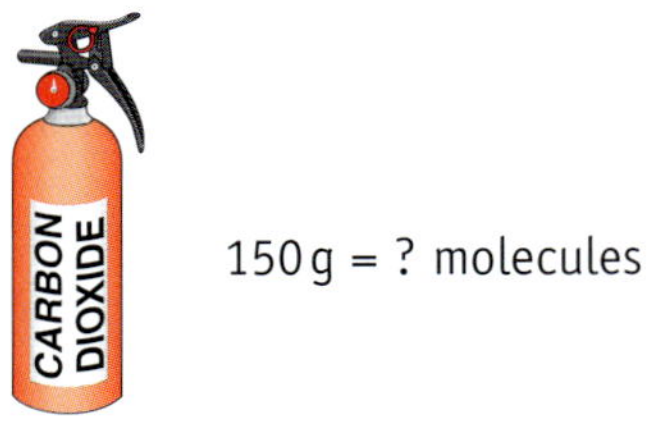

Fig 9.7
Calculate the number of carbon dioxide molecules in the fire extinguisher.

(C) Converting numbers of atoms or molecules to moles

Example 9.14

How many moles of aluminium atoms are there in 9×10^{22} atoms of aluminium?

Solution

6×10^{23} atoms of Al = 1 mole

1 atom of Al = $\dfrac{1}{6 \times 10^{23}}$ mole

9×10^{22} atoms of Al = $\dfrac{9 \times 10^{22}}{6 \times 10^{23}}$ mole

$= 0.15$ mole

Answer: 0.15 mole

Important note for students: It is vital that you know how to use your calculator to perform calculations involving powers. If you are unsure, study the instruction manual of your calculator. As an aid to solving problems like the above, it is useful to keep the Avogadro Constant, 6×10^{23}, in the memory of your calculator. To check that you are using your calculator correctly to perform calculations involving powers, carry out the calculations in the above worked examples using your own calculator.

(D) Converting numbers of atoms or molecules to grams

Example 9.15

How many grams of zinc are present in 3×10^{20} atoms of zinc?

Solution

1 mole of Zn has a mass of 65 grams, i.e. 6×10^{23} atoms of zinc have a mass of 65 g

1 atom of zinc has a mass of $\dfrac{65}{6 \times 10^{23}}$ g

3×10^{20} atoms have a mass of $\dfrac{3 \times 10^{23} \times 65}{6 \times 10^{23}}$ g

$= 0.0325$ g

Answer: 0.0325 g

Example 9.16

What is the mass (in grams) of 3×10^{22} molecules of ethanol, C_2H_5OH?

Solution

First calculate the relative molecular mass of ethanol

$2\ C = 2 \times 12$	$= 24$
$5\ H = 5 \times 1$	$= 5$
$1\ O = 1 \times 16$	$= 16$
$1\ H = 1 \times 1$	$= 1$
Rel Molecular Mass	$= 46$

i.e. 1 mole $C_2H_5OH =$ 46 g

$\therefore$ 6×10^{23} molecules of $C_2H_5OH = 46$ g

1 molecule of $C_2H_5OH = \dfrac{46}{6 \times 10^{23}}$ g

3×10^{22} molecules of $C_2H_5OH = \dfrac{3 \times 10^{22} \times 46}{6 \times 10^{23}}$ g

$= 2.3$ g

Answer: 2.3 g

Test Yourself:
Attempt questions 9.6 – 9.8 and W9.6 – W9.8.

9.5 MISCELLANEOUS QUESTIONS

In the Leaving Certificate examination paper, there may be some short questions on the mole concept. Some examples of this type of question now follow:

Example 9.17

How many moles of PO_4^{3-} ions are there in two moles of $Ca_3(PO_4)_2$? [LCH]

SOLUTION

1 molecule of $Ca_3(PO_4)_2$ contains 2 PO_4^{3-} ions (FIG 9.8).
i.e. 1 mole of $Ca_3(PO_4)_2$ contain 2 moles of PO_4^{3-} ions.
Therefore, 2 moles of $Ca_3(PO_4)_2$ contain 4 moles of PO_4^{3-} ions.

Answer: 4 moles.

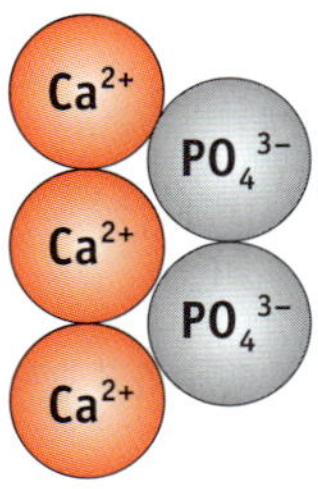

Fig 9.8

Example 9.18

How many moles of carbonate ions, CO_3^{2-}, are present in 2.76 grams of potassium carbonate, K_2CO_3?

SOLUTION

First calculate the relative molecular mass of K_2CO_3

$$\begin{aligned} 2\text{ K} &= 2 \times 39 &= 78 \\ 1\text{ C} &= 1 \times 12 &= 12 \\ 3\text{ O} &= 3 \times 16 &= 48 \\ \hline \text{Rel Molecular Mass} &= 138 \end{aligned}$$

$$\text{No. of moles of } K_2CO_3 = \frac{\text{Mass of } K_2CO_3}{\text{Rel Molecular Mass}} = \frac{2.76}{138} = 0.02 \text{ mole}$$

1 mole of K_2CO_3 contains 1 mole of CO_3^{2-} ions.
Therefore, 0.02 mole of K_2CO_3 contains 0.02 mole of CO_3^{2-} ions.

Answer: 0.02 mole

Example 9.19

What mass of chromium will contain the same number of atoms as 8 g of calcium? [LCH]

SOLUTION

$$\text{No of moles of calcium} = \frac{\text{mass}}{\text{rel atomic mass}} = \frac{8}{40} = 0.2 \text{ mole}$$

0.2 mole of chromium will contain the same number of atoms as 0.2 mole of calcium
0.2 mole chromium $= 0.2 \times 52$ g $= 10.4$ g

Answer: 10.4 grams

Test Yourself:
Attempt Revision Question W9.9

CHAPTER SUMMARY

- One mole of a substance is the amount of that substance which contains 6×10^{23} particles (atoms, ions or molecules) of that substance.
- The mass of one mole of an element is the relative atomic mass in grams.
- The mass of one mole of a compound is the relative molecular mass in grams.
- Mass (in grams) = Relative Molecular Mass × number of moles
- Number of moles of a substance = $\frac{\text{Mass of the substance}}{\text{Mass of one mole of the substance}}$
- Calculations involving Avogadro's number are based on the fact that one mole of a substance contain 6×10^{23} particles of that substance.

EXERCISES

When answering these questions, use the following values of relative atomic masses: H = 1, He = 4, Li = 7, C = 12, N = 14, O =16, F = 19, Na = 23, Mg = 24, Al = 27, Si = 28, P = 31, S = 32, Cl = 35.5, K = 39, Ca = 40, Cr = 52, Fe = 56, Cu = 63.5, Zn = 65, Br = 80, Ag = 108, Sn = 119, I = 127, Pb = 207.

9.1. A sulfur atom weighs twice as much as an oxygen atom. If a chemist has 16 grams of oxygen atoms, how many grams of sulfur would contain the same number of atoms as in the sample of oxygen atoms?

9.2. What is the mass of one mole of the following substances (a) hydrogen molecules, (b) nitrogen molecules, (c) oxygen molecules, (d) chlorine molecules, (e) fluorine molecules, (f) bromine molecules?

9.3. Calculate the relative molecular mass of each of the following substances (a) sulfuric acid, H_2SO_4, (b) carbon dioxide, CO_2, (c) hydrogen iodide, HI, (d) sodium chloride, NaCl, (e) aluminium bromide, $AlBr_3$, (f) sulfur dioxide, SO_2, (g) copper(II) carbonate, $CuCO_3$, (h) sodium carbonate, Na_2CO_3, (i) sodium hydroxide, NaOH, (j) nitrate ion, NO_3^-, (k) hydrated sodium carbonate, $Na_2CO_3.10H_2O$, (l) calcium phosphate $Ca_3(PO_4)_2$, (m) hydrated iron(II) sulfate, $FeSO_4.7H_2O$, (n) table sugar, $C_{12}H_{22}O_{11}$.

9.4. Calculate the number of moles of each element in (a) 69 g sodium, (b) 162.5 g zinc, (c) 6.5 g chromium, (d) 254 g copper, (e) 31.05 g lead.

9.5. Calculate the number of moles in (a) 192 g O_2, (b) 4 g Br_2, (c) 140 g N_2, (d) 56 g CO, (e) 0.7 g C_2H_4 , (f) 9.6 g sulfate ion, SO_4^{2-}.

9.6. How many atoms are present in (a) 0.54 moles carbon, (b) 1.24 moles tin, (c) 6.2 moles zinc, (d) 200 moles iron, (e) 5×10^{-5} moles silver.

9.7. How many atoms are present in (a) 54 g aluminium, (b) 168 g silver, (c) 96 g magnesium, (d) 3.36 g iron, (e) 0.12 g carbon.

9.8. How many moles of atoms are present in (a) 5×10^{24} atoms of carbon, (b) 8×10^{26} atoms of zinc, (c) 2.6×10^{12} atoms of gold, (d) 9.8×10^{27} atoms of silver, (e) 1.3×10^{6} atoms of magnesium.

10 Properties Of Gases

10.1 INTRODUCTION

You came across the term 'matter' before in your study of science. The word 'matter' is simply a scientific term for substances or materials. All substances exist as either solids, liquids, or gases. These are often referred to as the three **states of matter**. Gases have properties that distinguish them from solids and liquids. Many of these properties may be explained in terms of the particles of a gas having more freedom of movement than the particles of a solid or a liquid (FIG 10.1). In our study of bonding in CHAPTER 5, we saw that these particles may be either atoms, molecules or ions.

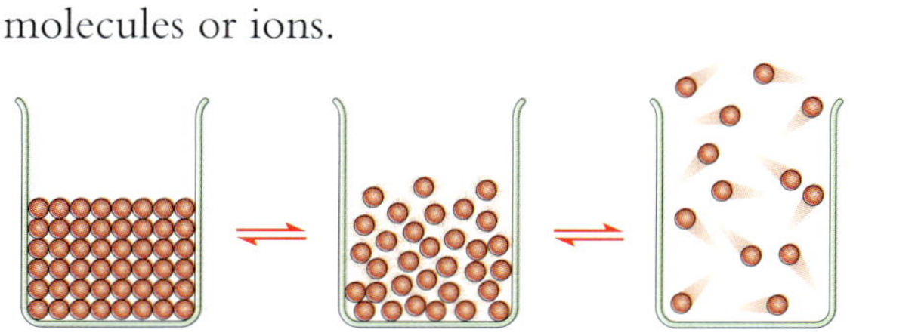

Fig 10.1
The particles in a gas have greater freedom of movement than those in a solid or liquid.

One example of a property of a gas is that it does not have a definite shape or size. It is found that a gas will spread throughout any container in which it is placed. This spreading of gases is called **diffusion**. For example, if a small amount of an irritating gas like chlorine is released into a room, it will be detected in all parts of the room after a short time. This property of a gas is often used to define what a gas is.

> A gas is defined as a substance that has no well-defined boundaries but diffuses rapidly to fill any container in which it is placed.

The ability of a gas to diffuse is illustrated in FIG 2.1. A white ring of ammonium chloride is formed where the ammonia gas meets with the hydrogen chloride gas. Note that the white ring is closer to one end showing that the lighter ammonia molecules diffused more quickly than the heavier hydrogen chloride molecules. Diffusion may also be shown by watching smoke particles travelling throughout the air in a room. Particles of a liquid also diffuse but not as quickly as the particles of a gas. For example, if a drop of a coloured substance like ink is added to a beaker of water, the colour spreads slowly throughout the water.

Gases also differ from solids and liquids in that gases do not have a fixed volume – they expand to fill the container in which they are placed. The volume of the gas is the volume of the container. Since a gas can be compressed and allowed to expand, the volume of a sample of gas depends on temperature and pressure. This dependence is studied in SECTION 10.2.

In this chapter we shall study the laws that gases obey and try to explain why they obey these laws. We shall also study how to measure the relative molecular mass of a volatile (low boiling point) liquid.

10.2 THE GAS LAWS

Before studying some of the laws that gases obey, it is important that we understand how to measure the three main properties of a fixed amount of gas: **temperature**, **pressure** and **volume**. Let us first consider what these terms mean and the units in which they are measured.

TEMPERATURE

Temperature is a measure of the degree of hotness of an object. There are two scales of temperature used in science: the **Celsius scale** (also called the

centigrade scale) and the **Kelvin scale** (also called the **absolute scale**).

The Celsius scale (named after Anders Celsius who proposed it in 1742) takes the freezing point of water as 0 °C and the boiling point of water as 100 °C. These two points are called the two **fixed points** of the Celsius scale.

The Kelvin scale is named after Lord Kelvin who proposed it in 1848. The zero of the Kelvin scale is taken as the temperature at which a gas would occupy no volume, if it could be cooled indefinitely without becoming a liquid or a solid. (This is covered in more detail later on in the chapter.) This temperature is called **absolute zero*** or zero kelvin and is given the symbol 0 K. Note that the unit does **not** have the degree symbol, °.

The relationship between the Celsius scale and the Kelvin scale is illustrated in Fig 10.2.

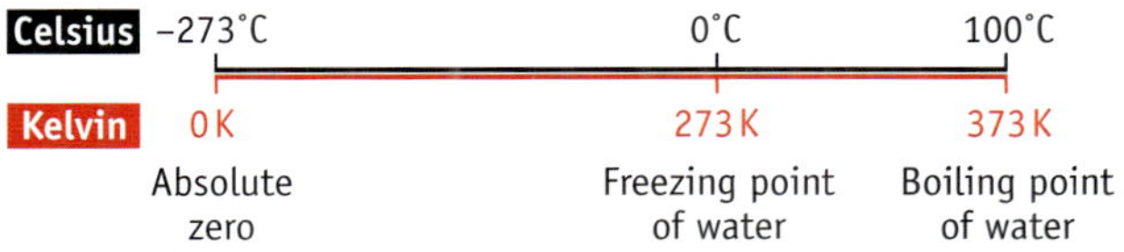

Fig 10.2
The Celsius and Kelvin scales of temperature.

> Temperature can be converted from the Celsius scale to the Kelvin scale by adding 273.

0 °C	=	0 + 273	=	273 K
50 °C	=	50 + 273	=	323 K
100 °C	=	100 + 273	=	373 K
-273 °C	=	-273 + 273	=	0 K

Note that a difference in temperature is the same for both scales as the size of a degree on the Celsius scale is the same as that on the Kelvin scale. For example, a rise in temperature of 10 °C is the same as a rise in temperature of 10 K. In the SI system, the unit of temperature is the kelvin.

* A temperature of absolute zero cannot be achieved. However, temperatures to within 0.001 K have been achieved in the laboratory. At this very low temperature the random movement of molecules almost vanishes. This causes substances to have very remarkable properties, e.g. metals cooled to near absolute zero lose their electrical resistance and become superconductors.

Pressure

The pressure of a gas is the force that the gas exerts on each unit area (e.g. one square metre) of its container. In the SI system, pressure is measured in **newtons per metre squared**, N m^{-2}. This is also called the **pascal** (Pa). Normal atmospheric pressure is **1×10^5 N m^{-2}** also written as **1×10^5 Pa** or **100,000 Pa**. As the pascal is a small unit of pressure, the **kilopascal**, symbol kPa, is also used. Thus, normal atmospheric pressure is 100 kPa.

> 1 kPa = 1000 Pa.

Note: The old method of expressing pressure of gases was in terms of millimetres of mercury or in atmospheres. Normal atmospheric pressure in this system was given as 760 mm Hg or 1 atmosphere (atm). One atmosphere is the pressure that supports a column of mercury which is 760 mm high at sea level. These units for measuring pressure are no longer used in the SI system.

Volume

The volume of a sample of gas is the same as the volume of the container in which the sample is held. The SI unit of volume is the **cubic metre** (m^3). However, this unit is too large for ordinary laboratory work. Therefore, we normally measure volume in cubic centimetres (cm^3) or **litres** (L). In this textbook we shall use the symbol L to represent litres rather than the small letter l which may be confused with the number 1. A litre is also called a **cubic decimetre** (dm^3). A cubic decimetre is exactly the same quantity as a litre. (A decimetre is one tenth of a metre, i.e. 10 cm. If you make a cube of length 10 cm, the volume of the cube will be 10 cm × 10 cm × 10 cm = 1000 cm^3, i.e. one litre or one cubic decimetre.)

> 1 litre = 1000 cm^3 = 1 cubic decimetre (dm^3)

1 cubic metre = 1 metre × 1 metre × 1 metre
= 100 cm × 100 cm × 100 cm
= 10^6 cm^3
(1,000,000 cm^3 or 1 million cm^3)

i.e. **1 m^3 = 10^6 cm^3**

Standard Temperature and Pressure (s.t.p.)

As we shall see later in this section, the volume of a gas varies with temperature and pressure. Therefore, in order to compare volumes of gases, the temperature and pressure at which they are measured must be known. Thus, when comparing volumes occupied by gases, it is normal to measure all volumes at the same temperature and pressure. For this reason, chemists decided to measure volumes of gases at **standard temperature and pressure (s.t.p.)**. The standards that have been chosen are:

> standard temperature = 273 K
> (i.e. 0 °C, the freezing point of water)

> standard pressure = 1×10^5 Pa = 100 kPa
> (i.e. normal atmospheric pressure = 1 atm or 760 mm Hg)

To be more precise, normal atmospheric pressure is actually 1.013×10^5 Pa. (A pascal may also be called a **newton per square metre**. Normal atmospheric pressure may be written as 1.013×10^5 N m^{-2}.) However, for all calculations in this course we may use the approximate value of 1×10^5 Pa or 100 kPa.

If one studies a wide variety of gases, it is observed that they differ from each other in many ways, e.g. some are flammable, some are coloured, some are vital to maintain life and others are poisonous. However, scientists have found that even though gases have different chemical properties, all gases obey certain laws. We now consider the five main laws that govern the behaviour of gases.

1. Boyle's Law

In 1657, the Irish chemist Robert Boyle (referred to in Chapter 4) built an air pump in order to study the properties of gases (Fig 10.3). He used this pump to investigate how the volume of a fixed mass of gas, kept at a constant temperature, varied with changes in pressure.

Fig 10.3
Robert Boyle's air pump. This air pump was used in experiments that led to Boyle's Law.

Consider a sample of gas trapped in a container with a moveable piston (Fig 10.4).

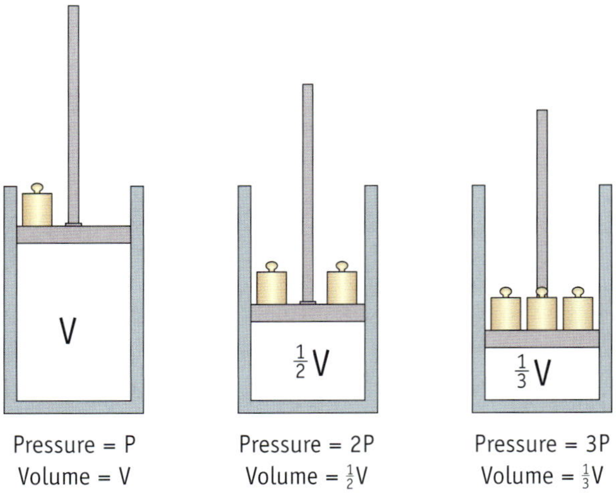

Fig 10.4
As the pressure is increased, the volume is decreased.

Boyle found that as the pressure increased, the volume decreased, i.e. the volume is inversely proportional to the pressure (Fig 10.5(a)). To check that this was correct, Boyle plotted the values of pressure against the inverse of volume as shown in (Fig 10.5(b)). The straight-line graph through the origin showed that the pressure and volume of the gas were inversely proportional.

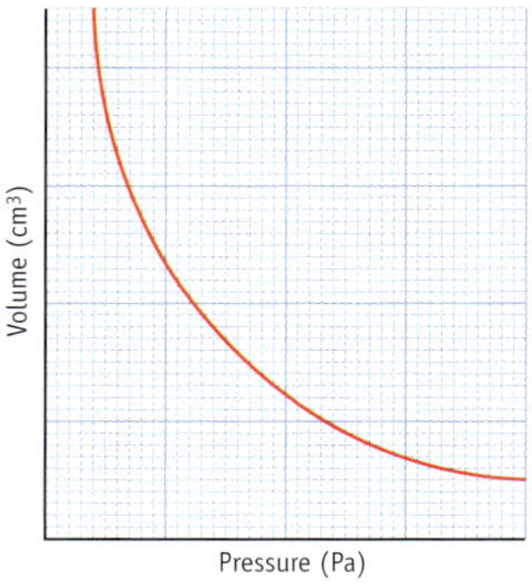

Fig 10.5 (a)
As the pressure increases, the volume decreases.

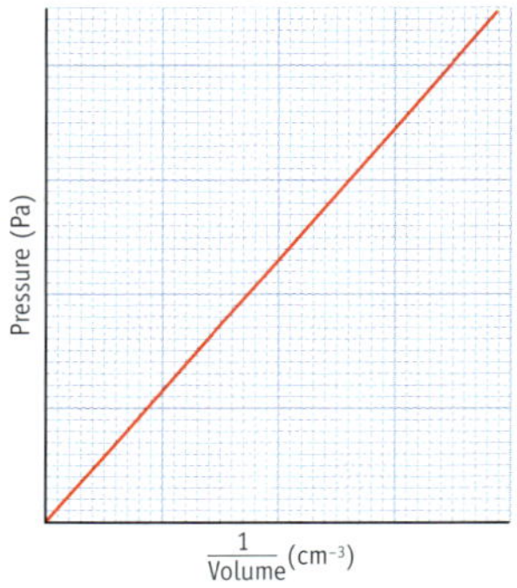

Fig 10.5 (b)
Pressure is inversely proportional to volume.

Boyle found that this relationship was true for all gases studied, regardless of their different chemical properties. He summarised his results in a statement that is now known as Boyle's Law.

> Boyle's Law states that, at constant temperature, the volume of a fixed mass of gas is inversely proportional to its pressure.

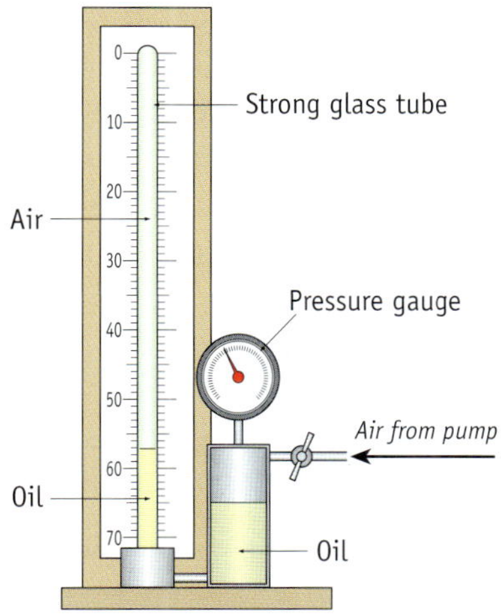

Fig 10.6
Apparatus to investigate Boyle's Law. Using a pump, the pressure on the air trapped in the glass tube is varied. The pressure is read on the pressure gauge. For each pressure value, the corresponding volume of air is read.

A modern apparatus for verifying Boyle's Law is shown in FIG 10.6.

Boyle's Law may be understood if one considers a gas as being made up of molecules which are continuously in motion. (This theory is called the 'kinetic theory of gases' and is covered more fully in SECTION 10.3.) Consider FIG 10.4. If the volume of the container is decreased, the molecules are now moving within a smaller space. This means that they collide more frequently with each other and with the walls of the container. The increased number of collisions causes an increase in the pressure.

Boyle's Law can be expressed mathematically as:

$$V \propto \frac{1}{p}$$

The symbol $\propto$ stands for 'is proportional to'.

Therefore, $V = k\frac{1}{p}$ where k is a constant of proportionality.
i.e.

$$pV = k$$

This gives another way of proving Boyle's Law: the pressure of the gas multiplied by its volume is always a constant value. This is shown in TABLE 10.1.

Pressure (p) (kPa)	Volume (V) (L)	Pressure × volume $p \times V$
40	0.100	4.0
67	0.060	4.0
80	0.050	4.0
100	0.040	4.0
120	0.033	4.0
160	0.025	4.0

Table 10.1
Data obtained from compressing a sample of air at 25 °C.

Since pV is always a constant value, if pV is plotted against p, a graph similar to that shown in FIG 10.7 will result, i.e. at different values of p, the value of pV will always be the same.

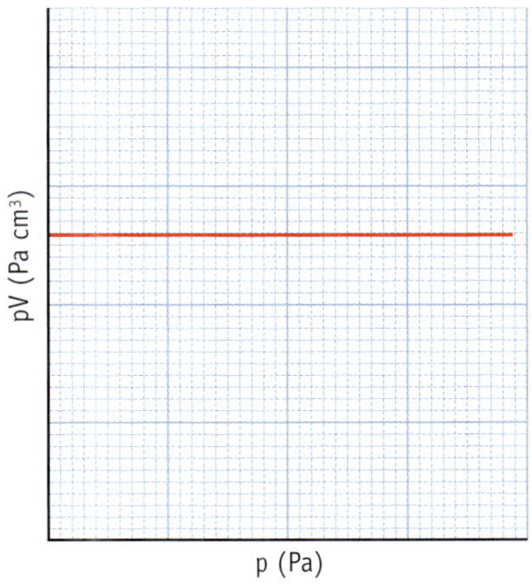

Fig 10.7
A plot of pV against p is a straight line parallel to the p axis.

2. CHARLES' LAW

We have seen that Robert Boyle studied the relationship between the volume and pressure of a fixed mass of gas at constant temperature. A French scientist, Jacques Charles (FIG 10.8), investigated the relationship between the volume and temperature of a fixed mass of gas at constant pressure. Imagine a fixed mass of gas trapped by a thin thread of concentrated sulfuric acid in a long glass tube as shown in FIG 10.9. As the water is heated, the volume of trapped air increases and the concentrated sulfuric acid is pushed up the tube.

Fig 10.8
Jacques Charles studied the relationship between the volume and temperature of a fixed mass of gas at constant pressure.

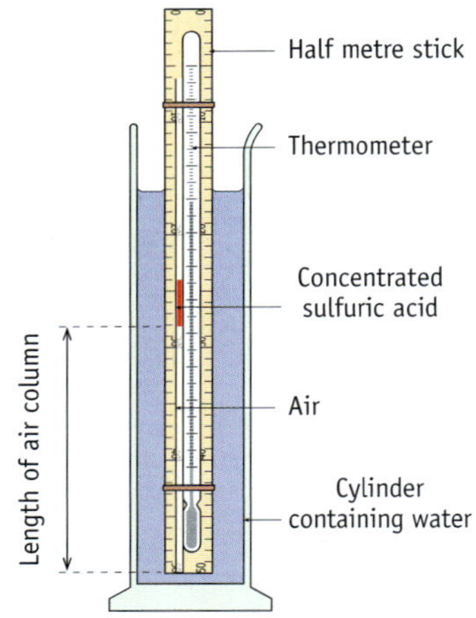

Fig 10.9
Apparatus to investigate Charles' Law. As the temperature of the trapped air increases, its volume also increases.

Charles summarised the results of his experiments on a graph similar to that shown in FIG 10.10.

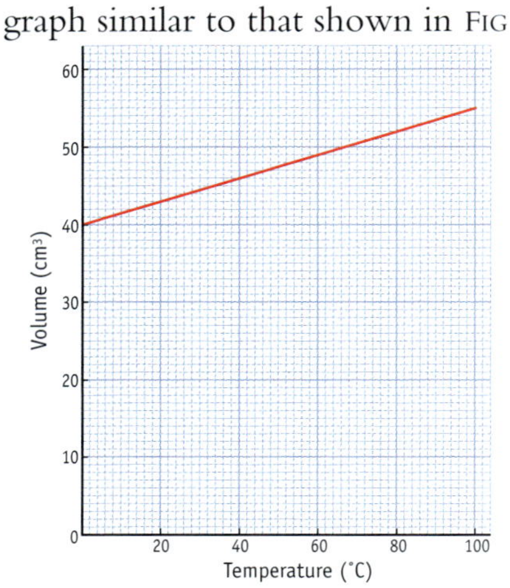

Fig 10.10
The volume of the gas increases as the temperature is increased.

You will observe that the straight line in FIG 10.10 does not pass through the origin. Therefore, one cannot say that the volume of the gas is directly proportional to the temperature measured in °C. However, if the line is continued backwards, it cuts the X axis at –273 °C (FIG 10.11).

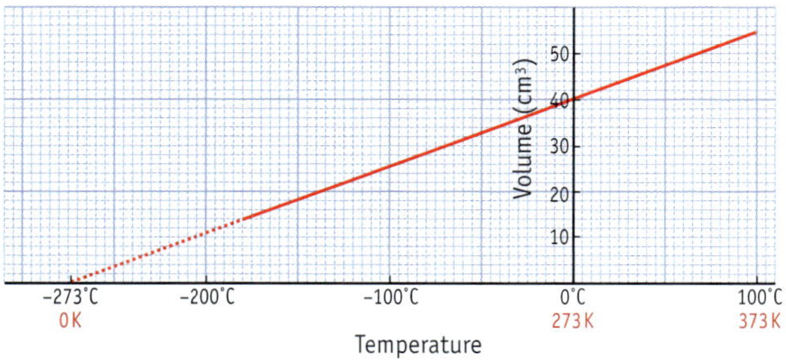

Fig 10.11
Volume is directly proportional to temperature on the Kelvin scale.

As already stated, this temperature of –273 °C is taken as the starting point for the Kelvin or absolute scale of temperature. As shown in FIG 10.11, using this scale of temperature there is now a direct relationship between volume and temperature measured on the Kelvin scale. This relationship is stated in Charles' Law:

> Charles' Law states that, at constant pressure, the volume of a fixed mass of gas is directly proportional to its temperature measured on the Kelvin scale.

Charles' Law may be understood if one considers a gas as being made up of molecules that are continually in motion. Consider FIG 10.12.

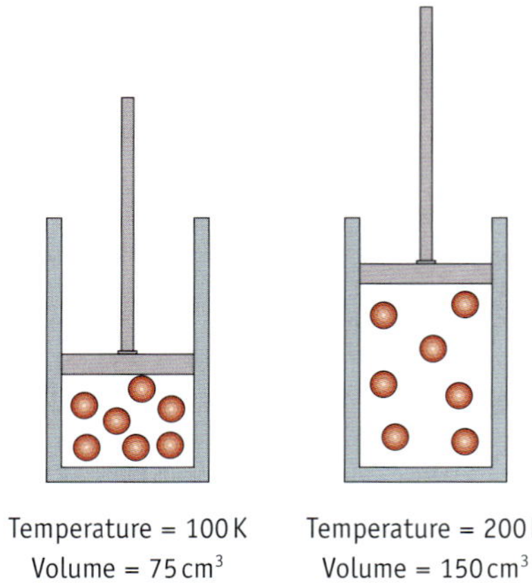

Fig 10.12
In order to keep the pressure constant as the container is heated, the volume of the container must increase.

If the temperature of the container is increased, the molecules are now moving more quickly due to the extra kinetic energy they obtain from the heat supplied. This means that they collide more frequently with each other and with the walls of the container. In order to prevent an increase in pressure, it is necessary for the volume of the container to increase.

Mathematically, Charles' Law may be stated as:

$V \propto T$ provided T is measured in kelvins.

$V = kT$ where k is a constant of proportionality.

i.e.

$$\frac{V}{T} = k$$

This gives another way of proving Charles' Law: the volume of the gas divided by its temperature (measured in kelvins) is always a constant value. This is shown in TABLE 10.2.

Volume (cm^3)	T (°C)	T (K)	$\frac{V}{T}$
75	-173	100	0.75
150	-73	200	0.75
225	27	300	0.75
300	127	400	0.75
375	227	500	0.75

Table 10.2

Data obtained by heating a sample of hydrogen at atmospheric pressure

Therefore, since $\frac{V}{T}$ is always a constant value, if $\frac{V}{T}$ is plotted against T, a straight-line graph as shown in FIG 10.13 should result, i.e. at different values of T, $\frac{V}{T}$ always has the same value.

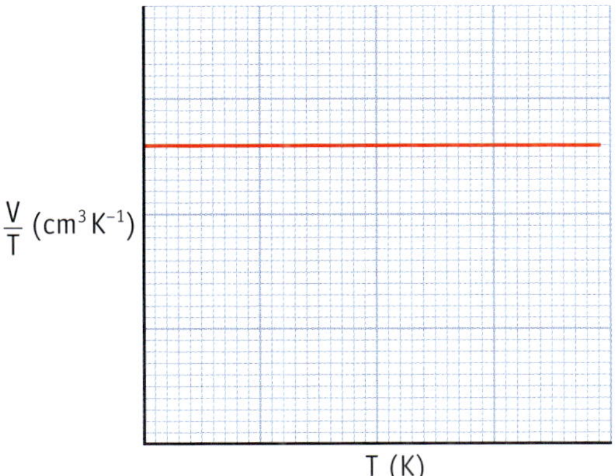

Fig 10.13

A plot of $\frac{V}{T}$ against T is a straight line parallel to the T axis.

3. THE COMBINED GAS LAW (THE GENERAL GAS LAW)

We have seen from Boyle's Law that $pV = k$.

Therefore, if the volume of a gas is allowed to change from V_1 to V_2 and its pressure changes from an initial value of p_1 to a final value of p_2 (keeping the temperature constant), then we can say:

$p_1V_1 = p_2V_2$

Similarly, from Charles' Law: $\frac{V_1}{T_1} = \frac{V_2}{T_2}$

Combining Boyle's Law and Charles' law, the following equation is obtained:

$$\frac{p_1V_1}{T_1} = \frac{p_2V_2}{T_2}$$

This is called the **Combined Gas Law** or the **General Gas Law**. It is very useful for converting the volume of a fixed mass of gas under one set of conditions of temperature and pressure to the volume of the gas under a different set of conditions.

This type of calculation is illustrated in EXAMPLE 10.1.

Example 10.1

A sample of nitrogen gas has a volume of 75 cm^3 at a temperature of 27 °C and a pressure of 95 kPa. Find the volume that the gas would have at s.t.p.

SOLUTION:

Given	s.t.p.
V_1 = 75 cm^3	V_2 = ?
p_1 = 95 kPa	p_2 = 100 kPa
T_1 = 27 °C + 273 = 300 K	T_2 = 0 °C + 273 = 273 K

Substitute into:

$$\frac{p_1V_1}{T_1} = \frac{p_2V_2}{T_2}$$

$$\frac{95 \times 75}{300} = \frac{100 \times V_2}{273}$$

$$V_2 = \frac{95 \times 75 \times 273}{300 \times 100}$$

$$= 64.84 \text{ cm}^3$$

Answer: 64.84 cm^3

Important points to note:

1. Since the Combined Gas Law is derived from Charles' Law, it is vital to convert all temperatures to the Kelvin scale. (Remember that Charles' Law is only true when temperature is measured on the Kelvin scale.)
2. The units on either side of the equation must be consistent, e.g. the above problem gives pressure in kilopascals. These units are used for pressure on the left-hand side of the equation. Therefore, standard pressure must be written in kilopascals on the right-hand side of the equation. This also applies to the units for volume.

Test Yourself:
Attempt questions 10.1 – 10.4 and W10.1 – W10.3.

4. GAY-LUSSAC'S LAW OF COMBINING VOLUMES

In 1805, Joseph Gay-Lussac, a French chemist (FIG 10.14) was studying reactions between various gases. He repeated an experiment first performed by Henry Cavendish (the discoverer of hydrogen) in 1781. He confirmed that, when hydrogen reacts with oxygen, two volumes of hydrogen always react with one volume of oxygen (all volumes being measured under the same conditions of temperature and pressure).

Fig 10.14
Joseph Gay-Lussac who performed some important experiments on the combining volumes of gases.

hydrogen	+	oxygen	⟶	steam
2 volumes		**1 volume**		**2 volumes**

This led him to study the reactions of other gases to see whether they also reacted in simple ratios. Some of the reactions which he studied were:

hydrogen	+	chlorine	⟶	hydrogen chloride
1 volume		**1 volume**		**2 volumes**
nitrogen monoxide	+	oxygen	⟶	nitrogen dioxide
2 volumes		**1 volume**		**2 volumes**

In 1808, Gay-Lussac summarised the results of his work in his Law of Combining Volumes. **Gay-Lussac's Law of Combining Volumes** states that:

> In a reaction between gases, the volumes of the reacting gases and the volumes of any gaseous products are in the ratio of small whole numbers provided the volumes are measured at the same temperature and pressure.

Special types of syringes made of glass called gas syringes are used to study the volumes of gases that react with each other (FIG 10.15).

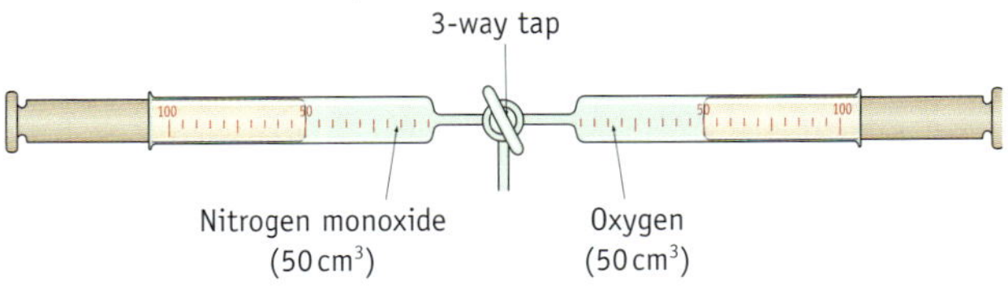

Fig 10.15
Apparatus to investigate the volume of nitrogen monoxide that reacts with oxygen.

Consider the reaction between nitrogen monoxide and oxygen (both colourless gases). These two gases react to form a brown gas called nitrogen dioxide. When the nitrogen dioxide is formed, it is found to occupy a smaller volume than the combined volume of the other two gases. Firstly, place 50 cm^3 of nitrogen monoxide in the gas syringe on the left and 50 cm^3 of oxygen in the gas syringe on the right. Push 5 cm^3 of oxygen from the syringe on the right into the nitrogen monoxide on the left. Allow time for the gases to react and for cooling to occur so that all volumes are measured at the same temperature and pressure. Note the total volume of gases left in the two syringes. Now add another 5 cm^3 of oxygen and keep doing this until the reaction is complete, i.e. until the total volume stops getting smaller. At this stage, the brown gas formed does not get any darker on adding the oxygen. The results of the experiment are plotted on a graph as shown in FIG 10.16.

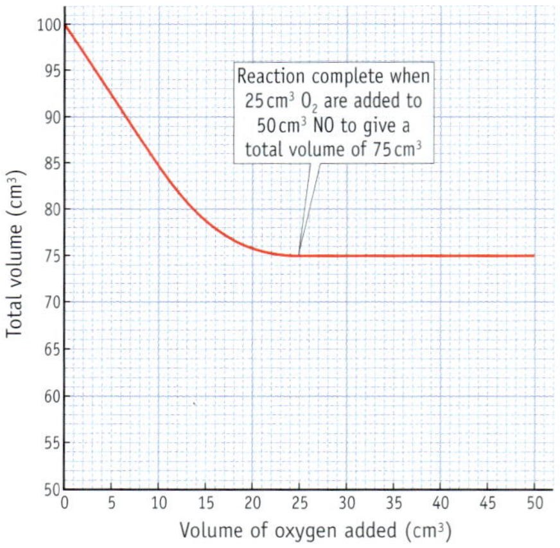

Fig 10.16
The reaction is complete when 25 cm^3 of oxygen are added to 50 cm^3 of nitrogen monoxide to give a total volume of 75 cm^3.

From the graph we see that the reaction is complete when approximately 25 cm^3 of oxygen have been added to the 50 cm^3 of nitrogen monoxide, i.e. one volume of oxygen reacts with two volumes of nitrogen monoxide.

nitrogen monoxide	+	oxygen	$\longrightarrow$	nitrogen dioxide
2 volumes		**1 volume**		**2 volumes**
(50 cm^3)		(25 cm^3)		(50 cm^3)

Note from the results of the experiment that the total volume of gas remaining is 75 cm^3. This consists of 25 cm^3 of unreacted oxygen (only 25 cm^3 of the original 50 cm^3 of oxygen reacted) and 50 cm^3 of nitrogen dioxide, i.e. two volumes of nitrogen dioxide are formed. This is exactly the type of result that Gay-Lussac obtained in a large number of experiments involving gases, i.e. the volumes of the gases which reacted and the volumes of the gaseous products formed were all in simple whole number ratios.

5. AVOGADRO'S LAW

After Gay-Lussac's discovery of the small whole number ratios in which gases react, scientists tried to develop a theory to explain the results of Gay-Lussac's experiments. It seemed likely that the explanation would rely on the idea that gases consist of particles. In this case, it would be the number of particles that would be important.

Gay-Lussac's results were published the same year as Dalton's Atomic Theory (1808). However, attempts to explain Gay-Lussac's Law using Dalton's atomic theory failed. Dalton imagined hydrogen and oxygen gas as being made up of **atoms** of hydrogen and oxygen. He could not explain why two volumes of hydrogen combine with one volume of oxygen. He suggested that Gay-Lussac was mistaken and that his observations were inaccurate. Three years later, the man with the answer to the problem came along – an Italian scientist called Amedeo Avogadro (Fig 10.17).

Fig 10.17
Amedeo Avogadro provided an explanation for Gay-Lussac's Law of Combining Volumes.

Avogadro reasoned as follows: by experiment, it is found that 100 cm^3 of hydrogen react with exactly 100 cm^3 of chlorine. This must mean that there is the **same number of molecules** of hydrogen and chlorine in each volume. Otherwise, some molecules of hydrogen or chlorine would be left over.

hydrogen	+	chlorine	$\longrightarrow$	hydrogen chloride
1 volume		**1 volume**		**2 volumes**

Suppose one volume of hydrogen and one volume of chlorine contain n molecules.

i.e.

n molecules hydrogen	+	n molecules chlorine	$\longrightarrow$	$2n$ molecules hydrogen chloride

i.e.

1 molecule hydrogen	+	1 molecule chlorine	$\longrightarrow$	2 molecules hydrogen chloride

i.e.

$$H_2 + Cl_2 \longrightarrow 2HCl$$

Similarly, the other reactions described when discussing Gay-Lussac's law may be written as:

$$2H_2 + O_2 \longrightarrow 2H_2O$$

$$2NO + O_2 \longrightarrow 2NO_2$$

In other words, the ratio in which the volumes of gases combine is the same as the ratio in which the molecules of gases combine. Thus, when dealing with gaseous reactions, and only gaseous reactions, the words volume and molecule can always be interchanged for each other. The reaction between hydrogen and chlorine is illustrated in Fig 10.18.

Avogadro's explanation of Gay-Lussac's Law

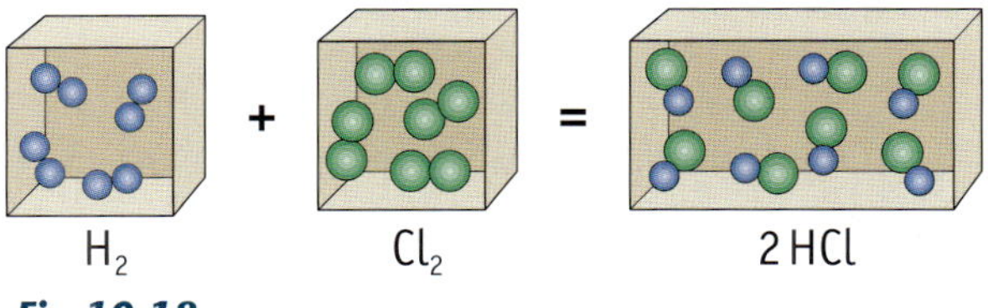

Fig 10.18
Avogadro assumed that equal volumes of gases at the same temperature and pressure contained equal numbers of molecules.

This assumption that Avogadro made is now called **Avogadro's Law** (also called **Avogadro's Theory** or **Avogadro's Hypothesis**) and is stated as follows:

> Avogadro's law states that equal volumes of gases contain equal numbers of molecules, under the same conditions of temperature and pressure.

Avogadro's theory was not accepted for nearly 50 years after he proposed it. This was because Dalton's Theory did not allow for the existence of molecules. Avogadro was the first to distinguish between atoms and molecules and he introduced the word *molecule* into the language of chemistry.

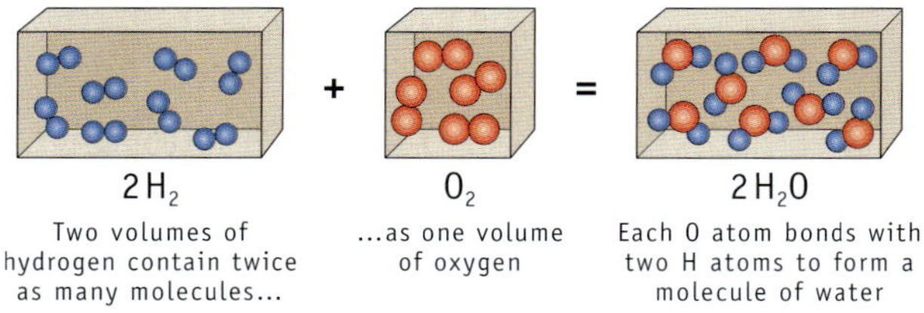

Fig 10.19
Avogadro's Law explains why two volumes of hydrogen react with one volume of oxygen to produce two volumes of steam.

Using his theory, he also proposed that the water molecule contained two hydrogen atoms for each oxygen atom. His explanation of Gay Lussac's results for the reaction of oxygen and hydrogen is summarised in FIG 10.19.

By experiment it is found that a mole of oxygen gas (i.e. 32 g O_2) occupies a volume of 22.4 litres at s.t.p. Since one mole of any gas contains 6×10^{23} molecules then, by Avogadro's law, one mole of any gas should occupy 22.4 litres at s.t.p.

> At s.t.p. one mole of any gas occupies a volume of 22.4 litres.

The volume occupied by one mole of any gas is called its **molar volume**. Thus, the molar volume of all gases is 22.4 litres. This value is normally given at the top of the examination paper. Since one mole of any gas occupies 22.4 L at s.t.p. the density of a gas is easily calculated. This is illustrated in EXAMPLE 10.2.

Example 10.2

Calculate the density of carbon dioxide at s.t.p. [Given C =12, O =16, molar volume at s.t.p. = 22.4 L.]

SOLUTION

Relative molecular mass CO_2 = 44

i.e. 1 mole CO_2 = 44 g

i.e. 44 g CO_2 occupy 22.4 L at s.t.p.

$$\text{Density} = \frac{\text{Mass}}{\text{Volume}} = \frac{44\text{g}}{22.4\text{L}} = 1.96 \text{ g/L}$$

Answer: 1.96 g/L

Example 10.3

Calculate the volume occupied by 20 g of sulfur dioxide gas at s.t.p. [Given O = 16, S = 32, molar volume at s.t.p. = 22.4 L.]

SOLUTION

1 mole of sulfur dioxide occupies 22.4 L at s.t.p.

i.e. 64 g SO_2 occupy 22.4 L (M_r SO_2 = 64)

1 g SO_2 occupies $\frac{22.4}{64}$ L

Therefore, 20 g SO_2 occupy $\frac{20 \times 22.4}{64}$ L = 7 L

Answer: 7 L

Example 10.4

200 cm^3 of a certain gas at s.t.p. have a mass of 0.25 g. Calculate the relative molecular mass of the gas.

SOLUTION

200 cm^3 have a mass of 0.25 g

1 cm^3 has a mass of $\frac{0.25}{200}$ g

22,400 cm^3 have a mass of $\frac{22{,}400 \times 0.25}{200}$ g = 28 g

i.e. 22.4 L of the gas have a mass of 28 g

Therefore, relative molecular mass of gas = 28

Answer: 28

Test Yourself:
Attempt questions 10.5 – 10.7 and W10.4 – W10.5.

We now investigate how we can use the gas laws to measure the relative molecular mass of a volatile liquid.

EXPERIMENT 10.1 MANDATORY STUDENT EXPERIMENT NO. 4

To measure the relative molecular mass of a volatile liquid.

Introduction

As discussed in CHAPTERS 4 AND 9, the relative molecular mass of a compound may be found to a very high degree of accuracy using a mass spectrometer. However, mass spectrometers are not usually available in school laboratories! Therefore, we shall use a method first developed by a German chemist called Victor Meyer to measure the relative molecular mass of a volatile liquid.

There are two methods of carrying out this experiment. In method A, we will fill a flask with the vapour of a volatile liquid at a known temperature. We will then find the mass of this vapour. We will measure the volume of the vapour and convert it to s.t.p. We will then calculate what the mass of 22,400 cm^3 of this vapour would be at s.t.p. This is the relative molecular mass of the liquid.

In method B (SECTION 10.4), we will use a gas syringe to measure the relative molecular mass of the liquid.

METHOD A

Apparatus required

Conical flask (250 cm^3), graduated cylinders (0 - 10 cm^3 and 0 - 100 cm^3), rubber band, Bunsen burner or hot plate, barometer, large beaker, aluminium foil, pin, retort stand, electronic balance.

Chemicals required

A volatile liquid e.g. propanone (acetone), cyclohexane, etc.

Procedure

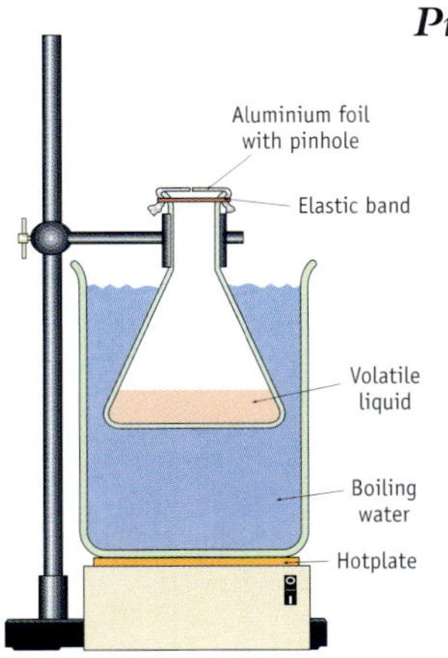

Fig 10.20
Apparatus to measure the relative molecular mass of a volatile liquid.

1. Take a clean, dry conical flask, a rubber band and a piece of aluminium foil (large enough to make a cap for the conical flask).
2. Place the conical flask, rubber band and aluminium foil on the balance and find their mass.
3. Using a graduated cylinder, pour 10 cm^3 approx. of the volatile liquid into the conical flask.
4. Put the aluminium foil over the mouth of the flask, press it down over the sides, and hold it in place **tightly** using the rubber band.
5. Using a pin, make a very small hole in the aluminium foil cap.
6. Clamp the flask in a large beaker of **boiling** water as shown in FIG 10.20. Try to ensure that as much as possible of the flask is covered by the boiling water. The purpose of the boiling water is to convert the volatile liquid to a vapour. The air is expelled from the flask which is then filled with the vapour of the liquid. Any excess vapour escapes from the flask through the pinhole.

7. Leave the flask in the boiling water for about eight minutes. This ensures that all the liquid has vaporised. The flask is now full of the vapour of the liquid at 100 °C (the temperature of the boiling water) and at the atmospheric pressure of the day.
8. Remove the flask from the boiling water and allow it to cool to room temperature. As it cools, you should see the vapour inside in the flask condensing back to a liquid.
9. Dry the outside of the flask and foil. Reweigh the flask, foil and band. Find the mass of the liquid whose vapour has filled the flask.
10. Find the total volume of the conical flask by filling it completely with water and pouring the water (a number of times) into a 100 cm^3 graduated cylinder.
11. Record the atmospheric pressure by taking a reading from a barometer.

Record your results as shown below. To assist you with calculations, some specimen results are shown.

Mass of empty flask, foil and band		= 105.21 g
Mass of flask, foil, band and condensed liquid		= 106.30 g
Therefore, mass of condensed liquid	= 106.30 - 105.21	= 1.09 g
Volume of flask = volume of vapour		= 326 cm^3
Atmospheric pressure		= 746 mm Hg*
Temperature of vapour in flask	= 100 °C + 273	= 373 K

Calculations

laboratory s.t.p

$$\frac{p_1V_1}{T_1} = \frac{p_2V_2}{T_2}$$

$$\frac{746 \times 326}{373} = \frac{760 \times V_2}{273}$$

(760: standard pressure; 273: standard temperature)

$$V_2 = \frac{746 \times 326 \times 273}{373 \times 760}$$

$$V_2 = 234 \text{ cm}^3 \text{ at s.t.p.}$$

i.e. at s.t.p., 234 cm^3 of the vapour have a mass of 1.09 g

1 cm^3 of the vapour has a mass of $\frac{1.09}{234}$ g

Therefore, 22,400 cm^3 of the vapour have a mass of $\frac{22{,}400 \times 1.09}{234}$ g

= 104 grams

i.e. Relative molecular mass of liquid = 104

* As most barometers are graduated in mm Hg, it is not necessary to convert this to pascals for the purpose of calculations. This is because the calculations merely involve a ratio of pressures and this ratio is the same whether it is measured in mm Hg or pascals.

10.3 THE KINETIC THEORY OF GASES

The kinetic theory of gases ('moving molecule theory') was developed in order to explain many of the gas laws discussed above. Scientists imagine all matter as being made up of tiny particles. In a solid these particles can only vibrate about a fixed position. In a liquid the particles possess larger amounts of energy and can move freely around each other. However, they do not have complete freedom as there are forces of attraction between the particles in the liquid. In gases the particles have much larger amounts of energy. The gas particles have sufficient energy to overcome the forces of attraction between them almost completely.

One of the first pieces of evidence to support the idea of particles in motion came from an observation made by a Scottish botanist, Robert Brown, in 1827. He was studying pollen grains suspended in a drop of water. Using his microscope, he observed that the tiny pollen grains were in continuous irregular motion. He could not explain his results but it is now accepted by scientists that the tiny pollen grains are being continuously bombarded by the moving water molecules. The molecules themselves are too tiny to be seen. This random movement of tiny particles suspended in a liquid or gas is called **Brownian motion**.

In order to help us understand the behaviour of gases, a number of assumptions were developed.

Assumptions of the Kinetic Theory Of Gases

1. Gases are made up of particles that are in continuous motion, colliding with each other and with the walls of the container.
2. There are no attractive or repulsive forces between the molecules of a gas, i.e. the molecules are independent of each other.
3. The gas molecules are so small and so widely separated that the actual volume of all the molecules is negligible compared with the space that they occupy.
4. When molecules collide, the collisions are perfectly elastic, i.e. there is no loss of kinetic energy in these collisions, although there may be a transfer of energy between the colliding particles (Fig 10.21). This means that, if one molecule loses energy in a collision, its partner in the collision gains this energy.
5. The average kinetic energy of the molecules in a sample of gas is proportional to the temperature measured on the absolute (Kelvin) scale.

If we examine the above assumptions, we see that there are certain limitations to the kinetic theory of gases. For example, in assumption 2, it is not valid to state that there are no attractive or repulsive

Fig 10.21
The collisions between molecules are similar to the collisions between the balls on a snooker table.

forces between the molecules of a gas. As discussed in Chapter 5, gas molecules that are **polar** (e.g. NH_3, HCl, etc) will have attractive forces because the partially negative end of one molecule will be attracted to the partially positive end of another molecule. Similarly, there are weak van der Waals forces between non-polar molecules (e.g. Cl_2).

A further limitation is that in assumption 3 it is not always valid to say that the volume of the molecules is negligible compared with the space that they occupy. Under high pressure, when molecules are crowded closer together, it is clear that their volume is not negligible compared with the distances between them.

Therefore, it is true to say that all of the assumptions of the kinetic theory do not hold for real gases but only hold for imaginary or **ideal** gases.

An ideal gas is one that obeys all the assumptions of the kinetic theory of gases under all conditions of temperature and pressure.

Ideal gases do not exist! However, a **real gas** behaves somewhat like an ideal gas at **low pressure** (when the molecules are widely spaced) and at **high temperatures** (when the molecules are moving rapidly and the forces between the molecules are small). At high pressures and low temperatures, gas molecules become much closer together. This means that we can no longer assume that the volume of the molecules can be ignored compared with the volume in which they move. Also, under these conditions, we cannot assume that there are no attractive forces between the molecules. At low temperatures, when the molecules are moving slowly, the attractive forces between the molecules are significant. Under ordinary conditions (e.g. atmospheric pressure and a temperature of 20 °C), real gases behave similarly to ideal gases and therefore the above assumptions of the kinetic theory are reasonably valid.

Real gases differ from ideal gases because of:
(i) forces of attraction and repulsion between the molecules and
(ii) the volume of the molecules is not negligible (especially at high pressures).

In terms of the gas laws, an ideal gas may be described as one that obeys the gas laws exactly, e.g. an ideal gas would have zero volume at -273 °C (see FIG 10.11). This never occurs as real gases become liquids long before they reach this temperature. It should be noted that, under the same conditions of temperature and pressure, non-polar molecules come closer to being ideal gases than polar molecules, since the attractive forces are less in the case of non-polar molecules.

Boyle's Law and Charles' Law have already been explained in terms of the kinetic theory of gases. The diffusion of gases can also be explained in terms of this theory. Diffusion occurs because gases consist of particles that have a great deal of freedom and are always on the move.

Example 10.5

Give two reasons why real gases depart from ideal behaviour. Under what conditions of temperature and pressure do real gases come closest to ideal behaviour? *[LCH]*

ANSWER

Real gases depart from ideal behaviour because:

(i) In an ideal gas, it is assumed that there are no forces of attraction between the molecules. In real gases, attractive forces exist between the molecules.

(ii) In an ideal gas, it is assumed that the molecules occupy negligible space. However, in real gases, the molecules occupy a small but definite space, which is not negligible at high pressure.

Real gases come closest to ideal behaviour at high temperature and low pressure.

10.4 THE EQUATION OF STATE FOR AN IDEAL GAS

Boyle's law, Charles' law and Avogadro's law may be combined to give a mathematical relationship between the volume (V), pressure (p), temperature (T) and number of moles (n) of a gas.

Boyle's Law: $V \propto \frac{1}{p}$ at constant temperature.

(The symbol $\propto$ stands for "is proportional to.")

Charles' Law: $V \propto T$ at constant pressure.

Avogadro's Law: $V \propto n$ at constant temperature and pressure.

Combining the above three relationships:

$$V \propto (\tfrac{1}{p})\ T\, n$$

i.e. $V = R\ (\frac{1}{p})\ T\ n$ where R is a constant of proportionality called the **Universal Gas Constant**.

Rearranging the above equation gives:

$$pV = nRT$$

This equation is called the **Equation of State for an Ideal Gas** or the **Ideal Gas Equation**. It is called the Equation of State because it tells us how, in going from one gaseous state to another, the four variables V, p, T and n change.

The numerical value for R can be found by substituting experimental values into the equation. For example, we know that at s.t.p. one mole of any gas occupies a volume of 22.4 litres.

At s.t.p $\quad T = 273 \text{ K}$

$p = 1.013 \times 10^5 \text{ N m}^{-2}$ (normal atmospheric pressure)

$n = 1$ (for 1 mole of gas)

$V = 22.4 \text{ L} = \dfrac{22.4}{1000} = 0.0224 \text{ m}^3$

Note: We must convert litres to cubic metres. (There are 1000 litres in a cubic metre.) This is because we use square metres in the units for pressure and we must therefore measure volume in cubic metres to keep the units consistent.

$$pV = nRT$$

$$R = \frac{pV}{nT}$$

$$= \frac{1.013 \times 10^5 \text{Nm}^{-2} \times 0.0224 \text{ m}^3}{1 \text{ mole } \times 273\text{K}}$$

$$= 8.31 \text{ Nm mol}^{-1} \text{ K}^{-1}$$

Since a newton metre is also called a joule (symbol J), the above is often written as:

$$R = 8.31 \text{ J mol}^{-1} \text{ K}^{-1}$$

When substituting into the equation $PV = nRT$, always ensure that the units of pressure are pascals, the units of volume are cubic metres, and the units of temperature are kelvins.

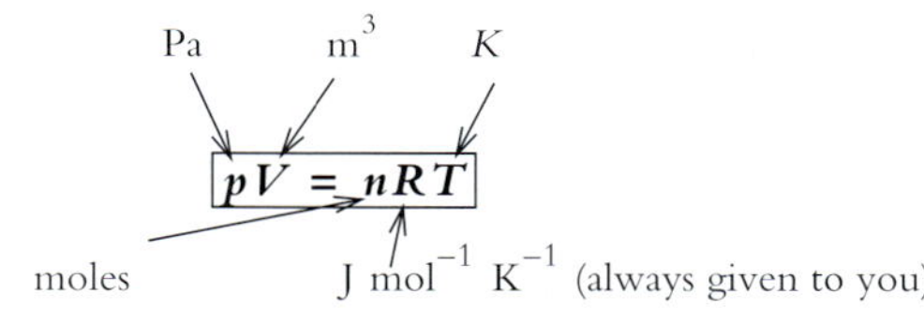

Example 10.6

Calculate the volume occupied by 10 g of oxygen gas at 25 °C and a pressure of 200 kPa. [R = 8.31 J mol^{-1} K^{-1}]

SOLUTION

$p = 200 \text{ kPa} = 200 \times 10^3 \text{ Pa} \qquad V = ?$

$$n = \frac{\text{Mass}}{\text{Rel Molecular Mass}} = \frac{10}{32} = 0.31 \text{ mole}$$

$R = 8.31 \text{ J mol}^{-1} \text{ K}^{-1}$

$T = 25 \text{ °C} + 273 = 298 \text{ K}$

$pV = nRT$

$$V = \frac{nRT}{P}$$

$$= \frac{0.31 \times 8.31 \times 298}{200 \times 10^3} = 0.0038 \text{ m}^3 = 0.0038 \times 1000 \text{ litres} = 3.8 \text{ L}$$

ALTERNATIVE METHOD

This problem could also be solved using the Combined Gas Law.

Given	s.t.p.
$V_1 = ?$	$V_2 = 22.4$ L
$p_1 = 200$ kPa	$p_2 = 100$ kPa
$T_1 = 25$ °C + 273 = 298 K	$T_2 = 0$ °C + 273 = 273 K

Substitute into $\dfrac{p_1 V_1}{T_1} = \dfrac{p_2 V_2}{T_2}$

$$\frac{200 \times V_1}{298} = \frac{100 \times 22.4}{273}$$

$$V_1 = \frac{100 \times 22.4 \times 298}{273 \times 200} = 12.23 \text{ L}$$

1 mole of O_2 occupies 12.23 L under the above conditions,

i.e. 32 g O_2 occupies 12.23 L

1g O_2 occupies $\dfrac{12.23}{32}$ L

10 g O_2 occupies $\dfrac{12.23 \times 10}{32} = 3.8\text{L}$

Answer: 3.8 L

Example 10.7

Calculate the relative molecular mass of a volatile liquid from the following information: 0.20 g of X on vaporisation gave 40 cm^3 of vapour measured at 100 °C and a pressure of 100 kPa.
[R = 8.31 J mol^{-1} K^{-1}]

SOLUTION

$p = 100 \text{ kPa} = 100 \times 10^3 \text{ N m}^{-2}$
$V = 40 \text{ cm}^3 = 40 \times 10^{-6} \text{ m}^3$
$n = ?$
$R = 8.31 \text{ J mol}^{-1} \text{ K}^{-1}$
$T = 100 \text{ °C} + 273 = 373 \text{ K}$

Note: A common error among students when substituting into the equation $pV = nRT$ is to forget to convert volumes to m^3. As already explained, if you are given the volume in litres, you must divide by 1000 (i.e. $\times 10^{-3}$) to convert to m^3. If you are given the volume in cm^3, you must divide by a million (i.e. $\times 10^{-6}$) since there are a million cubic centimetres in a cubic metre.

volume in cubic metres = volume in litres $\times 10^{-3}$
volume in cubic metres = volume in $cm^3 \times 10^{-6}$

$$pV = nRT$$
$$n = \frac{pV}{RT}$$
$$= \frac{100 \times 10^3 \times 40 \times 10^{-6}}{8.31 \times 373} = 0.0013 \text{ moles}$$

i.e. 0.0013 mole = 0.20 gram

$$1 \text{ mole} = \frac{0.2}{0.0013} \text{ grams}$$
$$= 153.85 \text{ grams}$$

i.e. Relative molecular mass = 153.85

Answer: 153.85

Test Yourself:
Attempt questions 10.8 – 10.12 and W10.6 – W10.9.

EXPERIMENT 10.1 MANDATORY STUDENT EXPERIMENT NO. 4

ALTERNATIVE METHOD

To measure the relative molecular mass of a volatile liquid using a gas syringe.

METHOD B

Introduction

In Method A, we saw how the relative molecular mass of a volatile liquid could be determined by filling a conical flask with the vapour of a liquid at a known temperature. An alternative method of carrying out this experiment is to use a gas syringe. Gas syringes are similar to ordinary plastic syringes except they are made of glass and usually have a capacity of 100 cm^3. They are manufactured to a very high degree of accuracy. In this experiment, a weighed quantity of a volatile liquid is injected into a gas syringe. The syringe is then warmed to vaporise the liquid. The volume taken up by the vapour is measured and the relative molecular mass of the liquid is calculated using the formula $PV = nRT$.

The apparatus required for this experiment is shown in FIG 10.22. This consists of an oven or furnace into which the gas syringe is placed. The oven heats the gas syringe to a temperature higher than the boiling point of the volatile liquid.

Apparatus required

Gas syringe (100 cm^3), small syringe (1 cm^3) and hypodermic needle, thermometer, self-sealing rubber cap, barometer, furnace for gas syringe or large beaker of hot water, electronic balance.

Chemicals required

A volatile liquid e.g. propanone (acetone), cyclohexane, etc.

Procedure

1. Draw about 5 cm^3 of air into the syringe. Place the self-sealing cap over the nozzle of the syringe. Place the syringe in the furnace.
2. Pass steam through the furnace until the reading of the thermometer and the volume of air in the gas syringe become steady.
3. Record the temperature and volume of air in the syringe.
4. Fill the small hypodermic syringe with about 1 cm^3 of the volatile liquid. Ensure that there is no air in the hypodermic syringe.
5. Find the mass of the hypodermic syringe and its contents.
6. Push the needle of the hypodermic syringe through the self-sealing cap of the gas syringe. Inject about 0.2 cm^3 of the liquid into the gas syringe. Note that when the liquid is injected into the gas syringe, the plunger of the gas syringe is pushed out as the liquid quickly vaporises.
7. Withdraw the hypodermic syringe and immediately reweigh the hypodermic syringe and its contents.
8. Record the final volume of air plus vapour in the gas syringe.
9. Record the atmospheric pressure.

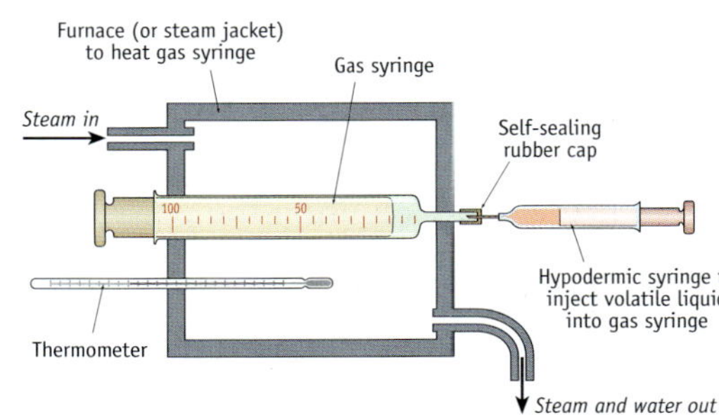

Fig 10.22
Apparatus to measure the relative molecular mass of a volatile liquid using a gas syringe.

Record your results as shown below. To assist you with calculations, some specimen results are shown.

Temperature of furnace	$= 95\ °C + 273$	$= 368$ K
Atmospheric pressure*	= 750 mm Hg	$= 9.997 \times 10^4\ Nm^{-2}$ (Pa)
Initial volume of air in gas syringe		$= 8\ cm^3$
Final volume of air + vapour in gas syringe		$= 74\ cm^3$
Therefore, volume of vapour	$= 74 - 8 = 66\ cm^3$	$= 66 \times 10^{-6}\ m^3$
Initial mass of hypodermic syringe + liquid		= 14.568 g
Final mass of hypodermic syringe + liquid		= 14.429 g
Therefore, mass of liquid injected	= 14.568 - 14.429	= 0.139 g

* As many barometers are marked in mm Hg, this reading is converted to Nm^{-2} as follows:

$760\ mm\ Hg = 1.013 \times 10^5\ Nm^{-2}$

$$1\ mm\ Hg = \frac{1.013 \times 10^5}{760}\ Nm^{-2}$$

$$750\ mm\ Hg = \frac{750 \times 1.013 \times 10^5}{760}\ Nm^{-2} = 9.997 \times 10^4\ Nm^{-2}$$

Calculations

Substitute the above values into the Equation of State

$pV = nRT \quad [R = 8.31 \text{ J mol}^{-1}\text{K}^{-1}]$

$$n = \frac{pV}{RT}$$

$$= \frac{9.997 \times 10^4 \times 66 \times 10^{-6}}{8.31 \times 368} = 0.00216 \text{ moles}$$

i.e. 0.00216 moles of liquid = 0.139 g

$$1 \text{ mole of liquid} = \frac{0.139}{0.00216} = 64.35 \text{ g}$$

i.e. relative molecular mass of liquid = 64.35

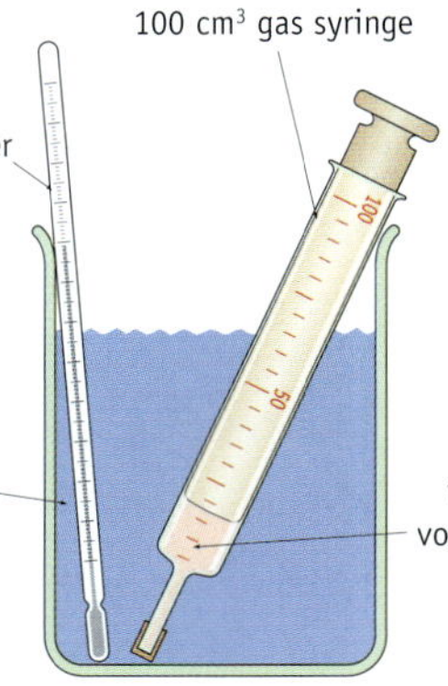

Fig 10.23
A large beaker of boiling water may be used instead of the furnace.

Note:

1. If a furnace is not available for the above experiment, a large beaker containing boiling water may be used instead. (FIG 10.23)
2. In Methods A and B of EXPERIMENT 10.1, there are a number of sources of experimental error. The main source lies in the accuracy of the measuring instruments. The gas syringe can only be read to an accuracy of 0.5 cm^3. The laboratory balance may only read to an accuracy of one or two decimal places. Another source of error is that there may be bubbles of air in the hypodermic syringe. When using the equation $PV = nRT$, we are assuming that the vapour is behaving as an ideal gas. This is not a perfectly valid assumption. Therefore, do not be disappointed if you do not get exactly the 'right' answer for the relative molecular mass!

Test Yourself:
Attempt questions W10.10 – W10.12.

CHAPTER SUMMARY

- Boyle's Law states that, at constant temperature, the volume of a fixed mass of gas is inversely proportional to its pressure.
- Charles' Law states that, at constant pressure, the volume of a fixed mass of gas is directly proportional to its temperature measured on the Kelvin scale.
- The Combined Gas Law (or General Gas Law) states that: $\frac{p_1V_1}{T_1} = \frac{p_2V_2}{T_2}$
- Gay-Lussac's Law of Combining Volumes states that the volumes of reacting gases and the volumes of any gaseous products are in the ratio of small whole numbers provided the volumes are measured at the same temperature and pressure.
- Avogadro's Law states that equal volumes of gases contain equal numbers of molecules, under the same conditions of temperature and pressure.
- An ideal gas is one that obeys all the assumptions of the kinetic theory of gases under all conditions of temperature and pressure.
- The equation $pV = nRT$ is called the Equation of State for an Ideal Gas or the Ideal Gas Equation.
- The relative molecular mass of a volatile liquid may be measured by filling a conical flask with its vapour or by using a gas syringe.

EXERCISES

Given: standard temperature = 273 K, standard pressure = 100 kPa = 1×10^5 Pa, R = 8.31 J mol^{-1} K^{-1}.

10.1. A student collected 75 cm^3 of hydrogen from the reaction between zinc and sulfuric acid at 20 °C and 900 kPa. What volume would this gas occupy at s.t.p.?

10.2. A sample of hydrogen of volume 100 cm^3 at a pressure of 1×10^5 Pa is compressed to 55 cm^3 at constant temperature. What is the new pressure of the gas?

10.3. A sample of helium gas occupies a volume of 5 litres at a pressure of 9.2×10^4 N m^{-2}. Calculate the volume that the gas will occupy if the pressure is increased to 1.01×10^5 N m^{-2} at constant temperature.

10.4. A chemical plant produced 2500 litres of ammonia at a pressure of 500 kPa and a temperature of 100 °C. A chemical engineer wished to store the ammonia at a pressure of 100 kPa and a temperature of 20 °C. What volume would the gas occupy at this new pressure and temperature?

10.5. 50 cm^3 of nitrogen gas combine with 150 cm^3 of hydrogen gas to form 100 cm^3 of ammonia. Are these experimental facts in keeping with Gay-Lussac's Law of Combining Volumes? Explain your answer.

10.6. Given that the molar volume is 22.4 L at s.t.p, calculate the density of each of the following gases at s.t.p. : (a) ammonia, NH_3; (b) nitrogen, N_2; (c) chlorine, Cl_2; (d) methane, CH_4; (e) oxygen, O_2; (f) ozone, O_3. [H = 1, C = 12, N = 14, O = 16, Cl = 35.5]

10.7. 200 cm^3 of a certain gas at s.t.p. have a mass of 1 g. Calculate the relative molecular mass of the gas.

10.8. A student carried out an experiment to measure the relative molecular mass of a volatile liquid by filling a conical flask with the vapour of the liquid.

(a) What is meant by the term 'volatile liquid'?

(b) Give an example of one volatile liquid that could be used in this experiment.

(c) Outline the main steps involved in carrying out this experiment.

(d) Calculate the relative molecular mass of the liquid from the following results:

Mass of empty flask, foil and band	= 108.41 g
Mass of flask, foil, band and condensed liquid	= 109.72 g
Volume of flask	= 332 cm^3
Atmospheric pressure	= 1.2×10^5 Pa
Temperature of vapour in flask	= 100 °C

(Take standard pressure as 1×10^5 Pa.)

10.9. State the assumptions of the kinetic theory of gases. What is an ideal gas? Under what conditions of temperature and pressure are deviations from ideal behaviour most marked? Which assumptions of the kinetic theory of gases are most likely to be false under these conditions?

10.10. Calculate the volume that 20 g of oxygen gas will occupy at 25 °C and 20 kPa pressure.

10.11. Calculate the relative molecular mass of a volatile liquid from the following information: 0.16 g of X on vaporisation gave 46 cm^3 of vapour measured at 100 °C and a pressure of 101.325 kPa.

10.12. A sample of chloroform of mass 0.495 g gave 127 cm^3 when vaporised at 98 °C at a pressure of 100.5 kPa. Calculate the relative molecular mass of chloroform.

11 Stoichiometry I

11.1 INTRODUCTION

Consider the reaction between hydrogen gas and oxygen gas to form water. This reaction may be represented by the following balanced equation:

$$2H_2 + 0_2 \rightarrow 2H_20$$

We can interpret this equation in a number of ways. Firstly, it tells us that **two** hydrogen molecules react with **one** oxygen molecule to form **two** molecules of water.

$2H_2$	+ O_2	$\longrightarrow$	$2H_2O$
2 molecules	1 molecule		2 molecules

Secondly, we can interpret it in terms of the Avogadro's number of molecules that react. If we multiply the number of molecules in the above balanced equation by the Avogadro constant, we now have:

$2H_2$	+ O_2	$\longrightarrow$	$2H_2O$
$2 \times 6 \times 10^{23}$ molecules	$1 \times 6 \times 10^{23}$ molecules		$2 \times 6 \times 10^{23}$ molecules

We have already seen that the amount of a substance which contains the Avogadro number of particles is called a **mole** of that substance. Therefore, the above equation may also be written as:

$2H_2$	+ O_2	$\longrightarrow$	$2H_2O$
2 moles	1 mole		2 moles

By converting moles to grams, we can also tell the masses of these substances that react and the mass of water that is formed.

$2H_2$	+ O_2	$\longrightarrow$	$2H_2O$
2 moles	1 mole		2 moles
i.e. (2×2) g	(1×32) g		(2×18) g
4 grams	**32 grams**		**36 grams**

The branch of chemistry that tells us the relationship between the amounts of reactants and products in a chemical reaction is called **stoichiometry.** (This comes from the Greek *stocheoin* meaning 'element' and *metron* meaning 'measure'.) A knowledge of stoichiometry is important for chemists because it enables us to decide on the correct quantities of substances to react together to form a certain amount of product.

We now consider the main types of calculations required in this section of the course.

11.2 TO CALCULATE THE PERCENTAGE OF EACH ELEMENT IN A COMPOUND

Very often, it is important to be able to calculate the percentage of an element in a compound. For example, a farmer may wish to know the percentage of nitrogen in a particular fertiliser or a chemical engineer may need to calculate the percentage of a metal in an ore. The method of performing this type of calculation is shown in the examples that follow.

Example 11.1

Calculate the percentage composition by mass of each element in calcium oxide, CaO.
(O = 16, Ca = 40)

SOLUTION

First calculate the relative 'molecular' mass of CaO.

Ca	= 40
O	= 16
Rel Molecular Mass	= 56

$$\text{Percentage of Ca} = \frac{40}{56} \times 100 = 71.43\%$$

$$\text{Percentage of O} = \frac{16}{56} \times 100 = 28.57\%$$

Answer:

The percentage composition of CaO by mass is 71.43% Ca and 28.57% O.

Example 11.2

Ammonium nitrate, NH_4NO_3, is often used in fertilisers to supply nitrogen to the soil. What is the percentage of nitrogen in ammonium nitrate?
(H = 1, N = 14, O = 16)

SOLUTION

First calculate the relative 'molecular' mass of ammonium nitrate. Note that in one molecule of ammonium nitrate, there are two atoms of nitrogen, four atoms of hydrogen and three atoms of oxygen:

$$\begin{aligned} 2\text{ N} &= 2 \times 14 &= 28 \\ 4\text{ H} &= 4 \times 1 &= 4 \\ 3\text{ O} &= 3 \times 16 &= 48 \\ \hline \text{Rel Molecular Mass} & &= 80 \end{aligned}$$

$$\text{Percentage of N} = \frac{28}{80} \times 100 \quad = 35\%$$

Answer:

The percentage of nitrogen is 35%

Example 11.3

Calculate the percentage of water of crystallisation in copper(II) sulfate crystals, $CuSO_4.5H_2O$. (H =1, O = 16, S = 32, Cu = 63.5)

Note: Water of crystallisation is water that is found chemically combined in some crystals. Substances containing water of crystallisation are usually referred to as **hydrated**. If there is no water of crystallisation present, the substance is said to be **anhydrous**.

SOLUTION

First calculate the relative molecular mass of $CuSO_4.5H_2O$. Note that you must include the five water molecules in your calculation.

$$\begin{aligned} 1\text{ Cu} &= 1 \times 63.5 &= 63.5 \\ 1\text{ S} &= 1 \times 32 &= 32 \\ 4\text{ O} &= 4 \times 16 &= 64 \\ 5\text{ H}_2\text{O} &= 5 \times 18 &= 90 \\ \hline \text{Rel Molecular Mass} & &= 249.5 \end{aligned}$$

$$\text{Percentage of water} = \frac{90}{249.5} \times 100 = 36.07\%$$

Answer:

The percentage of water of crystallisation is 36.07%

Test Yourself:
Attempt questions 11.1 – 11.2 and W11.1 – W11.2.

11.3 TO CALCULATE THE EMPIRICAL FORMULA OF A COMPOUND FROM ITS PERCENTAGE COMPOSITION

We have already seen that a compound is represented by a chemical formula, e.g. a molecule of water is represented by the formula H_2O. This is called the **molecular formula** of the compound. It tells us that a molecule of water consists of two atoms of hydrogen and one atom of oxygen, i.e. the molecular formula tells us the actual number of atoms of each element present.

The **empirical formula** of a compound is the formula showing the simplest whole number ratio of the numbers of different atoms present in the molecule. For example, the molecular formula of ethene is C_2H_4 but the empirical formula is CH_2. Similarly, the molecular formula of hydrogen peroxide is H_2O_2 but the empirical formula is HO. It should be clear that the molecular formula is a simple multiple of the empirical formula.

Molecular formula = empirical formula × n
where n = whole number.

Example 11.4

Glucose is the name of the sugar found in blood. The empirical formula of glucose is CH_2O. When the relative molecular mass of glucose is measured by a mass spectrometer, it is found to be 180. What is the molecular formula of glucose?

SOLUTION

$$\begin{aligned} \text{Molecular formula} &= \text{empirical formula} \times n \\ &= (CH_2O) \times n \\ \text{i.e.} \quad 180 &= (12 + 2 + 16) \times n \\ 180 &= 30 \times n \\ n &= 6 \end{aligned}$$

$$\begin{aligned} \text{Therefore, molecular formula} &= (CH_2O) \times 6 \\ &= C_6H_{12}O_6 \end{aligned}$$

Answer: $C_6H_{12}O_6$

Very often research chemists and chemists involved in quality control need to have compounds analysed. One of the methods of doing this is to determine the empirical formula of the compound using the combustion apparatus shown in FIG 11.1. This apparatus enables the chemist to determine accurately the percentage of each of the elements present in the compound. Using this information, it is possible to work out the empirical formula of the compound. The method is illustrated in EXAMPLE 11.5.

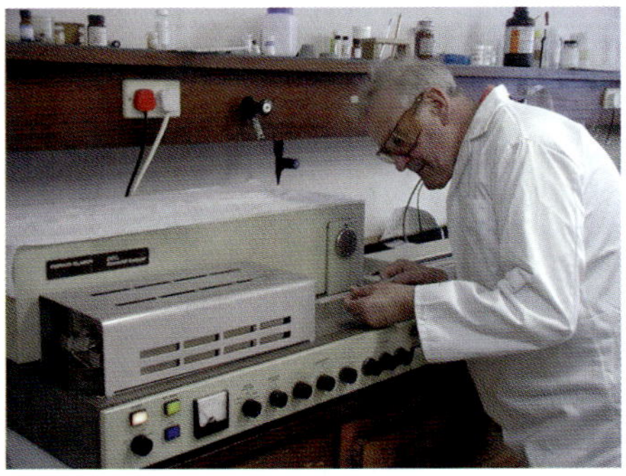

Fig 11.1
This apparatus is used to determine the empirical formula of a compound. A few milligrams of the material to be analysed is placed on a small platinum 'boat' inside the glass tube and is burned in an oven inside the instrument.

Example 11.5

A compound containing only carbon, hydrogen and oxygen was found on analysis to contain 52.17% carbon, 13.04% hydrogen and 34.79% oxygen. What is the empirical formula of the compound? Using a mass spectrometer, it was found that the relative molecular mass of the compound was 92. What is the molecular formula? [H = 1, C = 12, O = 16]

SOLUTION

The basic idea behind solving this problem is to realise that if we have 100 grams of the compound, 52.17 grams will be due to carbon atoms, 13.04 grams will be due to hydrogen atoms, and 34.79 grams will be oxygen atoms. If we convert each of these masses to moles of atoms (by dividing by the appropriate relative atomic mass), we will then have a mole ratio. Dividing by the **smallest** mole ratio will give us the simplest whole number ratio, i.e. the empirical formula. Lay out your work in the form of a table as follows:

Element	Percentage	$\frac{\text{Percentage}}{\text{Rel atomic mass}}$	Simplest ratio (divide by 2.17)
Carbon	52.17	$\frac{52.17}{12} = 4.35$	$\frac{4.35}{2.17} = 2$
Hydrogen	13.04	$\frac{13.04}{1} = 13.04$	$\frac{13.04}{2.17} = 6$
Oxygen	34.79	$\frac{34.79}{16} = 2.17$ ← *smallest number*	$\frac{2.17}{2.17} = 1$

i.e. empirical formula = C_2H_6O

We now find out the molecular formula as follows :

$$\text{empirical formula} \times n = \text{molecular formula}$$

$$\text{i.e.} \quad (C_2H_6O) \times n = 92$$

$$[2(12) + 6(1) + 1(16)] \times n = 92$$

$$46n = 92$$

$$n = 2$$

Therefore, molecular formula = $(C_2H_6O) \times 2 = C_4H_{12}O_2$

Answer: $C_4H_{12}O_2$

Note: In some cases, dividing by the smallest number obtained in the third column of the above table may not give whole numbers in the fourth column. To get whole numbers, it may be necessary to multiply by a certain factor. This is illustrated in EXAMPLE 11.6.

Example 11.6

A compound containing only carbon, hydrogen and oxygen was analysed and found to contain 65.11% carbon, 8.83% hydrogen and 26.06% oxygen. What is the empirical formula of the compound?

SOLUTION

Element	Percentage	$\frac{\text{Percentage}}{\text{Rel atomic mass}}$	Simplest ratio (divide by 1.63)	Multiply by 3 to get whole numbers
Carbon	65.11	$\frac{65.11}{12} = 5.43$	$\frac{5.43}{1.63} = 3.33$	10
Hydrogen	8.83	$\frac{8.83}{1} = 8.83$	$\frac{8.83}{1.63} = 5.42$	16
Oxygen	26.06	$\frac{26.06}{16} = 1.63$ (smallest number)	$\frac{1.63}{1.63} = 1$	3

i.e. empirical formula = $C_{10}H_{16}O_3$

Answer: $C_{10}H_{16}O_3$

Note: Multiplying the numbers 3.33 and 5.42 by 3 does not give whole numbers exactly. This is due to slight experimental error or rounding off in the percentages of the elements given. However, the deviation is so small that it may be ignored.

Test Yourself:
Attempt questions 11.3 – 11.6 and W11.3 – W11.5.

11.4 TO CALCULATE THE EMPIRICAL FORMULA OF A COMPOUND GIVEN THE MASSES OF REACTANTS AND PRODUCTS

In many cases, it is possible to calculate the formula of a compound by carefully measuring the masses of the substances involved in the reaction. Analysis of compounds based on the masses that react is usually referred to as **gravimetric analysis**. Some examples of this type of calculation now follow.

Example 11.7

1.02 g of copper powder was heated in a stream of oxygen gas. The brown copper powder turned black. It was found that the mass of black powder formed was 1.28 g. If the black powder is an oxide of copper, calculate the empirical formula of this oxide. (Cu = 63.5, O = 16)

SOLUTION

copper + oxygen ⟶ copper oxide
1.02 g   1.28 g

mass of copper = 1.02 g

mass of copper oxide formed = 1.28 g

∴ mass of oxygen in copper oxide = 1.28 - 1.02 = 0.26 g

No. of moles of copper atoms in copper oxide = $\frac{1.02}{63.5} = 0.016$

and no. of moles of oxygen atoms in copper oxide = $\frac{0.26}{16} = 0.016$

Ratio of Cu : O = 0.016 : 0.016 = 1 : 1

Therefore, empirical formula of copper oxide is CuO.

Thus, the following reaction has occurred:

$$Cu + \frac{1}{2}O_2 \longrightarrow CuO$$

Answer: CuO

Example 11.8

A student heated 1.61 g of hydrated sodium sulfate crystals, $Na_2SO_4.xH_2O$, in a crucible over a Bunsen burner until no further loss in mass occurred. It was found that 0.71 g of the anhydrous salt was obtained. Calculate the formula of the hydrated salt. (H = 1, O = 16, Na = 23, S = 32)

SOLUTION

Heating the crystals converts the hydrated sodium sulfate to anhydrous sodium sulfate.

$$\begin{array}{lll} Na_2SO_4.xH_2O & \longrightarrow Na_2SO_4 & + \ xH_2O \\ 1.61\text{ g} & \quad 0.71\text{ g} & \end{array}$$

Mass of hydrated crystals = 1.61 g

Mass of anhydrous salt = 0.71 g

Therefore, mass of water = 1.61 - 0.71 = 0.90 g

Rel Molecular Mass Na_2SO_4 = 2(23) + 32 + 4(16)
= 142

No. of moles of $Na_2SO_4 = \frac{0.71}{142} = 0.005$

No. of moles of water $= \frac{0.90}{18} = 0.05$

Therefore, ratio $Na_2SO_4 : H_2O = 0.005 : 0.05 = 1 : 10$

i.e. formula of hydrated salt is $Na_2SO_4.10H_2O$

Answer: $Na_2SO_4.10H_2O$

Example 11.9

When 1.2 g of metal M were reacted with oxygen, it was found that 2.0 g of the oxide MO were formed. Calculate the relative atomic mass of M.

SOLUTION

$$\begin{array}{lll} M & + \frac{1}{2}O_2 \longrightarrow & MO \\ 1.2\text{ g} & & 2.0\text{ g} \end{array}$$

Mass of metal = 1.2 g

Mass of metal oxide = 2.0 g

$\therefore$ mass of oxygen in metal oxide = 2.0 – 1.2 = 0.8 g

No. of moles of oxygen atoms $= \frac{0.8}{16} = 0.05$

Since in the compound MO, the ratio M : O = 1 : 1

then the number of moles of M must also be 0.05.

i.e. 0.05 moles of M = 1.2 g

1 mole of M $= \frac{1.2}{0.05} = 24$

i.e. relative atomic mass of M = 24 (The metal is magnesium).

Answer: Relative atomic mass of M = 24.

Test Yourself:
Attempt questions 11.7 – 11.10 and W11.6 – W11.8.

11.5 CALCULATING MASSES OF REACTANTS OR PRODUCTS FROM BALANCED CHEMICAL EQUATIONS

We have already seen in the introduction to this chapter that a balanced equation tells us the amounts of substances that react together and the amounts of products formed. This type of information is very useful for chemists and enables them to perform calculations to determine the amounts of reactants needed to produce a certain amount of product. We now consider some of these types of calculations.

Example 11.10

Magnesium reacts with oxygen to produce magnesium oxide according to the equation:

$2Mg + O_2 \longrightarrow 2MgO$

If a student burns 9 g of magnesium in excess oxygen (i.e. there is plenty of oxygen present to ensure that **all** of the magnesium reacts), what mass of magnesium oxide will be formed?

(Mg = 24, O = 16)

SOLUTION

Firstly, write down the given equation and circle what **we know** and what **we want to know**.

$\boxed{2Mg} + O_2 \longrightarrow \boxed{2MgO}$

Now make a shortened version of the equation and put in the amounts:

$2Mg \longrightarrow 2MgO$

2 moles $\longrightarrow$ 2 moles

2(24) g $\longrightarrow$ 2 (24 + 16) g

i.e. 48 g Mg $\longrightarrow$ 80 g MgO

Therefore, 1 g Mg $\longrightarrow \frac{80}{48}$ g MgO

Therefore, 9 g Mg $\longrightarrow \frac{9 \times 80}{48}$ g MgO

$= 15$ g MgO

Answer: 15 g of magnesium oxide will be formed from 9 g of Mg

Example 11.11

A student is told to make 100 g of magnesium oxide by burning magnesium in excess oxygen. How much magnesium should the student burn? Magnesium reacts with oxygen according to the equation:

$2Mg + O_2 \longrightarrow 2MgO$

(Mg = 24, O = 16)

SOLUTION

Write down the given equation and circle what **we know** and what **we want to know**.

$\boxed{2Mg} + O_2 \longrightarrow \boxed{2MgO}$

Now make a shortened version of the equation and put in the amounts:

$2Mg \longrightarrow 2MgO$

2 moles $\longrightarrow$ 2 moles

2(24) g $\longrightarrow$ 2 (24 + 16) g

i.e. 48 g Mg $\longrightarrow$ 80 g MgO

$\frac{48}{80}$ g Mg $\longrightarrow$ 1 g MgO

$\frac{48 \times 100}{80}$ g Mg $\longrightarrow$ 100 g MgO

i.e. 60 g Mg $\longrightarrow$ 100 g MgO

Answer: The student should burn 60 g of magnesium.

Example 11.12

Lime (calcium oxide) is manufactured in a lime kiln by heating limestone (calcium carbonate) to bring about the following reaction:

$CaCO_3 \longrightarrow CaO + CO_2$

Calculate the amount of lime that could theoretically be obtained by heating 5 kg of limestone.

(C = 12, Ca = 40, O = 16)

SOLUTION

This problem is fairly straightforward if one remembers to convert kilograms to grams.

$\boxed{CaCO_3} \longrightarrow \boxed{CaO} + CO_2$

Now make a shortened version of the equation and put in the amounts:

$CaCO_3 \longrightarrow CaO$
1 mole $\longrightarrow$ 1 mole
[40 + 12 + 3(16)] g $\longrightarrow$ [40 + 16] g
i.e. 100 g $CaCO_3 \longrightarrow$ 56 g CaO
1 g $CaCO_3 \longrightarrow \frac{56}{100}$ g CaO
Therefore, 5000 g $CaCO_3 \longrightarrow \frac{5000 \times 56}{100}$ g CaO
= 2800 g
= 2.8 kg.

Answer: 2.8 kg of lime will be obtained from 5 kg limestone.

Test Yourself:
Attempt questions 11.11 – 11.13 and W11.9 – W11.10.

11.6 CALCULATING VOLUMES OF GASES FROM BALANCED CHEMICAL EQUATIONS.

We have already learned that at s.t.p. one mole of any gas occupies a volume of 22.4 litres. We use this fact to calculate the volumes of gases used up or formed in chemical reactions. In calculating volumes of gases from balanced chemical equations, proceed as in EXAMPLES 11.13 – 11.15 but remember that, instead of measuring the **mass** of a gas, chemists usually measure the volume of a gas. The following examples will make this clear.

Example 11.13

Carbon dioxide is formed when carbon is burned in air according to the equation:
$C + O_2 \longrightarrow CO_2$
What volume of carbon dioxide (at s.t.p.) is formed when 30 g of carbon are burned in excess oxygen?

SOLUTION

$(C) + O_2 \longrightarrow (CO_2)$

Now make a shortened version of the equation and put in the amounts:
$C \longrightarrow CO_2$
1 mole $\longrightarrow$ 1 mole
12 g C $\longrightarrow$ 22.4 L of CO_2 at s.t.p.
1 g C $\longrightarrow \frac{22.4}{12}$ L of CO_2
30 g C $\longrightarrow \frac{30 \times 22.4}{12}$ L of CO_2
= 56 L of carbon dioxide.

Answer: 56 litres.

Example 11.14

Hydrogen gas may be prepared by reacting zinc metal with dilute sulfuric acid. The following reaction occurs:
$Zn + H_2SO_4 \longrightarrow ZnSO_4 + H_2$
What volume of hydrogen gas at s.t.p. is evolved when 7 g of Zn are added to excess dilute sulfuric acid?

SOLUTION

$(Zn) + H_2SO_4 \longrightarrow ZnSO_4 + (H_2)$

Now make a shortened version of the equation and put in the amounts:
$Zn \longrightarrow H_2$
1 mole $\longrightarrow$ 1 mole
65 g Zn $\longrightarrow$ 22.4 L of H_2 at s.t.p.
1 g Zn $\longrightarrow \frac{22.4}{65}$ L of H_2
7 g Zn $\longrightarrow \frac{7 \times 22.4}{65}$ L of H_2
= 2.41 L of H_2

Answer: 2.41 litres.

Example 11.15

Methane (natural gas) burns in air to form carbon dioxide and water according to the equation:

$CH_4 + 2O_2 \longrightarrow CO_2 + 2H_2O$

What volume of oxygen, measured at s.t.p. would be required for the complete combustion of 25 g of methane?

Solution

$CH_4 + 2O_2 \longrightarrow CO_2 + 2H_2O$

1 mole $= 16$ g　　2 moles $= 2 \times 22.4$ L $= 44.8$ L

i.e. 16 g of methane requires 44.8 litres of oxygen

Therefore, 1 g of methane requires $\frac{44.8}{16}$ L of O_2

25 g of methane requires $\frac{25 \times 44.8}{16}$ L of O_2

$= 70$ L of O_2

Answer: 70 litres.

Test Yourself:
Attempt questions 11.14 – 11.16 and W11.11 – W11.14.

In the Leaving Certificate examination, you may be asked a question on stoichiometry which contains a number of parts which draw on your knowledge of the previous types of questions. This is illustrated in Example 11.16.

Example 11.16

In a reaction 14.7 g of potassium dichromate(VI) reacted completely with concentrated hydrochloric acid according to the equation:

$K_2Cr_2O_7 + 14HCl \longrightarrow 2KCl + 2CrCl_3 + 7H_2O + 3Cl_2$

(i) How many moles of potassium dichromate(VI) were used?
(ii) How many moles of water were produced?
(iii) What volume of chlorine, measured at s.t.p., was produced?
(iv) How many molecules of chlorine did this volume contain?
(v) If the chlorine produced in the reaction were bubbled through a solution containing excess sodium bromide, calculate the mass of bromine liberated according to the equation:
$Cl_2 + 2NaBr \longrightarrow 2NaCl + Br_2$ [LCH]

Solution

(i) No. of moles of $K_2Cr_2O_7 = \frac{\text{Mass}}{\text{Rel Molecular Mass}} = \frac{14.7}{294} = 0.05$ mole

(ii) From the balanced equation,

1 mole $K_2Cr_2O_7 \longrightarrow 7$ moles H_2O

0.05 mole $K_2Cr_2O_7 \longrightarrow 7 \times 0.05$ mole H_2O

$= 0.35$ mole H_2O

(iii) From the balanced equation,

1 mole $K_2Cr_2O_7$ $\longrightarrow$ 3 moles Cl_2

0.05 mole $K_2Cr_2O_7$ $\longrightarrow$ 3×0.05 mole Cl_2

= 0.15 mole Cl_2

= 0.15×22.4 L

= 3.36 L

(iv) 22.4 L Cl_2 contains 6×10^{23} molecules

1 L Cl_2 contains $\frac{6 \times 10^{23}}{22.4}$ molecules

3.36 L Cl_2 contains $\frac{3.36 \times 6 \times 10^{23}}{22.4}$ molecules

= 9×10^{22} molecules

(v) $Cl_2 + 2NaBr \longrightarrow 2NaCl + Br_2$

From part (iii), 0.15 mole Cl_2 is produced

1 mole Cl_2 $\longrightarrow$ 1 mole Br_2

0.15 mole Cl_2 $\longrightarrow$ 0.15 mole Br_2

= 0.15×160 g Br_2 (M_r Br_2 = 160)

= 24 g Br_2

Answers: (i) 0.05 mole (ii) 0.35 mole (iii) 3.36 L (iv) 9×10^{22} molecules (v) 24 g

Test Yourself:
Attempt questions W11.15 – W11.20.

EXERCISES

When answering these questions, use the following values of relative atomic masses: H = 1, He = 4, Li = 7, C = 12, N = 14, O =16, F = 19, Na = 23, Mg = 24, Al = 27, Si = 28, P = 31, S = 32, Cl = 35.5, K = 39, Ca = 40, Cr = 52, Mn = 55, Fe = 56, Cu = 63.5, Zn = 65, Br = 80, Ag = 108, Sn = 119, I = 127, Pb = 207.

Percentage Composition

11.1. Calculate the percentage by mass of each element in the following compounds:

(a) magnesium oxide, MgO
(b) aluminium oxide, Al_2O_3
(c) magnesium sulfate, $MgSO_4$
(d) limestone, $CaCO_3$
(e) sodium chloride, NaCl
(f) ammonium sulfate, $(NH_4)_2SO_4$
(g) ethanedioic acid, $(COOH)_2$
(h) ethanoic acid, CH_3COOH
(i) lead nitrate, $Pb(NO_3)_2$
(j) potassium hydrogencarbonate, $KHCO_3$.

11.2. Which of the three fertilisers ammonium nitrate [NH_4NO_3], ammonium sulfate, [$(NH_4)_2SO_4$] and urea [$(NH_2)_2CO$] contains the highest percentage of nitrogen?

Empirical and Molecular Formulas

11.3. Write down the empirical formulas of the following compounds: C_6H_6, $C_2H_4O_2$, $C_3H_6O_3$.

11.4. Urea is the name of a substance found in urine. The empirical formula of urea is N_2H_4CO. The relative molecular mass of urea, when measured by a mass spectrometer, is found to be 60. What is the molecular formula of urea? Draw a possible 'electron dot' structure for a molecule of urea.

11.5. Calculate the empirical formulas of the compounds containing:

(a) nitrogen 47%, oxygen 53%
(b) nitrogen 30.5%, oxygen 69.5%
(c) carbon 75%, hydrogen 25%
(d) carbon 85.7%, hydrogen 14.3%
(e) iron 34.5%, chlorine, 65.5%
(f) carbon 92.3%, hydrogen 7.7%
(g) sulfur 40%, oxygen 60%
(h) sulfur 50%, oxygen 50%
(i) potassium 40.2%, chromium 26.9%, oxygen 32.9%
(j) potassium 26.6%, chromium 35.4%, oxygen 38%
(k) copper 34.6%, iron 30.5%, sulfur 34.9%

11.6. A sample of washing soda was analysed and found to have the following composition: carbon 4.2%, oxygen 16.8%, sodium 16.1%, water of crystallisation 62.9%. Determine the empirical formula of washing soda.

Gravimetric Analysis

11.7. When 2.0 g of calcium was completely burnt in oxygen, 2.8 g of calcium oxide was obtained. What is the empirical formula of calcium oxide?

11.8. When 2.70 g of aluminium combined with oxygen it was found that 5.10 g of aluminium oxide was formed. What is the empirical formula of aluminium oxide?

11.9. A student heated 4.10 g of hydrated magnesium sulfate, $MgSO_4.xH_2O$, and found that 2.00 g of the anhydrous salt remained. Calculate the degree of hydration in the hydrated salt, i.e. calculate the value of x.

11.10. When 5.30 g of a metal oxide, MO, was reduced, 4.24 g of the metal was formed. Calculate the relative atomic mass of M.

Calculations from balanced equations.

11.11. Iron reacts with chlorine to form iron(III) chloride according to the equation:

$2Fe + 3Cl_2 \rightarrow 2FeCl_3$

What mass of iron(III) chloride can be obtained from 20 g iron?

11.12. The following reaction occurs during the manufacture of sulfuric acid:

$2SO_2 + O_2 \longrightarrow 2SO_3$

Assuming that the sulfur dioxide is completely converted to sulfur trioxide, calculate the mass of sulfur trioxide that could be produced from 200 g of sulfur dioxide.

11.13. When glucose undergoes fermentation, ethanol and carbon dioxide are formed. This process may be represented by the equation:

$C_6H_{12}O_6 \longrightarrow 2C_2H_5OH + 2CO_2$

Assuming that the reaction goes to completion, calculate the mass of ethanol that could be obtained from 25 g of glucose.

Volumes of gases.

11.14. Calculate the volume of carbon dioxide formed at s.t.p. when 2.5 g of limestone (calcium carbonate) is heated. The reaction that occurs is:

$CaCO_3 \longrightarrow CaO + CO_2$

11.15. Baking powder contains sodium hydrogencarbonate, $NaHCO_3$. On heating this substance, it is decomposed to form carbon dioxide which causes the dough to rise. The equation for the reaction is:

$2NaHCO_3 \longrightarrow CO_2 + Na_2CO_3 + H_2O$.

Calculate the mass of sodium hydrogencarbonate that would be needed to form 800 cm^3 of carbon dioxide at s.t.p.

11.16. Oxygen is prepared by the decomposition of hydrogen peroxide according to the equation:

$2H_2O_2 \longrightarrow 2H_2O + O_2$

Calculate the volume of oxygen formed at s.t.p. by the decomposition of a solution containing 21 g of hydrogen peroxide.

12 Acids And Bases

12.1 INTRODUCTION

There are many chemical reactions that occur in our daily lives. Of these reactions, two types are of great importance: acid-base reactions and oxidation-reduction reactions. In this chapter we shall consider acid-base reactions and in CHAPTER 14 we shall study oxidation-reduction reactions.

Acid-base reactions occur when an antacid is taken to relieve an upset stomach, when a cake is baked, when hair conditioner is used, when lime is spread on soil, etc. In this chapter we will study the general properties of acids and bases and this will help us understand many of the chemical reactions that occur in our everyday lives.

12.2 ARRHENIUS' THEORY OF ACIDS AND BASES

Many people think of acids as dangerous substances. It is true that some acids can be dangerous and must be handled with care, e.g. the sulfuric acid in a car battery. However, many other acids are part of our diet, e.g. acetic acid (ethanoic acid) in vinegar, citric acid in oranges and lemons, carbonic acid in fizzy drinks, etc. Some acids are necessary to keep us healthy, e.g. vitamin C is an acid called ascorbic acid, soap contains the salt of an acid called stearic acid, benzoic acid is used as a food preservative, etc. The word 'acid' comes from the Latin word 'acidus' meaning 'sour' (FIG 12.1).

Fig 12.1
Many substances have a sharp taste due to the presence of acids in them.

As well as having a sour taste, acids are also known to turn moist blue litmus paper red. You may also be aware that acids react with many metals to liberate hydrogen gas, e.g.

$$\text{sulfuric acid} + \text{zinc} \rightarrow \text{zinc sulfate} + \text{hydrogen}$$
$$H_2SO_4 + Zn \rightarrow ZnSO_4 + H_2$$

We have already seen that acids react with carbonates and hydrogencarbonates to liberate carbon dioxide (see SECTION 6.4). But what exactly is an acid? One of the first theories about acids was put forward in 1884 by a Swedish chemist called Svante Arrhenius (FIG 12.2).

Fig 12.2
The Swedish chemist Arrhenius put forward one of the first definitions of acids. He received the Nobel Prize for chemistry in 1903 for his theory about the dissociation of certain substances in solution.

In his thesis for his Ph.D. degree, Arrhenius proposed the idea that some substances dissociate into ions when added to water. He defined an acid as follows:

> Arrhenius' definition of an acid: An acid is a substance that dissociates in water to produce H^+ ions.

For example, when hydrogen chloride is added to water, it dissociates or breaks up into H^+ ions and Cl^- ions:

$$HCl \rightarrow H^+ + Cl^-$$

Similarly, the dissociation of nitric acid and sulfuric acid may be represented as:

$$HNO_3 \rightarrow H^+ + NO_3^-$$
$$H_2SO_4 \rightarrow 2H^+ + SO_4^{2-}$$

A solution of hydrogen chloride in water is commonly called hydrochloric acid. Acids like hydrochloric acid and nitric acid, molecules of which donate **one** H^+ ion in solution, are called **monobasic** acids. Acids like sulfuric acid, a molecule of which donates **two** H^+ ions in solution, are called **dibasic** acids. An acid like phosphoric acid, H_3PO_4, a molecule of which can donate three H^+ ions, is called a **tribasic** acid.

Monobasic: $HCl \rightarrow H^+ + Cl^-$

Dibasic: $H_2SO_4 \rightarrow 2H^+ + SO_4^{2-}$

Tribasic: $H_3PO_4 \rightarrow 3H^+ + PO_4^{3-}$

Acids like hydrochloric acid, sulfuric acid and nitric acid are described as **strong acids** since they fully dissociate in water, i.e. almost every molecule breaks up to give H^+ ions. Acids like ethanoic acid (acetic acid) and methanoic acid (formic acid) are described as **weak acids** since they do not fully dissociate in water.

As knowledge about atomic structure developed, chemists realised that a H^+ ion (a hydrogen atom that had lost an electron) is simply a single proton. Arrhenius did not realise that it was not possible for a bare proton to exist independently in solution. In fact, using X-ray crystallography and diffraction studies, it was discovered that the H^+ ion actually reacts with water molecules to form the H_3O^+ ion. The formation of this ion is shown in Fig 12.3.

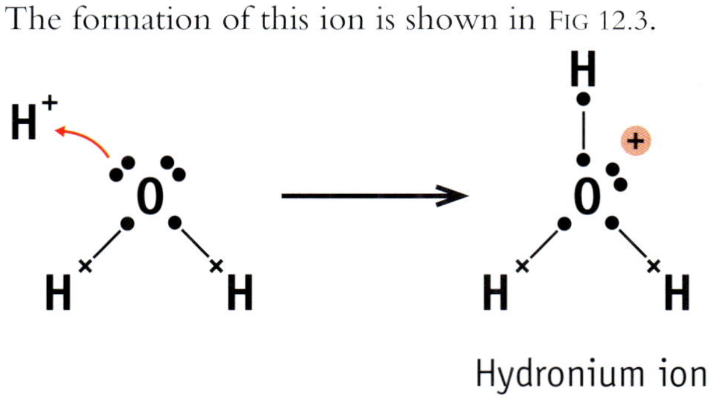

Fig 12.3
The H^+ ion reacts with a water molecule to form the H_3O^+ ion called the hydronium ion (or oxonium ion).

Note in this diagram that one of the lone pairs on the oxygen atom forms a covalent bond with the proton. Both electrons in this bond come from the one atom (the oxygen atom). Such a bond is often called a **dative** covalent bond. (The word 'dative' is a Latin word meaning 'giving' since the same atom gives both of the electrons in the bond.)

There are several names for the H_3O^+ ion, e.g. 'hydronium ion', 'hydroxonium ion', 'oxonium ion', etc. In this textbook we shall use the term 'hydronium ion' since this term is widely used. The hydronium ion is a very common ion – it is present in every acidic solution.

Thus, it is more accurate to represent the dissociation of acids in water (where HA represents an acid) as:

$$HA + H_2O \rightarrow H_3O^+ + A^-$$
$$\text{e.g.}\quad HCl + H_2O \rightarrow H_3O^+ + Cl^-$$

It should be realised that many chemists, for the sake of simplicity, use the term 'hydrogen ion' when talking about acidic solutions – even though hydrogen ions do not exist in solution.

The above equation recognises the essential role of the water molecules in the dissociation of acids. If substances like pure ethanoic acid (acetic acid) or pure hydrogen chloride are added to solvents like benzene or methylbenzene, no H^+ or H_3O^+ ions are found to be present. If a piece of **dry** blue litmus paper is then added, it does not show any colour change. Thus, substances like ethanoic acid and hydrogen chloride are only acidic when dissolved in water.

It is found that there is another group of substances that behave in the opposite way to acids, e.g. they turn litmus indicator from red to blue. Such substances are called bases (Fig 12.4).

Fig 12.4
All of these substances contain bases.

One example of a base is sodium hydroxide, NaOH, commonly called caustic soda, which is used in oven cleaners and to clear blocked drains. Another example of a base is magnesium hydroxide [$Mg(OH)_2$], found in *Milk of Magnesia*. A solution of calcium hydroxide [$Ca(OH)_2$], commonly called 'limewater' is a basic solution which is used to detect the presence of carbon dioxide. Many bases have a soapy feeling because they convert some of the oil in your skin into soap and this makes the skin slippery. Many household cleaners (e.g. oven cleaners, window cleaners, etc.) contain soluble bases as degreasing agents. They convert oil and grease into soluble soaps which are easily washed away.

In his study of bases, Arrhenius proposed the idea that bases dissociate into ions when added to water. He defined a base as follows:

> Arrhenius' definition of a base: A base is a substance that dissociates in water to produce OH^- ions.

The OH^- ion is commonly called the hydroxide ion. For example, when sodium hydroxide is added to water, it dissociates or breaks up into Na^+ ions and OH^- ions:

$$NaOH \rightarrow Na^+ + OH^-$$

Similarly, the dissociation of magnesium hydroxide and calcium hydroxide may be represented as:

$$Mg(OH)_2 \rightarrow Mg^{2+} + 2OH^-$$
$$Ca(OH)_2 \rightarrow Ca^{2+} + 2OH^-$$

Since sodium hydroxide dissociates almost completely in water to give OH^- ions it is referred to as a **strong base**. On the other hand, calcium hydroxide and magnesium hydroxide only partially dissociate in water and hence are referred to as **weak bases**.

Note: A base that dissolves in water is often referred to as an **alkali**.

Although Arrhenius' theory of acids and bases is a reasonably adequate one, it does have a number of shortcomings:

(i) As stated above, it is hydronium ions, rather than the bare H^+ ions suggested by Arrhenius, that exist in solution.

(ii) Arrhenius restricted his definitions to aqueous solutions, i.e. reactions occurring in water. Solvents like liquid ammonia, benzene, methylbenzene (toluene), etc. are excluded from his definitions.

(iii) Not all acid-base reactions require water. For example, when ammonia gas and hydrogen chloride gas come in contact, white fumes of ammonium chloride are formed:

NH_3 +	HCl	$\rightarrow NH_4Cl$
base	acid	ammonium chloride

However, the Arrhenius definitions of acids and bases would not allow us to classify the above reaction as an acid-base reaction since there is no water present.

In view of the limitations of Arrhenius' definitions, we now consider a more modern theory of acids and bases.

12.3 BRØNSTED-LOWRY THEORY OF ACIDS AND BASES

In 1923, Johannes Brønsted, a Danish chemist and Thomas Lowry, an English chemist, independently proposed new definitions of acid and base. Since the properties of acids and bases were, in many respects, opposed to each other, they believed it would make more sense to have definitions which were opposed to each other. These definitions are:

> Brønsted-Lowry definition of an acid: An acid is a proton donor.

> Brønsted-Lowry definition of a base: A base is a proton acceptor.

Let us now consider these definitions in a little more detail. When hydrogen chloride is added to water, we have already learned that the following reaction occurs:

HCl	+	H_2O	→	H_3O^+	+	Cl^-
donates a proton ∴B/L acid*		accepts a proton ∴B/L base				

In this reaction, we see that a molecule of HCl donates a proton to a molecule of water. Since HCl is a proton donor, it is classified as an **acid**. Since the molecule of water accepts a proton, it is classified as a **base**.

Consider what happens when ammonia gas is dissolved in water:

NH_3	+	H_2O	→	NH_4^+	+	OH^-
accepts a proton ∴B/L base		donates a proton ∴B/L acid				

Since the ammonia molecule accepts a H^+ ion from a water molecule, the ammonia molecule is acting as a base and the water molecule is acting as an acid. This reaction is shown in FIG 12.5.

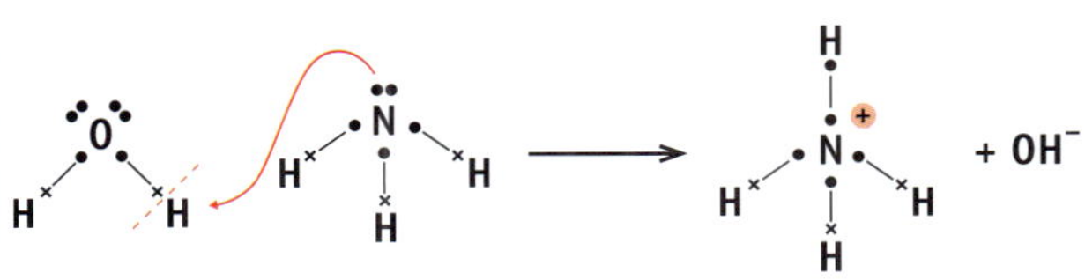

Fig 12.5
The water molecule donates a proton to the ammonia molecule. Thus, the water molecule is acting as an acid and the ammonia molecule is acting as a base.

It is clear from the above two examples that water can act either as an acid or a base. A substance which can act either as an acid or a base is said to be **amphoteric** or **amphiprotic**. The word 'amphi' comes from the Greek meaning 'on both sides'. The word 'protic' refers to protons. Therefore, an amphiprotic substance is one that can donate or accept protons.

The Brønsted-Lowry theory of acids and bases may be applied to acid-base reactions that do not involve water as the solvent. For example, in liquid ammonia, hydrogen chloride and ammonia react as follows:

* In this textbook we will use B/L to represent the term Brønsted-Lowry.

HCl	+	NH_3	→	NH_4^+	+	Cl^-
donates a proton ∴B/L acid		accepts a proton ∴B/L base				

The hydrogen chloride molecule acts as an acid by donating a proton to the ammonia molecule. Since there are no water molecules present, no hydronium ions are formed. Thus, whereas the Arrhenius theory is limited to aqueous solutions, the Brønsted-Lowry theory is not limited in this respect. Another advantage of the Brønsted-Lowry theory over the Arrhenius theory is that the Brønsted-Lowry definitions broaden the range of species that can be defined as acids and bases. Substances that would not be classified as acids or bases in the Arrhenius theory (e.g. NH_3) are classified as acids or bases in the Brønsted-Lowry theory.

Note in the above equations that an acid will only donate a proton if there is something else present to accept this proton from the acid. Similarly, a base can only accept a proton if there is something else present to donate the proton to the base.

Test Yourself:
Attempt questions 12.1 and W12.1.

12.4 CONJUGATE ACID-BASE PAIRS

Ethanoic acid (acetic acid), CH_3COOH, is the acid found in vinegar. When pure ethanoic acid is added to water, the following reaction occurs:

$$CH_3COOH + H_2O \rightarrow CH_3COO^- + H_3O^+$$

It is found that this reaction also takes place in the reverse direction:

$$CH_3COO^- + H_3O^+ \rightarrow CH_3COOH + H_2O$$

We use the symbol ⇌ to indicate that a reaction takes place in both directions:

CH_3COOH	+	H_2O	⇌	CH_3COO^-	+	H_3O^+
donates a proton ∴B/L acid		accepts a proton ∴B/L base				

Consider the reaction in the forward direction. In this reaction, the CH_3COOH molecule donates a proton to the H_2O molecule, i.e. the CH_3COOH molecule is behaving as an acid and the H_2O molecule is behaving as a base.

In the reverse direction, the CH_3COO^- ion accepts a proton from the H_3O^+ ion, i.e. the CH_3COO^- ion is behaving as a base and the H_3O^+ ion is behaving as an acid.

$$CH_3COOH + H_2O \rightleftharpoons CH_3COO^- + H_3O^+$$

		CH_3COO^-	H_3O^+
		accepts a proton	donates a proton
		∴B/L base	∴B/L acid

Note from the above equations that CH_3COOH changes into CH_3COO^- by donating a proton. We say that CH_3COO^- is the **conjugate base** of CH_3COOH.

An acid changes into a conjugate base when it donates a proton.

$$CH_3COOH \rightarrow CH_3COO^- + H^+$$

acid — conjugate base — proton

Also, CH_3COO^- changes into CH_3COOH by accepting a proton. CH_3COOH is said to be the **conjugate acid** of CH_3COO^-.

A base changes into a conjugate acid when it accepts a proton.

$$CH_3COO^- + H^+ \rightarrow CH_3COOH$$

base — proton — conjugate acid

Every acid has a conjugate base and every base has a conjugate acid.

Since CH_3COOH and CH_3COO^- only differ by a proton, we refer to this pair as a **conjugate acid-base pair**.

$$CH_3COOH + H_2O \rightleftharpoons CH_3COO^- + H_3O^+$$

acid — base — base — acid

(CH_3COOH / CH_3COO^-): conjugate acid-base pair

(H_2O / H_3O^+): conjugate acid-base pair

Similarly, H_2O and H_3O^+ are a conjugate acid-base pair since they differ only by one proton.

A conjugate acid-base pair is any pair consisting of an acid and a base which differ by one proton.

Using the formula HA to represent an acid, its conjugate base is represented as A^- and the equation relating them is then written as follows:

$$HA \rightleftharpoons A^- + H^+$$

Note: Since the definitions of a conjugate acid, conjugate base and conjugate acid-base pair all specify a **single** proton, it follows that the conjugate base of H_2SO_4 is HSO_4^- and not SO_4^{2-}.

Example 12.1

What is (i) the conjugate acid, (ii) the conjugate base of $HC_2O_4^-$? [LCH]

ANSWER

(i) A base changes into its conjugate acid by accepting a proton. Therefore, if $HC_2O_4^-$ is to change into its conjugate acid it must accept a proton, i.e. the following reaction must occur:

$$HC_2O_4^- + H^+ \rightarrow H_2C_2O_4$$

base — conjugate acid

Therefore, **the conjugate acid is $H_2C_2O_4$.**

(ii) An acid changes into its conjugate base by donating a proton. Therefore, if $HC_2O_4^-$ is to change into its conjugate base it must donate a proton, i.e. the following reaction must occur:

$$HC_2O_4^- \rightarrow H^+ + C_2O_4^{2-}$$

acid — conjugate base

Therefore, **the conjugate base is $C_2O_4^{2-}$.**

Example 12.2

Indicate which species are acting as acids and which are acting as bases:

$$HNO_3 + H_2F_2 \rightleftharpoons H_2NO_3^{+} + HF_2^{-}$$

Also indicate the conjugate acid-base pairs.

ANSWER

Working from left to right, we see that HNO_3 accepts a proton from H_2F_2 i.e. HNO_3 is acting as a base. The HNO_3 changes into $H_2NO_3^+$ (the conjugate acid). H_2F_2 is acting as an acid since it donates a proton to HNO_3.

Working from right to left, we see that $H_2NO_3^+$ donates a proton to HF_2^-. Therefore, $H_2NO_3^+$ is acting as an acid and HF_2^- is acting as a base.

We indicate the conjugate acid-base pairs by linking up the species that differ by a proton.

HNO_3 +	H_2F_2	$\rightleftharpoons$	$H_2NO_3^+$ +	HF_2^-
base	acid		acid	base

conjugate acid-base pair (HNO_3 / $H_2NO_3^+$)

conjugate acid-base pair (H_2F_2 / HF_2^-)

Note: In an examination it is very important that you clearly indicate the pairing as shown above.

12.5 NEUTRALISATION

When acids and bases react with each other in the right proportions, they 'cancel' each other out, i.e. both the acid and the base lose their characteristic properties. Since the solution formed has no effect on litmus (i.e. the solution is neutral), the process is called neutralisation. Consider the reaction between hydrochloric acid and sodium hydroxide:

$$HCl + NaOH \rightarrow NaCl + H_2O$$

Note that the reaction between the acid and the base produces a salt – in this case, sodium chloride or common table salt. The word '**salt**' is a general term to describe **the substance formed when the hydrogen in an acid is replaced by a metal or an ammonium ion**. Thus, when the hydrogen atom in HCl is replaced by sodium, the salt NaCl is formed. When the H atom in HCl is replaced by NH_4^+, the salt ammonium chloride, NH_4Cl, is formed.

Neutralisation is the reaction between an acid and a base to form a salt and water.

acid + base → salt + water

In order to understand what is happening in a neutralisation reaction, consider the ions which are present in solution.

$$HCl + NaOH \rightarrow NaCl + H_2O$$

i.e. in solution:

$$H^+ + Cl^- + Na^+ + OH^- \rightarrow Na^- + Cl^- + H_2O$$

If we ignore the ions that are present on each side of the equation ('spectator ions'), the above equation becomes:

$$H^+ + \cancel{Cl^-} + \cancel{Na^+} + OH^- \longrightarrow \cancel{Na^+} + \cancel{Cl^-} + H_2O$$

i.e. $$H^+ + OH^- \longrightarrow H_2O$$

Thus, neutralisation occurs because the hydrogen ions (or more precisely the hydronium ions) from the acid react with the hydroxide ions from the base to form water.

There are many examples of neutralisation in our everyday lives:

1. **Medicine**. Hydrochloric acid is produced by the stomach to help digest food. Sometimes the stomach produces too much HCl as a result of over-eating, stress, etc. If this condition persists, the lining of the stomach may be damaged and this gives rise to ulcers. However, relief from excess HCl may be obtained by taking an antacid preparation which contains a base to neutralise the excess HCl. (FIG 12.6)

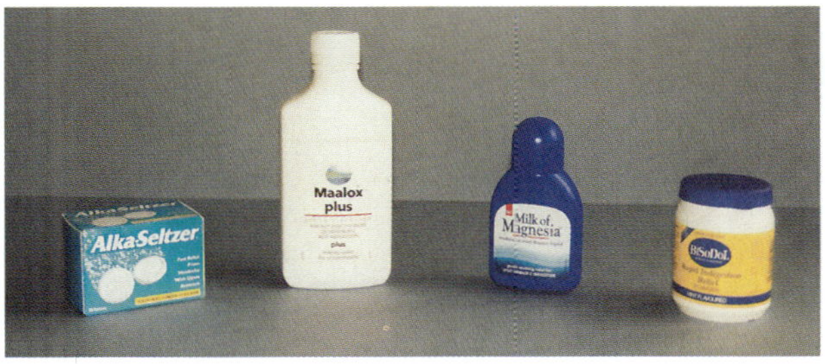

Fig 12.6
Alka-Seltzer and Bisodol contain the base sodium hydrogencarbonate, $NaHCO_3$. Milk of Magnesia and Maalox contain the base magnesium hydroxide, $Mg(OH)_2$.

Alka-Seltzer and *Bisodol* contain sodium hydrogencarbonate which reduces the acidity in the stomach by the following neutralisation reaction:

hydrochloric acid + sodium hydrogen-carbonate → sodium chloride + water + carbon dioxide

$$HCl + NaHCO_3 \rightarrow NaCl + H_2O + CO_2$$

Similarly, *Milk of Magnesia* and *Maalox* contain magnesium hydroxide which reacts with the hydrochloric acid as follows:

hydrochloric acid + magnesium hydroxide → magnesium chloride + water

$$2HCl + Mg(OH)_2 \rightarrow MgCl_2 + 2H_2O$$

2. **Agriculture**. If soil is too acidic, crop yields tend to be low, e.g. sugarbeet grows best in soils which are approximately neutral or slightly basic. Therefore, farmers often spread lime on soil to neutralise the acidity in the soil. Lime is calcium oxide, CaO, also known as quicklime which reacts with water to form calcium hydroxide, $Ca(OH)_2$, commonly called slaked lime.

calcium oxide + water → calcium hydroxide

$$CaO + H_2O \rightarrow Ca(OH)_2$$

Calcium hydroxide neutralises the acidity in the soil as follows:

sulfuric acid + calcium hydroxide → calcium sulfate + water

$$H_2SO_4 + Ca(OH)_2 \rightarrow CaSO_4 + 2H_2O$$

3. **Environmental protection**. In areas that suffer from acid rain, limestone is often added to lakes to neutralise the acidity.

sulfuric acid + limestone → calcium sulfate + water + carbon dioxide

$$H_2SO_4 + CaCO_3 \rightarrow CaSO_4 + H_2O + CO_2$$

Limestone is often used in the tall chimneys of coal burning power plants. The limestone neutralises the acidic substances formed as a result of burning coal thereby reducing harmful emissions.

4. **Miscellaneous**. Toothpaste is slightly basic to neutralise the acids in the food which cause tooth decay. Baking soda (sodium hydrogencarbonate) is used to neutralise the acidic sting of bees. Vinegar is used to neutralise the alkaline sting of wasps. Shampoo is slightly basic and this can cause the scales that coat the surface of each hair to open out. Hair conditioner is slightly acidic to neutralise the base in shampoo. This causes the scales on each hair to close down and leaves the hair more manageable and shiny. (FIG 12.7)

Fig 12.7
The reaction between hair conditioner and shampoo is an example of a neutralisation reaction.

Test Yourself:
Attempt questions 12.2 – 12.6 and W12.2 – W12.4.

CHAPTER SUMMARY

- The Arrhenius definition of an acid is that it is a substance which dissociates in water to produce H^+ ions.
- The Arrhenius definition of a base is that it is a substance which dissociates in water to produce OH^- ions.
- The Brønsted-Lowry definition of an acid is that an acid is a proton donor.
- The Brønsted-Lowry definition of a base is that a base is a proton acceptor.
- An acid changes into a conjugate base when it donates a proton.
- A base changes into a conjugate acid when it accepts a proton.
- A conjugate acid-base pair is defined as any pair consisting of an acid and a base which differ by one proton.
- Neutralisation is the reaction between an acid and a base to form a salt and water.

EXERCISES

12.1. (a) How do acids and bases differ with regard to their effect on litmus? ....

(b) A solution of hydrogen chloride in water is called..

(c) Explain why bases tend to have a soapy feeling...

(d) Write down the Arrhenius definition of an acid...

(e) Write an equation to show how the HSO_4^- ion can act as an Arrhenius acid.....................

(f) Write an equation to show how the HSO_4^- ion can act as a Brønsted-Lowry base.............

(g) Give two limitations of the Arrhenius acid-base theory..

(h) The formula for the hydronium ion is ....

(i) Many toothpastes contain a base called aluminium hydroxide. What is the purpose of this base in toothpaste? ....

(j) Name the acid found in car batteries ....

(k) The Brønsted-Lowry definition of a base is...

(l) Give one advantage of the Brønsted-Lowry theory compared to the Arrhenius theory. .... *[LCH]*

(m) An amphoteric substance is ....

(n) What is the purpose of an antacid?..

12.2. Write down the conjugate base of each of the following Brønsted-Lowry acids:

(a) HCl (b) H_3O^+ (c) HSO_4^- (d) $H_2PO_4^-$

(e) H_2S (f) HSO_3^- (g) $HClO_2$ (h) HNO_2.

12.3. Write down the conjugate acid of each of the following Brønsted-Lowry bases:

(a) CH_3COO^- (b) $H_2PO_4^-$ (c) HPO_4^{2-} (d) HSO_3^-

(e) H_2O (f) CH_3NH^- (g) SO_3^{2-} (h) NH_2^-

12.4. Identify the species acting as bases in the dissociation of nitrous acid in aqueous solution.
$HNO_2 + H_2O \rightleftharpoons H_3O^+ + NO_2^-$ *[LCH]*

12.5. Orthophosphoric acid, H_3PO_4, is a tribasic acid. The equation for the first dissociation of the acid in aqueous solution is:
$H_3PO_4 + H_2O \rightleftharpoons H_3O^+ + H_2PO_4^-$
Write an equation for the second dissociation in aqueous solution. Indicate the conjugate pairs present in the second dissociation. *[LCH]*

12.6. Write down each of the following equations and indicate in each case the acids, bases and conjugate acid-base pairs present.

(a) $NH_3 + H_2O \rightleftharpoons NH_4^+ + OH^-$ (b) $HSO_4^- + HNO_2 \rightleftharpoons H_2NO_2^+ + SO_4^{2-}$

(c) $H_2O + H_2O \rightleftharpoons H_3O^+ + OH^-$ (d) $HSO_4^- + OH^- \rightleftharpoons SO_4^{2-} + H_2O$ *[LCH]*

(f) $HOCl + OH^- \rightleftharpoons H_2O + ClO^-$ (e) $O^{2-} + H_2O \rightleftharpoons OH^- + OH^-$ *[LCH]*

(g) $H^- + H_2O \rightleftharpoons OH^- + H_2$ (h) $HCN + H_2SO_4 \rightleftharpoons H_2CN^+ + HSO_4^-$

(i) $H_2SO_4 + HF \rightleftharpoons H_2F^+ + HSO_4^-$ (j) $CH_3COOH + HNO_3 \rightleftharpoons CH_3COOH_2^+ + NO_3^-$

13

Volumetric Analysis: Acid-Base

13.1 INTRODUCTION

In CHAPTER 11 we studied how the composition of substances could be determined by careful weighing. We called this method of analysis **gravimetric analysis**. Gravimetric analysis is one of the branches of quantitative analysis, i.e. investigating the quantities or amounts of materials present. (This is not to be confused with qualitative analysis which is concerned with identifying substances as studied in the tests for anions, SECTION 6.4.) The other main branch of quantitative analysis is **volumetric analysis**. This method of analysis, as the name suggests, involves reacting together volumes of solutions.

Volumetric analysis has many applications not only in the school laboratory but also in industry and in research. In this course, we shall use volumetric analysis to tell us the amount of ethanoic acid in a sample of vinegar, to analyse bleach, to determine the level of pollution in a sample of water, etc. In FIG 13.1, the laboratory technician is using volumetric analysis to determine the amount of dissolved oxygen in water.

Fig 13.1
Volumetric analysis is being used by this laboratory technician in Inniscarra Water Treatment Plant, Co. Cork, to determine the amount of dissolved oxygen in water.

Before learning the techniques of volumetric analysis, we must first study how chemists express the concentrations of solutions.

13.2 EXPRESSING THE CONCENTRATIONS OF SOLUTIONS

If we add copper sulfate to water, the copper sulfate is referred to as the **solute** and the water is referred to as the **solvent**. The solute and solvent together form a perfect mixture which is called a **solution.** If there is a large amount of solute relative to the quantity of solvent, we say that the solution is **concentrated**. If there is only a small amount of solute relative to the quantity of solvent, the solution is said to be **dilute**.

In many cases, chemists need to know the exact amount of solute present in a certain volume of solution, e.g. a chemist analysing polluted water must very accurately measure the quantity of oxygen dissolved in the water. The precise amount of solute in a given amount of solution is referred to as the **concentration** of the solution.

> The concentration of a solution is the amount of solute that is dissolved in a given volume of solution.

There are a number of different ways of expressing the concentration of a solution:

(A) PERCENTAGE OF SOLUTE

This method of expressing concentration is usually used in many household solutions and in the areas of biochemistry and medicine (FIG 13.2).

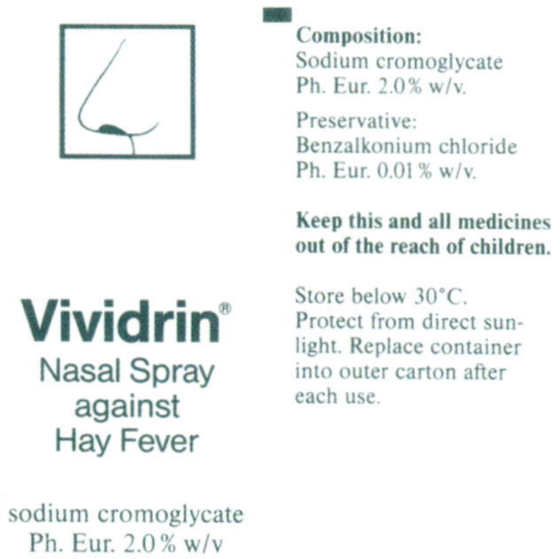

Fig 13.2
Percentage concentrations are often used in medicine.

There are three ways in which the percentage of solute in a solution is expressed:

(i) **Weight per weight (w/w)**. A solution labelled 10% w/w NaCl means that there are 10 g of sodium chloride per 100 g of solution. To make up this solution, you would weigh out 10 g of the salt, dissolve this amount in water and then keep adding water until the total weight of the solution becomes 100 grams.

(ii) **Weight per volume (w/v)**. A solution labelled 10% w/v NaCl means that there are 10 g of sodium chloride per 100 cm^3 of solution. To make up this solution, you would weigh out 10 g of the salt and keep adding water until the total volume of the solution becomes 100 cm^3. The concentration of glucose soltion is usually stated as %w/v (FIG 13.3).

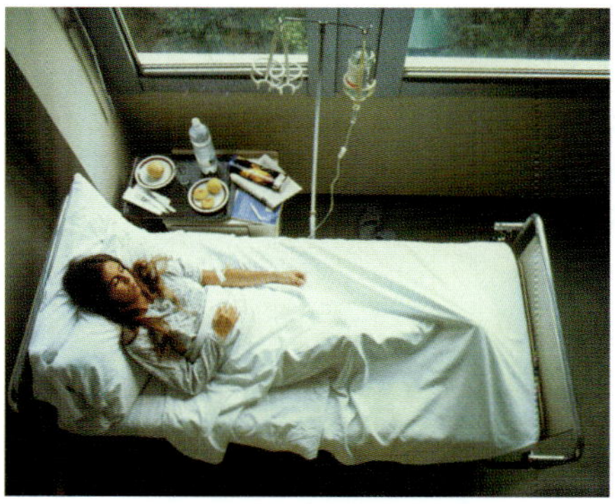

Fig 13.3
Patients are often fed glucose intravenously. The glucose solution is normally about 4% w/v concentration.

(iii) **Volume per volume (v/v)**. Since this method of expressing the concentration refers to the **volume** of solute, we must be dealing with a solute which is a liquid, i.e. the solution is made up from two liquids. For example, a solution marked 10% v/v ethanol means that there are 10 cm^3 of ethanol per 100 cm^3 of solution. To make up this solution, you would measure out 10 cm^3 of ethanol, add it to some water and keep adding water until the volume of solution is 100 cm^3. The amount of alcohol in alcoholic drinks is usually given as a percentage v/v, e.g. beer and cider are about 7–8% v/v, wine is 10–12% v/v, sherry is 20% v/v and spirits (whiskey, gin, vodka, etc.) have an alcohol content of 40% v/v.

Example 13.1

Some illnesses can upset the salt balance in the body and it may be necessary for a doctor or nurse to administer salt intravenously. The solution of salt that is injected is marked 0.85% w/v. What weight of sodium chloride is needed to make up 250 cm^3 of this solution?

SOLUTION

0.85% w/v means that there are 0.85 g NaCl per 100 cm^3 solution.

Therefore, there are $0.85 \times 2.5 = 2.125$ g NaCl per 250 cm^3 solution.

Answer: 2.125 g NaCl

Example 13.2

65 g of a solution contains 8 g of solute. Express the concentration of this solution as % w/w.

SOLUTION

65 g of solution contains 8 g solute.

$$\text{Therefore, \% w/w} = \frac{8}{65} \times 100$$
$$= 12.31\ \%$$

Answer: Concentration of solution is 12.31% w/w.

(b) Parts per million (ppm)

This method of expressing the concentration of a solution is only used for very dilute solutions. It is particularly useful in water analysis (Chapter 19) where one is dealing with very low concentrations of certain substances, i.e. in terms of a few milligrams of a substance per litre of water (Fig 13.4).

1 mg per litre = 1 mg per 1000 grams of water (density water is 1g/cm^3)
= 1 mg per 1000 × 1000 mg
= 1 mg per million mg
= 1 ppm

In other words, **1 ppm = 1 mg/L**

For example, a chemist may report that the concentration of chlorine in water is 2 ppm. This means that there are two milligrams of chlorine per litre of water.

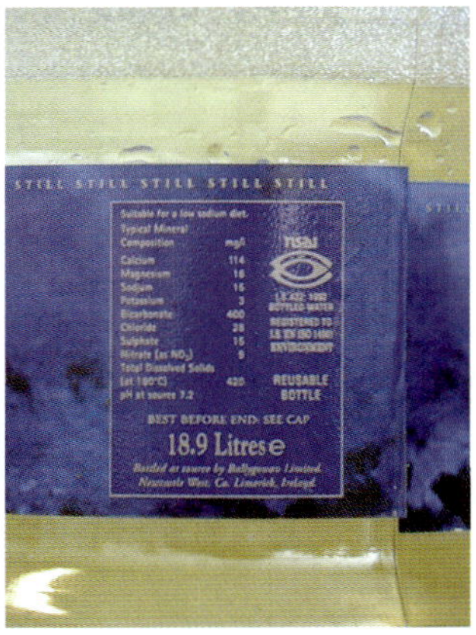

Fig 13.4
Expressing concentrations in parts per million is used when the amounts present are very small.

Example 13.3

On analysing a sample of water, a chemist finds that 100 cm^3 of it contain 0.005 g of calcium carbonate. What is the concentration of calcium carbonate in ppm?

Solution

0.005 g per 100 cm^3 = 0.005 × 10 g per litre
= 0.05 g /L
= 0.05 × 1000 mg/L
= 50 mg/L
= 50 ppm

Answer: 50 ppm

Test Yourself:
Attempt questions 13.1 – 13.2 and W13.1 – W13.2.

(c) Moles of solute per litre of solution (molarity)

The most important method of expressing the concentration of a solution is in terms of moles of solute per litre of solution. This is often referred to as the **molarity** of the solution. The molarity of a solution is the number of moles of solute per litre of solution. The symbol is **M** or **mol/L** (also written as **mol L^{-1}**), i.e. moles per litre. It is more common to use the symbol M. Thus, if a solution contains one mole of solute per litre of solution, we say that the concentration of the solution is 1 M or 1 mol/L. In the S.I. system of units, the symbol mol/L is recommended rather than M. However, since the term molarity and its symbol M are used so widely in industry and at third level, it is important for students to realise that the two have the same meaning. Whenever the symbol M is used, always make sure that it is used in conjunction with a formula, e.g. a 1 M NaOH solution means a sodium hydroxide solution which has a concentration of 1 mole per litre.

1 M NaOH = 1 mol/L NaOH

A 1 molar solution is one that contains one mole of the solute dissolved in one litre of solution.

Thus a 1 mol/L solution of sodium hydroxide contains 40 grams of sodium hydroxide (M_r NaOH = 40) per litre of solution. Since it is important in volumetric analysis to have solutions whose concentrations are **accurately** known, we usually use a flask called a **volumetric flask** when making up solutions (Fig 13.5). The correct procedure for using a volumetric flask to prepare a definite volume of solution will be studied at a later stage.

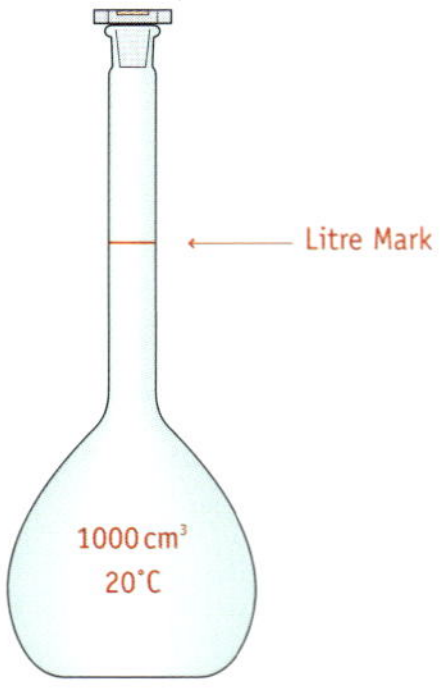

Fig 13.5
If one mole of a substance is dissolved in water and the solution is made up to the litre mark, we have a 1 mol/L (1 molar = 1M) solution.

Suppose that two moles of sodium hydroxide are dissolved in water and the solution is made up to the calibration mark in a 1 litre flask. The concentration of this solution is 2 mol/L or 2 M. Similarly, a solution which is marked 0.5 mol/L or 0.5 M will contain only half a mole of sodium hydroxide (20 g) per litre of solution. A solution that contains one tenth of a mole per litre is often said to be **decimolar** (0.1 M), e.g. a decimolar solution of sodium hydroxide contains 4 g (one tenth of a mole) per litre of solution.

13.3 CALCULATIONS INVOLVING SOLUTIONS

(A) Converting moles per litre to grams per litre

In many cases of analysing substances, chemists find it necessary to convert from concentrations expressed in moles per litre to concentrations expressed in grams per litre. The method of carrying out this conversion is shown in the following examples.

Example 13.4

How many grams of NaCl per litre are present in a solution marked 0.25 M NaCl. (Na = 23, Cl = 35.5)

Solution

Rel Molecular Mass of NaCl = 23 + 35.5 = 58.5

A 1 M NaCl solution contains 58.5 grams per litre.

∴ a 0.25 M NaCl solution contains 58.5×0.25
$= 14.625$ g/L

Answer: 14.625 grams NaCl per litre

Example 13.5

Calculate the concentration in grams per litre of bench dilute sulfuric acid whose concentration is 1.5 mol/L. (H = 1, O = 16, S = 32)

Solution

Rel Molecular Mass of H_2SO_4 = 2(1) + 32 + 4(16)
= 98

A 1 M H_2SO_4 solution contains 98 grams per litre.

∴ a 1.5 M H_2SO_4 solution contains 98×1.5
$= 147$ g/L

Answer: 147 grams H_2SO_4 per litre

(B) Converting grams per volume to moles per litre (molarity)

This type of calculation is met very often in volumetric analysis. Study the following examples carefully.

Example 13.6

What is the molarity of a solution that contains 3.68 g of NaOH per litre of solution?
(H = 1, O = 16, Na = 23)

Solution

Rel Molecular Mass of NaOH = 23 + 16 + 1 = 40

40 g in 1 litre of solution $\rightarrow$ 1 M NaOH solution.

1 g in 1 litre of solution $\rightarrow \frac{1}{40}$ M NaOH solution.

3.68 g in 1 litre of solution $\rightarrow \frac{3.68}{40}$ M NaOH solution.
= 0.092 M NaOH solution

Answer: 0.092 M NaOH.

Example 13.7

Calculate the concentration in moles per litre of a solution containing 45 grams of sulfuric acid per 250 cm^3 of solution. (H = 1, O = 16, S = 32)

SOLUTION

Rel Molecular Mass of H_2SO_4 = 2(1) + 32 + 4 (16) = 98

98 g H_2SO_4 in 1 litre of solution $\rightarrow$ 1 M solution

1 g H_2SO_4 in 1 litre of solution $\rightarrow$ $\frac{1}{98}$ M solution

45 g H_2SO_4 in 1 litre of solution $\rightarrow$ $\frac{45}{98}$ M solution

45 g H_2SO_4 in 250 cm^3 of solution $\rightarrow$ $\frac{4 \times 45}{98}$ M solution

= 1.84 M

Answer: The concentration of the solution is 1.84 M.

Study the last step of EXAMPLE 13.7 carefully. Since the sulfuric acid is actually dissolved in only 250 cm^3 of solution, the concentration of the solution is four times higher. This is an important idea in volumetric analysis that we shall be coming across many times in this course.

If you find the above difficult to understand, studying the dissolving of copper sulfate crystals in water (FIG 13.6) may help. One mole of copper sulfate crystals is dissolved in water and the resulting solution is made up to the calibration mark. Note that, as the volume becomes smaller, the concentration increases. In the case of a coloured solution, as the concentration increases, the intensity of the colour also increases.

(c) Calculating the number of moles given a volume and molarity

Chemists, having prepared a solution in a volumetric flask, usually take a smaller quantity (e.g. 25 cm^3) of that solution out of the volumetric flask in order to perform various experiments in volumetric analysis. When using this smaller volume, it is necessary to be able to answer questions like that posed in EXAMPLE 13.8.

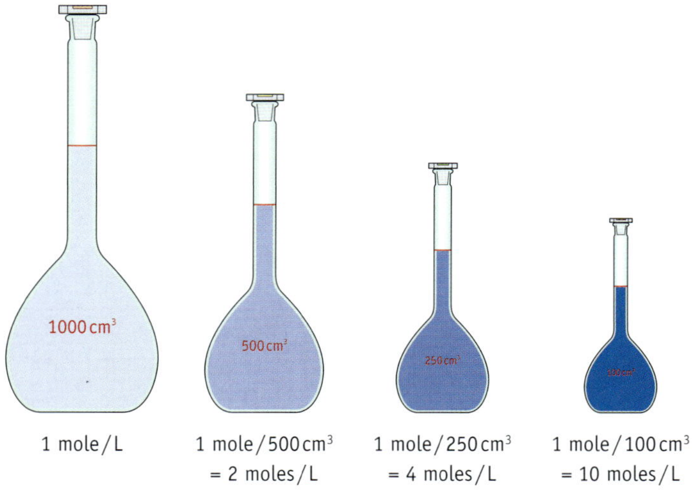

Fig 13.6
If the same mass of solute is dissolved in different quantities of water, the smaller the volume of water, the higher the concentration.

Example 13.8

How many moles of NaOH are present in 25 cm^3 of 0.55 M NaOH.

SOLUTION

We shall first work out the answer from first principles. Having done this we will then deduce a general expression for answering this type of question.

By definition:

1 litre of 0.55 M NaOH contains 0.55 moles NaOH.

i.e. 1000 cm^3 of 0.55 M NaOH contains 0.55 moles NaOH

$\therefore$ 1 cm^3 of 0.55 M NaOH contains $\frac{0.55}{1000}$ moles NaOH

$\therefore$ 25 cm^3 of 0.55 M NaOH contains $\frac{25 \times 0.55}{1000}$ moles NaOH

$= 0.014$ moles NaOH.

Answer: 0.014 moles

In general, if the volume is given in **cm^3** and the concentration in **molarity** (or **moles per litre**), the number of moles may be calculated from the expression:

$$\text{Number of moles} = \frac{\text{volume} \times \text{molarity}}{1000}$$

This is a very useful expression and is worth learning off.

Example 13.9

How many moles of hydrochloric acid are present in 30 cm^3 of 0.2 M HCl.

SOLUTION

$$\text{Number of moles of HCl} = \frac{\text{volume} \times \text{molarity}}{1000}$$

$$= \frac{30 \times 0.2}{1000}$$

$$= 0.006 \text{ moles}$$

Answer: 0.006 moles.

Example 13.10

What mass of sodium hydroxide is contained in 25 cm^3 of a 0.75 M solution of sodium hydroxide? (H = 1, O = 16, Na = 23)

SOLUTION

First calculate the number of moles present and then convert this number of moles to grams.

$$\text{Number of moles} = \frac{\text{volume} \times \text{molarity}}{1000}$$
$$= \frac{25 \times 0.75}{1000}$$
$$= 0.019 \text{ moles}$$
$$= 0.019 \times 40 \text{ g } (M_r \text{ NaOH} = 40)$$
$$= 0.76 \text{ g}$$

Answer: 0.76 grams

Example 13.11

What volume of 0.15 M sodium hydroxide solution will contain 5 grams of sodium hydroxide. (H = 1, O = 16, Na = 23)

SOLUTION

First convert 5 grams to moles.

$$\text{No. of moles of NaOH} = \frac{\text{Mass}}{\text{Rel Molecular Mass}}$$
$$= \frac{5}{40}$$
$$= 0.125 \text{ moles}$$

Calculate the volume using the formula:

$$\text{Number of moles} = \frac{\text{volume} \times \text{molarity}}{1000}$$
$$0.125 = \frac{\text{volume} \times 0.15}{1000}$$
$$\frac{0.125 \times 1000}{0.15} = \text{volume}$$
$$\text{i.e. volume} = 833.33 \text{ cm}^3$$

Answer: 833.33 cm^3

Test Yourself:
Attempt questions 13.3 – 13.5 and W13.3 – W13.5.

13.4 REACTION BETWEEN A SOLUTION AND A SOLID

In a number of chemical reactions, solids react with solutions, e.g. many metals dissolve in acidic solutions. EXAMPLE 13.12 shows how we can calculate the mass of metal which reacts with a given volume of acid.

Example 13.12

What mass of magnesium will react with 50 cm^3 of 0.5 M H_2SO_4? (Mg = 24). The balanced equation for the reaction is:

$$Mg + H_2SO_4 \rightarrow MgSO_4 + H_2$$

SOLUTION

$$\text{No. of moles of acid} = \frac{\text{volume} \times \text{molarity}}{1000}$$
$$= \frac{50 \times 0.5}{1000}$$
$$= 0.025 \text{ moles}$$

From the balanced equation, one mole of sulfuric acid reacts with one mole of magnesium.

$$\text{Therefore, moles of magnesium which react} = 0.025$$
$$= 0.025 \times 24 \text{ g}$$
$$= 0.6 \text{ g}$$

Answer: 0.6 g magnesium.

Example 13.13

Sodium carbonate, Na_2CO_3, reacts with dilute hydrochloric acid according to the equation:

$Na_2CO_3 + 2HCl \rightarrow 2NaCl + CO_2 + H_2O$

What volume of hydrochloric acid of concentration 0.75 M would be needed to neutralise 7.5 g of anhydrous sodium carbonate? (C = 12, O = 16, Na = 23)

Solution

Rel Molecular Mass of Na_2CO_3 = 2(23) + 12 + 3(16) = 106

No. of moles of $Na_2CO_3 = \dfrac{\text{Mass}}{\text{Rel Molecular Mass}} = \dfrac{7.5}{106} = 0.071$ moles

From the balanced equation, two moles of HCl react with one mole of Na_2CO_3.
Therefore, no. of moles of HCl which react = $2 \times 0.071 = 0.142$ moles

$$\text{No. of moles of HCl} = \frac{\text{volume} \times \text{molarity}}{1000}$$

$$\text{i.e. } 0.142 = \frac{\text{volume} \times 0.75}{1000}$$

$$\text{volume} = \frac{0.142 \times 1000}{0.75}$$

$$= 189.33 \text{ cm}^3$$

Answer: 189.33 cm^3 of HCl are needed.

Test Yourself:
Attempt questions 13.6 – 13.7 and W13.6 – W13.7.

13.5 DILUTION OF SOLUTIONS

Many common laboratory reagents are purchased as solutions of high concentration, e.g. the concentration of sulfuric acid supplied to laboratories is 18 M. Some of the common laboratory reagents and their concentrations as supplied are given in Table 13.1.

SUBSTANCE	CONCENTRATION
Sulfuric acid	18 M
Nitric Acid	16 M
Hydrochloric Acid	12 M
Ammonia	16 M
Ethanoic acid (acetic acid)	17 M

Table 13.1
The concentrations of common reagents as supplied to laboratories.

However, for most laboratory experiments, these reagents must be used in their diluted state, e.g. the concentrations of 'bench dilute' hydrochloric acid and 'bench dilute' ammonia are only 3 M approx. When a solution is diluted, more solvent is added but the quantity of solute is unchanged. Since the volume of the solution increases, and the same number of moles are present, then the concentration of the solution must decrease (Fig 13.7).

Fig 13.7
Adding water does not affect the amount of solute present but does reduce the concentration.

In order to dilute a concentrated solution to give a solution of known concentration, it is important to remember that the number of moles of solute in the diluted solution is the same as the number of moles of solute in the concentrated solution.

No. of moles of solute in diluted solution = No. of moles of solute in concentrated solution

$$\text{i.e. } \frac{Vol_{dil} \times M_{dil}}{1000} = \frac{Vol_{conc} \times M_{conc}}{1000}$$

We use this relationship to solve problems on dilution.

Note: It does not matter what units are used to express the concentration as long as the same units are used on each side of the equation.

Example 13.14

What volume of 12 M HCl solution is required to make up 500 cm^3 of 3 M HCl solution?

SOLUTION

No. of moles of solute in diluted solution = No of moles of solute in concentrated solution

$$\text{i.e. } \frac{Vol_{dil} \times M_{dil}}{\cancel{1000}} = \frac{Vol_{conc} \times M_{conc}}{\cancel{1000}}$$

$$500 \times 3 = Vol_{conc} \times 12$$

$$Vol_{conc} = \frac{500 \times 3}{12}$$

$$Vol_{conc} = 125 \text{ cm}^3$$

Answer: 125 cm^3 of the concentrated acid would have to be diluted to 500 cm^3.

Example 13.15

A student takes 15 cm^3 of 2 M HNO_3 solution and dilutes it to 250 cm^3 in a volumetric flask. What is the new concentration of the acid?

SOLUTION:

No. of moles of solute in diluted solution = No of moles of solute in concentrated solution

$$\text{i.e. } \frac{Vol_{dil} \times M_{dil}}{\cancel{1000}} = \frac{Vol_{conc} \times M_{conc}}{\cancel{1000}}$$

$$250 \times M_{dil} = 15 \times 2$$

$$M_{dil} = \frac{15 \times 2}{250} = 0.12 \text{ M}$$

Answer: The concentration of the diluted solution is 0.12 M.

Test Yourself:
Attempt questions 13.8 – 13.9 and W13.8 – W13.10.

13.6 STANDARD SOLUTIONS

A **standard solution** is a solution of precisely known concentration, e.g. a solution containing 4.65 grams of sodium chloride per litre is a standard solution. A standard solution is prepared by dissolving a precisely known mass of solute in a suitable solvent (usually deionised water) to give a definite known volume of solution.

The concentration of a solution may be determined either by preparing the solution ourselves or by performing an experiment to analyse the solution. Because many substances cannot be obtained in a high degree of purity, it is difficult for chemists to make up a wide range of solutions by simply dissolving a weighed amount of solute in a solvent. For example, it is not possible to weigh out a precise quantity of sulfuric acid because sulfuric acid absorbs moisture from the air. Similarly, an accurately known quantity of iodine cannot be weighed out because iodine tends to sublime (change to a vapour) at room temperature.

In order to make up standard solutions, chemists need substances which can be obtained in a highly pure state and which are stable in air. Such substances are called **primary standards.**

A primary standard is a substance that meets the following criteria:

1. It must be available in a highly pure state.
2. It must be stable in air.
3. It must dissolve easily in water.
4. It should have a fairly high relative molecular mass (so that the precision in weighing out a known mass is as high as possible).

5. When used in volumetric analysis, it should undergo complete and rapid reaction.
6. It should not be hydrated.

Thus, a primary standard may be defined as follows:

> A primary standard is a substance which can be obtained in a stable, pure and soluble solid form so that it can be weighed out and dissolved in water to give a solution of accurately known concentration.

Because primary standards must satisfy so many conditions, only a small number of common substances are classed as primary standards, e.g. anhydrous sodium carbonate, Na_2CO_3, sodium chloride, NaCl, potassium dichromate, $K_2Cr_2O_7$, etc. As standard solutions are widely used in industry and in research, it is important that you know how to prepare one.

Experiment 13.1 Mandatory Student Experiment No. 5

To prepare a standard solution of sodium carbonate.

Introduction

In this experiment, we weigh out accurately a known amount of sodium carbonate, dissolve it in water and make the solution up to a known volume. Since sodium carbonate is a primary standard, the concentration of the solution will then be known precisely.

Apparatus required

Clock glass, wash bottle, glass rod, beaker, funnel, 250 cm^3 volumetric flask, spatula, laboratory balance, adhesive label, dropper.

Chemicals required

Anhydrous sodium carbonate, deionised water.

Procedure

1. Place a clean, dry clock glass on the pan of the laboratory balance.
2. Adjust the balance so that the weight of the clock glass is 'tared off', i.e. so that the balance reads zero.
3. Using a clean, dry spatula, add about 1.30 grams of sodium carbonate to the clock glass. You must ensure that the weight of sodium carbonate taken is known precisely (to within 0.01 g or better). However, it need not be the exact weight specified – it could be 1.31 g, 1.32 g, 1.33 g etc.). Do not waste time adding or removing small amounts of solid to get exactly 1.30 g. Make a note of the exact amount of sodium carbonate weighed out. It is important that you know the exact amount in order to calculate the concentration of the solution you are about to make up.
4. Carefully add the carbonate to a beaker containing about 100 cm^3 of deionised water. (Deionised water must always be used when making up a standard solution as tap water contains impurities that could interfere with the solution.)
5. Using a wash bottle of deionised water, wash all traces of the sodium carbonate into the beaker.
6. Stir the mixture with a clean, glass rod until the solid dissolves completely. It may be necessary to break up the lumps of sodium carbonate using the glass rod.
7. Using a glass rod and funnel, transfer the solution carefully to a clean 250 cm^3 volumetric flask as shown in stage 4 (Fig 13.8). A detailed description of a volumetric flask is given in the next section.

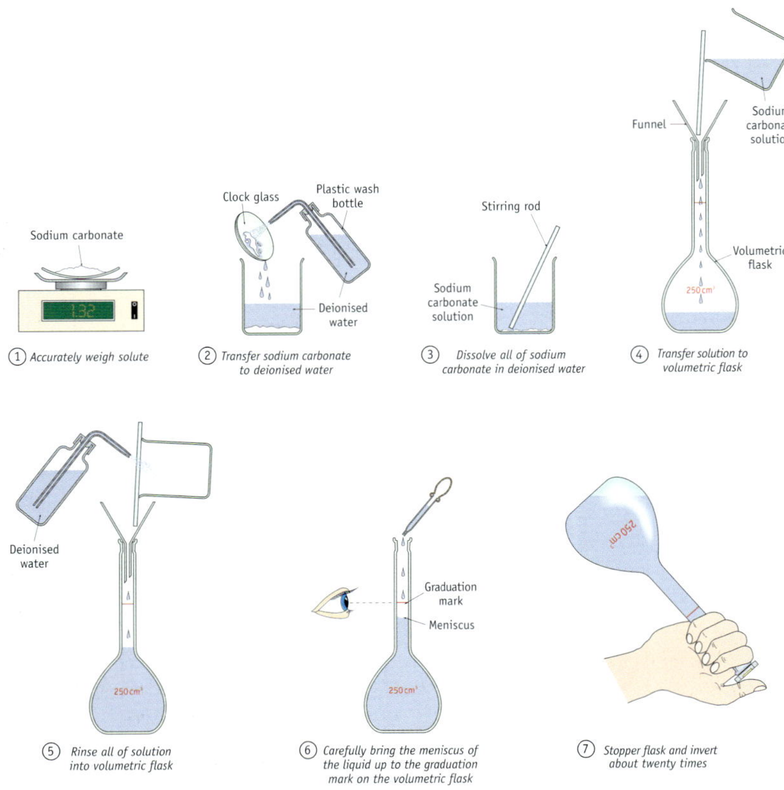

Fig 13.8
Making up a standard solution of sodium carbonate.

8. Using the wash bottle, wash all traces of the solution from the beaker, rod and funnel into the volumetric flask.
9. Stand the volumetric flask on a level surface. Add deionised water to the flask to within about 1 cm of the graduation mark. Remove the funnel.
10. Using a dropper, add deionised water to the flask until the bottom of the meniscus of the liquid is exactly on the graduation mark (FIG 13.9). Ensure that the graduation mark is at eye-level to avoid any error in viewing the meniscus level.
11. Place the stopper on the volumetric flask and, keeping one hand on the stopper, invert the flask about 20 times to mix the contents thoroughly. (If you simply invert the flask once or twice, the contents are not mixed properly and hence the solution will not be uniform. This is a common error.)
12. Calculate the concentration (in moles per litre) of the solution you have made up. (The calculation involved is similar to that in EXAMPLE 13.7.)
13. Label the flask with the name of the contents, the concentration of the solution, your name and the date. You now have a solution whose concentration is known precisely. Store the flask in a safe place as you will be using this solution in the next experiment to find out the concentration of another solution.

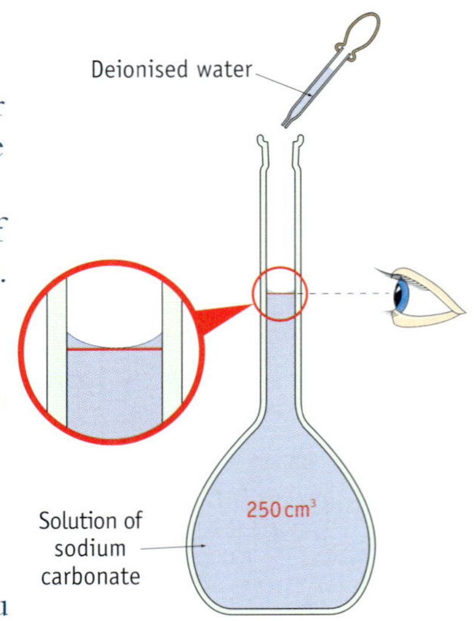

Fig 13.9
The meniscus must be brought up to the graduation mark using a dropper. The volumetric flask must be placed on a level surface.

The main steps are summarised in FIG 13.8.

Note: All experiments should be written up under the headings: Title, Date, Background Theory, Procedure, Results, Calculations, Conclusions. A labelled diagram should accompany all reports of experiments.

13.7 VOLUMETRIC ANALYSIS. TITRATIONS

We have already studied gravimetric analysis and have seen that it involves careful and accurate weighing. However, another type of analysis called **volumetric analysis** is more often used by chemists. As the name suggests, volumetric analysis involves reacting together volumes of solutions. In volumetric analysis, the concentration of a solution is determined by reacting the solution with another solution of known concentration. A definite volume (usually 25 cm^3) of one of the solutions is taken and the other solution is added slowly until the two solutions have reacted completely. This procedure is known as a **titration**. The stage when the two solutions just react completely with each other is called the **equivalence point** of the titration. When the equivalence point is shown by an indicator, it is called the **end point** of the titration.

A titration is a laboratory procedure where a measured volume of one solution is added to a known volume of another solution until the reaction is complete.

Before studying the correct procedure for performing a titration, we shall first study the apparatus which is used in volumetric analysis.

APPARATUS USED IN VOLUMETRIC ANALYSIS

One of the advantages of volumetric analysis is that the apparatus required is fairly simple. The main apparatus used in volumetric analysis consists of the graduated cylinder, the volumetric flask, the pipette, the burette and the conical flask. A brief note on each of these now follows.

1. The Graduated Cylinder

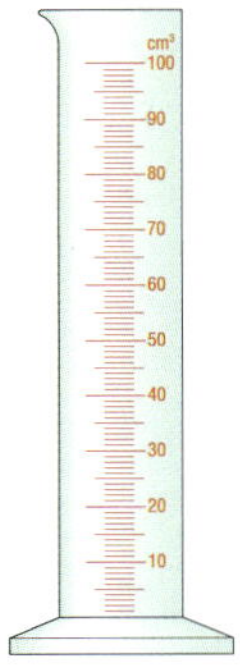

Fig 13.10
Graduated cylinders may be used for measuring approximate volumes of liquids.

The graduated cylinder is not particularly accurate and should be used only for measuring out approximate volumes of liquids, i.e. correct to within 0.5 cm^3 (Fig 13.10). Before using a graduated cylinder, it should always be rinsed out with deionised water.

2. The Volumetric Flask

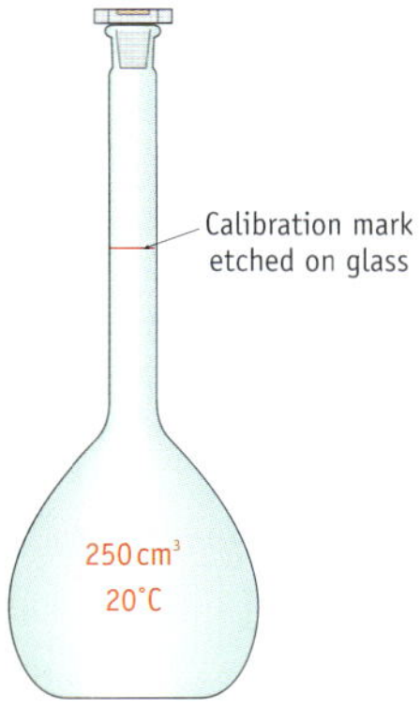

Fig 13.11
A volumetric flask holds an exact volume of liquid.

A volumetric flask is a vessel that is designed to contain a definite volume of solution. There is an engraved graduation mark on the narrow neck of the flask (Fig 13.11). This allows the volume of solution in the flask to be adjusted accurately. The capacity of the volumetric flask and the temperature at which the stated volume is most accurately measured are usually marked on the flask. (This is necessary as the volume of a liquid changes with temperature.)

Before using a volumetric flask, it should always be rinsed out with deionised water. There are two important points to remember when using a volumetric flask:

(a) The last few cm^3 of water must be added carefully so that the meniscus rests on the calibration mark. A small dropper is used for this purpose.

(b) After making the solution up to the mark, the contents must be well mixed by inverting the stoppered flask about 20 times. This ensures a homogeneous mixture, i.e. that the solution is uniform throughout. Otherwise the solution in the narrow neck could be a different concentration than that in the body of the volumetric flask.

3. The Pipette

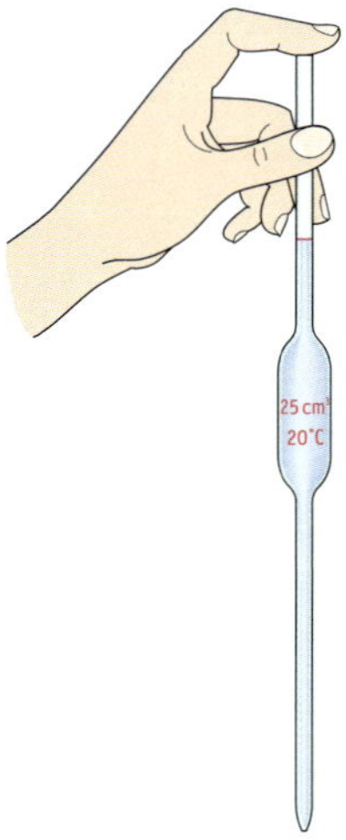

Fig 13.12
A pipette is designed to deliver the exact volume of liquid that is indicated on the bulb.

A pipette is designed to deliver only an exact volume of solution, usually 20 or 25 cm^3 (Fig 13.12).

A pipette is filled by drawing the solution into it by suction. The liquid is drawn into the pipette above the graduation mark etched on its surface. This is done using a pipette filler (Fig 13.13).

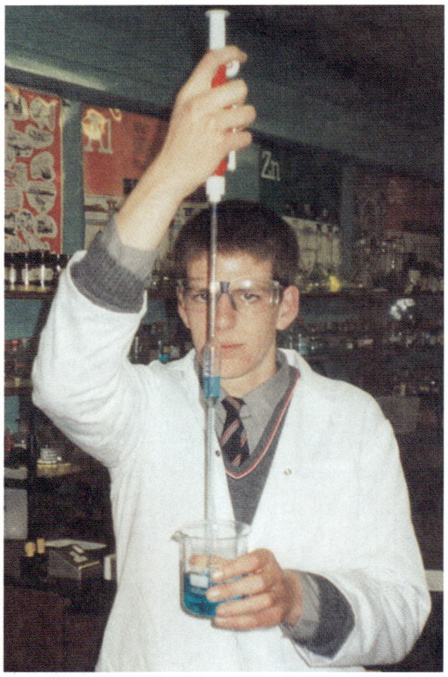

Fig 13.13
Using a pipette filler.

The liquid is then allowed to run into a beaker (for waste) until the bottom of the meniscus is level with the graduation mark. When the bottom of the meniscus is level with the graduation mark, the tip of the pipette is touched against the side of a glass beaker to remove any drop adhering to the tip. The pipette is then allowed to discharge slowly into a conical flask, keeping the tip of the pipette in contact with the side of the flask. This avoids splashing of the liquid. When the discharge is complete, touch the tip of the pipette against the side of the flask and wait a few seconds before removing the pipette from the flask (draining time). Having removed the pipette from the flask, it will be observed that a small amount of liquid remains in the tip. Do NOT blow out this liquid into the conical flask. It has been taken into account when the pipette was calibrated at the manufacturing stage.

Before using a pipette, it should always be rinsed with deionised water and then with the liquid which it will contain. Rinsing with water will remove any impurities in it. Rinsing with the liquid that that it will contain is necessary in order to remove any drops of water in the pipette. This water would dilute the solution placed in the pipette.

After use, the pipette should be rinsed out with deionised water and allowed to drain.

4. The Burette

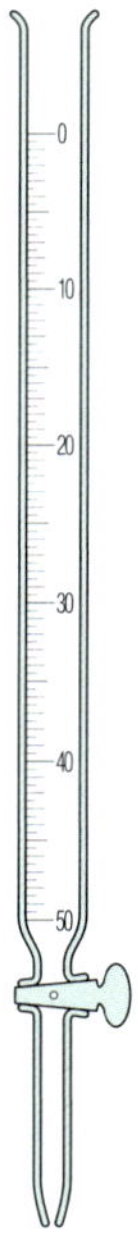

Fig 13.14
The burette.

A burette consists of a long narrow glass tube that is closed at the bottom by means of a tap (Fig 13.14).

Note that the graduations on the burette read downwards. The burette usually used in the school laboratory delivers variable amounts of liquid up to 50 cm^3. The burette is graduated in divisions of 0.1 cm^3. The volume of liquid is normally read to 0.1 cm^3. A second place of decimals can be obtained by estimation.

Before using the burette, the tap should be checked to see that it turns freely. Modern burette taps are made of plastic and are not likely to stick. Older burettes have taps made of ground glass and, if this sticks, it may be necessary to grease the tap slightly.

The burette should be rinsed out with deionised water and then with the liquid to be used in it. Otherwise, the water in the burette would dilute the solution placed in it. A small funnel should be used when filling the burette. The burette is then filled almost completely with the liquid, the tap opened and the liquid allowed to fill the part of the burette below the tap (the nozzle). The level of the liquid in the burette is then adjusted until the bottom of the meniscus is on the zero mark. The nozzle of the burette should be checked to ensure that there are no air bubbles trapped in it. Any drop of liquid adhering to the tip of the burette should be removed by touching it against the side of a glass beaker. The burette is now ready for use and will deliver any volume of liquid up to 50 cm^3.

When using a burette is should always be clamped vertically. In order to read the volume of liquid in the burette, the eye should be level with the bottom of the meniscus as shown in FIG 13.15.

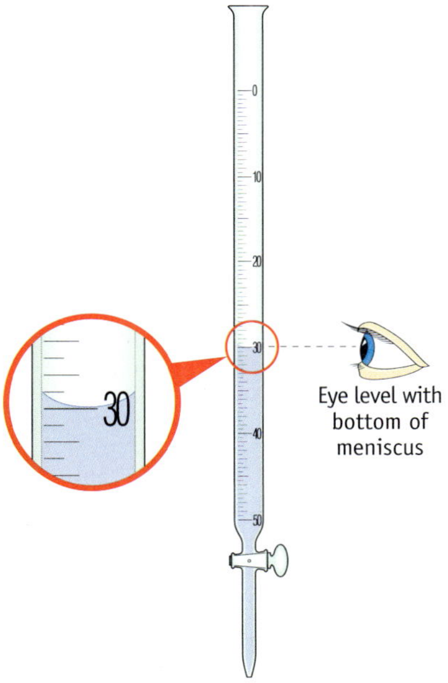

Fig 13.15
When reading the volume of liquid in the burette, the eye should be level with the bottom of the meniscus.

As a help to reading the meniscus, it is useful to hold a piece of white paper behind the burette and near to the meniscus. (If the liquid is highly coloured, as in the case of potassium permanganate solution, readings are normally taken from the **top** of the meniscus.)

After use, the burette should be rinsed out with deionised water and clamped in an inverted position, with the tap open, in a burette stand.

5. The Conical Flask

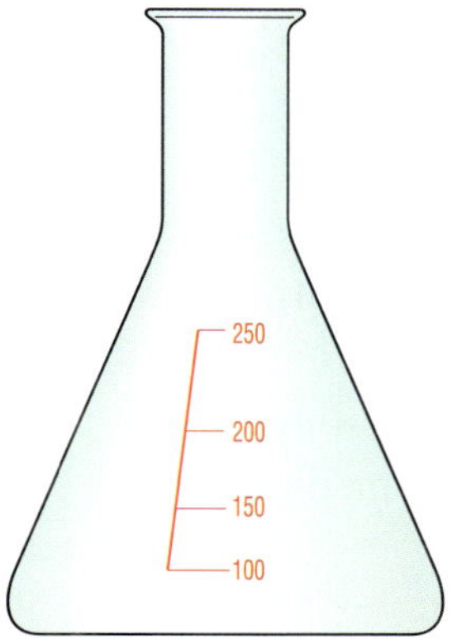

Fig 13.16
The conical flask is designed to prevent liquid from splashing out.

A conical flask (also called a titration flask) is designed so that it may be swirled without the liquid splashing out (FIG 13.16). The shape of the conical flask also allows any droplets of solution adhering to the side of it to be easily washed down.

Before use, the conical flask should always be washed out with deionised water. It should NOT be rinsed with the solution it is to contain. The reason for this arises from the fact that a pipette is used to place an **exact volume** of solution in the conical flask. If drops of the solution were already present in the conical flask, these would be additional to the volume added from the pipette. Thus, the actual volume of solution in the conical flask would not be known.

PROCEDURE IN CARRYING OUT A TITRATION

Because of the importance of volumetric analysis, it is vital that the student masters the technique of carrying out a titration accurately. This technique is described in detail in EXPERIMENT 13.2.

EXPERIMENT 13.2 MANDATORY EXPERIMENT NO. 6

To use the standard solution of sodium carbonate solution prepared in the previous experiment to standardise a given hydrochloric acid solution.

Introduction

Sodium carbonate is a base that reacts with hydrochloric acid according to the equation:

$Na_2CO_3 + 2HCl \rightarrow 2NaCl + H_2O + CO_2$

Essentially, what is happening is that the carbonate ion is accepting two protons from the hydrochloric acid to form carbon dioxide and water:

$CO_3^{2-} + 2H^+ \rightarrow H_2O + CO_2$

By titrating the given HCl solution against a known volume of Na_2CO_3 solution, whose concentration is accurately known, we can determine the concentration of the HCl solution. An indicator (methyl orange in this case) is used to determine the end point of the titration.

Note: In acid-base titrations, the acid is usually placed in the burette and the base in the conical flask. One of the main reasons for this is because some common bases (e.g. sodium hydroxide) if placed in a burette would cause the tap to stick. Sodium hydroxide solution attacks glass, especially the ground surfaces in some burette taps. Also, it absorbs carbon dioxide from the air and forms a crust of sodium carbonate around the tap – this happens even with the modern plastic taps. Both of these actions cause the burette taps to become seized up.

Apparatus required

Pipette (25 cm^3), pipette filler, wash bottle, retort stand, spatula, labaratory balance, white tile, funnel, dropper, burette, three beakers (plastic), conical flask.

Chemicals required

Sodium carbonate solution (prepared in EXPERIMENT 13.1), hydrochloric acid solution, methyl orange indicator, deionised water.

Procedure

The detailed procedure given for this experiment should be followed in all future titrations. The main stages in carrying out a titration are summarised in FIG 13.17.

1. Wash out the pipette, burette and conical flask with deionised water.
2. Pour about 100 cm^3 of the sodium carbonate solution from the volumetric flask into a clean, dry beaker. Also pour about 100 cm^3 of the hydrochloric acid solution into another clean, dry beaker. Label both beakers.
3. Rinse out the pipette with some of the sodium carbonate solution and the burette with some of the hydrochloric acid solution.
4. Pipette 25 cm^3 of the sodium carbonate solution into a conical flask (FIG 13.17 (a)). (It is bad practice to pipette directly from the volumetric flask as any impurities on the outside of the pipette could contaminate the entire solution of sodium carbonate. This is the reason why you first pour the sodium carbonate solution into a beaker.) When using the pipette do not forget to take all the precautions outlined earlier.

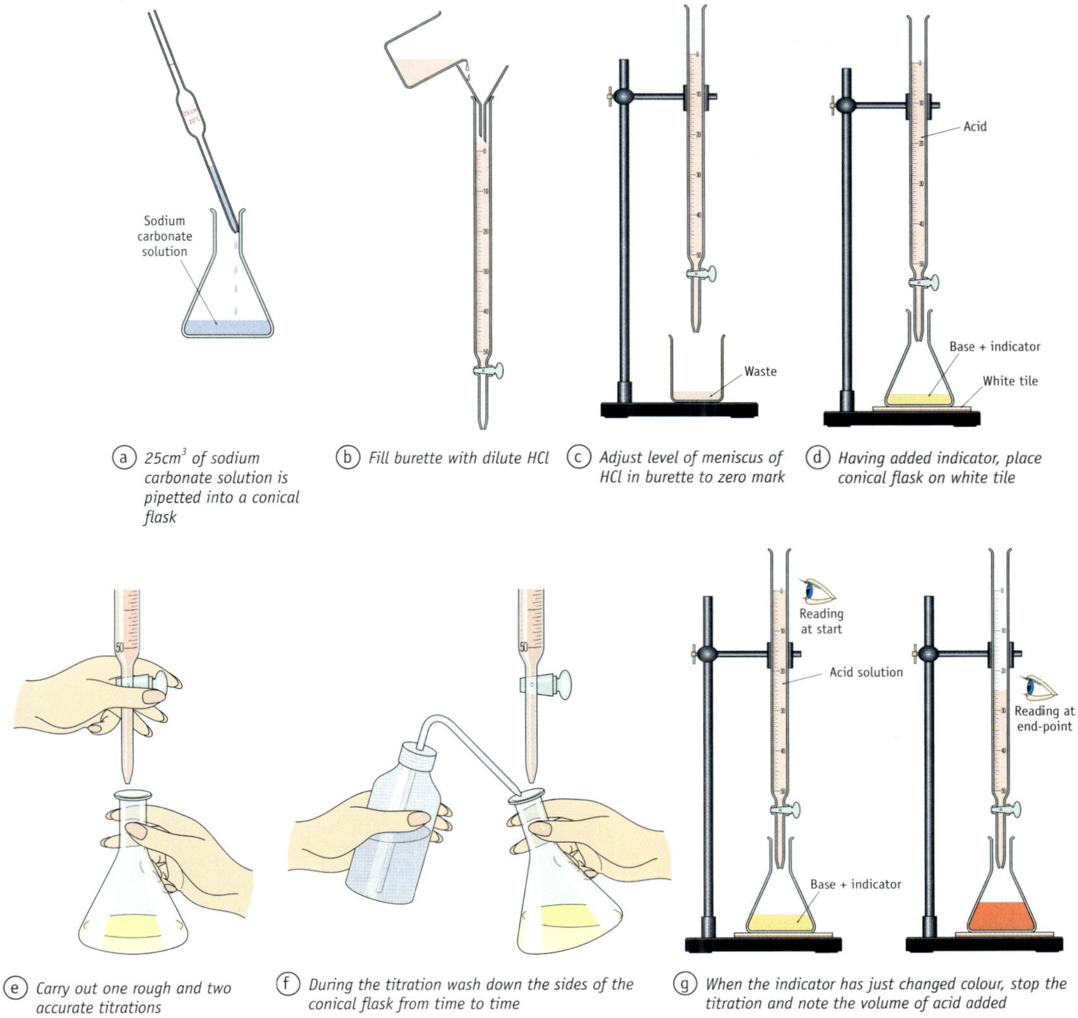

Fig 13.17
The main stages in carrying out a titration.

5. Add about three drops of methyl orange indicator to the sodium carbonate solution in the conical flask. Note the yellow colour. Methyl orange is yellow in neutral or basic solution and red in acidic solution. Do not use too much indicator as an intense colour may make it difficult to see the colour change at the end point. (In addition, as we shall see later, indicators themselves are either weak acids or weak bases; adding too much indicator could affect the result.)
6. Clamp the burette vertically using the retort stand.
7. Using a small funnel, fill the burette almost completely with the dilute HCl solution. Open the tap briefly and allow some of the liquid to run out into a beaker marked 'waste'. This is done in order to fill the tap and the space below the tap.
8. Remove the funnel from the burette. If you forget to do this, you may introduce an error into your readings as drops of liquid may fall from the funnel in the course of the titration.
9. Using the thumb and first two fingers of the left hand, open the tap of the burette. Allow the solution to run slowly into the beaker marked 'waste' until the bottom of the meniscus is on the zero graduation mark. Check that there are no air bubbles in the nozzle.
10. Place a white tile on the base of the retort stand and put the conical flask on the white tile. The purpose of the white tile is to enable us to detect colour changes more easily.

11. Using the thumb and first two fingers of the left hand, open the tap of the burette. Allow the acid to run fairly quickly into the conical flask. The conical flask should be continuously swirled with the right hand. Stop the titration from time to time and, using the wash bottle, wash down the sides of the conical flask. The reason for doing this is to ensure that no drops of acid are stuck to the side of the flask. It is important that all of the acid added from the burette reacts with the sodium carbonate.
Note: Deionised water may be added to the conical flask at any stage of the titration. It does not affect the amount of sodium carbonate in the conical flask. We have pipetted 25 cm^3 of the sodium carbonate solution into the conical flask. This volume of sodium carbonate contains a certain number of moles of sodium carbonate. Whether we now add 10 cm^3 or 100 cm^3 of deionised water to the conical flask, will not affect the number of moles of sodium carbonate present in the conical flask. Note the distinction between washing out the pipette and burette with the solutions which they contain and **not** washing out the conical flask with the solution it is to contain. In the case of adding water to the conical flask during the titration, the dilution occurs **after** the amount has been measured.
12. When the colour of the indicator has turned pink, stop the titration and note the volume of acid added. A piece of white paper behind the burette will help you to take the reading more accurately. This rough titration tells you approximately where the end point is.
13. You must now perform two further **accurate** titrations, adding the acid drop by drop as you come near the end point. First, bring the level of hydrochloric acid back up to the zero mark in the burette. (Although it is not essential to have the burette reading at the zero mark for each titration, it is less likely to cause errors for the student if the meniscus is brought back to the zero mark. This means that the volume may be read directly and there is less likelihood of an error when subtracting the initial reading from the final reading.) The conical flask should be emptied and rinsed out well with deionised water before each titration is carried out. When performing each accurate titration, one drop of acid should turn the solution in the conical flask a faint pink colour at the end point. (Turning the tap through a revolution of 180° allows one drop into the conical flask.) The results of these two accurate titrations should agree to within 0.1 cm^3. If they do not agree, a further titration should be carried out until agreement to within 0.1 cm^3 is obtained.
14. The titration result (also called the titre) for this experiment is taken as the average of the two readings which agree to within 0.1 cm^3. The rough titration, performed to give you an approximate idea where the end point is, should be neglected. For example, if the three titration readings are 22.9 cm^3, 22.4 cm^3 and 22.5 cm^3, the titration figure to be used in calculations is the average of 22.4 and 22.5, i.e. 22.45 cm^3.

Calculations

The method of calculating the concentration of the hydrochloric acid solution is shown in Example 13.16 in the next section.

13.8 SOLVING VOLUMETRIC PROBLEMS

In solving volumetric problems involving acids and bases, there are two main types of calculations: 1. calculating the unknown concentration of a solution from titration data, and 2. calculating the relative molecular mass and the amount of water of crystallisation in a compound and the relative molecular mass from titration data. We now consider these two types of calculations.

1. CALCULATING THE UNKNOWN CONCENTRATION OF A SOLUTION FROM TITRATION DATA

The most common type of problem encountered in volumetric analysis is where we are required to measure the concentration of a solution of a given substance. For example, in EXPERIMENT 13.2 we are required to work out the concentration of the hydrochloric acid solution. The type of calculation involved is shown in the following example:

Example 13.16

1.45 g of sodium carbonate was dissolved in water and the solution was made up to 250 cm^3 in a volumetric flask. 25 cm^3 of this solution were titrated against a solution of hydrochloric acid using methyl orange as indicator. One rough and two accurate titrations were performed. The titration results are shown in the following table.

Titration	1	2	3
Volume (cm^3)	19.8	19.5	19.6

The equation for the reaction is:

$Na_2CO_3 + 2HCl \rightarrow 2NaCl + H_2O + CO_2$

Calculate the concentration of the hydrochloric acid solution in (a) moles per litre and (b) grams per litre. (H = 1, C = 12, O = 16, Na = 23, Cl = 35.5)

SOLUTION

Since we are not given the molarity of the sodium carbonate solution, the first step is to calculate this before using the results of the titration.

1 mole Na_2CO_3 in 1 litre of solution $\rightarrow$ 1M Na_2CO_3 solution

i.e. 106 g Na_2CO_3 in 1 litre of solution $\rightarrow$ 1 M Na_2CO_3 solution

1g Na_2CO_3 in 1 litre of solution $\rightarrow \frac{1}{106}$ M Na_2CO_3 solution

1.45 g Na_2CO_3 in 1 litre of solution $\rightarrow \frac{1 \times 1.45}{106}$ M Na_2CO_3 solution

1.45 g Na_2CO_3 in 250 cm^3 of solution $\rightarrow \frac{4 \times 1.45}{106}$ M Na_2CO_3 solution

= 0.055 M Na_2CO_3 solution

The next step is to use the concentration of the sodium carbonate solution to calculate the concentration of the hydrochloric acid solution.

To do this, we will first deduce a general expression for answering this type of question.

Consider the reaction between sodium carbonate and hydrochloric acid:

$Na_2CO_3 + 2HCl \rightarrow 2NaCl + H_2O + CO_2$

From the balanced equation, we see that two moles of acid react with one mole of base , i.e. the ratio of the number of moles of acid to the number of moles of base is 2 : 1

$$\frac{\text{number of moles of acid}}{\text{number of moles of base}} = \frac{2}{1}$$

If, in a given titration, V_a is the volume of acid, M_a is the molarity of the acid, n_a is the number of moles of acid, V_b is the volume of base, M_b is the molarity of the base and n_b is the number of moles of base, then the above expression becomes:

$$\frac{\frac{V_a \times M_a}{1000}}{\frac{V_b \times M_b}{1000}} = \frac{n_a}{n_b}$$

i.e. $$\frac{V_a \times M_a}{V_b \times M_b} = \frac{n_a}{n_b}$$

Rearranging this expression gives:

$$\frac{V_a \times M_a}{n_a} = \frac{V_b \times M_b}{n_b}$$

molarity of acid → M_a; volume of base (cm^3) → V_b; vol of acid (cm^3) → V_a; molarity of base → M_b; no. of moles of acid → n_a; no. of moles of base → n_b

This is a very important relationship that is very useful for solving many volumetric problems. We now substitute into this equation the values for the particular problem we are solving.

(HCl) (Na$_2$CO$_3$)

GIVEN:

$V_a = 19.55$*

$M_a = ?$

$n_a = 2$

$V_b = 25$

$M_b = 0.055$

$n_b = 1$

$$\frac{V_a \times M_a}{n_a} = \frac{V_b \times M_b}{n_b}$$

$$\frac{19.55 \times M_a}{2} = \frac{25 \times 0.055}{1}$$

$$M_a = \frac{25 \times 0.055 \times 2}{1 \times 19.55}$$

= 0.141 moles per litre.

= 0.141 × 36.5 grams per litre (M_r HCl = 36.5)

= 5.147 g/L.

* The average of 19.5 and 19.6.

Answer: Concentration of HCl solution is (a) 0.141 moles per litre, (b) 5.147 grams per litre.

Example 13.17

Hydrochloric acid and sodium hydroxide react according to the equation:

$HCl + NaOH \rightarrow NaCl + H_2O$

25 cm^3 of a sodium hydroxide solution was titrated against a 0.2 M HCl solution. The average titration figure was 23.5 cm^3. Calculate the concentration of the sodium hydroxide solution in (a) moles per litre and (b) grams per litre. (H = 1, O = 16, Na = 23)

SOLUTION

From the balanced equation, we see that one mole of acid reacts with one mole of base, i.e. $n_a = 1$ and $n_b = 1$.

GIVEN:

$V_a = 23.5$

$M_a = 0.2$

$n_a = 1$

$V_b = 25$

$M_b = ?$

$n_b = 1$

HCl | NaOH

$$\frac{V_a \times M_a}{n_a} = \frac{V_b \times M_b}{n_b}$$

$$\frac{23.5 \times 0.2}{1} = \frac{25 \times M_b}{1}$$

$$M_b = \frac{23.5 \times 0.2}{25}$$

= 0.188 moles per litre.

= 0.188 × 40 grams per litre (M_r NaOH = 40)

= 7.52 g/L

Answer: Concentration of NaOH solution is (a) 0.188 M, (b) 7.52 g/L.

IMPORTANT NOTE FOR STUDENTS:
One of the most common errors in solving volumetric problems is substituting the wrong values for n_a and n_b in the formula. As you will always be given the balanced equation in the question, the values for n_a and n_b will be obvious, i.e. n_a is the number in front of the acid and n_b is the number in front of the base. Writing down all the values in the form of a list as shown in the above examples will help to avoid errors.

Example 13.18

A chemist is asked to determine the percentage purity of a sample of potassium hydroxide. 3.0 g of the impure sample were dissolved in water and made up to 250 cm^3 in a volumetric flask. 25 cm^3 portions of the base were titrated against 0.1 M H_2SO_4 solution. The average titration figure was 22.9 cm^3. The equation for the reaction is:

$H_2SO_4 + 2KOH \rightarrow K_2SO_4 + 2H_2O$

Use the above information to calculate the percentage purity of the sample of potassium hydroxide. (H = 1, O = 16, K = 39)

SOLUTION

First calculate the concentration of the potassium hydroxide solution. From the balanced equation, we see that one mole of acid reacts with two moles of base, i.e. $n_a = 1$ and $n_b = 2$.

H_2SO_4 KOH

GIVEN:

$V_a = 22.9$

$M_a = 0.1$

$n_a = 1$

$V_b = 25$

$M_b = ?$

$n_b = 2$

$$\frac{V_a \times M_a}{n_a} = \frac{V_b \times M_b}{n_b}$$

$$\frac{22.9 \times 0.1}{1} = \frac{25 \times M_b}{2}$$

$$M_b = \frac{22.9 \times 0.1 \times 2}{25}$$

$= 0.183$ moles per litre.

$= 0.183 \times 56$ grams per litre (M_r KOH = 56)

$= 10.248$ grams per litre

$= \frac{10.248}{4}$ grams per 250 cm^3

$= 2.562$ grams per 250 cm^3

Note that the above formula always gives us the concentration in moles per litre. Since the volume of solution that we have is only 250 cm^3, this means that there would be four times less solid present – hence we must divide the answer in grams per litre by 4.

Thus, in the volumetric flask we have only 2.562 grams of potassium hydroxide. The rest of the 3 g of the sample is due to impurities.

$$\% \text{ purity} = \frac{\text{mass of pure KOH}}{\text{mass of impure sample}} \times 100$$

$$= \frac{2.562}{3.0} \times 100 = 85.4\%$$

Answer: 85.4%

Test Yourself:
Attempt questions 13.10 - 13.12 and W13.11 - W13.13.

EXPERIMENT 13.3 MANDATORY STUDENT EXPERIMENT 6A

To make up an approximate 0.1 M solution of sodium hydroxide, to standardise it (i.e. obtain the exact concentration) with a standard hydrochloric acid solution, and hence to prepare a sample of sodium chloride.

Introduction

In EXPERIMENT 13.1, we used a standard solution of sodium carbonate to determine the exact concentration of a hydrochloric acid solution. The sodium carbonate solution was said to be a primary standard. The hydrochloric acid solution is called a secondary standard. This hydrochloric acid solution may now be used to standardise other solutions, e.g. a solution of sodium hydroxide. Since sodium hydroxide absorbs water vapour from the air, we cannot weigh out a fixed amount of it and make up a solution of exact concentration. In this experiment, we will make up an approximate 0.1 M NaOH solution and standardise it by titrating it against a HCl solution of known concentration. The equation for the reaction is:

$HCl + NaOH \rightarrow NaCl + H_2O$

Apparatus required

Pipette (25 cm^3), pipette filler, burette, three beakers (plastic), spatula, laboratory balance, conical flask, wash bottle, retort stand, white tile, funnel, dropper, evaporating dish, hotplate or Bunsen burner.

Chemicals required

Sodium hydroxide pellets, standard hydrochloric acid solution, methyl orange indicator, deionised water.

Procedure

1. Weigh out about one gram of sodium hydroxide on a clock glass. **CAUTION! DO NOT ALLOW THE SODIUM HYDROXIDE TO COME IN CONTACT WITH THE SKIN AS IT CAUSES SEVERE BURNS!** Since we will be finding the concentration of the sodium hydroxide solution by titration, the mass of sodium hydroxide need not be accurately known.
2. Taking the usual precautions as outlined in EXPERIMENT 13.1, transfer the sodium hydroxide to a beaker containing about 100 cm^3 of deionised water. Stir the sodium hydroxide with a glass rod until it is dissolved.
3. Transfer the solution to a 250 cm^3 volumetric flask and make up to the graduation mark with deionised water. Ensure that the solution is homogeneous by inverting the flask about 20 times.
4. Pour about 100 cm^3 of the sodium hydroxide solution from the volumetric flask into a clean, dry beaker. Also pour about 100 cm^3 of the hydrochloric acid solution into another clean, dry beaker. Label both beakers.
5. Rinse out the pipette with deionised water and then with some of the sodium hydroxide solution. Rinse out the burette with water and then with some of the hydrochloric acid solution.
6. Pipette 25 cm^3 of the sodium hydroxide solution into a conical flask that has been previously rinsed with deionised water.

7. Add about three drops of methyl orange indicator to the sodium hydroxide solution in the conical flask.
8. Using a small funnel, fill the burette with the dilute HCl solution. Remove the funnel from the burette.
9. Adjust the bottom of the meniscus to the zero graduation mark.
10. Place a white tile on the base of the retort stand and titrate the sodium hydroxide against the HCl taking all the usual precautions. For methyl orange indicator, the colour change is from yellow to red.
11. Perform one rough and two accurate titrations agreeing to within 0.1 cm^3. Do not dispose of the sodium hydroxide solution left over as it will be required in EXPERIMENT 13.4.
12. If time permits, a sample of sodium chloride may be isolated by repeating the above titration. When repeating the titration, leave out the indicator and add the same volume of hydrochloric acid as obtained in the above titration to the sodium hydroxide in the conical flask. Pour the contents of the conical flask into an evaporating dish and evaporate almost to dryness over a hotplate or Bunsen burner. Allow the solution to cool. Note the formation of white crystals of sodium chloride in the dish. These crystals may be left to dry on a piece of filter paper.

Calculations

From your titration results, work out the exact concentration of the sodium hydroxide solution. Express the concentration of the sodium hydroxide solution in (a) moles per litre and (b) grams per litre.

EXPERIMENT 13.4 MANDATORY STUDENT EXPERIMENT NO. 7

To determine the percentage of ethanoic acid (acetic acid) in vinegar.

Introduction

Vinegar is a dilute solution of an acid called ethanoic acid, CH_3COOH, the old name of which was acetic acid. The amount of ethanoic acid in vinegar can be determined by titrating the vinegar against a standard sodium hydroxide solution. The equation for the reaction is:

$$CH_3COOH + NaOH \rightarrow CH_3COONa + H_2O$$

To detect the end point, we use phenolphthalein indicator. Phenolphthalein is colourless in an acid and pink in a base. (Phenolphthalein is suitable because we are titrating a weak acid against a strong base. This will be explained in more detail in CHAPTER 18.)

As the concentration of sodium hydroxide we are using is quite small (0.1 M approx.), we will dilute the vinegar so that the end point will not be too low.

Apparatus required

Pipette (25 cm^3), pipette filler, burette, conical flask, wash bottle, retort stand, three beakers (plastic), white tile, funnel, dropper.

Chemicals required

Sample of vinegar, standard sodium hydroxide solution, phenolphthalein indicator, deionised water.

Procedure

1. Dilute the sample of vinegar as follows: pipette two 25 cm^3 samples of vinegar into a clean 250 cm^3 volumetric flask and make up to the mark with deionised water. Note that we have diluted the vinegar from 50 cm^3 to 250 cm^3, i.e. five times.
2. Wash out the pipette with deionised water and then with the standard sodium hydroxide solution.
3. Pipette 25 cm^3 of the sodium hydroxide solution into a conical flask that has been previously rinsed with deionised water.
4. Add about three drops of phenolphthalein indicator to the sodium hydroxide solution in the conical flask.
5. Rinse out the burette with water and then with some of the diluted vinegar.
6. Using a small funnel, fill the burette with the diluted vinegar.
7. Remove the funnel from the burette. Adjust the meniscus to the zero graduation mark.
8. Place a white tile on the base of the retort stand and titrate the sodium hydroxide solution against the diluted vinegar taking all the usual precautions. For phenolphthalein, the colour change is from pink to colourless.
9. Perform one rough and two accurate titrations agreeing to within 0.1 cm^3.

Note: When using phenolphthalein indicator, it is sometimes difficult to see the colour change from pink to colourless. It is easier to see a colour change from colourless to pink, i.e. it is easier to see a colour appearing in the conical flask rather than disappearing. For this reason, you could also perform the above titration by putting the acid in the conical flask and the base in the burette. However, if you put sodium hydroxide in the burette, it is vital that you wash out the burette thoroughly with deionised water after use. Why is this washing out of the burette of particular importance?

Calculations

From the results of your titrations, work out the concentration of ethanoic acid in the original vinegar in (a) moles per litre, (b) grams per litre and (c) % w/v. A sample calculation is given in EXAMPLE 13.19.

Example 13.19

A 20 cm^3 sample of vinegar was diluted to 100 cm^3 in a volumetric flask. This diluted solution was then titrated against 20 cm^3 of 0.12 M NaOH solution. The balanced equation for the reaction is:
$CH_3COOH + NaOH \rightarrow CH_3COONa + H_2O$
The average titration figure was 14.2 cm^3. Calculate the concentration of ethanoic acid in the original vinegar in (a) mol/L, (b) g/L and (c) % w/v. (H = 1, C = 12, O = 16)

SOLUTION

From the balanced equation, we see that one mole of acid reacts with one mole of base, i.e. $n_a = 1$ and $n_b = 1$.

GIVEN:
$V_a = 14.2$
$M_a = ?$
$n_a = 1$
$V_b = 20$
$M_b = 0.12$
$n_b = 1$

$(CH_3COOH) \qquad (NaOH)$

$$\frac{V_a \times M_a}{n_a} = \frac{V_b \times M_b}{n_b}$$

$$\frac{14.2 \times M_a}{1} = \frac{20 \times 0.12}{1}$$

$$M_a = \frac{20 \times 0.12}{14.2}$$

$= 0.169$ moles per litre.

Since the solution was diluted by a factor of five, concentration of ethanoic acid in original vinegar $= 0.169 \times 5$ moles per litre
$= 0.845$ mol/L........................(a)
$= 0.845 \times 60$ g/L $\quad$ (M_r $CH_3COOH = 60$)
$= 50.7$ g/L..............................(b)

Since there are 50.7 g of ethanoic acid in a litre of vinegar, there are 5.07 g of ethanoic acid in 100 cm^3 of vinegar.
Therefore, % w/v = 5.07% ....(c)

Answer: Concentration of ethanoic acid in vinegar is (a) 0.845 M (b) 50.7 g/L (c) 5.07% w/v.

Test Yourself:
Attempt questions 13.13 – 13.14 and W13.14 – W13.15.

2. CALCULATING THE AMOUNT OF WATER OF CRYSTALLISATION IN A COMPOUND AND THE RELATIVE MOLECULAR MASS OF THE COMPOUND FROM TITRATION DATA

We now consider how the results of a titration may be used to determine the percentage of water of crystallisation in washing soda crystals. (Water of crystallisation is water that makes up part of the crystal structure of a compound.) Crystals that contain water of crystallisation are said to be **hydrated**. Washing soda (FIG 13.18) consists of hydrated crystals of sodium carbonate, $Na_2CO_3.xH_2O$. EXPERIMENT 13.5 shows how the value of x is determined.

Fig 13.18
Washing soda consists of crystals of hydrated sodium carbonate.

EXPERIMENT 13.5 MANDATORY STUDENT EXPERIMENT NO. 8

To determine the percentage of water of crystallisation in hydrated sodium carbonate (washing soda).

Introduction

In this experiment, we will make up a solution of washing soda and titrate it against a hydrochloric acid solution of known concentration. From the results of our titration, we will work out the percentage of water of crystallisation in the washing soda.

Apparatus required

Pipette (25 cm^3), pipette filler, spatula, laboratory balance, burette, conical flask, wash bottle, retort stand, three beakers (plastic), white tile, funnel, dropper, clock glass.

Chemicals required

Washing soda crystals, standardised hydrochloric acid solution (0.1 M or 0.2 M), methyl orange indicator, deionised water.

Procedure

1. Wash out the pipette, burette and conical flask with deionised water.
2. Using a clock glass, weigh out accurately about 5 g of washing soda crystals.
3. Taking the precautions outlined earlier, dissolve the crystals in about 100 cm^3 of deionised water. Stirring with a glass rod helps the crystals to dissolve.
4. Transfer the solution to a 250 cm^3 volumetric flask taking all the usual precautions to avoid loss of solution. Make the solution up to the mark in the usual manner and invert the volumetric flask about 20 times.
5. Rinse out the pipette with some of the sodium carbonate solution and the burette with some of the hydrochloric acid solution.
6. Pipette 25 cm^3 of the sodium carbonate solution into a conical flask.
7. Add about three drops of methyl orange indicator to the sodium carbonate solution in the conical flask.
8. Clamp the burette vertically using the retort stand and fill the burette to the mark with the HCl solution (FIG 13.19).
9. Remove the funnel from the burette and fill the part of the burette below the tap. Adjust the bottom of the meniscus to the zero graduation mark. With the white tile in place, titrate in the usual manner until the yellow colour of the indicator changes to pink.
10. Note the titration figure and repeat the experiment to get two more readings that agree to within 0.1 cm^3.

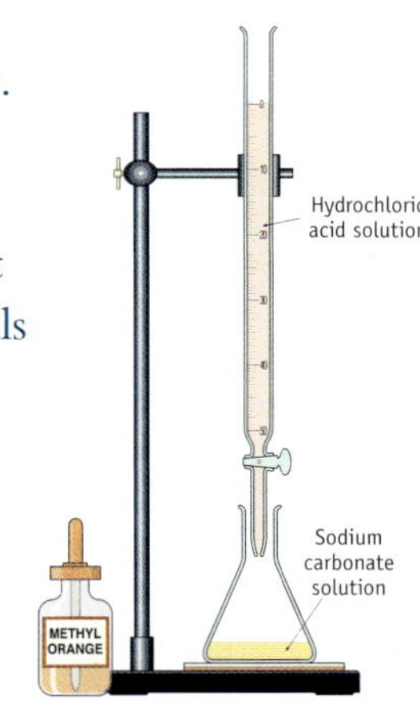

Fig 13.19
Titration to determine the percentage of water of crystallisation in washing soda crystals.

Calculations

The method of calculating the percentage of water of crystallisation in hydrated sodium carbonate is shown in EXAMPLE 13.20.

Example 13.20

A student is asked to analyse a sample of washing soda, $Na_2CO_3.xH_2O$, in order to find the number of moles of water per mole of sodium carbonate. He weighs out 8.61g of the washing soda, dissolves this amount in water and makes the solution up to 250 cm^3 in a volumetric flask. 25 cm^3 of this solution required an average of 28.7 cm^3 of 0.21 M HCl for neutralisation. Calculate (i) the percentage of water of crystallisation in the washing soda crystals and (ii) the value of x in the above formula. The balanced equation for the reaction is:

$$Na_2CO_3 + 2HCl \rightarrow 2NaCl + H_2O + CO_2$$

SOLUTION

In solving problems of this type, first use the titration data to calculate the concentration of the sodium carbonate solution in moles per litre.

From the balanced equation, we see that two moles of acid react with one mole of base, i.e. $n_a = 2$ and $n_b = 1$.

GIVEN:

$V_a = 28.7$

$M_a = 0.21$

$n_a = 2$

$V_b = 25$

$M_b = ?$

$n_b = 1$

$$\underset{(HCl)}{\frac{V_a \times M_a}{n_a}} = \underset{(Na_2CO_3)}{\frac{V_b \times M_b}{n_b}}$$

$$\frac{28.7 \times 0.21}{2} = \frac{25 \times M_b}{1}$$

$$M_b = \frac{28.7 \times 0.21}{2 \times 25}$$

$$= 0.121 \text{ moles per litre.}$$

Since the concentration of the sodium carbonate solution is 0.121 moles per litre, there must be $\frac{0.121}{4} = 0.03$ moles of sodium carbonate in 250 cm^3 solution.

$1 \text{ mole } Na_2CO_3 = 106 \text{ g}$

$\therefore 0.03 \text{ mole } Na_2CO_3 = 0.03 \times 106 = 3.18 \text{ g}$

In the 8.61 g of $Na_2CO_3.xH_2O$ crystals, there are only 3.18 g of Na_2CO_3.

$\therefore$ mass of water of crystallisation = 8.61 - 3.18

= 5.43 g

$$\% \text{ water of crystallisation} = \frac{5.43}{8.61} \times 100$$

$$= 63.07\% \quad \text{..(i)}$$

We now calculate the value of x in the formula $Na_2CO_3.xH_2O$. We note from the above that, by dissolving 8.61 g of washing soda in 250 cm^3 of solution, we obtain a solution which contains 0.03 moles per 250 cm^3.

Therefore, 0.03 moles = 8.61 g

$$1 \text{ mole} = \frac{8.61}{0.030}$$

$$= 287$$

i.e. Rel Molecular Mass of $Na_2CO_3.xH_2O$ = 287

$$106 + 18x = 287$$
$$18x = 287 - 106$$
$$= 181$$
$$x = 10 \text{ (x must be a whole number)......(ii)}$$

(The mathematical calculations give us a value for x of 10.06. The slight deviation is probably due to experimental error.)

Thus, the formula of hydrated sodium carbonate is $Na_2CO_3.10H_2O$

Answer: (i) 63.07%, (ii) x = 10

Example 13.21

A student is asked to analyse a sample of a metal carbonate whose formula is M_2CO_3. She does this by dissolving 2.11 g of the carbonate in water and diluting it to 250 cm^3 in a volumetric flask. 25 cm^3 samples of this carbonate are titrated against 0.12 M HCl solution. The average titration figure is 25.4 cm^3. The balanced equation for the reaction is:

$M_2CO_3 + 2HCl \rightarrow 2MCl + H_2O + CO_2$

Calculate the relative molecular mass of the carbonate and hence identify the metal M.

SOLUTION

In solving problems of this type, we first use the titration data to calculate the concentration of the substance in moles per litre. We then relate this number of moles to the mass of the substance dissolved in the water.

From the balanced equation, we see that two moles of acid react with one mole of base, i.e. $n_a = 2$ and $n_b = 1$.

GIVEN:

$V_a = 25.4$

$M_a = 0.12$

$n_a = 2$

$V_b = 25$

$M_b = ?$

$n_b = 1$

$$\underset{(HCl)}{\frac{V_a \times M_a}{n_a}} = \underset{(Na_2CO_3)}{\frac{V_b \times M_b}{n_b}}$$

$$\frac{25.4 \times 0.12}{2} = \frac{25 \times M_b}{1}$$

$$M_b = \frac{25.4 \times 0.12}{2 \times 25} = 0.061 \text{ moles per litre.}$$

Therefore, in the 250 cm^3 volumetric flask there must be $\frac{0.061}{4} = 0.0153$ moles of the carbonate, i.e. by dissolving 2.11 g of the carbonate in 250 cm^3 of solution, we obtain a solution which contains 0.0153 moles in 250 cm^3.

Therefore, 0.0153 moles = 2.11g

$$1 \text{ mole} = \frac{2.11}{0.0153}$$
$$= 137.91$$

i.e. M_2CO_3 has a relative molecular mass of 137.91

2M + 12 + 3(16) = 137.91

Solving for M, M = 39

The metal that has a relative atomic mass of 39 is potassium.

Answer: Relative molecular mass is 137.91. M is potassium.

Test Yourself:
Attempt questions 13.15 – 13.16 and W13.16 – W13.17.

CHAPTER SUMMARY

- The concentration of a solution is the amount of solute that is dissolved in a given volume of solution.
- Concentration may be expressed in terms of percentage of solute or in parts per million or in moles per litre (molarity).
- A 1 M solution is one which contains one mole of the solute dissolved in one litre of solution.
- To calculate the number of moles, given the volume (in cm^3) and the molarity, use the formula: number of moles $= \frac{\text{volume} \times \text{molarity}}{1000}$
- For calculations involving dilution of solutions use the formula: $\frac{Vol_{dil} \times M_{dil}}{1000} = \frac{Vol_{conc} \times M_{conc}}{1000}$
- A standard solution is a solution of precisely known concentration.
- A primary standard is a substance which can be obtained in a stable, pure and soluble solid form so that it can be weighed out and dissolved in water to give a solution of accurately known concentration.
- A titration is a laboratory procedure where a measured volume of one solution is added to a known volume of another solution until the reaction is complete.
- Many volumetric problems may be solved using the formula: $\frac{V_a \times M_a}{n_a} = \frac{V_b \times M_b}{n_b}$

EXERCISES

When answering these questions use the following values of relative atomic masses: H = 1, He = 4, Li = 7, C = 12, N = 14, O =16, F = 19, Na = 23, Mg = 24, Al = 27, Si = 28, P = 31, S = 32, Cl = 35.5, K = 39, Ca = 40, Cr = 52, Fe = 56, Cu = 63.5, Zn = 65, Br = 80, Ag = 108, Sn = 119, I = 127, Pb = 207.

Concentrations of solutions.

13.1. A sample of sea water has a mass of 1.3 kg. On evaporating the water, 148 g of salt was recovered from it. Express the concentration of the salt as % w/w.

13.2. Convert the following concentrations to ppm:
(a) 0.75 g/L
(b) 0.06 g/L
(c) 0.017 g/100 cm^3
(d) 0.0003 g/100cm^3.

13.3. Calculate the concentration in grams per litre of:
(a) 1 M NaOH
(b) 2 mol/L KOH
(c) 0.125 M H_2SO_4
(d) 0.01 mol/L Na_2CO_3
(e) 0.005 mol/L HNO_3.

13.4. What is the concentration in moles per litre of the following solutions:
(a) 1 g NaOH in 1 litre of solution
(b) 24 g H_2SO_4 in 500 cm^3 of solution
(c) 22 g HCl in 200 cm^3 of solution
(d) 20 g NaOH in 100 cm^3 of solution
(e) 4 g NaOH in 750 cm^3 of solution.

13.5. Calculate the number of moles of sulfuric acid in:

(a) 25 cm^3 of 0.01 M H_2SO_4
(b) 18 cm^3 of 2.5 M H_2SO_4
(c) 8.7 cm^3 of 0.5 M H_2SO_4
(d) 100 cm^3 of 0.005 M H_2SO_4
(e) 50 cm^3 of 0.09 M H_2SO_4.

Reaction between a solution and a solid.

13.6. Iron reacts with dilute sulfuric acid according to the equation:
$Fe + H_2SO_4 \rightarrow FeSO_4 + H_2$
What mass of iron will react with 75 cm^3 of 0.5 M H_2SO_4?

13.7. It is required to neutralise 25 cm^3 of a 2 M H_2SO_4 solution using some sodium hydroxide pellets. Sodium hydroxide reacts with sulfuric acid according to the equation:
$2NaOH + H_2SO_4 \rightarrow Na_2SO_4 + 2H_2O$
What mass of sodium hydroxide pellets is required?

Dilution of solutions.

13.8. What volume of 10 M HCl would be required to make 500 cm^3 of 0.25 M solution?

13.9. What volume of 18 M H_2SO_4 would be required to prepare 10 litres of 1 M H_2SO_4?

Titrations. Volumetric Problems.

13.10. With regard to the procedure for carrying out a titration, answer each of the following:

(a) Why is the tap of the burette opened briefly before the titration is begun?
(b) Why is it necessary to remove the funnel from the burette before the titration is begun?
(c) What is the purpose of placing the conical flask on a white tile?
(d) Why is a volumetric flask preferable to a graduated cylinder for making up a solution?
(e) Why is a conical flask preferable to a beaker for carrying out a titration?
(f) Which of the pieces of apparatus, the pipette, the burette or the conical flask, should **not** be rinsed with the solution it is to contain?
(g) What is the purpose of carrying out a rough titration?
(h) Why should the conical flask be swirled during the titration?
(i) Would any error be introduced if the conical flask contained some droplets of deionised water before the base is pipetted into it?
(j) Sodium hydroxide is not normally placed in burettes. What is the reason for this?
(k) When reading the volume of liquid in a burette, what precautions should be taken to ensure that the reading is taken correctly?
(l) Why is it necessary to wash down the sides of the conical flask during a titration?
(m) Why is it not advisable to insert a pipette into a volumetric flask in order to transfer some solution directly to a conical flask?

13.11. It was found that 11.5 cm^3 of a 0.5 M sulfuric acid solution neutralised 25 cm^3 of a given solution of sodium hydroxide. The balanced equation for the reaction is:
$H_2SO_4 + 2NaOH \rightarrow Na_2SO_4 + 2H_2O$
Calculate the concentration of the sodium hydroxide solution in (a) mol/L and (b) g/L.

13.12. In carrying out an experiment, 2.5 g of an impure sample of caustic soda (sodium hydroxide), NaOH, were dissolved in water and made up to 250 cm^3 in a volumetric flask. 25 cm^3 portions of the base were titrated against 0.15 M H_2SO_4 solution. The titration figures obtained were 17.1 cm^3, 16.7cm^3, 16.8 cm^3. Calculate the percentage purity of the sample of caustic soda.

13.13. A 20 cm^3 sample of vinegar was diluted to 100 cm^3 in a volumetric flask. This diluted solution was then titrated against 25 cm^3 of 0.11 M NaOH solution. The balanced equation for the reaction is:

$CH_3COOH + NaOH \rightarrow CH_3COONa + H_2O$

The average titration figure was 15.8 cm^3. Calculate the concentration of ethanoic acid in the original vinegar in (a) mol/L, (b) g/L and (c) % w/v.

13.14. A 25 cm^3 sample of vinegar was diluted to 250 cm^3 in a volumetric flask. This diluted solution was then titrated against 25 cm^3 of 0.09 M NaOH solution. The balanced equation for the reaction is:

$CH_3COOH + NaOH \rightarrow CH_3COONa + H_2O$

The average titration figure was 26.55 cm^3. Calculate the concentration of ethanoic acid in the original vinegar in (a) mol/L, (b) g/L and (c) % w/v.

Water of crystallisation. Relative Molecular Mass.

13.15. In an experiment, 7.16 g of hydrated sodium carbonate, $Na_2CO_3.xH_2O$, were weighed out, dissolved in water, and the solution was made up to 250 cm^3 in a volumetric flask. 25 cm^3 of this solution were titrated against 0.17 M hydrochloric acid solution. The average titration figure was 29.38 cm^3. The balanced equation for the reaction is:

$Na_2CO_3 + 2HCl \rightarrow 2NaCl + H_2O + CO_2$

Calculate (a) the percentage of water of crystallisation in the compound and (b) the value of x in the formula $Na_2CO_3.xH_2O$.

13.16. In carrying out an experiment, 3.55 g of hydrated sodium carbonate, $Na_2CO_3.xH_2O$, were weighed out, dissolved in water, and the solution was made up to 500 cm^3 in a volumetric flask. 25 cm^3 of this solution were titrated against 0.11 M hydrochloric acid solution. The average titration figure was 22.8 cm^3. The balanced equation for the reaction is:

$Na_2CO_3 + 2HCl \rightarrow 2NaCl + H_2O + CO_2$

Calculate (a) the percentage of water of crystallisation in the compound and (b) the value of x in the formula $Na_2CO_3.xH_2O$.

Oxidation and Reduction

14

14.1 INTRODUCTION

Oxygen is the most abundant element in the Earth's crust and in the human body. It is also one of the most reactive elements – it combines with the vast majority of the other elements. Oxygen is involved in many of the most important chemical reactions in our lives, e.g. burning, respiration, rusting, etc., (Fig 14.1). Therefore, it is not surprising that chemists classify many chemical reactions in terms of the reaction with oxygen.

Fig 14.1
Burning of fuels is an example of an oxidation reaction.

In the early days of chemistry, **oxidation** was defined as simply **the addition of oxygen to a substance**. For example, when coal is burned, the chemical reaction that takes place may be described as the combining of carbon in the coal with oxygen in the air to form the gas carbon dioxide.

carbon + oxygen → carbon dioxide

$$C + O_2 \rightarrow CO_2$$

C gains oxygen
∴ oxidation

A similar oxidation reaction occurs when magnesium metal is burned in air.

magnesium + oxygen → magnesium oxide

$$2Mg + O_2 \rightarrow 2MgO$$

Mg gains oxygen
∴ oxidation

One of the most commonly observed oxidation reactions is the rusting of iron. This may be represented in a simplified form as:

$$4Fe + 3O_2 \rightarrow 2Fe_2O_3$$

Fe gains oxygen
∴ oxidation

Chemists also discovered that it was possible to remove oxygen from some substances. For example, when hydrogen gas is passed over heated copper(II) oxide, it is found that copper metal is obtained from the copper(II) oxide.

copper(II) oxide + hydrogen → copper + water

$$CuO + H_2 \rightarrow Cu + H_2O$$

CuO loses oxygen
∴ reduction

Since this type of reaction was often used to extract metals from their ores, it was often referred to as a **reduction** reaction since a large amount of ore was converted or reduced to a smaller mass of metal. Note in the above reaction that the copper(II) oxide has lost oxygen. Thus, reduction could be said to be the opposite to oxidation. **Reduction** may be defined as **the removal of oxygen from a substance**. Since hydrogen was often used to remove the oxygen, **reduction** was also defined as **the addition of hydrogen**.

14.2 OXIDATION AND REDUCTION IN TERMS OF ELECTRON TRANSFER

After the discovery of the electron, chemists began to take a closer look at various chemical reactions. It became clear that many chemical reactions involved the transfer of electrons. Let us again consider the oxidation of magnesium to magnesium oxide.

$$2Mg + O_2 \rightarrow 2Mg^{2+}O^{2-}$$

Mg loses two electrons

oxidation

We see that what is actually happening in this reaction is that the magnesium is losing electrons and the oxygen is gaining electrons. In fact, each magnesium atom is losing two electrons. Therefore, oxidation may also be defined in terms of electron loss.

> Oxidation of an element takes place when it loses electrons.

Consider now the reduction of copper(II) oxide to copper metal.

$$Cu^{2+}O^{2-} + H_2 \rightarrow Cu + H_2O$$

Cu^{2+} gains two electrons

reduction

We see that what is actually happening is that the Cu^{2+} ion is gaining two electrons to become a copper atom. Thus, reduction may also be defined in terms of electron transfer.

> Reduction of an element takes place when it gains electrons.

The words **OIL RIG** ('Oxidation Is Loss of electrons, Reduction Is Gain of electrons') may be used to help you remember these definitions.

If we think of oxidation and reduction in terms of electrons being transferred, it becomes clear that oxidation and reduction must always occur together. If one substance loses electrons (oxidation), there must be some other substance to accept these electrons (reduction). This may easily be seen in the reaction between sodium and chlorine to form sodium chloride.

Each Na atom loses one electron

oxidation

$$2Na + Cl_2 \rightarrow 2Na^+Cl^-$$

reduction

Each Cl atom gains one electron

In other words, oxidation reactions and reduction reactions always occur together. Thus, it is more correct to refer to these types of reactions as **oxidation-reduction** reactions or **redox** reactions. Note also that in the above reaction neither oxygen nor hydrogen is present. Therefore, it is clear that the definitions of oxidation and reduction in terms of electron transfer are broader than the previous definitions that described oxidation and reduction in terms of oxygen and hydrogen.

If we take a closer look at the reaction between magnesium and oxygen to form magnesium oxide, we see that this is another example of an oxidation-reduction reaction.

Each Mg atom loses two electrons

oxidation

$$2Mg + O_2 \rightarrow 2Mg^{2+}O^{2-}$$

reduction

Each oxygen atom gains two electrons

An interesting experiment in which the results of an oxidation-reduction reaction are visible to the observer is the reaction between zinc metal and copper ions (FIG 14.2).

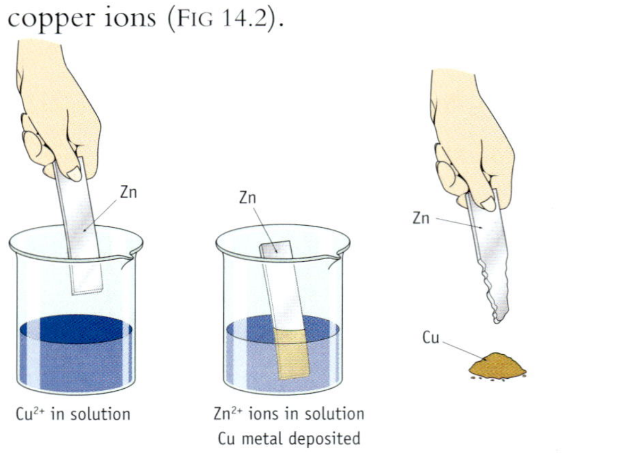

Fig 14.2
When zinc metal is left standing in a solution containing copper ions, an oxidation-reduction reaction takes place. This reaction causes copper metal to be deposited and zinc goes into solution as Zn^{2+} ions.

If zinc metal is left standing in copper sulfate solution, it is observed that the zinc becomes covered with a reddish deposit which, on analysis, is found to be copper metal. It is also observed that the blue colour of the solution (caused by the presence of Cu^{2+} ions) begins to fade. This fading implies that the Cu^{2+} ions are being used up. On analysing the solution in the beaker, it is found that Zn^{2+} ions are present. The above evidence led chemists to conclude that the following chemical reaction was taking place:

$$Zn + Cu^{2+} \rightarrow Zn^{2+} + Cu$$

If we study this equation closely, we see that the Zn is oxidised and the Cu^{2+} is reduced:

Each Zn atom loses two electrons

oxidation

$$Zn + Cu^{2+} \rightarrow Zn^{2+} + Cu$$

reduction

Each Cu^{2+} ion gains two electrons

In this reaction, the zinc atom could not lose electrons (oxidation) without the presence of something to accept these electrons (i.e. the Cu^{2+} ion). We say that the Cu^{2+} ion is an **oxidising agent**.

> An oxidising agent is a substance that brings about oxidation in other substances.

If we look at the above equation, we see that the Cu^{2+} ion (the oxidising agent) accepts two electrons from the Zn atom, i.e. the Cu^{2+} ion is itself reduced. This is a general characteristic of oxidising agents – **the oxidising agent is always reduced**. The substance that causes the Cu^{2+} ion to be reduced is Zn. We say that Zn is the **reducing agent**.

> A reducing agent is a substance that brings about reduction in other substances.

In order for a substance to act as a reducing agent, it must donate electrons to some other substance, e.g. in the above equation Zn gives two electrons to the Cu^{2+} ion. When it does this, the Zn itself is oxidised. This is a general characteristic of reducing agents – **the reducing agent is always oxidised**.

We can now indicate the oxidising and reducing agents in the above equation.

Each Zn atom loses two electrons

Reducing agent → Zn; oxidation

$$Zn + Cu^{2+} \rightarrow Zn^{2+} + Cu$$

Oxidising agent → Cu^{2+}; reduction

Each Cu^{2+} ion gains two electrons

One of the most common oxidising agents is hydrogen peroxide, H_2O_2, which is used to bleach hair. The hydrogen peroxide converts the coloured pigment in the hair to a colourless hair 'pigment' which leaves the hair blonde – hence the expression 'peroxide blonde' (FIG 14.3).

coloured hair pigment + H_2O_2 → colourless hair pigment + H_2O

(H_2O_2: oxidising agent)

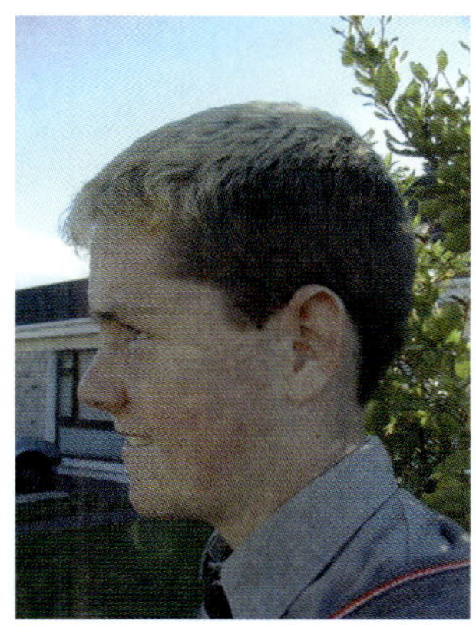

Fig 14.3
Hydrogen peroxide is a very useful oxidising agent. It oxidises the coloured pigments in hair to a blonde colour.

Another example of an oxidising agent is 'tincture of iodine' which is often found in first aid boxes. This is a solution of I_2 in ethanol and is used to prevent infection of cuts.

live germ + $I_2 \rightarrow 2I^-$ + dead germ

(I_2: oxidising agent)

Similarly, small amounts of chlorine are added to drinking water and swimming pool water to oxidise chemicals in the cells of germs. This kills the germs and disinfects the water.

live germ + $Cl_2 \rightarrow 2Cl^-$ + dead germ

(Cl_2: oxidising agent)

Common household bleach, e.g. *Parazone*, *Domestos*, etc. contain a compound of chlorine called sodium hypochlorite, NaClO. The hypochlorite ions, ClO^-, oxidise the coloured material in fabrics to a colourless material.

$$\text{coloured fabric} + \underset{\substack{\uparrow \\ \text{oxidising agent}}}{ClO^-} \rightarrow Cl^- + \text{colourless fabric}$$

A very useful reducing agent in industry is carbon monoxide. This gas is used to remove the oxygen from iron ore in order to convert it to pure iron.

$$Fe_2O_3 + \underset{\substack{\uparrow \\ \text{reducing agent}}}{3CO} \rightarrow 2Fe + 3CO_2$$

Another useful reducing agent is sulfur dioxide which is used in the bleaching of wood pulp in the manufacture of paper. It is also used in the bleaching of textiles and straw. It is used in making straw hats; it converts the yellow colour of the dye in straw to a colourless form of the dye (FIG 14.4).

Fig 14.4
Sulfur dioxide is a useful reducing agent. It is used to bleach straw and paper. The reducing action of sulfur dioxide is slowly reversed by the oxygen in the air. Hence, paper and straw often turn yellow with age.

$$\text{coloured form of dye} + \underset{\substack{\uparrow \\ \text{reducing agent}}}{SO_2} \rightarrow \text{colourless form of dye}$$

Example 14.1

Define oxidation in terms of electron transfer. In the following example indicate clearly, by showing electron loss or gain, which species is being oxidised and which is being reduced:

$2KI + Cl_2 \rightarrow 2KCl + I_2$

What are the oxidising and reducing agents in the above reaction? [LCH]

SOLUTION

Oxidation is the loss of electrons.

The electron loss and electron gain may be more easily seen if we ignore the K^+ ion ('spectator ion') on each side of the equation. All of the required information may then be summarised as follows:

Each I^- atom loses one electron — oxidation ($2I^- \rightarrow I_2$)

Reducing agent: $2I^-$

$$2I^- + Cl_2 \rightarrow 2Cl^- + I_2$$

Oxidising agent: Cl_2

reduction ($Cl_2 \rightarrow 2Cl^-$) — Each Cl atom gains one electron

Note: Once you have established what is oxidised and what is reduced, it is easy to pick out the oxidising and reducing agents. In EXAMPLE 14.1, since the I^- is oxidised, the oxidising agent must be the Cl_2. Similarly, since the Cl_2 is reduced, the reducing agent must be I^-. The question merely asks you to indicate the various items so full marks are obtained for laying out your work as shown above. There is no need for you to give any further explanations.

Test yourself:
Attempt questions 14.1 and 14.2.

14.3 OXIDATION NUMBERS

As chemistry developed, it became clear that the definitions of oxidation and reduction needed to be widened still further. We have already seen that oxidation-reduction reactions need not involve oxygen at all. In addition, it was realised that not all oxidation-reduction reactions involved electron transfer, i.e. ionic compounds were not formed in all chemical reactions. For example, consider the reaction between hydrogen and oxygen:

$$2H_2 + O_2 \rightarrow 2H_2O$$

This reaction is very different from the reaction of magnesium with oxygen as no ions are formed. Water is a covalent compound. In order to keep track of electrons during chemical reactions involving covalent compounds, chemists introduced the idea of **oxidation numbers** (also called **oxidation states**). These are simply numbers that are assigned to atoms to help us decide how much they are oxidised or reduced.

> Oxidation number is the charge that an atom has or appears to have when electrons are distributed according to certain rules.

Oxidation number is also described as the charge which an atom appears to have in a compound if the bonding is assumed to be completely ionic. For example, in the water molecule the two electrons in each of the single covalent bonds are counted with the more electronegative element, i.e. the oxygen atom (Fig 14.5).

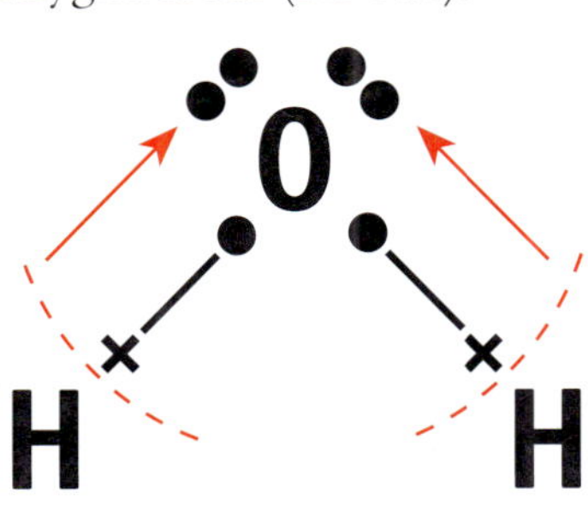

Fig 14.5
When assigning oxidation numbers, the electrons are assigned to the more electronegative element.

Since we have 'transferred ownership' (on paper) of two electrons from hydrogen to oxygen, we assign an oxidation number of -2 to the oxygen atom. Since each hydrogen atom has formally lost an electron, we assign an oxidation number of +1 to each of the hydrogen atoms. This may be indicated by writing:

$$\underset{+1\ -2}{H_2O}$$

Since oxidation numbers are not actual charges, they are usually written under the symbol of the corresponding element. This avoids confusion when dealing with charged species.

In order to make it easy for us to assign oxidation numbers, chemists have developed the following rules.

Rules For Assigning Oxidation Numbers

1. **The oxidation number of any uncombined element is zero.**
   An uncombined element is an element that is not bonded to another element, e.g. Zn, Fe, O_2, Cl_2, etc. This means that in H_2 the oxidation number (O.N.) of hydrogen is 0, in sodium metal (Na) the O.N. of Na is 0, in F_2 the O.N. of F is 0, etc.
2. **The oxidation number of an ion of an element is the same as its charge.**
   This means that the O.N. of Br in the Br^- ion is −1, the O.N. of Fe in Fe^{2+} is +2, the O.N. of Na in Na^+Cl^- is +1, the O.N. of Cl in Na^+Cl^- is -1, the O.N. of Mg in $Mg^{2+}O^{2-}$ is +2, etc.
   It follows that some elements nearly always have the same oxidation number in their compounds. Such elements are the alkali metals (Li, Na, K, etc.) which have an oxidation number of +1. Similarly, the alkaline earth metals (Be, Mg, Ca, etc.) have an oxidation number of +2. You will find that the oxidation numbers of these two groups of elements will be very useful as reference points for assigning oxidation numbers to elements in compounds.
3. **The sum of the oxidation numbers of all the elements in a compound must add up to zero.**
   For example, in NaCl the sum of the oxidation numbers is +1 (for sodium) added to -1 (for chlorine) which equals zero. Similarly, in H_2O the sum is 2(+1) for the hydrogen atoms added to -2 for the oxygen atom giving a total of zero.

4. **Oxygen always has an oxidation number of -2 except in peroxides (when it has an oxidation number of -1) and in the compound OF_2 (when it has an oxidation number of +2).**

For example, in magnesium oxide, $Mg^{2+}O^{2-}$, the oxidation number of oxygen is -2. However, in hydrogen peroxide, H_2O_2, the oxygen is assigned an oxidation number of -1. This is done so that the sum of all of the oxidation numbers in H_2O_2 will add up to zero, i.e. 2(+1) for the two hydrogen atoms and 2(-1) for the two oxygen atoms.

A similar reason holds in the case of oxygen difluoride, OF_2. Fluorine is the only element that is more electronegative than oxygen. Therefore, in the OF_2 molecule (Fig 14.6), the electrons in the two covalent bonds are assigned to the fluorine atoms. Therefore, in order that the sum of all of the oxidation numbers adds up to zero, the oxygen atom must be assigned an oxidation number of +2.

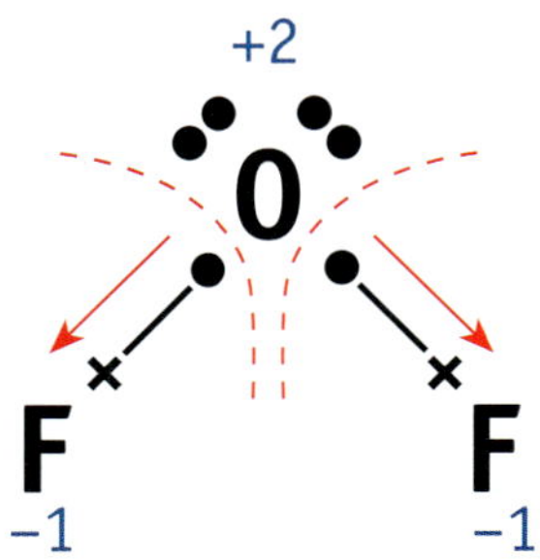

Fig 14.6
Oxygen has an oxidation number of +2 in oxygen difluoride.

5. **Hydrogen is always assigned an oxidation number of +1 in its compounds, except in the metal hydrides when it is assigned an oxidation number of -1.**

For example, in NH_3 each H atom is assigned an oxidation number of +1. (It follows from rule 3 above that the oxidation number of the N atom must be -3.) However, in a metal hydride (e.g. NaH), the H atom is assigned an oxidation number of -1. The reason for this is because sodium hydride is an ionic compound, Na^+H^-. The oxidation number of -1 arises from the one negative charge on the hydrogen atom (Rule 2).

6. **The halogens always have an oxidation number of -1 in their binary compounds unless bonded to a more electronegative atom.**

For example, since fluorine is the most electronegative element in the Periodic Table, its oxidation number will always be -1. Chlorine usually has an oxidation number of -1 also, except when it is bonded to oxygen or fluorine (both of which are more electronegative than chlorine). For example, in Cl_2O the oxidation number of each chlorine atom is +1. In the compound Cl_2O_7, each chlorine atom has an oxidation number of +7.

Cl_2O	Cl_2O_7
2(+1) -2	2(+7) 7(-2)
O.N. of Cl = +1	O.N. of Cl = +7

Thus, whereas the O.N. of fluorine is always -1, the O.N. of the other halogens can vary depending on the particular compound.

Note the difference in the way that the formulas OF_2 and Cl_2O are written. As a general rule, the more electronegative element is placed second.

7. **The sum of the oxidation numbers of all the elements in a complex ion must equal the charge on the ion.**

Examples of complex ions are NO_3^-, SO_4^{2-}, CO_3^{2-}, etc. (Ions like Cl^-, Na^+, Mg^{2+}, etc. are often referred to as simple ions.) The following examples will illustrate the usefulness of this rule.

Example 14.2

What is the oxidation number of the N atom in the NO_3^- ion?

Solution

We know that the oxidation number of the O atom is -2. Therefore, the negative charge due to three oxygen atoms is $3 \times (-2) = -6$

But overall charge on ion = -1

From rule 7, the oxidation number of the N atom added to -6 must equal -1.

Let x = oxidation number of the N atom.

$$x + (-6) = -1$$
$$x = -1 + 6 = 5$$

$$NO_3^-$$
+5 3(-2)

O.N. of the N atom = 5

Example 14.3

What is the oxidation number of each of the elements in $CuSO_4$?

SOLUTION

Note that of the three elements in $CuSO_4$, only one (oxygen) is mentioned in the rules for assigning oxidation numbers. To work out the oxidation numbers of Cu and S, we need to know that there are two negative charges on the sulfate ion, i.e. SO_4^{2-} This must mean that, in order to balance the charges, there must be two positive charges on the copper, i.e. Cu^{2+}. Thus, the oxidation number of copper is +2.

Now consider the SO_4^{2-} ion. The four oxygen atoms give a total negative charge of 4(-2) = -8. The overall charge is -2. Therefore, the oxidation number of the sulfur atom must be +6.

$$\underset{+2}{Cu}\,\underset{+6}{S}\,\underset{4(-2)}{O_4}$$

For examination purposes, it is always best to clearly state the oxidation numbers of the elements. In the above notation, it is not clear if the oxidation number of oxygen is 4 or -2. To avoid any confusion in an examination, it is best to set out your work as follows:

$$\underset{+2}{Cu}\,\underset{+6}{S}\,\underset{4(-2)}{O_4}$$

O.N. of Cu = +2 O.N. of S = +6 O.N. of O = -2

In the above example, copper has an oxidation number of +2. Since transition metals exhibit variable valency, this oxidation number also varies, e.g. in CuCl the oxidation number of Cu is +1. The oxidation number (written in Roman numerals) is used in naming transition metal compounds, e.g. CuCl is called copper(I) chloride, $CuCl_2$ is called copper(II) chloride, FeO is called iron(II) oxide, Fe_2O_3 is called iron(III) oxide, etc. An alternative way of stating this is to say that iron exists in an oxidation state of two and an oxidation state of three. Thus, the Roman numeral tells us the oxidation state of the element. Note when writing the name that there is no space between the name of the element and the number.

Example 14.4

What is the oxidation number of sulfur in $Na_2S_2O_3$?

SOLUTION

Each sodium atom has an oxidation number of +1.
Each oxygen atom has an oxidation number of -2.

$$\underset{2(+1)}{Na_2}\,\underset{2(x)}{S_2}\,\underset{3(-2)}{O_3}$$

Let x = oxidation number of sulfur. The sum of all the oxidation numbers must be zero.

$$2(+1) + 2x + 3(-2) = 0$$
$$2x = 6 - 2 = 4$$
$$x = 2$$

Oxidation number of sulfur = +2

Since oxidation numbers are assigned to help us 'keep track' of electrons on paper, we sometimes find that when we apply the rules, the oxidation number of an element may not work out to be a whole number. This is the case in EXAMPLE 14.5.

Example 14.5

What is the oxidation number of sulfur in $Na_2S_4O_6$?

SOLUTION

Each sodium atom has an oxidation number of +1.
Each oxygen atom has an oxidation number of -2.

$$Na_2S_4O_6$$

2(+1) 4(x) 6(-2)

Let x = oxidation number of sulfur. The sum of all the oxidation numbers must be zero.

$$2(+1) + 4x + 6(-2) = 0$$
$$4x = 12 - 2 = 10$$
$$x = 2.5$$

Oxidation number of sulfur = +2.5

You may wonder what an oxidation number of 2.5 means. It does not mean that some atoms have gained two and a half electrons! The 2.5 can be thought of as an average value of oxidation number. A more detailed explanation is outside the range of the syllabus.

Note: There are cases where an element may have an oxidation number of 0 even when it is in a compound, e.g. C has an oxidation number of 0 in sucrose, $C_{12}H_{22}O_{11}$.

Test yourself:
Attempt questions 14.3 – 14.5 and W14.1 – W14.8.

14.4 OXIDATION AND REDUCTION IN TERMS OF OXIDATION NUMBERS

Oxidation numbers may be used to give very comprehensive definitions of oxidation and reduction.

When an element is oxidised its oxidation number increases, i.e. **oxidation is an increase in oxidation number.**

When an element is reduced, its oxidation number decreases, i.e. **reduction is a decrease in oxidation number.**

The use of oxidation numbers to determine the species that are oxidised and reduced is shown in EXAMPLES 14.6 AND 14.7.

Example 14.6

Using oxidation numbers, indicate the species oxidised and reduced in the reaction:
$2H_2 + O_2 \rightarrow 2H_2O$

SOLUTION

Assign oxidation numbers and indicate the elements that change oxidation number.

Increase in oxidation number
oxidation

$$2H_2 + O_2 \rightarrow 2H_2O$$

0 0 +1 -2

reduction
Decrease in oxidation number

Example 14.7

Use oxidation numbers to determine whether or not the following is an oxidation/reduction reaction and, if so, indicate which species is being oxidised and which is being reduced:
$Cu + 2H_2SO_4 \rightarrow CuSO_4 + SO_2 + 2H_2O$ *[LCH]*

SOLUTION

When answering questions like the above, it is always best to write under each element the oxidation numbers of that element (using a pencil).

Cu +	$2H_2SO_4$	$\rightarrow CuSO_4$	+ SO_2	+ $2H_2O$
0	2(+1) +6 4(-2)	+2 +6 4(-2)	+4 2(-2)	2(+1)-2

Now mark (in ink) those elements that show a change in oxidation number.

Cu +	$2H_2SO_4$	$\rightarrow CuSO_4$	+ SO_2	+ $2H_2O$
0	+6	+2	+4	

It now becomes clearer which elements are oxidised and which are reduced. (Note that in this particular example, one of the sulfur atoms is neither oxidised nor reduced and the other sulfur atom is reduced.)

Increase in oxidation number
oxidation (Cu 0 → $CuSO_4$ +2)

$Cu + 2H_2SO_4 \rightarrow CuSO_4 + SO_2 + 2H_2O$
0 +6 +2 +4

reduction (S +6 → SO_2 +4)
Decrease in oxidation number

In this example, sulfuric acid is the oxidising agent (and is therefore reduced) and copper is the reducing agent (and is therefore oxidised).

Test yourself:
Attempt questions 14.6 – 14.7 and W14.9 – W14.13.

EXPERIMENT 14.1 MANDATORY STUDENT EXPERIMENT NO. 9

To study some oxidation-reduction reactions.

Introduction

In this experiment, we shall study some oxidation-reduction reactions involving the halogens (Group VII) as oxidising agents. We shall also study some displacement reactions of metals.

All of the halogens act as oxidising agents because of the high electronegativity values of these elements: F = 4.0, Cl = 3.0, Br = 2.8, I = 2.5, At = 2.2. In other words, these elements have such a strong attraction for electrons that they tend to remove electrons from other substances i.e. they oxidise other substances. As fluorine is too reactive to be found in school laboratories and astatine is a radioactive substance, our study will be confined to the properties of chlorine, bromine and iodine.

Note: For examination purposes, you need only remember the half reactions involved in the reactions relating to this experiment:

Apparatus required

Test-tube rack, test-tubes, stoppers.

Chemicals required

Chlorine water, bromine water, iodine solution, potassium iodide solution (0.1 M), potassium bromide solution (0.1 M), sodium sulfite solution (0.1 M), barium chloride solution (0.25 M), hydrochloric acid (3 M), iron(II) sulfate (0.5 M), zinc metal, magnesium metal, copper sulfate solution (0.25 M).

PART A: HALOGENS AS OXIDISING AGENTS

Procedure

1. Add about one fifth of a test-tube of chlorine water to the same quantity of potassium bromide solution. Note that the colourless solution of KBr turns orange - implying the formation of Br_2. The Cl_2 is oxidising the Br^- ion to Br_2 according to the equation:

$$Cl_2 + 2Br^- \rightarrow 2Cl^- + Br_2$$

Oxidising agent: Cl_2

-1 → 0: oxidation

It is also clear that the chlorine (O.N. = 0) is reduced to chloride (O.N. = -1). (This reaction is used to extract bromine from bromide salts in sea water.)

Oxidation: $2Br^- \rightarrow Br_2 + 2e^-$
Reduction: $Cl_2 + 2e^- \rightarrow 2Cl^-$

2. Add about one fifth of a test-tube of chlorine water to the same quantity of potassium iodide solution. Note the appearance of the reddish/brown colour of iodine. The chlorine is oxidising the iodide ions to iodine according to the equation:

$$Cl_2 + 2I^- \rightarrow 2Cl^- + I_2$$

Oxidising agent: Cl_2

-1 → 0: oxidation

Once again, note that the chlorine (O.N. = 0) is reduced to chloride (O.N. = -1). The half reactions are:
Oxidation: $2I^- \rightarrow I_2 + 2e^-$
Reduction: $Cl_2 + 2e^- \rightarrow 2Cl^-$

3. Add about one fifth of a test-tube of bromine water to the same quantity of potassium iodide solution. Note the appearance of the brown colour of the iodine. The bromine is oxidising the iodide ions to iodine according to the equation:

$$Br_2 + 2I^- \rightarrow 2Br^- + I_2$$

Oxidising agent: Br_2

-1 → 0: oxidation

Once again, note that the bromine (O.N. = 0) is reduced to bromide (O.N. = -1). The half reactions are:
Oxidation: $2I^- \rightarrow I_2 + 2e^-$
Reduction: $Br_2 + 2e^- \rightarrow 2Br^-$

From the above three experiments, we may conclude that (a) chlorine displaces bromine from a solution of a bromide salt, (b) chlorine displaces iodine from a solution of an iodide salt and (c) bromine displaces iodine from a solution of an iodide salt. This makes sense if we look at the electronegativity values of the halogens.

F = 4.0
Cl = 3.0 Decreasing
Br = 2.8 Electronegativity
I = 2.5 ↓

The reasons why the electronegativity values decrease down the group have already been discussed in CHAPTER 7. Thus, chlorine being more electronegative than bromine, forces the bromide ion to give up its outermost electron to chlorine, leaving a bromine atom behind. Two bromine atoms then combine to form Br_2 as shown in the above equation. Similar reactions occur when chlorine reacts with iodide ions and when bromine reacts with iodide ions. In short, a halogen will oxidise the ions of any other halogen below it in the above listing, i.e. chlorine will oxidise bromide ions and iodide ions and bromine will oxidise iodide ions. However, bromine will not oxidise chloride ions and iodine will not oxidise either bromide or chloride ions.

4. In this test, we will demonstrate the oxidising ability of the halogens by using solutions of chlorine, bromine and iodine to oxidise the sulfite ion (SO_3^{2-}) to the sulfate ion (SO_4^{2-}).
   To about one fifth of a test-tube of sodium sulfite solution, add the same quantity of chlorine water. Gently shake the test-tube and test for the presence of the sulfate ion using barium chloride solution. If the white precipitate does **not** dissolve in dilute HCl, then the sulfate ion is present. We conclude that the chlorine has oxidised the sulfite ion (O.N. of S atom = +4) to the sulfate ion (O.N. of S atom = +6) as shown by the following equation:

$$Cl_2 + \underset{+4}{SO_3^{2-}} + H_2O \rightarrow 2Cl^- + \underset{+6}{SO_4^{2-}} + 2H^+$$

oxidation (+4 → +6)

Oxidising agent (Cl_2)

The half equations are:

Oxidation: $SO_3^{2-} + H_2O \rightarrow SO_4^{2-} + 2H^+ + 2e^-$

Reduction: $Cl_2 + 2e^- \rightarrow 2Cl^-$

A similar reaction occurs if you use bromine water or iodine solution. In each case, it is found that the sulfite ion is oxidised to the sulfate ion.

5. The final test to demonstrate the oxidising properties of the halogens is to show the conversion of Fe^{2+} to Fe^{3+}. A useful source of Fe^{2+} ions is an acidified solution of $FeSO_4$. The solution is acidified with dilute sulfuric acid as this prevents the Fe^{2+} ion being oxidised to Fe^{3+} ion by the air.
   To about one fifth of a test-tube of acidified iron(II) sulfate solution, add about the same quantity of chlorine water. To test for the presence of the Fe^{3+} ion, add some dilute sodium hydroxide solution. A greenish/brown precipitate of Fe_2O_3 is formed if the solution contains Fe^{3+} ions.

$$Cl_2 + \underset{+2}{2Fe^{2+}} \rightarrow \underset{+3}{2Fe^{3+}} + 2Cl^-$$

oxidation (+2 → +3)

Oxidising agent (Cl_2)

Note that the chlorine (O.N. = 0) is reduced to chloride (O.N. = -1).
The half equations are:
Oxidation: $2Fe^{2+} \rightarrow 2Fe^{3+} + 2e^-$
Reduction: $Cl_2 + 2e^- \rightarrow 2Cl^-$
A similar reaction occurs if you use Br_2 or I_2 instead of Cl_2.

Part B: Displacement Reactions of Metals

1. Oxidation-reduction reactions also occur between one metal and another. For example, we have already referred to the fact that a piece of zinc metal dipped into copper sulfate solution displaces Cu^{2+} ions as copper metal, (Fig 14.2). You may verify this for yourself by adding pieces of granulated zinc to copper sulfate solution. Note that the pieces of zinc become covered with copper metal. This is because Zn has a greater tendency than copper to lose electrons and the zinc forces the Cu^{2+} ions to accept electrons:
Oxidation: $Zn \rightarrow Zn^{2+} + 2e^-$
Reduction: $Cu^{2+} + 2e^- \rightarrow Cu$
The ionic equation is:

oxidation (Zn → Zn^{2+})

$$Zn + Cu^{2+} \rightarrow Zn^{2+} + Cu$$
$$0 \quad\quad +2 \quad\quad +2 \quad\quad 0$$

reduction (Cu^{2+} → Cu)

2. Another example that demonstrates an oxidation-reduction reaction between metals is the reaction between magnesium metal and copper sulfate (Fig 14.7). As in the previous case, magnesium has a much greater tendency than copper to lose electrons and forces the Cu^{2+} ions to accept electrons.
The half equations are:
Oxidation: $Mg \rightarrow Mg^{2+} + 2e^-$
Reduction: $Cu^{2+} + 2e^- \rightarrow Cu$
The ionic equation is:

oxidation (Mg → Mg^{2+})

$$Mg + Cu^{2+} \rightarrow Mg^{2+} + Cu$$
$$0 \quad\quad +2 \quad\quad +2 \quad\quad 0$$

reduction (Cu^{2+} → Cu)

Fig 14.7
Magnesium displaces Cu^{2+} ions out of solution as copper metal.

The tendency of metals to take part in oxidation-reduction reactions will be studied in more detail at a later stage (Chapter 20).

14.5 BALANCING REDOX EQUATIONS

In Section 6.3 we learned how to balance equations by inspection. This trial-and-error approach is often quite adequate for fairly simple equations. However, for more complicated equations, it is preferable to balance the equations in a more systematic manner. In the case of oxidation-reduction reactions, we use oxidation numbers to help us balance the equations. The method is illustrated in the following examples.

(In all of the following examples, ionic equations are given as these are the ones specified on the syllabus.)

Example 14.8

Using oxidation numbers, balance the following equation:
$MnO_4^- + Fe^{2+} + H^+ \rightarrow Mn^{2+} + Fe^{3+} + H_2O$

SOLUTION

Step 1. Assign oxidation numbers to all atoms.

$MnO_4^- + Fe^{2+} + H^+ \rightarrow Mn^{2+} + Fe^{3+} + H_2O$

+7 4(-2) +2 +1 +2 +3 2(+1) -2

Step 2. Note which elements change oxidation number.

$MnO_4^- + Fe^{2+} + H^+ \rightarrow Mn^{2+} + Fe^{3+} + H_2O$

+7 +2 +2 +3

Step 3. Show the number of electrons lost and gained.
An increase in oxidation number represents a loss of electrons. A decrease in oxidation number represents a gain of electrons.

loses 1 electron

$MnO_4^- + Fe^{2+} + H^+ \rightarrow Mn^{2+} + Fe^{3+} + H_2O$

+7 +2 +2 +3

gains 5 electrons

Step 4. Work out the ratio of oxidising agent to reducing agent.
In step 3, we see that each MnO_4^- ion must gain five electrons from the Fe^{2+} ion. However, each Fe^{2+} ion can only lose one electron. Therefore, there must be five Fe^{2+} ions for every one MnO_4^- ion, i.e. MnO_4^- and Fe^{2+} react in the ratio of 1 : 5. Therefore, we can write:

$1MnO_4^- + 5Fe^{2+} + H^+ \rightarrow Mn^{2+} + Fe^{3+} + H_2O$
(The '1' before the MnO_4^- is usually omitted. It is written in here just to emphasise the ratio.)

Step 4 is a vital step. Therefore, it is best to just concentrate on getting this ratio correct and forget about everything else in the equation.

Step 5. Balance the remaining items in the equation by the inspection method.
As a general rule, leave oxygen and hydrogen atoms as the last items to be balanced. From the above, we see that there is one Mn atom on the left-hand side (L.H.S) of the equation. Therefore, there must be one Mn atom on the right-hand side (R.H.S) of the equation.

$1MnO_4^- + 5Fe^{2+} + H^+ \rightarrow 1Mn^{2+} + Fe^{3+} + H_2O$

Similarly, there are five iron atoms on the L.H.S. Therefore, there must be five iron atoms on the R.H.S.

$1MnO_4^- + 5Fe^{2+} + H^+ \rightarrow 1Mn^{2+} + 5Fe^{3+} + H_2O$

There are four oxygen atoms on the L.H.S. Therefore, there must be four oxygen atoms on the R.H.S.

$1MnO_4^- + 5Fe^{2+} + H^+ \rightarrow 1Mn^{2+} + 5Fe^{3+} + 4H_2O$

Finally, there are eight hydrogen atoms on the R.H.S. Therefore, there must be eight H atoms on the L.H.S.

$1MnO_4^- + 5Fe^{2+} + 8H^+ \rightarrow 1Mn^{2+} + 5Fe^{3+} + 4H_2O$

Always double check that the equation is balanced

1. For atoms:

	L.H.S.	R.H.S.
Mn	1	1 ✔
Fe	5	5 ✔
O	4	4 ✔
H	8	8 ✔

2. For charges:

$1MnO_4^- + 5Fe^{2+} + 8H^+ \rightarrow 1Mn^{2+} + 5Fe^{3+} + 4H_2O$	
Charge on L.H.S = 1(-1) + 5(+2) +8(+1) = +17	Charge on R.H.S. = 1(+2) +5(+3) +4(0) = +17

Note: It is important that you use oxidation number to balance oxidation-reduction reactions. Otherwise, it would be very unlikely that the correct ratio of oxidising agent to reducing agent would be obtained if you simply attempted to balance the entire equation by inspection.

Example 14.9

Using oxidation numbers, balance the following equation:

$Cu + HNO_3 + H^+ \rightarrow Cu^{2+} + NO + H_2O$

SOLUTION

Step 1. Assign oxidation numbers to all atoms.

$Cu + HNO_3 + H^+ \rightarrow Cu^{2+} + NO + H_2O$

0   +1+5 3(-2)   +1   +2   +2-2   2(+1) -2

Step 2. Note which elements change oxidation number.

$Cu + HNO_3 + H^+ \rightarrow Cu^{2+} + NO + H_2O$

0   +5   +2   +2

Step 3. Show the number of electrons lost and gained.

loses 2 electrons

$Cu + HNO_3 + H^+ \rightarrow Cu^{2+} + NO + H_2O$

0   +5   +2   +2

gains 3 electrons

Step 4. Work out the ratio of oxidising agent to reducing agent.

In order to have the same number of electrons lost and gained, bring this number up to six, i.e. 3Cu and $2HNO_3$.

$3Cu + 2HNO_3 + H^+ \rightarrow Cu^{2+} + NO + H_2O$

Step 5. Balance the remaining items in the equation by the inspection method.

$3Cu + 2HNO_3 + 6H^+ \rightarrow 3Cu^{2+} + 2NO + 4H_2O$

Double check that the equation is balanced.

Example 14.10

Using oxidation numbers, balance the following equation:

$Fe^{2+} + Cl_2 \rightarrow Fe^{3+} + Cl^-$

SOLUTION

Step 1. Assign oxidation numbers to all atoms.

$Fe^{2+} + Cl_2 \rightarrow Fe^{3+} + Cl^-$

+2 0 +3 -1

Step 2. Note which elements change oxidation number.

In this equation, all of the reactants change oxidation number when forming the products.

Step 3. Show the number of electrons lost and gained.

loses 1 electron

$Fe^{2+} + Cl_2 \rightarrow Fe^{3+} + Cl^-$

+2 0 +3 -1

Each Cl atom gains 1 electron

$\therefore$ Cl_2 gains 2 electrons

Step 4. Work out the ratio of oxidising agent to reducing agent.

Note that each Cl_2 molecule gains 2 electrons. Therefore, two Fe^{2+} ions must react with one Cl_2 molecule in order to balance the electron gain and electron loss.

$2Fe^{2+} + Cl_2 \rightarrow Fe^{3+} + Cl^-$

Step 5. Balancing the remaining items in the equation by the inspection method.

$2Fe^{2+} + Cl_2 \rightarrow 2Fe^{3+} + 2Cl^-$

Double check that the equation is balanced.

Test Yourself:
Attempt question 14.8 and W14.14.

CHAPTER SUMMARY

- Oxidation is the addition of oxygen to a substance.
- Reduction is the removal of oxygen from a substance or the addition of hydrogen to a substance.
- Oxidation of an element takes place when it loses electrons.
- Reduction of an element takes place when it gains electrons.
- An oxidising agent is a substance that brings about oxidation in other substances.
- A reducing agent is a substance that brings about reduction in other substances.
- Oxidation number is the charge that an atom has or appears to have when electrons are distributed according to certain rules.
- Oxidation is an increase in oxidation number.
- Reduction is a decrease in oxidation number.
- Oxidation numbers may be used to balance redox equations.

EXERCISES

14.1. (a) Define oxidation in terms of electron transfer .... *[LCH]*
(b) An oxidising agent is defined as ....
(c) Household bleaches usually contain an oxidising agent called ....
(d) What would you observe when a piece of zinc is left for some hours in an aqueous solution of copper(II) sulfate? .... *[LCH]*
(e) Chlorine is an example of an .... agent.
(f) Hydrogen peroxide is a commonly used .... agent.
(g) Tincture of iodine is a solution of iodine in ....
(h) Small amounts of .... are usually added to drinking water to kill germs.
(i) When a peeled apple is left exposed to the air it begins to turn brown. This is because one of the substances in the apple is being oxidised. What is the oxidising agent?......................

14.2. In the following examples indicate clearly, by showing electron loss or gain, which species is being oxidised and which is being reduced.
Show the oxidising and reducing agents in each of the reactions.
(a) $2Ca + O_2 \rightarrow 2CaO$ (b) $2K + Br_2 \rightarrow 2KBr$
(c) $Zn + I_2 \rightarrow ZnI_2$ (d) $2Cu + S \rightarrow Cu_2S$
(e) $2Fe^{2+} + Cl_2 \rightarrow 2Fe^{3+} + 2Cl^-$ (f) $Br_2 + 2I^- \rightarrow 2Br^- + I_2$
(g) $Zn + CuSO_4 \rightarrow ZnSO_4 + Cu$ (h) $2Br^- + Cl_2 \rightarrow Br_2 + 2Cl^-$

14.3. Write down the oxidation number of each element in each of the following: I_2; KCl; MgF_2; H_2SO_4; H_2O_2; $CaCO_3$; Fe; H_2S; IO^-; IO_3^-; PCl_5; $NiSO_4$; CS_2; MgH_2; $Fe_2(SO_4)_3$.

14.4. Write down the oxidation number of Mn in each of the following compounds: $MnSO_4$, Mn_2O_3, MnO_2, $KMnO_4$.

14.5. Define oxidation number. What is the oxidation number of chromium in the following:
(i) $HCrO_4^-$, (ii) Cr_2O_3 , (iii) $Cr_2O_7^{2-}$?
What is the oxidation number of oxygen in OF_2 ? Explain why the oxidation number you have given is assigned to oxygen in this case. *[LCH]*

14.6. Use oxidation numbers to determine whether or not each of the following is an oxidation-reduction reaction and if so, state which species is being oxidised and which is being reduced:
$H_2SO_3 + OH^- \rightarrow HSO_3^- + H_2O$
$Cu_2O + 2H^+ \rightarrow Cu + Cu^{2+} + H_2O$
$3Ag + 4H^+ + NO_3^- \rightarrow 3Ag^+ + NO + 2H_2O$ *[LCH]*

14.7. In the case of the following reactions, indicate clearly by showing the changes in the oxidation numbers, which species is being oxidised and which is being reduced.
(i) $2HgCl_2 + SnCl_2 \rightarrow Hg_2Cl_2 + SnCl_4$
(ii) $NH_4NO_3 \rightarrow N_2O + 2H_2O$
(iii) $3Cl_2 + 6OH^- \rightarrow 5Cl^- + ClO_3^- + 3H_2O$ *[LCH]*

14.8. Using oxidation numbers, balance each of the following equations.
(a) $MnO_4^- + H_2S + H^+ \rightarrow Mn^{2+} + S + H_2O$ *[LCH]*
(b) $Cr_2O_7^{2-} + Cl^- + H^+ \rightarrow Cr^{3+} + H_2O + Cl_2$ *[LCH]*
(c) $Zn + NO_3^- + H^+ \rightarrow Zn^{2+} + N_2H_4 + H_2O$ *[LCH]*
(d) $MnO_4^- + H^+ + Cl^- \rightarrow Mn^{2+} + Cl_2 + H_2O$ *[LCH]*
(e) $ClO_3^- + I^- + H^+ \rightarrow I_2 + Cl^- + H_2O$
(f) $MnO_4^- + I^- + H^+ \rightarrow Mn^{2+} + H_2O + I_2$ *[LCH]*
(g) $Cr_2O_7^{2-} + H^+ + S \rightarrow SO_2 + OH^- + Cr_2O_3$
(h) $MnO_4^- + H_2O_2 + H^+ \rightarrow Mn^{2+} + O_2 + H_2O$

Volumetric Analysis: Oxidation-Reduction

15

15.1 INTRODUCTION

In CHAPTER 13 we studied how volumetric analysis could be used to determine the concentrations of solutions of acids and bases. In CHAPTER 14 we saw that many reactions in chemistry may be classified as oxidation-reduction reactions. In this chapter we shall study how volumetric analysis may be used to determine the concentrations of oxidising and reducing agents.

The most important oxidising agent that you will come across in this course is **potassium permanganate** which has the formula **$KMnO_4$**. The more modern name for this substance is **potassium manganate(VII)**. However, since the name 'potassium permanganate' is so well known, we shall use this term throughout this chapter. Before studying redox titrations, it is important to be familiar with some key points regarding potassium permanganate. These points may be summarised as follows:

(i) Potassium permanganate is a purple coloured solid. It is not a primary standard as it cannot be obtained in a state of very high purity and it decomposes in the presence of sunlight. Therefore, solutions of potassium permanganate must be standardised by titration against a primary standard solution.

(ii) Potassium permanganate acts as an oxidising agent by gaining five electrons in acidic solution. The following is the reaction which takes place in acidic solution:

$$\underset{\text{purple}}{MnO_4^-} + 8H^+ + 5e^- \rightarrow \underset{\text{colourless}}{Mn^{2+}} + 4H_2O$$

The above reaction is illustrated in FIG 15.1.

Fig 15.1
As the acidified potassium permanganate solution is being added to the reducing agent, the manganese atom changes from an oxidation state of +7 (purple) to an oxidation state of +2 (colourless).

As a general rule, when dealing with redox reactions, the transfer of electrons is emphasised by writing the half reactions rather than the fully balanced equations. The above half reaction is so important and occurs so frequently in this course, that we recommend to the student that it be learned off by heart.

(iii) In order to get the permanganate to undergo the above reaction, it is necessary to add some dilute acid to supply the H^+ ions. When acidifying $KMnO_4$ solution, it is usual to use **sulfuric acid**. If hydrochloric acid were used, the Cl^- ions would be oxidised to Cl_2 by the $KMnO_4$. If nitric acid were used, it could interfere with the reaction as nitric acid is itself an oxidising agent.

(iv) In neutral or basic solution, the above reaction does not take place. Instead of the manganese ion changing from the oxidation state of +7 (purple) to +2 (colourless), an intermediate oxidation state of +4 is formed. This oxidation state appears as a brown colour and is due to the formation of manganese dioxide, MnO_2 (FIG 15.2).

(The manganese dioxide is formed according to the reaction:
$MnO_4^- + 2H_2O + 3e^- \rightarrow MnO_2 + 4OH^-$)

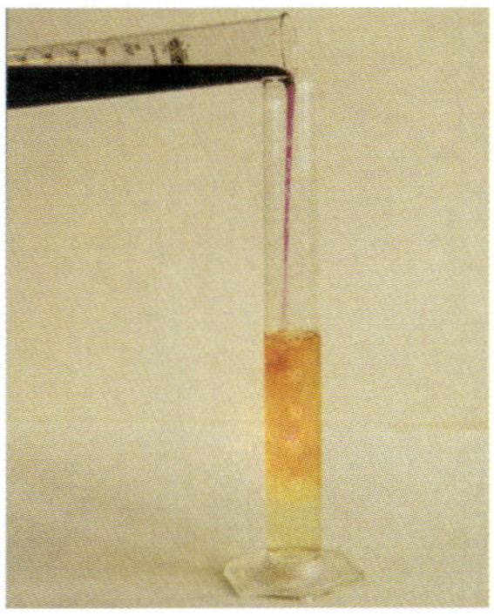

Fig 15.2
When permanganate reacts with a reducing agent in neutral or basic solution, the manganese atom changes from an oxidation state of +7 (purple) to an oxidation state of +4 (brown).

(v) Potassium permanganate acts as its own indicator. As explained in (ii) above, it changes colour from purple to colourless when acting as an oxidising agent in acidic solution. This colour change enables us to detect the end point. (This will be covered in more detail at a later stage.)

(vi) Because of the intense purple colour of potassium permanganate solution, it is usual to read the top rather than the bottom of the meniscus (FIG 15.3).

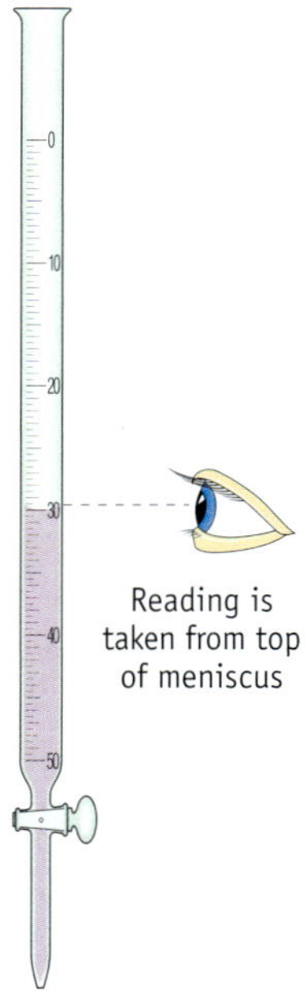

Fig 15.3
In the case of potassium permanganate, it is normal to take the readings from the top of the meniscus.

Due to the intense purple colour, its fairly low solubility and its power as an oxidising agent, it is normal practise to use fairly dilute solutions of $KMnO_4$ in redox titrations, e.g. 0.02 M solutions will frequently be used in the redox titrations on this course.

In the Leaving Certificate chemistry course, we shall study two examples of redox titrations: (i) the reaction of MnO_4^- and Fe^{2+} ions and (ii) the reaction of iodine and thiosulfate ions. We now take a closer look at both of these titrations.

15.2 THE REACTION OF MnO_4^- AND Fe^{2+} IONS

As already stated, $KMnO_4$ is not a primary standard, i.e. it is not possible to make up a solution of $KMnO_4$ whose concentration is accurately known, by simply weighing out a certain amount of the solid and dissolving it in a known volume of solution. However, it is possible to determine the concentration of a $KMnO_4$ solution by titrating it against a solution containing Fe^{2+} ions.

$KMnO_4$ is an oxidising agent and, when added to an acidified solution containing Fe^{2+} ions, converts the Fe^{2+} ions to Fe^{3+} ions, i.e. the following two reactions occur:

Oxidising agent:
$MnO_4^- + 8H^+ + 5e^- \rightarrow Mn^{2+} + 4H_2O$ (a)

Reducing agent:
$Fe^{2+} \rightarrow Fe^{3+} + e^-$ (b)

If we examine equation (a) above, we see that each MnO_4^- ion can accept **five** electrons. However, each Fe^{2+} ion can lose only **one** electron. Therefore, you will require five Fe^{2+} ions to react with one MnO_4^- ion, i.e. the ions react in the ratio of one MnO_4^- ion to five Fe^{2+} ions. Think carefully about this. To balance the electron transfer we may write the above two equations as follows:

Oxidising agent:
$MnO_4^- + 8H^+ + 5e^- \rightarrow Mn^{2+} + 4H_2O$

Reducing agent:
$5Fe^{2+} \rightarrow 5Fe^{3+} + 5e^-$

Adding:
$MnO_4^- + 8H^+ + 5Fe^{2+} \rightarrow Mn^{2+} + 5Fe^{3+} + 4H_2O$

The important point to note from this equation is that $\mathbf{MnO_4^-}$ ions react with $\mathbf{Fe^{2+}}$ ions in the ratio of **1 : 5**. (We have already used this idea in the last chapter to balance redox equations.) You will need to know this ratio for subsequent calculations.

You might think that Fe^{2+} ions for the titration would be obtained by dissolving crystals of $FeSO_4$ in water. However, it is not possible to use $FeSO_4$ as a primary standard because (i) the crystals tend to be oxidised slightly by the air and (ii) the crystals tend to lose their water of crystallisation when exposed to the air. (The spontaneous loss of water of crystallisation to the air is known as *efflorescence*.) Instead of using $FeSO_4$, we use a substance called **ammonium iron(II) sulfate** as a source of Fe^{2+} ions. This can be obtained in a high degree of purity and is not affected by the air, i.e. it can be used as a primary standard. The formula of ammonium iron(II) sulfate is written as $(NH_4)_2SO_4.FeSO_4.6H_2O$. This is referred to as a 'double salt'. To remember its formula, write down the formula of iron(II) sulfate and put the formula of ammonium sulfate in front of it. The formula of ammonium iron(II) sulfate is normally given on the examination paper. (The old name for ammonium iron(II) sulfate was 'ferrous ammonium sulfate'. This name is still widely used and is probably the name that appears on the container of the salt in your school laboratory.) The high relative molecular mass of ammonium iron(II) sulfate (392) ensures a high degree of accuracy when weighing.

Note: When making up a solution of ammonium iron(II) sulfate, it is important that it is dissolved in water which contains dilute sulfuric acid. The sulfuric acid prevents the Fe^{2+} ions reacting with the water, i.e. if the sulfuric acid were not present, instead of the ammonium iron(II) sulfate crystals simply dissolving in the water, they would actually react with the water to form a new substance. This type of reaction where a substance added to water reacts with the water is called **hydrolysis**. Thus, the purpose of adding the dilute sulfuric acid is to prevent hydrolysis. In addition, the sulfuric acid helps to prevent the Fe^{2+} ions being oxidised to Fe^{3+} ions by the oxygen in the air or oxygen dissolved in the water.

EXPERIMENT 15.1 MANDATORY STUDENT EXPERIMENT NO. 10

To prepare a standard solution of ammonium iron(II) sulfate and to use this solution to standardise a solution of $KMnO_4$ by titration.

Introduction

In this experiment, we shall make up a standard solution of ammonium iron(II) sulfate. We shall then use this standard solution to find out the exact concentration of a solution of $KMnO_4$ by titration.

Apparatus required

Pipette (25 cm^3), burette, pipette filler, conical flask, wash bottle, clock glass, retort stand and clamp, three beakers, graduated cylinder, white tile, funnel, glass rod, volumetric flask (250 cm^3), spatula.

Chemicals required

Potassium permanganate solution (0.02 M approx.), dilute sulfuric acid solution (1M), ammonium iron(II) sulfate crystals, deionised water.

Procedure

1. Weigh out accurately on a clock glass about 9.8 g ammonium iron(II) sulfate crystals.
2. Using a graduated cylinder, add 20 cm^3 approx. of dilute sulfuric acid to a beaker containing 100 cm^3 approx. of deionised water.

3. Taking all the precautions outlined previously (CHAPTER 13), transfer the crystals to the beaker containing the dilute sulfuric acid.
4. Stir to dissolve the crystals and, when dissolved, transfer the solution to a clean 250 cm^3 volumetric flask.
5. Transfer all the washings to the volumetric flask and make up to the mark with deionised water. Invert the stoppered flask about 20 times.
6. Wash out the pipette, burette and conical flask with deionised water.
7. Wash out the burette with some of the potassium permanganate solution and fill the burette above the zero mark with the solution. Remove the funnel from the burette and adjust the level of the solution to the zero mark. Remember to read from the top of the meniscus. Ensure that the tap and space below the tap are also filled.
8. Pour 100 cm^3 approx. of the ammonium iron(II) sulfate solution into a clean, dry beaker. Wash out the pipette with a small amount of this solution and then pipette 25 cm^3 of the solution into the conical flask. (FIG 15.4)

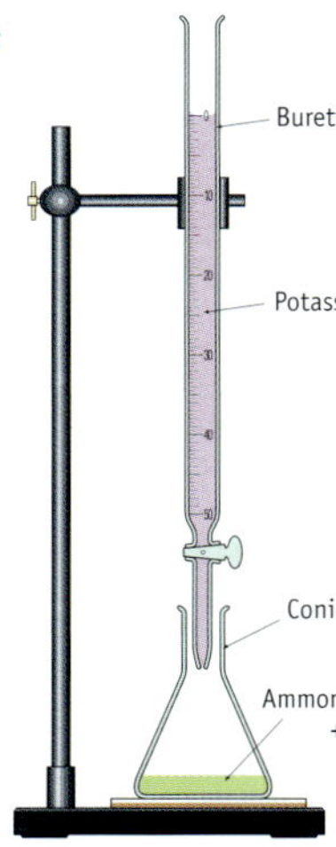

Fig 15.4
Titration of $KMnO_4$ solution against a solution of Fe^{2+} ions.

9. Using a graduated cylinder, add 20 cm^3 approx. of dilute sulfuric acid to the conical flask. Recall that the purpose of adding the acid is to supply the H^+ ions in order for the following reaction to occur:

   $MnO_4^- + 8H^+ + 5e^- \rightarrow Mn^{2+} + 4H_2O$

   If you forget to add the acid, a dark brown precipitate of MnO_2 will be formed in the conical flask during the titration. If this occurs it is best to begin the titration again.
10. Titrate in the usual manner until a faint permanent pink colour is observed at the end point. Consider what is happening at the end point: as long as Fe^{2+} ions are present in the conical flask, the $KMnO_4$ solution which is run into the flask loses its colour. Once all the Fe^{2+} ions have reacted, the next drop of $KMnO_4$ solution gives a permanent pink colour. Thus, at the end point one single drop of $KMnO_4$ from the burette will turn the solution in the conical flask a faint pink colour. Thus, $KMnO_4$ acts as its own indicator. Perform one rough and two accurate titrations. The two accurate titrations should agree to within 0.1 cm^3.
11. Ensure that all glassware is washed out when you are finished the titration as $KMnO_4$ stains glassware.

Note: The first few drops of $KMnO_4$ added to the conical flask are decolourised slowly but subsequent drops are decolourised more rapidly. This is due to an occurrence called **autocatalysis** (CHAPTER 16), i.e. the reaction is catalysed by the Mn^{2+} ions formed in the reaction. As soon as the Mn^{2+} ions are formed, the rate of the reaction increases.

Calculations

Using the average titration figure, we must now calculate the exact concentration of the $KMnO_4$ solution. The method of calculation is shown in EXAMPLE 15.1.

15.3 SOLVING VOLUMETRIC PROBLEMS IN REDOX REACTIONS

The general expression derived for solving volumetric problems involving acid-base reactions may also be used for solving volumetric problems involving redox reactions. (The derivation of this formula is the same as outlined for acid-base reactions.) For redox reactions we write this expression as follows:

molarity of oxidising agent — volume of reducing agent (cm^3)

volume of oxidising agent → $\frac{V_o \times M_o}{n_o} = \frac{V_r \times M_{red}}{n_r}$ ← molarity of reducing agent

no. of moles of oxidising agent — no. of moles of reducing agent

V_o = volume of oxidising agent (cm^3)

M_o = molarity of oxidising agent

n_o = number of moles of oxidising agent (from balanced equation)

V_r = volume of reducing agent (cm^3)

M_{red} = molarity of reducing agent*

n_r = number of moles of reducing agent (from balanced equation).

The use of this equation is illustrated in the following examples.

* M_{red} is used instead of M_r to avoid confusion with the symbol for relative molecular mass.

Example 15.1

A student made up a 0.12 M ammonium iron(II) sulfate solution and used this to standardise a solution of potassium permanganate. 25 cm^3 of the ammonium iron(II) sulfate solution was titrated against the potassium permanganate solution. The equation for the reaction is:

$$MnO_4^- + 8H^+ + 5Fe^{2+} \rightarrow Mn^{2+} + 5Fe^{3+} + 4H_2O$$

The average titration figure was 23.75 cm^3. Calculate the concentration of the potassium permanganate solution in (a) moles per litre and (b) grams per litre.

SOLUTION

From the balanced equation, we see that one mole of MnO_4^- reacts with 5 moles of Fe^{2+}, i.e. $n_o = 1$ and $n_r = 5$.

GIVEN:

$V_o = 23.75$

$M_o = ?$

$n_o = 1$

$V_r = 25$

$M_{red} = 0.12$

$n_r = 5$

(MnO_4^-) (Fe^{2+})

$$\frac{V_o \times M_o}{n_o} = \frac{V_r \times M_{red}}{n_r}$$

$$\frac{23.75 \times M_o}{1} = \frac{25 \times 0.12}{5}$$

$$M_o = \frac{25.5 \times 0.12}{5 \times 23.75}$$

= 0.0253 moles per litre

= 0.0253 × 158 grams per litre (Rel molecular mass $KMnO_4$ = 158)

= 3.997 g/L.

Answer: Concentration of $KMnO_4$ solution is (a) 0.0253 moles per litre, (b) 3.997 grams per litre.

Example 15.2

A sample of iron(II) sulfate crystals $FeSO_4.xH_2O$ was analysed in order to find out the value of x. 6.42 g of the crystals were dissolved in deionised water to which some dilute sulfuric acid had been added and the resulting solution was made up to 250 cm^3 in a volumetric flask. 25 cm^3 samples of this solution were titrated against 0.018 M $KMnO_4$ solution. The balanced equation for the reaction is:

$$MnO_4^- + 8H^+ + 5Fe^{2+} \rightarrow Mn^{2+} + 5Fe^{3+} + 4H_2O$$

The average titration figure was 25.5 cm^3. Calculate the value of x in the above formula.

Solution

When solving problems of this type, proceed as we did in the case of acid-base volumetric problems. First calculate the concentration of the iron(II) sulfate solution in moles per litre. Then relate this number of moles to the mass of substance dissolved and hence calculate the relative molecular mass of the hydrated salt. Once this is known it is easy to calculate the value of x.

(MnO_4^-) (Fe^{2+})

GIVEN:

$V_o = 25.5$

$M_o = 0.018$

$n_o = 1$

$V_r = 25$

$M_{red} = ?$

$n_r = 5$

$$\frac{V_o \times M_o}{n_o} = \frac{V_r \times M_{red}}{n_r}$$

$$\frac{25.5 \times 0.018}{1} = \frac{25 \times M_{red}}{5}$$

$$M_{red} = \frac{25.5 \times 0.018 \times 5}{25}$$

$$= 0.0918 \text{ moles/litre}$$

$$= 0.023 \text{ moles/250 cm}^3$$

i.e. by dissolving 6.42 g of the crystals in 250 cm^3 of solution, we obtain a solution which contains 0.023 moles.

Therefore, 0.023 moles $FeSO_4.xH_2O$ = 6.42 g

$$1 \text{ mole } FeSO_4.x\,H_2O = \frac{6.42}{0.023} \text{ g}$$

$$= 279.13 \text{ g}$$

$$1 \text{ mole } FeSO_4.xH_2O = 279.13 \text{ g}$$

$$152 + 18x = 279.13$$

$$18x = 279.13 - 152 = 127.13$$

$$x = \frac{127.13}{18} = 7 \quad \text{(x must be a whole number).}$$

Answer: x = 7. Formula is $FeSO_4.7H_2O$

Test Yourself:
Attempt questions 15.1 – 15.6 and W15.1 – W15.6.

Our bodies can easily become deficient in iron. This is particularly true for young teenagers who are growing rapidly. Iron deficiency (also called anaemia) is usually remedied with a course of iron tablets and an alteration in diet. Some foods like liver, kidney, egg yolk, etc. are rich in iron. Iron is an important element in our bodies because it is part of the haemoglobin molecule in our blood. Haemoglobin is the molecule that is responsible for the transport of oxygen in our blood. The oxygen actually becomes attached to the iron atom in the haemoglobin molecule. The unit that contains iron in haemoglobin is called haem (FIG 15.5). You are not expected to remember the structure of this molecule!

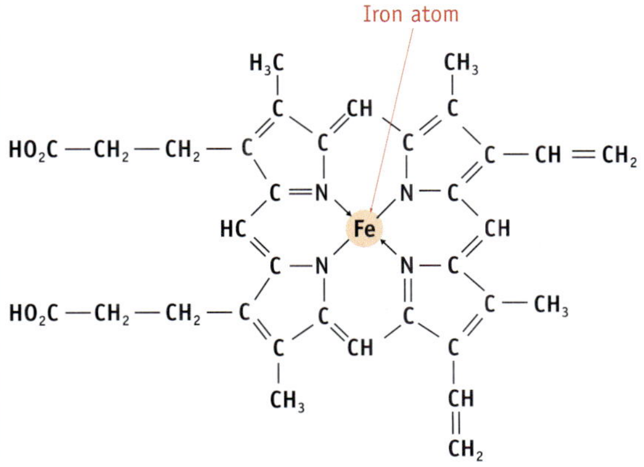

Fig 15.5
Iron is part of the haem unit of the haemoglobin molecule in our blood.

Since anaemia is caused by a shortage of haemoglobin in the blood, people suffering from anaemia are often prescribed 'iron tablets' (FIG 15.6). These tablets usually contain anhydrous iron(II) sulfate, $FeSO_4$. In EXPERIMENT 15.2, we will attempt to find the actual quantity of iron(II) sulfate in the tablets.

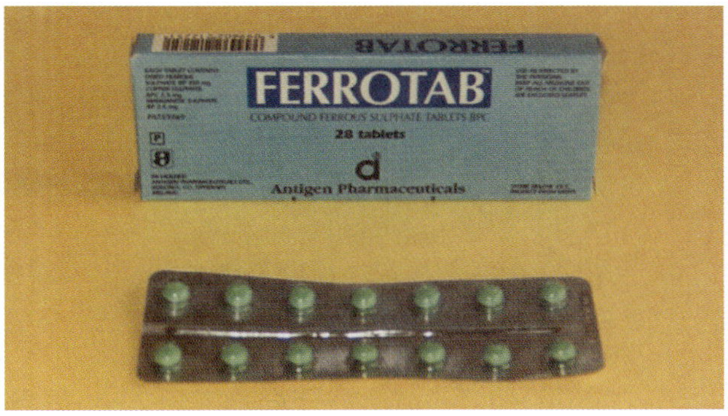

Fig 15.6
Iron tablets usually contain anhydrous iron(II) sulfate, $FeSO_4$.

EXPERIMENT 15.2 MANDATORY STUDENT EXPERIMENT NO. 11

To determine the amount of iron in an iron tablet.

Introduction

In this experiment, we shall measure the amount of iron(II) sulfate, $FeSO_4$, in an iron tablet and then compare our result with the quantity stated on the packet. The quantity on the packet is usually stated in terms of milligrams of the anhydrous salt per tablet, e.g. 200 mg $FeSO_4$ per tablet.

The amount of iron(II) sulfate is measured by dissolving some tablets in dilute sulfuric acid and titrating the solution against standardised potassium permanganate solution.

Apparatus required

Pipette (25 cm^3), burette, pipette filler, mortar and pestle, conical flask, wash bottle, clock glass, retort stand and clamp, three beakers, graduated cylinder, white tile, funnel, glass rod, volumetric flask (250 cm^3).

Chemicals required

Potassium permanganate solution (0.01 M approx.), sulfuric acid solution (1 M), deionised water, iron tablets, e.g. *Ferrotab* (Antigen Pharmaceuticals Ltd. Roscrea).

Procedure

1. Using a clock glass and electronic balance, accurately find the mass of five iron tablets. If the iron tablets you have been given are in the form of capsules, empty the contents of five capsules on to the clock glass and find their mass.
2. Grind up the tablets using a mortar and pestle. It will make it easier to grind them if you add about 5 cm^3 of 1 M sulfuric acid to the mortar.
3. Taking all the precautions outlined previously, transfer the paste of acid and tablets to a beaker containing about 100 cm^3 of 1 M sulfuric acid. Do not forget to rinse the mortar and pestle to ensure that all traces of the tablets have been transferred to the beaker.
4. Stir to dissolve the paste. You may find that the outer coating of the tablets will not dissolve and this may give a cloudy appearance to your solution. Do not worry about this, as it will not affect your results. (Iron tablets often have a coating to slow down the rate at which they dissolve in the stomach. This reduces the chance of stomach upsets.)
5. Taking all the precautions outlined in previous experiments, transfer the solution to a clean 250 cm^3 volumetric flask. Transfer all the washings to the volumetric flask and make up to the mark with deionised water. Invert the stoppered flask about 20 times.
6. Wash out the pipette, burette and conical flask with deionised water.
7. Wash out the burette with some of the potassium permanganate solution and fill the burette above the zero mark with the solution. Remove the funnel from the burette and adjust the level of the solution to the zero mark (FIG 15.7). Remember to read from the top of the meniscus. Ensure that the tap and space below the tap are also filled.
8. Pour 100 cm^3 approx. of the iron tablet solution into a clean, dry beaker. Wash out the pipette with a small amount of this solution and then pipette 25 cm^3 of the solution into the conical flask.
9. Using a graduated cylinder, add 20 cm^3 approx. of dilute sulfuric acid to the conical flask. Recall that the purpose of adding the acid is to supply the H^+ ions for the following reaction to occur:

   $MnO_4^- + 8H^+ + 5e^- \rightarrow Mn^{2+} + 4H_2O$

10. Titrate in the usual manner until a faint permanent pink colour is observed at the end point. Perform one rough and two accurate titrations. The two accurate titrations should agree to within 0.1 cm^3.
11. Ensure that all glassware is washed out when you are finished the titration as $KMnO_4$ stains glassware.

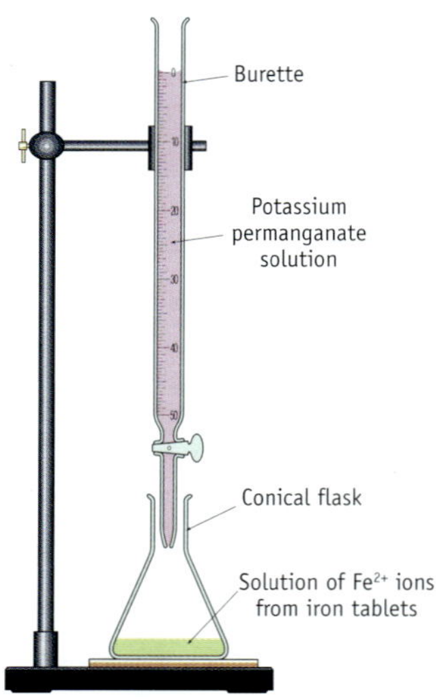

Fig 15.7
Apparatus to measure the amount of iron(II) sulfate in iron tablets.

Calculations

Using the average titration figure, it is now possible to calculate the mass of iron(II) sulfate in each tablet. The method of calculation is shown in EXAMPLE 15.3.

Example 15.3

A student was asked to analyse an iron tablet to determine if the amount of iron sulfate per tablet stated on the packet was correct. He dissolved five tablets whose total mass was 1.2 grams in dilute sulfuric acid and made the solution up to 250 cm^3 in a volumetric flask. He then titrated 25 cm^3 portions of the resulting solution against 0.015 M potassium permanganate. The average titration figure was 5.75 cm^3. Calculate (a) the mass of anhydrous $FeSO_4$ in each tablet, (b) the mass of iron in each tablet and (c) the percentage of $FeSO_4$ in each tablet.
The equation for the reaction is:

$$MnO_4^- + 8H^+ + 5Fe^{2+} \rightarrow Mn^{2+} + 5Fe^{3+} + 4H_2O$$

SOLUTION

From the balanced equation, we see that one mole of MnO_4^- reacts with five moles of Fe^{2+}, i.e. $n_o = 1$ and $n_r = 5$.

GIVEN:
$V_o = 5.75$
$M_o = 0.015$
$n_o = 1$
$V_r = 25$
$M_{red} = ?$
$n_r = 5$

$$\underset{(MnO_4^-)}{\frac{V_o \times M_o}{n_o}} = \underset{(Fe^{2+})}{\frac{V_r \times M_{red}}{n_o}}$$

$$\frac{5.75 \times 0.015}{1} = \frac{25 \times M_{red}}{5}$$

$$M_{red} = \frac{5.75 \times 0.015 \times 5}{25}$$

$= 0.0173$ moles per litre
$= 0.0173 \times 152$ g/L (Rel molecular mass $FeSO_4 = 152$)
$= 2.63$ g/L.

Therefore, in the 250 cm^3 volumetric flask there are $\frac{2.63}{4} = 0.658$ grams of $FeSO_4$.

Thus, since we used 5 tablets, 1 tablet contains $\frac{0.658}{5}$ grams of $FeSO_4 = 0.132$ grams (a)

The percentage of iron in $FeSO_4$ is $\frac{56}{152} \times 100 = 36.84\%$

Therefore, mass of iron in each tablet is 36.84% of 0.132 g = 0.049 g (b)

Mass of one tablet = $\frac{1.2}{5} = 0.24$ g

% $FeSO_4$ in each tablet = $\frac{0.132}{0.24} \times 100 = 55\%$ (c)

Answer: (a) 0.132 g $FeSO_4$ (b) 0.049 g Fe (c) 55% $FeSO_4$

Test Yourself:
Attempt questions 15.7 – 15.8 and W15.7 – W15.8.

15.4 THE REACTION OF IODINE AND SODIUM THIOSULFATE

Sodium thiosulfate, $Na_2S_2O_3$, is an important reducing agent used in chemistry. It is commonly called 'hypo' and is used in developing photographs and in analytical chemistry (FIG 15.8).

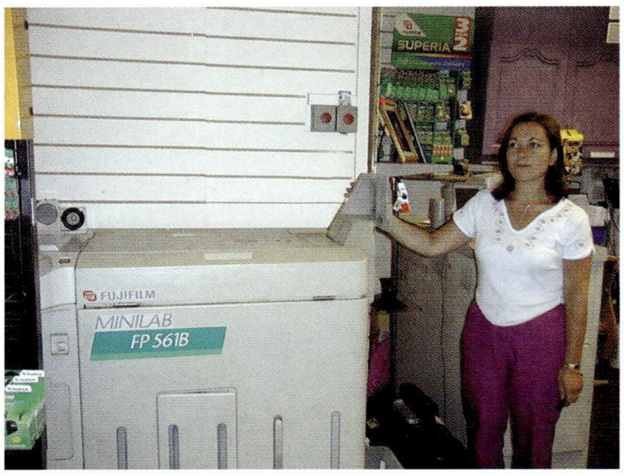

Fig 15.8
Sodium thiosulfate is used in processing photographic film. It is used to 'fix' the image when developing the photographic film.

The thiosulfate ion, $S_2O_3^{2-}$, has a structure similar to that of the sulfate ion, SO_4^{2-}, except that one of the oxygen atoms in the sulfate ion is replaced by a sulfur atom in the thiosulfate ion (FIG 15.9). (The prefix 'thio' means 'sulfur'.)

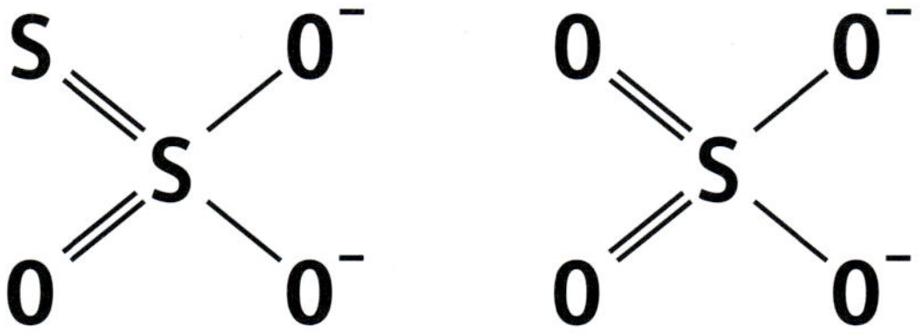

Fig 15.9
The thiosulfate and sulfate ions.

Sodium thiosulfate is commonly encountered as a colourless crystalline solid that has the formula $Na_2S_2O_3.5H_2O$. (The water content of sodium thiosulfate crystals is variable but the pentahydrate is the most common one.) The main points about the use of sodium thiosulfate in volumetric analysis may be summarised as follows:

(i) Sodium thiosulfate is a reducing agent and reacts with iodine molecules, I_2, to convert them to iodide ions, I^-.
The equation for the reaction is :

I_2	$+ 2S_2O_3^{2-} \rightarrow$	$S_4O_6^{2-}$	$+ 2I^-$
iodine	thiosulfate ion	tetrathionate ion	iodide ion
↑	↑		
oxidising agent	reducing agent		

Note: The $S_4O_6^{2-}$ ion formed in the above reaction is called the tetrathionate ion. Its structure may be represented as shown in FIG 15.10.

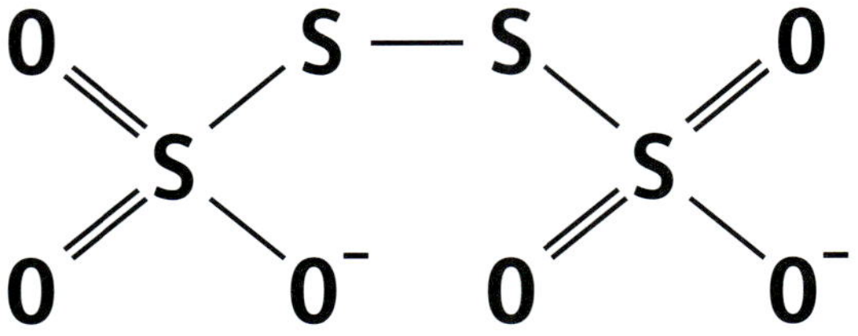

Fig 15.10
The tetrathionate ion, $S_4O_6^{2-}$.

The thiosulfate ion (the reducing agent) is oxidised to the tetrathionate ion. This may be easily seen by assigning oxidation numbers.

$$I_2 + \underset{+2}{2S_2O_3^{2-}} \rightarrow \underset{+2.5}{S_4O_6^{2-}} + 2I^-$$

(ii) Sodium thiosulfate is not a primary standard, i.e. it cannot be obtained in a sufficiently pure state so that a solution of known concentration may be made up by weighing out a certain amount of the solid and dissolving it in a known volume of solution. A sodium thiosulfate solution is standardised by titrating it against a standard iodine solution.

(iii) A standard solution of iodine cannot be made up by direct weighing. This is because iodine vaporises slightly even at room temperature. In addition, iodine does not dissolve in water. For these two reasons, iodine is not a primary standard. How therefore do we make up a standard solution of iodine? Stay tuned for point (iv)!

(iv) A standard solution of iodine is obtained by reacting a standard solution of acidified potassium permanganate with excess potassium iodide solution. Consider the reaction of acidified permanganate ions and iodide ions:

$$MnO_4^- + 8H^+ + 5e^- \rightarrow Mn^{2+} + 4H_2O \quad (a)$$

$$2I^- \rightarrow I_2 + 2e^- \quad (b)$$

In equation (a) above, we see that five electrons are gained. In equation (b), two electrons are lost. We can balance the electron gain and loss by multiplying (a) by 2 and (b) by 5. This gives us:

(a) × 2 $2MnO_4^- + 16H^+ + 10e^- \rightarrow 2Mn^{2+} + 8H_2O$

(b) × 5 $10I^- \rightarrow 5I_2 + 10e^-$

Add $2MnO_4^- + 10I^- + 16H^+ \rightarrow 2Mn^{2+} + 5I_2 + 8H_2O$

The I^- ions are kept in excess and, therefore, the amount of iodine formed in the above reaction depends on the amount of permanganate present. Thus, if we use a standard solution of $KMnO_4$, we can easily determine the amount of iodine formed. This known amount of iodine formed is then titrated against a sodium thiosulfate solution to standardise the thiosulfate solution. Details of the method of carrying out this experiment are given in EXPERIMENT 15.3. **Note:** Iodine (non-polar) is only sparingly soluble in water (polar solvent) but is readily soluble in potassium iodide solution. When iodine is added to KI solution, the I_2 molecule reacts with the iodide ion to form an ion called the tri-iodide ion, I_3^-. The I_3^- ion, because it is charged, dissolves easily in water.

$I_2 + I^- \rightarrow I_3^-$

The I_3^- ion is identical in its chemical behaviour to I_2. Thus, solutions of I_3^- may be considered as solutions of I_2, i.e. a solution of iodine in potassium iodide may be treated as if it were a solution of iodine in water.

(v) How is the end point detected in an iodine/thiosulfate titration?

As already stated, thiosulfate reacts with iodine according to the equation:

$I_2 + 2S_2O_3^{2-} \rightarrow S_4O_6^{2-} + 2I^-$

reddish brown … colourless

As thiosulfate is added to iodine, the reddish/brown colour gradually turns yellow which then becomes paler and paler until the solution turns colourless (FIG 15.11).

Fig 15.11
As the thiosulfate uses up the iodine, the colour in the conical flask becomes lighter.

Since there is no sudden colour change, this would make the end point very difficult to detect. Therefore, it is necessary to use an indicator to detect the end point accurately (FIG 15.12). The indicator used is **starch**, which forms a blue-black colour with iodine. When all of the iodine has reacted with the thiosulfate, the blue/black colour suddenly disappears to leave a colourless solution.

Fig 15.12
Starch solution forms a blue/black complex with iodine.

Note: When performing an iodine/thiosulfate titration, the starch is only added when most of the iodine has disappeared, i.e. when the reddish/brown colour of the iodine has become pale yellow. The reason for this is because if starch is added at an early stage in the titration, the iodine present may become strongly adsorbed on to the starch and this makes the titration less accurate. In addition, when the pale yellow colour is observed, it tells us that the end point is quite near. Therefore, when the starch is added at this stage, adding the thiosulfate drop by drop helps to obtain an accurate titration figure.

EXPERIMENT 15.3 MANDATORY STUDENT EXPERIMENT NO. 12

To prepare a solution of sodium thiosulfate and standardise it by titration against a solution of iodine.

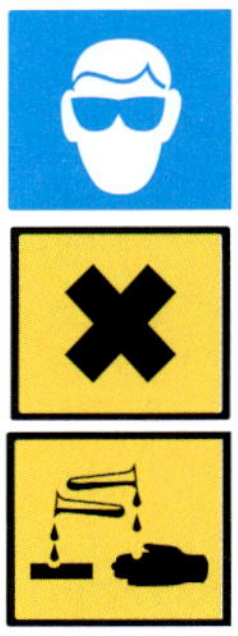

Introduction

In this experiment, we shall make up an approximate 0.1 M solution of sodium thiosulfate. We shall then find the exact concentration of this solution by titrating it against a solution of iodine. The iodine solution is made by the reaction between acidified $KMnO_4$ solution (of known concentration) and excess iodide ions. The permanganate ions and iodide ions react in the conical flask as follows:

$$2MnO_4^- + 10I^- + 16H^+ \rightarrow 2Mn^{2+} + 5I_2 + 8H_2O$$

The iodine formed is then reacted with sodium thiosulfate, which is added from a burette. The following reaction occurs:

$$I_2 + 2S_2O_3^{2-} \rightarrow S_4O_6^{2-} + 2I^-$$

The experimental set-up is illustrated in FIG 15.13.

It is important to use deionised water in this experiment as tap water contains small quantities of chlorine which may interfere with the result.

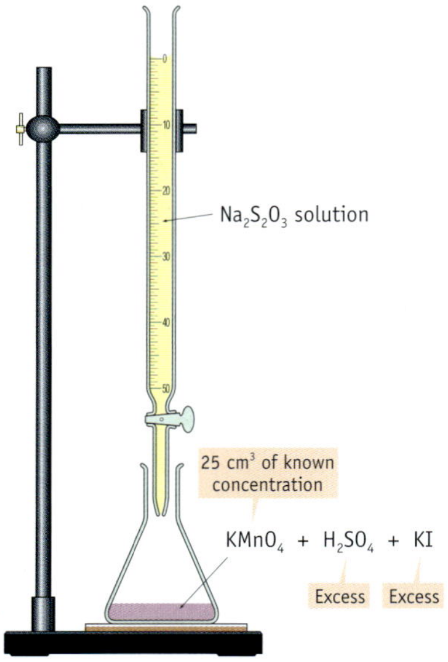

Fig 15.13
Experiment to standardise a solution of sodium thiosulfate.

Apparatus required

Pipette (25 cm^3), pipette filler, burette, conical flask, wash bottle, retort stand, three beakers, white tile, funnel, graduated cylinder, stirring rod, volumetric flask (250 cm^3), spatula.

Chemicals required

Sodium thiosulfate crystals, potassium permanganate solution (0.02 M), dilute sulfuric acid (1M), potassium iodide solution (0.5 M), starch indicator, deionised water.

Procedure

1. Weigh out on a clock glass about 6.25 g of sodium thiosulfate crystals.
2. Taking all the precautions outlined previously, transfer the crystals to a beaker containing 100 cm^3 approx. of deionised water.
3. Stir to dissolve the crystals and when dissolved transfer the solution to a clean 250 cm^3 volumetric flask.
4. Transfer all the washings to the volumetric flask and make up to the mark with deionised water. Invert the stoppered flask about 20 times.
5. Wash out the pipette, burette and conical flask with deionised water.
6. Wash out the burette with some of the sodium thiosulfate solution and fill the burette above the zero mark with the solution. Remove the funnel from the burette and adjust the level of the solution to the zero mark. Ensure that the tap and space below the tap are also filled.
7. Pour 100 cm^3 approx. of standardised $KMnO_4$ solution into a clean, dry beaker. Wash out the pipette with a small amount of this solution and, using a pipette filler, pipette 25 cm^3 of the solution into the conical flask.

8. Using a graduated cylinder, add 20 cm^3 approx. of dilute sulfuric acid, followed by 10 cm^3 of 0.5 M potassium iodide solution to the conical flask. Note the reddish/brown colour of the liberated iodine.
9. Titrate in the usual manner until the solution in the conical flask becomes pale yellow (straw colour). Now add a few drops of starch indicator until a blue/black colour is observed.
10. Continue the titration, adding the thiosulfate solution drop by drop, until the blue/black colour disappears to give a colourless solution. Note the titration figure.
11. Perform one rough and two accurate titrations. The two accurate titration figures should agree to within 0.1 cm^3.

Calculations

The method of calculating the concentration of the sodium thiosulfate solution in iodine/thiosulfate titrations is shown in EXAMPLES 15.4 AND 15.5.

Example 15.4

25 cm^3 of a 0.045 M iodine solution (previously standardised) was titrated against a solution of sodium thiosulfate. The average titration figure was 27.45 cm^3. Calculate the concentration of the sodium thiosulfate solution in (a) moles per litre and (b) grams per litre of crystalline $Na_2S_2O_3.5H_2O$. The reaction between iodine and thiosulfate may be represented as:

$$I_2 + 2S_2O_3^{2-} \rightarrow S_4O_6^{2-} + 2I^-$$

SOLUTION

From the balanced equation, we see that two moles of thiosulfate (the reducing agent) react with one mole of iodine (the oxidising agent), i.e. $n_o = 1$ and $n_r = 2$.

GIVEN:
$V_o = 25$
$M_o = 0.045$
$n_o = 1$
$V_r = 27.45$
$M_{red} = ?$
$n_r = 2$

$$\underset{(I_2)}{\frac{V_o \times M_o}{n_o}} = \underset{(S_2O_3^{2-})}{\frac{V_r \times M_{red}}{n_r}}$$

$$\frac{25 \times 0.045}{1} = \frac{27.45 \times M_{red}}{2}$$

$$M_{red} = \frac{25 \times 0.045 \times 2}{27.45}$$

$= 0.082$ moles per litre

$= 0.082 \times 248$ grams (Rel molecular mass of $Na_2S_2O_3.5H_2O = 248$)

$= 20.336$ grams per litre

Answer: Concentration of sodium thiosulfate solution is:
(a) 0.082 moles per litre
(b) 20.336 grams $Na_2S_2O_3.5H_2O$ per litre.

Example 15.5

25 cm^3 of 0.02 M potassium permanganate is pipetted into a conical flask. Some dilute sulfuric acid and excess potassium iodide solution are then added and the liberated iodine is titrated against a solution of sodium thiosulfate of unknown concentration. The average titration figure is 18.7 cm^3. Calculate the concentration of the sodium thiosulfate solution in (a) moles per litre and (b) grams per litre of crystalline $Na_2S_2O_3.5H_2O$. The reactions may be represented as:

$$2MnO_4^- + 10I^- + 16H^+ \rightarrow 2Mn^{2+} + 5I_2 + 8H_2O$$

$$I_2 + 2S_2O_3^{2-} \rightarrow S_4O_6^{2-} + 2I^-$$

Solution

From the balanced equations, we see that two moles of MnO_4^- produce five moles of I_2 and these five moles of I_2 would then react with ten moles of $Na_2S_2O_3$, i.e. $2MnO_4^- \equiv 5I_2 \equiv 10S_2O_3^{2-}$. This means that $n_o = 2$ and $n_r = 10$.

GIVEN:

$V_o = 25$

$M_o = 0.02$

$n_o = 2$

$V_r = 18.7$

$M_{red} = ?$

$n_r = 10$

$$\underset{(MnO_4^-)}{\frac{V_o \times M_o}{n_o}} = \underset{(S_2O_3^{2-})}{\frac{V_r \times M_{red}}{n_r}}$$

$$\frac{25 \times 0.02}{2} = \frac{18.7 \times M_{red}}{10}$$

$$M_{red} = \frac{25 \times 0.02 \times 10}{2 \times 18.7}$$

= 0.1337 moles per litre.

= 0.1337×248 g/L (Rel molecular mass of $Na_2S_2O_3.5H_2O$ = 248)

= 33.158 g/L

Answer: Concentration of sodium thiosulfate solution is:

(a) 0.1337 moles per litre

(b) 33.158 grams $Na_2S_2O_3.5H_2O$ per litre

Experiment 15.4 Mandatory Student Experiment No. 13

To determine the percentage (w/v) of sodium hypochlorite in household bleach.

Introduction

Most household bleaches (e.g. *Parazone*, *Domestos*, etc.) contain chlorine in the form of sodium hypochlorite, NaClO (Fig 15.14). This substance contains the hypochlorite ion, ClO^-. The more modern name for this ion is the chlorate(I) ion but since the term hypochlorite is more widely used, we shall use this term in the following discussion. The hypochlorite ion is an oxidising agent and converts iodide ions to iodine according to the following equation:

$$ClO^- + 2I^- + 2H^+ \rightarrow Cl^- + I_2 + H_2O$$

To determine the strength of household bleach, a sample of the bleach is reacted with a solution of iodide ions. The liberated iodine is then titrated against a solution of sodium thiosulfate of known concentration. From the results of the titration, we can then calculate the concentration of sodium hypochlorite in the bleach. As household bleach is usually fairly concentrated, it must be diluted before it is used in this experiment. Otherwise, an excessive amount of reagents would be used in the titration.

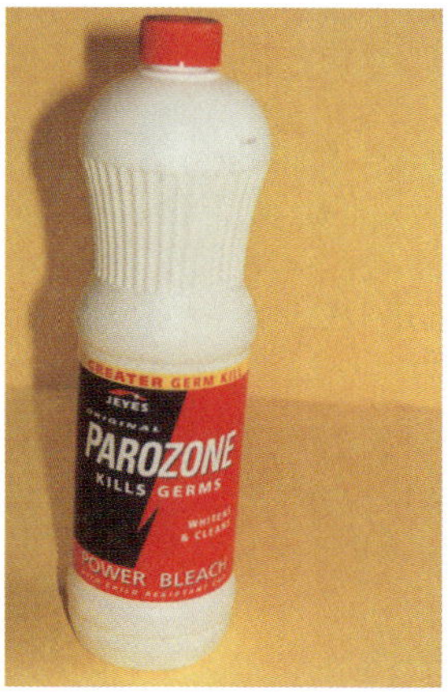

Fig 15.14 Sodium hypochlorite, NaClO, is the active ingredient in bleach.

Apparatus required

Pipette (25 cm^3), pipette filler, burette, conical flask, wash bottle, retort stand, three beakers, white tile, funnel, graduated cylinder, volumetric flask (250 cm^3).

Chemicals required

Sodium thiosulfate (0.1 M), dilute sulfuric acid (1M), potassium iodide solution (0.5 M), starch indicator, bleach, deionised water.

Procedure

1. Using a pipette filler, pipette 25 cm^3 of the concentrated household bleach into a clean 250 cm^3 volumetric flask. (Do not use the 'thick' type of bleach as the gelling agent in this bleach may interfere with the end point.)
2. Add deionised water to the volumetric flask and make up to the mark. Note that we have diluted the bleach 10 times. Invert the stoppered flask about 20 times.
3. Wash out the pipette, burette and conical flask with deionised water.
4. Wash out the burette with some of the 0.1 M sodium thiosulfate solution (previously standardised) and fill the burette above the zero mark with the solution. Remove the funnel from the burette and adjust the level of the solution to the zero mark. Ensure that the tap and space below the tap are also filled.
5. Pour 100 cm^3 approx. of the diluted bleach solution into a clean, dry beaker. Wash out the pipette with a small amount of this solution and, using a pipette filler, pipette 25 cm^3 of the solution into the conical flask.
6. Using a graduated cylinder, add 20 cm^3 approx. of dilute sulfuric acid, followed by 10 cm^3 of 0.5 M potassium iodide solution to the conical flask (Fig 15.15). Note the reddish/brown colour of the liberated iodine.
7. Titrate in the usual manner until the solution in the conical flask becomes pale yellow. Now add a few drops of starch indicator. A blue/black colour should be observed.
8. Continue the titration, adding the thiosulfate solution drop by drop, until the blue/black colour disappears to give a colourless solution. Note the titration figure.
9. Perform one rough and two accurate titrations. The two accurate titration figures should agree to within 0.1 cm^3.

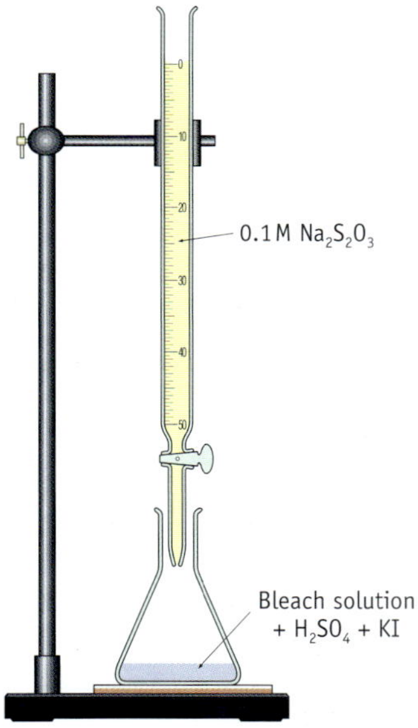

Fig 15.15 Experiment to determine the percentage (w/v) of hypochlorite in bleach.

Calculations

The method of calculating the amount of sodium hypochlorite in the bleach is shown in Example 15.6.

Example 15.6

25 cm^3 of household bleach was diluted to 250 cm^3. A 25.0 cm^3 portion of the diluted solution was added to an excess of acidified potassium iodide solution and titrated against 0.12 M sodium thiosulfate solution (previously standardised). The average titration figure was 32.1 cm^3. Calculate the concentration of sodium hypochlorite in the bleach in (a) moles per litre, (b) grams per litre and (c) % w/v.

The equations for the reactions are:

$ClO^- + 2I^- + 2H^+ \rightarrow Cl^- + I_2 + H_2O$

$2S_2O_3^{2-} + I_2 \rightarrow S_4O_6^{2-} + 2I^-$

SOLUTION

From the first balanced equation, we see that one mole of hypochlorite (the oxidising agent) produces one mole of iodine. Each mole of iodine reacts with two moles of thiosulfate (the reducing agent).

$ClO^- \equiv I_2 \equiv 2S_2O_3^{2-}$

i.e. $n_o = 1$ and $n_r = 2$

GIVEN:

$V_o = 25$

$M_o = ?$

$n_o = 1$

$V_r = 32.1$

$M_{red} = 0.12$

$n_r = 2$

$$\underset{(ClO^-)}{\frac{V_o \times M_o}{n_o}} = \underset{(S_2O_3^{2-})}{\frac{V_r \times M_{red}}{n_r}}$$

$$\frac{25 \times M_o}{1} = \frac{32.1 \times 0.12}{2}$$

$$M_o = \frac{32.1 \times 0.12}{2 \times 25}$$

$$= 0.077 \text{ mole per litre}$$

$\therefore$ concentration of original solution = 0.77 mole per litre (ten times more concentrated) (a)

= 0.77×74.5 g NaClO per litre (Rel mol mass of NaClO = 74.5)

= 57.37 g NaClO per litre (b)

= 5.74 g NaClO per 100 cm^3

= 5.74 % w/v (c)

Answer: Concentration of bleach solution is:

(a) 0.77 mole NaClO per litre

(b) 57.37 grams NaClO per litre

(c) 5.74 % w/v NaClO.

Test Yourself:
Attempt questions 15.9 - 15.12 and W15.9 - W15.12.

CHAPTER SUMMARY

- Potassium permanganate is an oxidising agent. It is not a primary standard.
- Potassium permanganate acts as its own indicator when it is used in titrations.
- Potassium permanganate oxidises Fe^{2+} ions to Fe^{3+} ions in acidified solution.
- Volumetric problems involving redox reactions may be solved using the formula:

$$\frac{V_o \times M_o}{n_o} = \frac{V_r \times M_{red}}{n_r}$$

- Sodium thiosulfate is a reducing agent. It reduces iodine molecules to iodide ions.
- Starch is used to detect the end point in an iodine/thiosulfate titration.
- Household bleaches contain chlorine in the form of sodium hypochlorite.
- The concentration of sodium hypochlorite in bleach may be determined by means of an iodine/thiosulfate titration.

EXERCISES

Use the following values of relative atomic masses in answering these questions: H = 1, C = 12, N = 14, O = 16, Na = 23, S = 32, Cl = 35.5, K = 39, Mn = 55, Fe = 56, I = 127.

15.1. A student made up a 0.15 M ammonium iron(II) sulfate solution and used this to standardise a solution of potassium permanganate. 25 cm^3 of the ammonium iron(II) sulfate solution was titrated against the potassium permanganate solution. The average titration figure was 22.85 cm^3. Calculate the concentration of the potassium permanganate solution in (a) moles per litre and (b) grams per litre.

15.2. It was found that 25 cm^3 of a solution of iron(II) sulfate reacted with 23.45 cm^3 of 0.02 M $KMnO_4$ solution. Calculate the concentration of the iron(II) sulfate solution in (a) moles per litre and (b) grams of crystalline $FeSO_4.7H_2O$ per litre. What other substance, apart from the iron(II) sulfate solution, should be added to the conical flask? Give the reason for adding this other substance.

15.3. A student weighed out 10.52 g of ammonium iron(II) sulfate crystals, $(NH_4)_2SO_4.FeSO_4.6H_2O$, dissolved them in dilute sulfuric acid and made the solution up to 250 cm^3 in a volumetric flask. 25 cm^3 of this solution reacted with 23.8 cm^3 of a solution of potassium permanganate. Calculate the concentration of the potassium permanganate solution in (a) moles per litre and (b) grams of crystalline $KMnO_4$ per litre.

15.4. A mass of 6.08 grams of iron(II) sulfate crystals ($FeSO_4.xH_2O$) was dissolved in deionised water which had been acidified with dilute sulfuric acid. The resulting solution was made up to 250 cm^3 in a volumetric flask. 25 cm^3 of this solution required 21.8 cm^3 of 0.02 M $KMnO_4$ solution for complete reaction.

(a) Outline the procedure for transferring the crystals to the volumetric flask.

(b) Why was it necessary to acidify the deionised water used to dissolve the crystals?

(c) In carrying out the titration, it was also necessary to add sulfuric acid to the conical flask. What is the reason for this? What would be observed if sulfuric acid were not added to the conical flask?

(d) Calculate the value of x in the above formula.

15.5. A mass of 9.80 g of ammonium iron(II) sulfate crystals, $(NH_4)_2SO_4.FeSO_4.xH_2O$, was dissolved in dilute sulfuric acid and the resulting solution was made up to 250 cm^3 in a volumetric flask. 25 cm^3 of this solution were titrated against 0.021 M $KMnO_4$ solution. The average titration figure was 23.85 cm^3. Calculate the value of x in the above formula.

15.6. A chemist dissolved 7.84 g of ammonium iron(II) sulfate, $(NH_4)_2SO_4.FeSO_4.xH_2O$, in dilute sulfuric acid and made up the solution to 250 cm^3 with distilled water. 25 cm^3 of this solution required 20 cm^3 of 0.02 M potassium manganate(VII) (potassium permanganate) for complete reaction according to the equation:

$$5Fe^{2+} + MnO_4^- + 8H^+ \rightarrow 5Fe^{3+} + Mn^{2+} + 4H_2O$$

(i) Why was dilute sulfuric acid used in making up the iron(II) salt solution?
(ii) Find the concentration, in moles per litre, of Fe^{2+} ions in the solution.
(iii) Find the value of x.
(iv) Show clearly any changes in oxidation numbers which have occurred in the reaction. *[LCH]*

15.7. Five tablets of total mass 1.69 g and containing iron(II) sulfate were dissolved in dilute sulfuric acid and made up to 250 cm^3 in a volumetric flask. 20 cm^3 portions of the resulting solution were titrated against 0.01 M potassium permanganate. The average titration figure was 9.15 cm^3.

(i) Name the piece of apparatus used to measure (a) 25 cm^3 of the iron(II) sulfate solution and (b) to measure approximately 20 cm^3 of dilute sulfuric acid.
(ii) Describe how the end point is detected in this titration.
(iii) Calculate (a) the mass of anhydrous $FeSO_4$ in each tablet, (b) the mass of iron in each tablet and (c) the percentage of $FeSO_4$ in each tablet. The equation for the reaction is:

$$MnO_4^- + 8H^+ + 5Fe^{2+} \rightarrow Mn^{2+} + 5Fe^{3+} + 4H_2O$$

15.8. In an analysis of iron tablets to determine the amount of iron sulfate per tablet, four tablets whose total mass was 0.96 g were dissolved in dilute sulfuric acid and made up to 250 cm^3 in a volumetric flask. 25 cm^3 portions of the resulting solution were titrated against 0.015 M potassium permanganate. The average titration figure was 4.51 cm^3.

(i) Why are the tablets dissolved in dilute sulfuric acid?
(ii) Why is dilute sulfuric acid added to the tablet solution before titrating?
(iii) What is observed as the potassium permanganate solution is added to the iron(II) sulfate solution?
(iv) Calculate (a) the mass of anhydrous $FeSO_4$ in each tablet, (b) the mass of iron in each tablet and (c) the percentage of $FeSO_4$ in each tablet. The equation for the reaction is: $MnO_4^- + 8H^+ + 5Fe^{2+} \rightarrow Mn^{2+} + 5Fe^{3+} + 4H_2O$

15.9. 25 cm^3 of a 0.05 M iodine solution (previously standardised) was titrated against a solution of sodium thiosulfate. The average titration figure was 18.55 cm^3. Calculate the concentration of the sodium thiosulfate solution in (a) moles per litre and (b) grams per litre of crystalline $Na_2S_2O_3.5H_2O$. The reaction between iodine and thiosulfate may be represented as:

$$2S_2O_3^{2-} + I_2 \rightarrow S_4O_6^{2-} + 2I^-$$

15.10. 25 cm^3 of 0.018 M potassium permanganate is pipetted into a conical flask. Some dilute sulfuric acid and excess potassium iodide solution are then added and the liberated iodine is titrated against a solution of sodium thiosulfate of unknown concentration. The average titration figure is 15.6 cm^3. The reactions may be represented as:

$$2MnO_4^- + 10I^- + 16H^+ \rightarrow 2Mn^{2+} + 5I_2 + 8H_2O$$

$$I_2 + 2S_2O_3^{2-} \rightarrow S_4O_6^{2-} + 2I^-$$

(i) The potassium permanganate solution was made up by dissolving crystals of $KMnO_4$ in water. How would you ensure that all of the crystals were transferred from the clock glass to the volumetric flask?
(ii) How was the solution made up to the correct volume in the volumetric flask?
(iii) Explain how you would standardise the potassium permanganate solution.

(iv) Why was sulfuric acid added to the potassium permanganate solution before adding the potassium iodide solution? Could hydrochloric acid be used? Explain your answer.

(v) Calculate the concentration of the sodium thiosulfate solution in (a) moles per litre and (b) grams per litre of crystalline $Na_2S_2O_3.5H_2O$.

15.11. An experiment was carried out by a student in order to determine the percentage (w/v) of sodium hypochlorite, NaClO, in a sample of household bleach. 25 cm^3 of the bleach was diluted to 250 cm^3. A 25.0 cm^3 portion of the solution was added to an excess of potassium iodide solution and titrated against 0.22 M sodium thiosulfate solution (previously standardised). The average titration figure was 22.4 cm^3.
The equations for the reactions are:

$$ClO^- + 2I^- + 2H^+ \rightarrow Cl^- + I_2 + H_2O$$
$$2S_2O_3^{2-} + I_2 \rightarrow S_4O_6^{2-} + 2I^-$$

(i) Why was it necessary to dilute the sample of bleach?

(ii) Describe the procedure for diluting the bleach to an accurate volume of 250 cm^3.

(iii) Name the instrument used by the student to measure out 25 cm^3 of the diluted bleach. What precautions should be observed when using this instrument to measure this volume of the diluted bleach?

(iv) The student added 20 cm^3 approx. of dilute sulfuric acid and 10 cm^3 approx. of 0.5 M potassium iodide to the conical flask containing the diluted bleach. What instrument is usually used to measure out these volumes of liquids?

(v) What colour change is observed in the conical flask when the sulfuric acid and potassium iodide are added to the bleach?

(vi) Name the indicator used in the above titration and describe how the end point is detected.

(vii) Calculate the concentration of sodium hypochlorite in the bleach in (a) moles per litre, (b) grams per litre and (c) % w/v.

15.12. When carrying out an experiment to analyse domestic bleach (NaClO solution), exactly 25 cm^3 of a bleach solution were placed in a 250 cm^3 volumetric flask and diluted to the mark with deionised water. 25 cm^3 of this solution were placed in a titration flask and dilute sulfuric acid and excess potassium iodide were added. The resulting solution was then titrated against 0.1 M sodium thiosulfate solution. The following results were obtained for the three titrations:
25.5 cm^3, 24.9 cm^3, 25.0 cm^3.
The equations for the reactions are:

$$ClO^- + 2I^- + 2H^+ \rightarrow Cl^- + I_2 + H_2O$$
$$2S_2O_3^{2-} + I_2 \rightarrow S_4O_6^{2-} + 2I^-$$

(i) Name the instrument used to measure the 25 cm^3 of bleach solution.

(ii) Name the indicator used in the titration. At what stage in the titration is the indicator added.

(iii) Name two operations carried out on the conical flask during the titration.

(iv) Comment on the shape of the conical flask.

(v) Calculate the concentration of sodium hypochlorite in the bleach in (a) moles per litre, (b) grams per litre and (c) % w/v.

16 Rates of Reactions

16.1 INTRODUCTION

Chemical reactions occur at different speeds. If a lighted taper is put near a balloon filled with hydrogen, a chemical reaction takes place very quickly and a loud bang is heard. This reaction is extremely fast. Explosions are usually extremely fast reactions in which a large volume of gas is produced in a very short period of time (Fig 16.1). On the other hand, the chemical reactions that take place when an apple ripens or when iron rusts are very slow and may take several months. It is important for chemists to know how quickly a particular chemical reaction occurs and how to change the speed of the reaction. If a chemical reaction is being used in an industrial process, economic problems may arise if the reaction occurs too slowly.

Fig 16.1
Explosions occur very quickly. The explosive nitroglycerine gives off a very large volume of gas in a small period of time.

To illustrate the different rates at which chemical reactions occur, consider the experiment shown in Fig 16.2. A sample of acidified potassium permanganate is added to three conical flasks containing ammonium iron(II) sulfate solution, sodium sulfite solution and ethanedioic acid (an acid found in rhubarb) respectively. The reaction occurs extremely fast in the case of the ammonium iron(II) sulfate solution and the sodium sulfite solution but occurs very slowly in the case of the ethanedioic acid (oxalic acid) – it takes about 30 minutes for the purple colour to disappear.

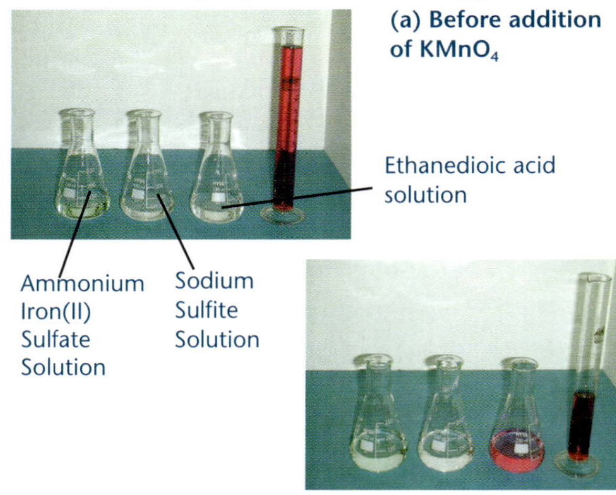

Fig 16.2
The reactions between acidified potassium permanganate and ammonium iron(II) sulfate and sodium sulfite are very rapid. However, in the case of the ethanedioic acid (oxalic acid) the reaction takes about half an hour to go to completion.

Chemists use the term **rate of reaction** to describe how quickly chemical changes occur. You may have already heard the word rate used to describe the speed of a car. The speed of a car is the rate at which the car is travelling. It is calculated by dividing the distance travelled by the time taken for the car to travel that distance.

$$\text{Speed} = \text{Rate of change of distance with time} = \frac{\text{Distance travelled}}{\text{Time taken}}$$

Thus, a car which travels 20 km in 10 minutes has an average speed of 2 km per minute.

Chemists measure the rate of a chemical reaction by measuring how the concentration of any one reactant or product changes with time. For example, if a chemical reaction produces a gas, the rate of reaction may be measured by dividing the volume of gas produced by the time taken to produce that particular volume.

The rate of reaction is defined as the change in concentration per unit time of any one reactant or product.

For example, when hydrogen peroxide decomposes to give off oxygen gas, the rate of the reaction may be determined by measuring the volume of oxygen given off as the reaction proceeds. This experiment is studied in the next section.

16.2 MEASURING RATES OF REACTIONS

In this section, we will investigate how chemists measure rates of reactions in the laboratory. We first consider a reaction that can be monitored by collecting the gas evolved in the reaction.

EXPERIMENT 16.1 MANDATORY STUDENT EXPERIMENT NO. 14

To monitor the rate of production of oxygen from hydrogen peroxide using manganese dioxide as a catalyst.

Introduction

In this experiment, we will investigate the rate at which oxygen is released from a sample of hydrogen peroxide. Hydrogen peroxide decomposes slowly into oxygen and water. However, the rate at which it decomposes may be increased by adding a small amount of manganese dioxide. Manganese dioxide is a black powder manufactured by Mitsui Denman in Little Island, Co. Cork for use in dry cells (batteries). In this experiment, we will use it as a catalyst to speed up the decomposition of the hydrogen peroxide.

$$H_2O_2 \xrightarrow{MnO_2} H_2O + \tfrac{1}{2}O_2$$

Apparatus required

Conical flask, one-hole rubber stopper, rubber or plastic tubing, glass tubing, trough, 250 cm^3 graduated cylinder, spatula, retort stand and clamp, stopwatch. (A second 250 cm^3 graduated cylinder may be necessary depending on the concentration of the hydrogen peroxide solution being used.)

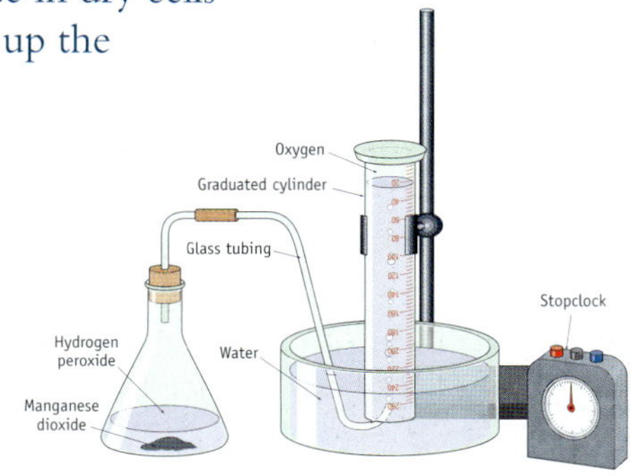

Fig 16.3 Apparatus to monitor rate of production of oxygen from hydrogen peroxide.

Chemicals required

Hydrogen peroxide ('20 volume'), manganese dioxide.

Procedure

1. Fill a trough to about half its capacity with water and invert a 250 cm^3 graduated cylinder (filled with water) over it as shown in FIG 16.3.
2. Place 50 cm^3 of '20 volume' hydrogen peroxide solution in the conical flask.
3. Place the end of the delivery tube under the graduated cylinder. One person will take charge of reading the volume of gas given off and the other person will take charge of starting the reaction and measuring the time.
4. When you are ready to begin the experiment, quickly add a small amount of manganese dioxide (enough to fit on the tip of a spatula) to the hydrogen peroxide in the conical flask, quickly put on the rubber stopper and immediately start the stopwatch.

5. At half-minute intervals, tell your partner to read the volume of gas in the graduated cylinder. Enter the results in a table, similar to that shown, in your practical notebook.

Time (mins)	0	0.5	1.0	1.5	2.0	2.5	3.0	3.5	4.0 etc.
Volume O_2 (cm^3)	0								

6. Continue measuring the volume of gas until the reaction is over, i.e. when the volume of gas evolved remains constant.
7. Using graph paper, plot a graph of volume of oxygen (y-axis) against time (x-axis). When drawing a graph always make sure that you use at least half the page and that you label both axes. Each point on the graph should be marked with a dot and a small circle drawn around it. A graph of similar shape to that in FIG 16.4 should be obtained.

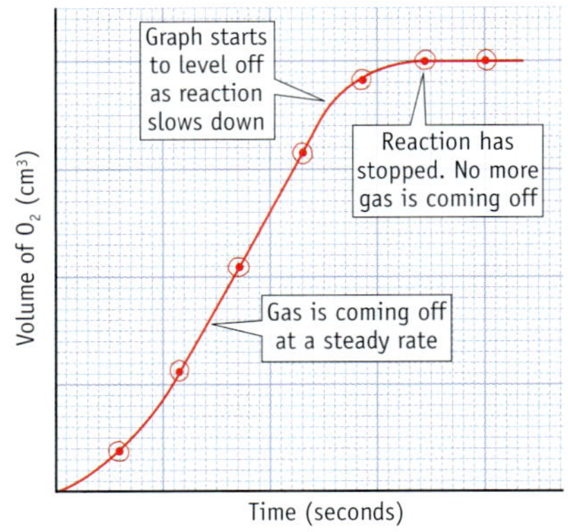

Fig 16.4
When the volume of oxygen given off is plotted against the time, a graph similar to this one is obtained.

8. Now calculate the average rate of evolution of oxygen.

$$\text{Average rate} = \frac{\text{Total volume of oxygen given off } (cm^3)}{\text{Total time for reaction to go to completion (secs)}}$$

Your answer should be given in cm^3 per second.

When carrying out the above experiment, we saw that the reaction proceeds quite rapidly at the beginning of the experiment and slows down later on. In this experiment, we have measured the **average rate** of evolution of oxygen gas. However, in some cases chemists need to measure the rate of reaction at a particular instant in time. To understand this, consider the analogy with the speed of a car. A car may travel a distance of 20 km in 20 minutes, i.e. an average speed of 1 km per minute. However, this number does not tell us anything about the speed of the car at any particular instant – it may in fact be travelling very fast on a straight stretch of roadway (FIG 16.5).

Fig 16.5
The Gardaí are more interested in the instantaneous speed than in the average speed!

The rate of reaction at a particular instant in time is called the **instantaneous rate**. The method of measuring the instantaneous rate of reaction is shown in the following example.

Example 16.1

In an experiment, 25 cm^3 of 0.1 M sulfuric acid were added to excess granulated zinc in a reaction vessel. The hydrogen evolved in the reaction was collected and its volume (at room temperature and pressure) was measured at regular intervals of time. The results obtained are shown in the following table:

Time (mins)	0	1	2	3	4	5	6	7	8	9	10	11
Volume of hydrogen	0	30	45	52.5	56.3	58.2	60	61	61.5	62	62	62

(i) Draw a diagram of the apparatus you would use in this experiment.

(ii) Plot on graph paper the results in the above table and comment on the shape of the graph.

(iii) Find, using the graph, the rate of reaction after two minutes (i.e. the instantaneous rate) expressing the rate in terms of cm^3 of hydrogen liberated per minute. [LCH]

Solution

(i) The gas may be collected over water in a graduated cylinder or it may be collected in a gas syringe (Fig 16.6). When answering questions of this type, always ensure you label the diagram. A common error among students is to forget to show the graduation marks on the graduated cylinder (and hence the graduated cylinder could be mistaken for a gas jar).

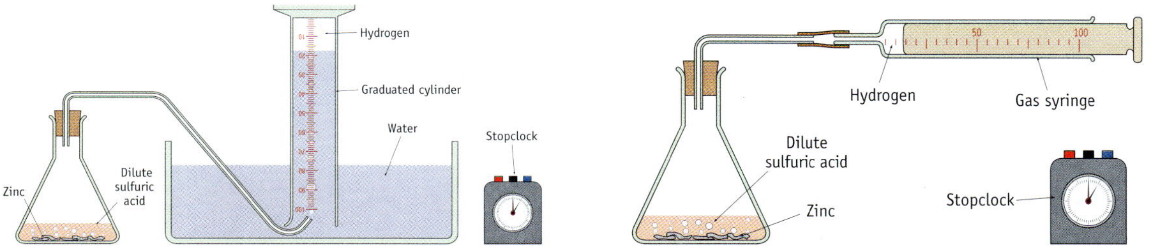

Fig 16.6
The reaction between zinc and sulfuric acid may be monitored using either apparatus.

(ii) The data is plotted in Fig 16.7.

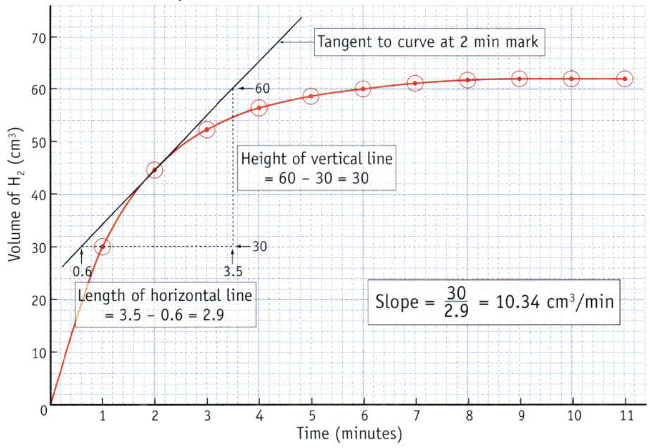

Fig 16.7
The results of the experiment are plotted on a graph.

The shape of the graph tells us that the reaction is fairly fast initially – the graph rises steeply. In fact, the steepness of the curve at the beginning of the experiment gives us a good measure of the initial reaction rate. However, as time proceeds, the reaction becomes slower and slower as indicated by the fact that the slope of the curve becomes less and less steep and the graph eventually levels off.

(iii) To measure the instantaneous rate after two minutes, we must measure the slope of the tangent to the curve at the two minute mark on the graph, i.e. the slope of this tangent tells us the rate of production of hydrogen at that instant in time. To measure the slope of the line, draw a tangent at the two-minute mark on the graph and then draw any vertical line and any horizontal line as shown in FIG 16.7. The slope of the line is then the length of the vertical line divided by the length of the horizontal line.

Instantaneous rate at 2 mins = Slope of line = $\frac{30}{2.9}$ = 10.34 cm^3/min.

Note: A slight variation in the value of the slope is always obtained depending on where you draw the vertical and horizontal lines. This variation is always taken into account in examination questions. As a general rule-of-thumb, choose points on the tangent at least half a unit of time on the right and half a unit of time on the left of the given time, e.g. in the above example, the vertical line should begin at some time **after** 2.5 minutes and the horizontal line some time **before** 1.5 minutes.

16.3 FACTORS AFFECTING RATES OF REACTIONS

By experiment, it is found that the rate of a reaction depends on five factors:

(i) Nature of reactants
(ii) Particle size
(iii) Concentration
(iv) Temperature
(v) Catalysts

Let us now consider each of the above.

(I) NATURE OF REACTANTS

When carrying out experiments to test for the presence of various anions in solution (CHAPTER 6), we saw that all of the precipitation reactions were instantaneous, i.e. a precipitate was immediately formed on mixing the reagents. Thus, on adding a solution containing some Ag^+ ions to a solution containing Cl^- ions, a white precipitate was formed immediately. The reason why the reaction takes place so quickly is because all that is involved is simply the coming together of oppositely charged ions.

$$Ag^+ + Cl^- \rightarrow AgCl\downarrow$$

A similar instantaneous reaction occurs if we add some acidified sodium dichromate solution to some ammonium iron(II) sulfate solution (FIG 16.8). Both of these compounds are ionic. However, if acidified sodium dichromate is added to a compound called ethanal (acetaldehyde), a covalent compound, the reaction occurs much more slowly. This is because it takes time for the covalent bonds to be broken and new bonds to be formed in the reaction. For the same reason, the acidified potassium permanganate in FIG 16.2 takes a long time to become decolorised when added to ethanedioic acid (oxalic acid). Ethanedioic acid is a covalent compound.

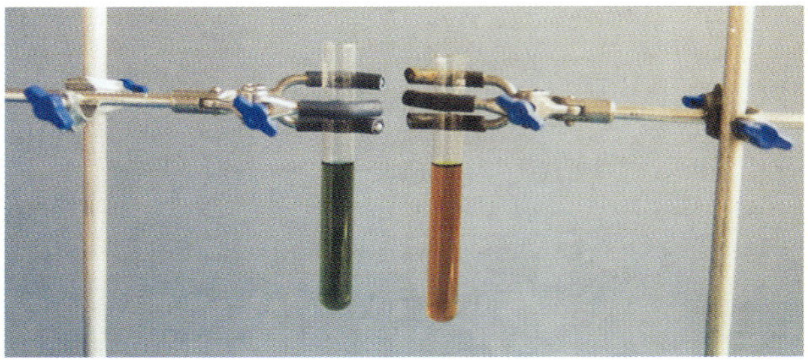

Fig 16.8
The reaction between acidified sodium dichromate and ammonium iron(II) sulfate is instantaneous (test-tube on left). However, the reaction between acidified sodium dichromate and ethanal (acetaldehyde) occurs much more slowly (test-tube on right).

As a general rule, reactions that do not involve bond breaking, but merely the coming together of oppositely charged species in solution, are usually fast at room temperature. Reactions that involve breaking of covalent bonds are usually slow at room temperature.

(II) PARTICLE SIZE

When potatoes are being cooked in boiling water, a chemical reaction occurs. One way of increasing the rate at which the potatoes cook is to cut them into smaller pieces. Similarly, it is easier to light a fire using small pieces of wood rather than a large log. The effect of particle size on reaction rates may be easily demonstrated in the laboratory. Marble reacts with dilute hydrochloric acid to release carbon dioxide (FIG 16.9). However, if the marble is powdered, there is a very large increase in the rate. Prove it for yourself!

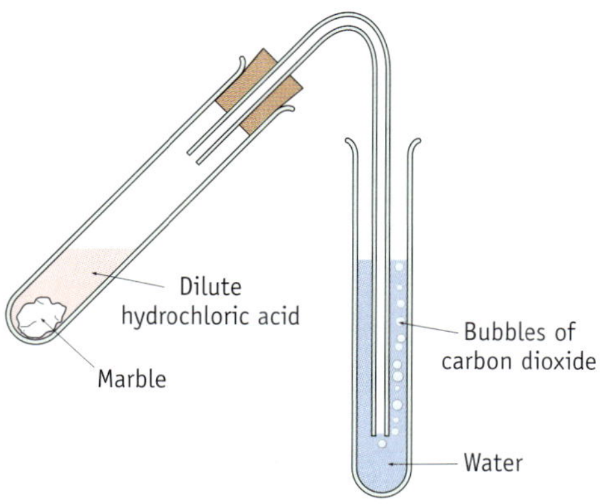

Fig 16.9
A lump of marble reacts with the hydrochloric acid to give off bubbles of carbon dioxide at a steady rate. Powdered marble would react at a much faster rate.

An experiment to quantify the effect of particle size on the rate of reaction is shown in FIG 16.10. A conical flask containing about 20 g of **large** marble chips and a graduated cylinder containing about 50 cm^3 of dilute hydrochloric acid are placed on the pan of the electronic balance. A cotton wool plug is placed in the mouth of the conical flask to prevent any loss of spray during the experiment.

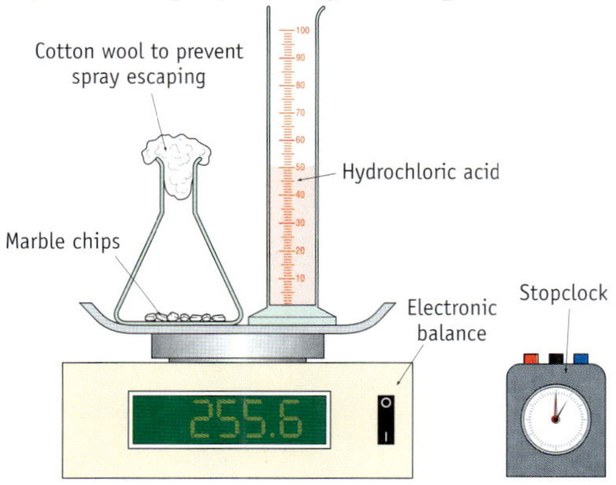

Fig 16.10
Apparatus to study the effect of particle size on the rate of reaction. As the hydrochloric acid reacts with the marble chips, there is a loss in mass as carbon dioxide is evolved.

The total mass is read from the balance. The dilute hydrochloric acid is then added to the conical flask with swirling, the cotton wool plug is replaced, the stop clock is started and the conical flask and graduated cylinder are replaced on the balance. The mass is recorded every 30 seconds for about 10 minutes. The loss in mass is understood if we consider the reaction which is taking place:

$$CaCO_3 + 2HCl \rightarrow CaCl_2 + H_2O + CO_2$$

calcium carbonate + hydrochloric acid → calcium chloride + water + carbon dioxide

The loss in mass is due to the fact that the carbon dioxide escapes through the cotton wool plug. The cotton wool plug prevents any loss of mass due to the escape of the acid spray formed as the marble reacts with the hydrochloric acid. The results are summarised in the form of a table.

Time (mins)	0	0.5	1.0	1.5	2.0	2.5	3.0	3.5, etc
Total Mass (g)								
Loss in mass (g)	0							

The loss in mass is calculated by subtracting each value of the mass from the initial mass before the reaction began.

The experiment is repeated with 20 g of **small** marble chips keeping all other conditions the same. It is found that the rate of loss of carbon dioxide is greatly increased with the smaller-sized chips.

A graph of mass against time will give a graph similar to that shown in FIG 16.11. Alternatively, a graph of loss in mass against time will give a graph similar to that shown in FIG 16.12.

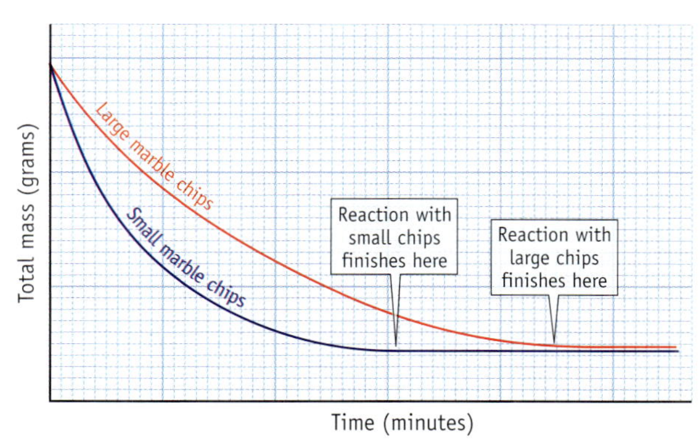

Fig 16.11
There is a faster rate of reaction with the smaller marble chips.

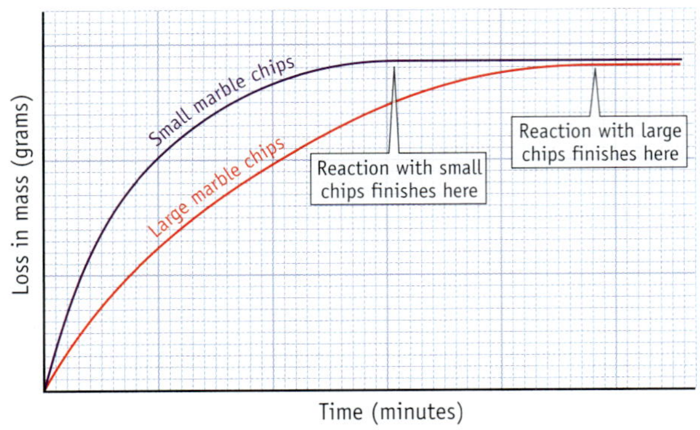

Fig 16.12
The rate of loss of mass is much greater with the small marble chips.

Very finely divided particles may cause a 'dust explosion'. This is a particular problem in flour mills, grain silos and coal mines where a build-up of very fine dust in the air can lead to an explosion. This may be demonstrated in the laboratory using a very fine powder called lycopodium powder (a type of pollen). The powder is blown over a candle in a large coffee tin on which the lid has been firmly placed (Fig 16.13). The force of the explosion is sufficient to shoot the lid high into the air (Fig 16.14).

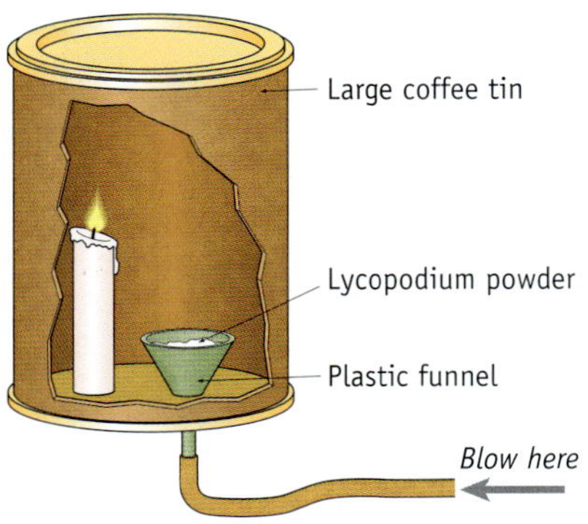

Fig 16.13
Experiment to illustrate the fact that materials in a powdered state undergo very fast reactions.

Fig 16.14
The lycopodium bomb works very well! The lid blew off too quickly to be caught by the camera!

(iii) Concentration

By experiment, it is found that changing the concentrations of the reactants in a chemical reaction alters the rate of the reaction. For example, consider the reaction in which magnesium metal reacts with hydrochloric acid.

Mg	+	$2HCl$	→	$MgCl_2$	+	H_2
magnesium	+	hydrochloric acid	→	magnesium chloride	+	hydrogen

Suppose that we react 0.1 g of magnesium with 50 cm^3 of 1.0 M HCl in the apparatus shown in Fig 16.15. (Instead of using a gas syringe we could also, of course, collect the gas over water in a graduated cylinder as illustrated in Fig 16.3.) In this experiment, the hydrochloric acid is in excess so that the reaction comes to an end when the magnesium metal is used up. The volume of gas evolved is measured at regular intervals. When the reaction ceases, the flask is washed out and the experiment repeated with 50 cm^3 of less concentrated acid, e.g. 0.5 M HCl.

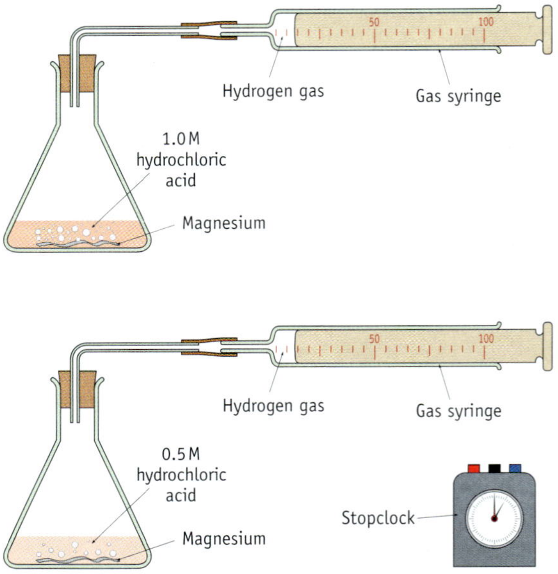

Fig 16.15
Studying the rate of reaction between magnesium and excess hydrochloric acid.

The results of the experiment are plotted in Fig 16.16.

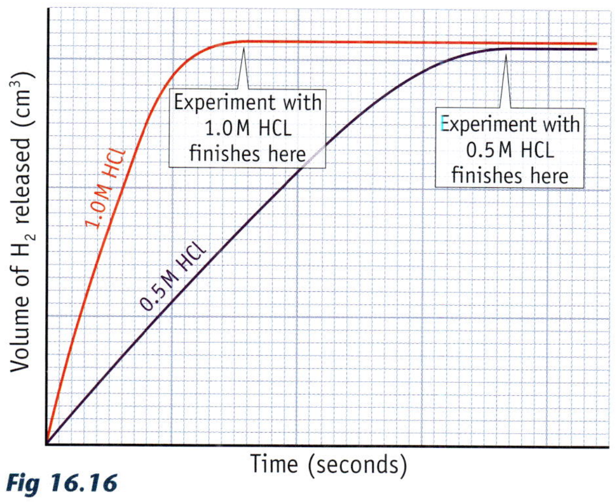

Fig 16.16
The more concentrated the acid the steeper the curve i.e. the faster the rate. The experiment with the 1.0 M HCl is completed in a much shorter time than the experiment using the 0.5 M HCl.

Note from the graph that the experiment with the more concentrated acid takes place more quickly.

Note also that the line for the more concentrated acid shows a steep climb upwards, i.e. a large amount of gas is coming off in a short time. In the experiment with the less concentrated acid, there is a less steep climb upward, i.e. the gas is coming off more slowly. Thus, the more concentrated acid gives rise to a faster rate of reaction. Note that the same amount of gas is given off in both experiments. This is because the same mass of magnesium was used in both experiments. Since the hydrochloric acid is present in excess, the magnesium is the limiting reactant and determines the volume of hydrogen given off.

An alternative experiment to study how the rate of reaction depends on concentration is the reaction between sodium thiosulfate solution and hydrochloric acid. This reaction is investigated in EXPERIMENT 16.2.

EXPERIMENT 16.2 MANDATORY STUDENT EXPERIMENT NO. 15(I)

To study the effect of concentration on the rate of reaction using sodium thiosulfate and hydrochloric acid.

Introduction

In this experiment, we will take a number of solutions of different concentration of sodium thiosulfate. We will then react each of these solutions with a certain quantity of dilute hydrochloric acid. Sodium thiosulfate and hydrochloric acid react together to produce a precipitate of sulfur:

$$Na_2S_2O_3 + 2HCl \rightarrow S\downarrow + 2NaCl + SO_2 + H_2O$$

The sulfur does not settle but forms a cloudy yellow suspension. In order to follow the rate at which sulfur is formed, we will time how long it takes until there is enough sulfur to block the view of a cross drawn on a piece of paper under a flask.

Apparatus required

Beaker, conical flask, paper, two graduated cylinders (100 cm^3), stopclock.

Chemicals required

Sodium thiosulfate (0.2 M), hydrochloric acid (3 M), deionised water.

Procedure

1. Using a graduated cylinder, pour 100 cm^3 of 0.2 M sodium thiosulfate into a conical flask. Place the conical flask on a sheet of white paper on which a mark has been made (FIG 16.17). Using the other graduated cylinder, pour 10 cm^3 of the hydrochloric acid solution into the conical flask, swirl the flask and start the stopclock. Look at the cross through the solution. When the yellow precipitate of sulfur ('milkiness') reaches such a stage that the cross is no longer visible, stop the stopclock. Note the time taken in seconds and fill in the result in the table.

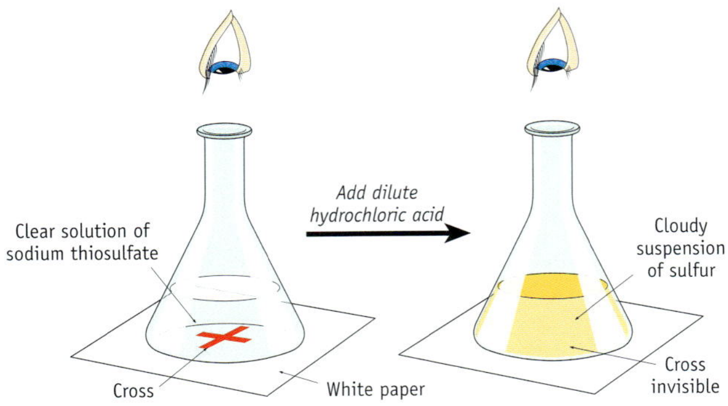

Fig 16.17
The sulfur formed by the reaction of sodium thiosulfate and hydrochloric acid obscures the cross under the conical flask.

Conc. Of $Na_2S_2O_3$ (moles/L)	0.2	0.16	0.12	0.08	0.04	0.02
Time (secs)						

2. Rinse out the conical flask and the graduated cylinder that contained the sodium thiosulfate solution. Now place 80 cm^3 of the 0.2 M sodium thiosulfate solution and 20 cm^3 of deionised water in the graduated cylinder. This gives us a solution of concentration 0.16 M (i.e. 80% of 0.2 M). Pour this solution into the conical flask, swirl the conical flask and place it on the cross as in step 1. Add 10 cm^3 of the dilute hydrochloric acid as in step 1, swirl the flask, start the stopclock and note how long it takes for the cross to become obscured. Enter your result in the table.
3. Repeat step 2 but this time use a solution of sodium thiosulfate containing 60 cm^3 of the 0.2 M sodium thiosulfate solution made up to 100 cm^3 with deionised water in the graduated cylinder. This gives a concentration of 0.12 M (i.e. 60% of 0.2 M).
4. Repeat step 2 but this time use a solution of sodium thiosulfate containing 40 cm^3 of the 0.2 M sodium thiosulfate solution made up to 100 cm^3 with deionised water. This gives a concentration of 0.08 M (i.e. 40% of 0.2 M).
5. Repeat step 2 but this time use a solution of sodium thiosulfate containing 20 cm^3 of the 0.2 M sodium thiosulfate solution made up to 100 cm^3 with deionised water. This gives a concentration of 0.04 M (i.e. 20% of 0.2 M).
6. Repeat step 2 but this time use a solution of sodium thiosulfate containing 10 cm^3 of the 0.2 M sodium thiosulfate solution made up to 100 cm^3 with deionised water. This gives a concentration of 0.02 M (i.e. 10% of 0.2 M).
7. Plot a graph of time taken (y-axis) against concentration (x-axis). A graph similar to that shown in Fig 16.18 should be obtained. Note from the graph that the cross is obscured most quickly when the solution is most concentrated. Another way of stating this is that, as the concentration increases, the time for the cross to be obscured decreases, i.e. the rate of the reaction increases with increased concentration.

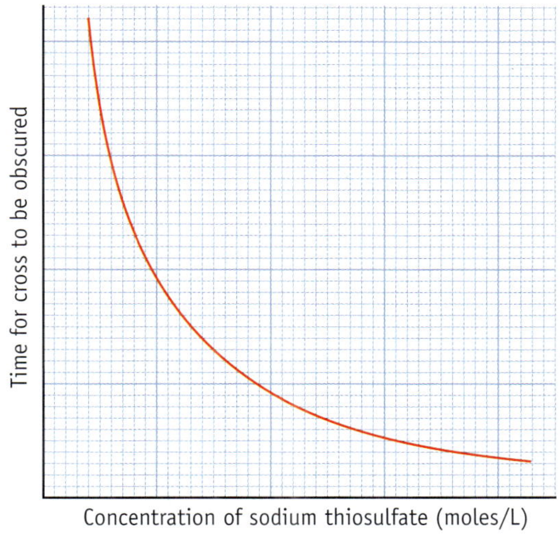

Fig 16.18

This graph shows that the more concentrated the solution, the less the time needed to obscure the cross.

8. In order to get an idea of the initial rate of a reaction, chemists often use the reciprocal of the time:

$$\text{rate} \propto \frac{1}{\text{time}}$$

9. Time and rate are inversely proportional. The shorter the time for the cross to be obscured, the faster the rate of reaction. The longer the time it takes the cross to be obscured, the slower the rate of reaction. To indicate more clearly that the rate of reaction increases with the concentration of thiosulfate ion, a graph similar to that shown in Fig 16.19 could be drawn. Note that a straight line through the origin is obtained showing that the rate of reaction is directly proportional to the concentration of the sodium thiosulfate solution. (You may find it easier to plot the graph involving $\frac{1}{t}$ if you multiply all values of $\frac{1}{t}$ by 1000 to remove the decimal points in front of the values.)

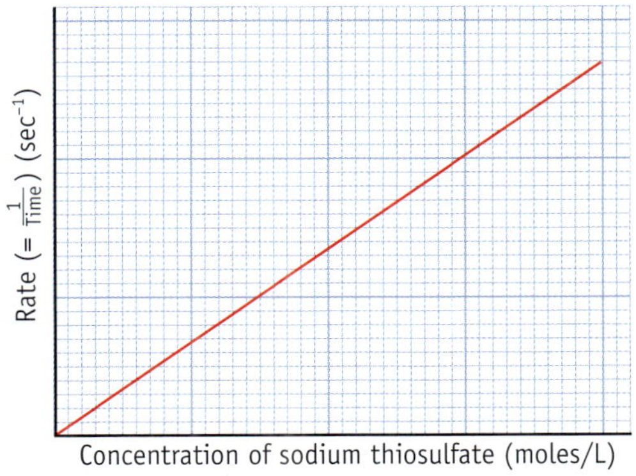

Fig 16.19

This graph shows that the rate of the reaction is directly proportional to the concentration of sodium thiosulfate

(IV) TEMPERATURE

We are all aware that food keeps fresh for longer if stored in a fridge. The reason for this is that the chemical reactions which cause the food to go bad take place more slowly at the lower temperature. In order to study the effect of temperature on the rate of a reaction, we use the same reaction between sodium thiosulfate and dilute hydrochloric acid as in EXPERIMENT 16.2.

EXPERIMENT 16.3 MANDATORY STUDENT EXPERIMENT NO. 15 (II)

To study the effect of temperature on the rate of reaction using sodium thiosulfate and hydrochloric acid.

Introduction

In this experiment, we will use the same chemical reaction studied in EXPERIMENT 16.2. However, we will use just **one concentration** of sodium thiosulfate solution but at **different temperatures.** We will use quite a dilute solution of sodium thiosulfate in this experiment so that the measured times will not be too short at the higher temperatures.

Apparatus required

Conical flask, thermometer, paper, Bunsen burner or hot plate, Pyrex beaker (50 cm^3 or 100 cm^3), two graduated cylinders (100 cm^3), stopclock.

Chemicals required

Sodium thiosulfate (0.05 M), hydrochloric acid (3 M), deionised water.

Procedure

1. Pour 100 cm^3 of the 0.05 M sodium thiosulfate into a conical flask. Place the conical flask on a sheet of white paper on which a mark has been made. Pour 10 cm^3 of the hydrochloric acid solution into the conical flask, swirl the flask and start the stopclock.
2. Look at the cross through the solution and when the degree of 'milkiness' reaches such a stage that the cross is no longer visible, stop the stopclock. Note the time taken in seconds, measure the temperature of the solution (e.g. 15 °C) and fill in the result in the table.

Temp. (°C)	15	20	30	40	50	60	70
Time (secs)							

3. Wash out the conical flask, pour 100 cm^3 of the 0.05 M sodium thiosulfate solution into the flask and warm the flask using a hot plate or a Bunsen burner to a temperature of about 20 °C. Warm a beaker containing 10 cm^3 of hydrochloric acid to the same temperature. Place the conical flask on top of the cross, add in the 10 cm^3 of dilute hydrochloric acid to the conical flask and measure the time it takes to obscure the cross. While doing this, measure the temperature of the mixture as some cooling may have occurred on adding the dilute hydrochloric acid.
4. Repeat the experiment for temperatures of around 30 °C, 40 °C, 50 °C, 60 °C and 70 °C, (FIG 16.20.)

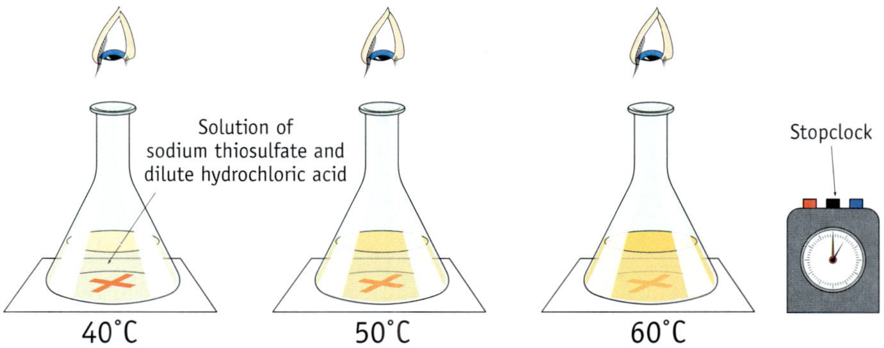

Fig 16.20
Investigating how temperature affects the rate of reaction.

You should find that the hotter solution takes less time to obscure the cross, i.e. increased temperature causes an increased reaction rate. We can represent the results of the experiments on two types of graphs. First, we can plot the time taken against the temperature (Fig 16.21). Second, we can plot the reciprocal of the time (i.e. the rate) against the temperature (Fig 16.22). This graph shows clearly that the rate of the reaction increases with temperature. Note that the rate of reaction is not directly proportional to the temperature, since we do not get a straight line passing through the origin. Chemists have discovered that, for many reactions, the rate of the reaction is approximately doubled for every 10 °C rise in temperature.

We shall see at a later stage that, because a rise in temperature causes the rate of a reaction to increase, many industrial processes are carried out at high temperatures. For example, the process for manufacturing ammonia is carried out at around 500 °C and nitric acid is manufactured at a temperature of around 900 °C.

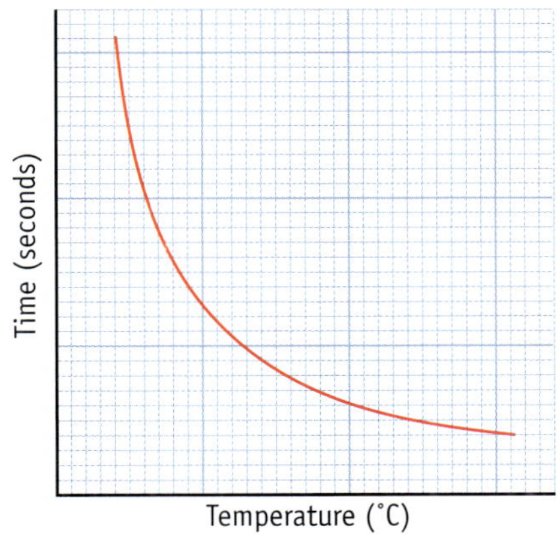

Fig 16.21
The higher the temperature, the less time it takes for the cross to be obscured.

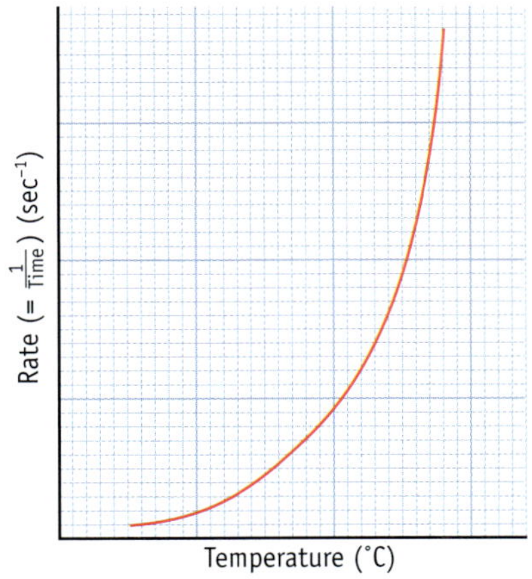

Fig 16.22
An increase in temperature causes an increase in the rate of the reaction. Note that the graph is not a straight line, i.e. the rate of reaction is not directly proportional to the temperature.

(v) Catalysts

We have already seen in Experiment 16.1 that a substance called manganese dioxide, when added to hydrogen peroxide, helps the hydrogen peroxide to decompose more rapidly.

$$H_2O_2 \xrightarrow{MnO_2} H_2O + \tfrac{1}{2}O_2$$

If you were to measure the mass of manganese dioxide present at the beginning of the experiment, you would find that the same mass of manganese dioxide would be present at the end of the experiment, i.e. the manganese dioxide is not used up in the reaction. We say that the manganese dioxide is a **catalyst**.

> A catalyst is a substance that alters the rate of a chemical reaction but is not consumed in the reaction.

Note that the definition uses the phrase 'alters' rather than 'speeds up'. The reason for this is that while most catalysts are used to speed up reactions, some can be used to slow down reactions. These latter substances are called **negative catalysts** or **inhibitors**. Glycerine is an example of a negative catalyst. It is sometimes added to hydrogen peroxide to slow down its rate of decomposition when it is being stored. Similarly, if a substance called calcium propionate is added to dough when making bread, the bread will keep fresher a lot longer as this substance slows down the chemical reactions which cause the bread to become stale.

Our study of catalysts will be considered under the following headings:

(a) General properties of catalysts.
(b) Types of catalysis.
(c) Mechanisms of catalysis.
(d) Catalytic converters.

(a) General properties of catalysts

1. Catalysts are recovered chemically unchanged at the end of a reaction. For example, if the manganese dioxide used to catalyse the decomposition of hydrogen peroxide is examined before and after the decomposition reaction has taken place, it is found to have exactly the same chemical properties.

2. Catalysts tend to be specific, i.e. a catalyst may catalyse one reaction but may not have any effect on a similar reaction. Biological catalysts called **enzymes** are produced by living cells and tend to be extremely specific. For example, liver contains an enzyme called **catalase.** This enzyme decomposes hydrogen peroxide in living things. Hydrogen peroxide is highly toxic to living cells and is constantly being produced as a result of chemical reactions in the cells. It is essential that hydrogen peroxide be decomposed as soon as it is formed. The enzyme catalase acts so quickly to break down the hydrogen peroxide that it is the fastest enzyme known. The effect of catalase on hydrogen peroxide is shown in Fig 16.23.

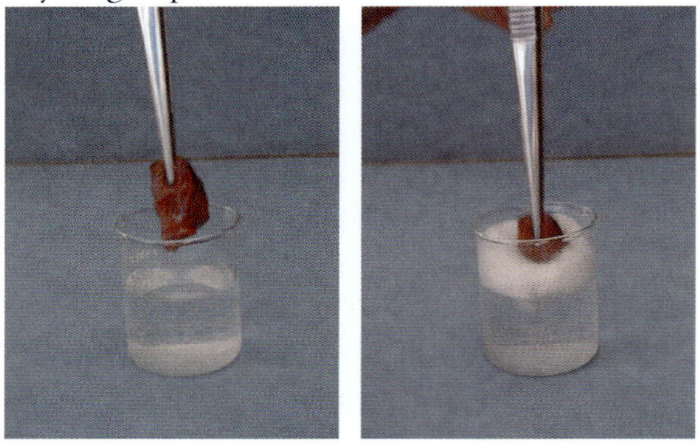

Fig 16.23
Liver contains an enzyme called catalase which acts as a catalyst in the decomposition of hydrogen peroxide.

Another example of an enzyme is **lysozyme** which catalyses the breakdown of the cell walls of bacteria and helps protect us from infection. Large amounts of enzymes are used in the manufacture of biological washing powders. These washing powders contain enzymes which catalyse the breaking down of the chemicals in food and other stains on clothing. Almost all of the enzymes in washing powders are **hydrolases –** these are enzymes which catalyse the breaking down (hydrolysis) of fats, proteins and carbohydrates. An enzyme called **protease** is used in washing powders for its ability to digest protein-based stains, such as blood, at low temperatures. Enzymes are also used in the food-processing industry to manufacture bread, cheese, wine, etc. In the digestive system, the enzyme **lipase** helps to break down fats, **pepsin** helps break down proteins, **amylase** helps break down starch, **sucrase** helps break down sucrose, etc. (For your examination you need to know just two examples of enzymes.)

3. Catalysts need only be present in very small amounts to function as catalysts. You may assume that in the reactions studied in this course, increasing the amount of catalyst does not significantly alter the rate of the reaction. (The effect on the rate of altering the amount of catalyst varies depending on the reaction itself. The effect can only be determined by experiment.)
4. In the case of equilibrium reactions, a catalyst helps equilibrium to be achieved more quickly. However, it does not affect the position of equilibrium or the composition of the final mixture in an equilibrium reaction. This will be studied in more detail in CHAPTER 17.
5. The action of a catalyst may be destroyed by certain substances called **catalytic poisons**, e.g. the lead in petrol destroys the catalysts in the catalytic converters in cars. Another example of a catalyst poison is arsenic. This is very dangerous because it inhibits the actions of certain enzymes in the body. We shall meet more examples of catalytic poisons in our study of industrial chemistry at a later stage (CHAPTER 25).

(b) Types of catalysis

Chemists have discovered that there are three types of catalysis –homogeneous catalysis, heterogeneous catalysis and autocatalysis.

1. **Homogeneous catalysis.** This is catalysis in which both the reactants and the catalyst are in the **same phase**, i.e. there is no boundary between the reactants and the catalyst. An example of this type of catalysis is the 'iodine snake' experiment. In this experiment, some concentrated potassium iodide solution is added to washing-up liquid and some hydrogen peroxide (100 volume) is then added to the container. Potassium iodide catalyses the decomposition of hydrogen peroxide into water and oxygen, i.e. the catalyst and the reactant are both liquids. The oxygen is quickly released and carries foam with it out of the lemonade bottle as shown in FIG 16.24

Fig 16.24
The iodine snake is an example of homogeneous catalysis!

2. **Heterogeneous catalysis**. This is catalysis in which the reactants and the catalyst are in **different phases**, i.e. there is a boundary between the reactants and the catalyst. An example of this type of catalysis is the decomposition of hydrogen peroxide (liquid) catalysed by manganese dioxide (solid). Another example of this type of catalysis is the catalytic oxidation of methanol (liquid) by platinum (solid). When hot methanol vapour passes over the platinum catalyst, so much heat is given out that the platinum glows as shown in FIG 16.25.

Fig 16.25
The catalytic oxidation of methanol by platinum is an example of heterogeneous catalysis.

3. **Autocatalysis**. We have already studied the reaction between permanganate ions and Fe^{2+} ions in CHAPTER 15.

$$MnO_4^- + 8H^+ + 5Fe^{2+} \rightarrow Mn^{2+} + 5Fe^{3+} + 4H_2O$$

In performing EXPERIMENT 15.1, you may have observed that the first few drops of potassium permanganate were decolorised slowly but subsequent drops were decolorised more rapidly. This is because the reaction is catalysed by the Mn^{2+} ions formed in the reaction. This occurrence, whereby **one of the products of the reaction catalyses the reaction**, is called autocatalysis.

Autocatalysis may be demonstrated in the laboratory using the reaction between potassium permanganate and ethanedioic acid (oxalic acid). If some acidified potassium permanganate solution is added to a solution of ethanedioic acid, the purple colour disappears very slowly. If more permanganate is added, it disappears more quickly. However, if a small amount of manganese(II) sulfate solution ($MnSO_4$) is added to another sample of the ethanedioic acid solution before adding the acidified potassium permanganate solution, the rate of reaction is greatly increased by the presence of the Mn^{2+} ions.

(c) Mechanisms of catalysis

A knowledge of catalysis is very important for chemists as it helps them to speed up or slow down chemical processes in industry. A lot of research takes place to determine the mechanisms of catalysis, i.e. how catalysts work. It has been discovered that there are two general mechanisms to explain how catalysts work.

1. The Intermediate Formation Theory

Arrhenius suggested that a catalyst works by forming an intermediate compound. If one considers a general reaction W + X → Y + Z, Arrhenius suggested that one (or sometimes more than one) of the reactants combined with the catalyst to form an intermediate compound. This intermediate compound may only exist for a very short time and may be difficult to detect.

$$\underset{\text{reactant}}{W} + \underset{\text{catalyst}}{C} \rightarrow \underset{\text{intermediate}}{[WC]}$$

This intermediate then reacts with the other reactant (X) to give the final product and regenerate the catalyst.

$$[WC] + X \rightarrow Y + Z + \underset{\text{catalyst}}{C}$$

Overall reaction: $W + X \rightarrow Y + Z$

The decomposition of hydrogen peroxide catalysed by the presence of I^- ions ('iodine snake' reaction) illustrates the formation of an intermediate.

$$2H_2O_2 \xrightarrow{I^-} 2H_2O + O_2$$

Step 1: One of the hydrogen peroxide molecules reacts with an I^- ion to form the IO^- intermediate.

$$H_2O_2 + I^- \rightarrow H_2O + \underset{\text{intermediate}}{IO^-}$$

Step 2: The intermediate then reacts with another hydrogen peroxide molecule to form the products and regenerate the catalyst.

$$H_2O_2 + IO^- \rightarrow H_2O + O_2 + I^-$$

Adding the equations in steps 1 and 2 gives us the overall equation:

$$2H_2O_2 \xrightarrow{I^-} 2H_2O + O_2$$

An experiment that provides visual evidence for the Intermediate Formation Theory is the oxidation of a compound called potassium sodium tartrate (don't worry about the formula!) using hydrogen peroxide. This reaction is catalysed by the presence of Co^{2+} ions. Some potassium sodium tartrate is dissolved in hot water (nearly boiling) and some crystals of cobalt(II) chloride are added. The solution is pink due to the presence of the catalyst. The beaker is placed on a tripod standing in a trough (FIG 16.26a). Some '100-volume' hydrogen peroxide is then added and, after a short time, the colour changes to green (the colour of the intermediate) and there is a vigorous evolution of carbon dioxide and steam (FIG 16.26b). The pink colour is restored at the end, suggesting that the Co^{2+} ions have not been used up.

(a) (b)

Fig 16.26
The green colour is evidence for the existence of an intermediate compound!

2. The Surface Adsorption Theory

Although some examples of heterogeneous catalysis may be explained in terms of the Intermediate Formation Theory, most heterogeneous catalysis reactions are explained using a theory called the **Surface Adsorption Theory**.

We must be careful to distinguish between the words a**d**sorption and a**b**sorption. The term **absorption** is used to describe what happens when one substance moves into another, e.g. a sponge absorbing water. However, the word **adsorption** refers to the accumulation of substances only at the surface of another substance, e.g. gas molecules accumulating on the surface of a solid. A good example of adsorption occurs in the reaction between hydrogen and oxygen to form water. At room temperature the two gases, hydrogen and oxygen, do not react but, if we add some finely divided platinum to the mixture, the reaction proceeds rapidly.

$$2H_2 + O_2 \xrightarrow{\text{Platinum}} 2H_2O$$

The first step in the reaction occurs when the hydrogen and oxygen molecules settle on (are adsorbed on) the surface of the pieces of platinum. The increased concentration of the molecules on the surface of the catalyst makes it more likely that the molecules will collide with each other and that a chemical reaction will occur. The oxygen and hydrogen molecules are held on the surface of the platinum by temporary bonds formed between these molecules and the platinum. (Platinum, being a transition metal, has vacant d orbitals that help the molecules to form temporary bonds with it.) The oxygen and hydrogen molecules then react to form water molecules on the surface of the platinum. This reaction on the surface of the platinum is the second stage of the reaction (Fig 16.27).

The final stage (called the **desorption** stage) is when the products leave the surface of the catalyst. More reactants are then adsorbed on to the surface of the catalyst and the whole process repeats over and over again.

The more finely divided the catalyst, the better is its performance. This is because there is a greater surface area on which adsorption can take place. When a catalyst is 'poisoned', a lot of the surface of the catalyst is blocked off due to permanent bond

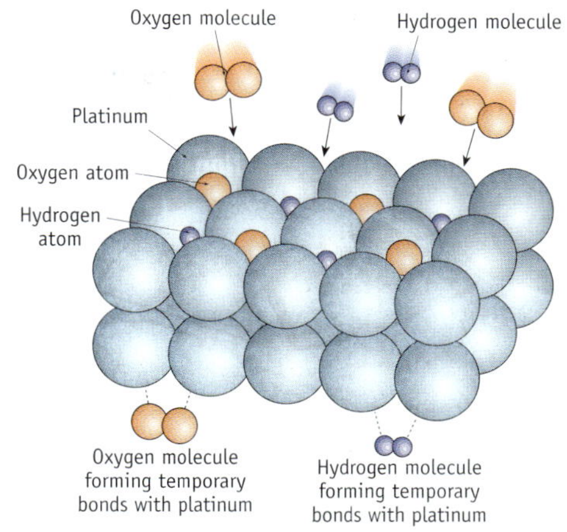

Fig 16.27
Hydrogen and oxygen molecules are adsorbed on to the surface of the platinum. The adsorbed atoms form weak bonds with the metal atoms. Transition metals can act as catalysts because they have vacant d orbitals to help molecules to form these temporary bonds with the metal atoms.

formation between the metal and the substance acting as the catalytic poison.

(d) Catalytic converters

Exhaust fumes from motor vehicles are a major source of air pollution. These exhaust fumes contain carbon monoxide (CO), nitrogen monoxide (NO), nitrogen dioxide (NO_2) and unburned hydrocarbons (i.e. compounds containing carbon and hydrogen). Carbon monoxide comes from the incomplete combustion of hydrocarbons in the fuel. The oxides of nitrogen in the exhaust fumes are commonly referred to as NO_x compounds. These oxides of nitrogen are formed because, when any fuel is burned in air at high temperatures, part of the nitrogen in the air is oxidised to form nitrogen monoxide, NO, which reacts with more oxygen to form nitrogen dioxide, NO_2. If the petrol being used is leaded petrol, then lead compounds are also emitted in the exhaust fumes.

It is very important to reduce these emissions as much as possible. Carbon monoxide is a colourless, odourless and tasteless gas that is highly poisonous. The oxides of nitrogen, as well as being toxic, contribute to acid rain. The unburned hydrocarbons arise because some of the fuel is not burned in the engine and the hydrocarbons are emitted in the exhaust fumes. These unburned hydrocarbons can give rise to smog.

A catalytic converter, fitted to the exhaust system of a car, converts these environmentally harmful gases into harmless gases. From the outside, the catalytic converter looks just like an extra silencer box on the car's exhaust system (FIG 16.28).

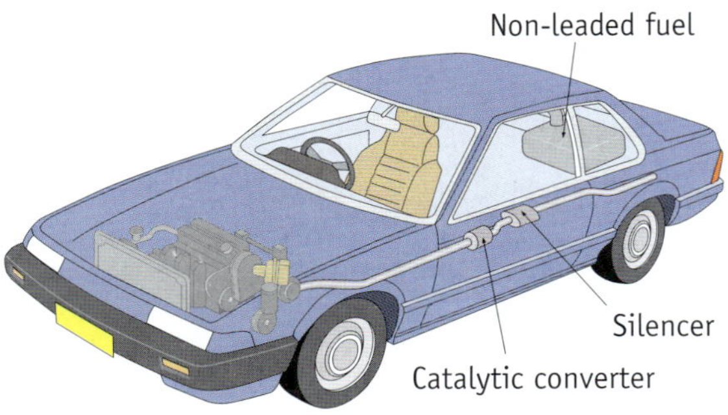

Fig 16.28
A catalytic converter is fitted into the exhaust system of a car to convert the harmful gases into harmless or less harmful ones.

A catalytic converter consists of a thin coating of platinum, palladium and rhodium (the catalysts) on a ceramic or metal 'honeycomb' inside a stainless steel case (FIG 16.29).

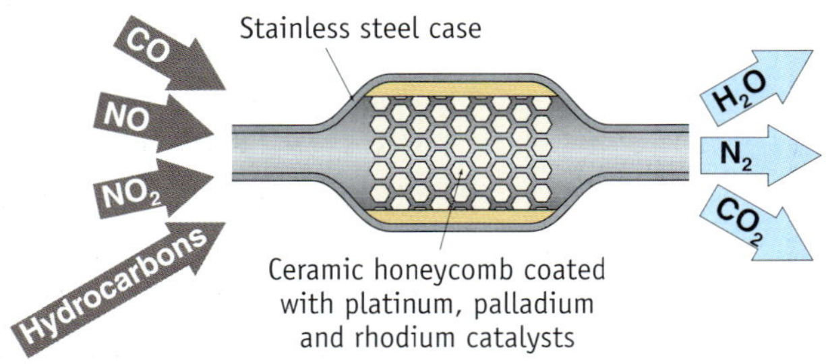

Fig 16.29
A typical catalytic converter.

The catalysts are spread over the surface of the honeycomb, which has a huge surface area – as large as two or three football fields! The mixture of hot harmful gases (at a temperature of around 300 °C) is then converted into harmless products. The main reaction that takes place in the catalytic converter is the reaction between the polluting gases themselves (carbon monoxide and nitrogen monoxide) to form carbon dioxide and nitrogen:

$$2CO + 2NO \xrightarrow{Pt/Pd/Rh} 2CO_2 + N_2$$

(The other pollutants, i.e. the unburned hydrocarbons, react with the oxides of nitrogen to form carbon dioxide, nitrogen and water.) Note that the pollutants are being converted to CO_2, N_2 and H_2O, all of which are naturally present in the air. Note also that the above reaction is a good example of heterogeneous catalysis.

Catalytic converters will last for about 50,000 miles. They need to be replaced as a result of poisoning of the catalyst and disintegration of the catalysts and ceramic surfaces as a result of engine vibration and wear from fast moving hot gases. However, if leaded petrol is used, the catalytic converter has a very short life. It is ruined as the lead poisons the surface of the catalysts. The lead compounds are more strongly adsorbed on to the surface of the catalyst than the other compounds and thus 'poison' or prevent the catalyst from doing its job.

When the converter is changed, it is possible to recover most of the catalysts and recycle them. Each catalytic converter contains only a few grams of the catalysts. Since 1993, all new cars in EU countries must be fitted with catalytic converters (FIG 16.30).

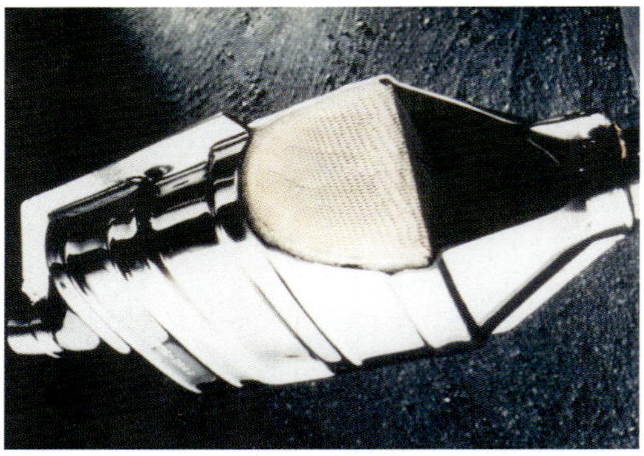

Fig 16.30
This photograph shows the ceramic honeycombs used in catalytic converters.

Pollution from exhaust fumes is a serious problem when engines are starting from cold, particularly on short city journeys when the engine does not have time to warm up fully. The lower the temperature at which the catalytic converters start working the better and a lot of research is taking place to have their operation temperature as low as possible.

16.4 COLLISION THEORY AND ACTIVATION ENERGY

Many of the observed facts about rates of reactions are interpreted using a theory called the **Collision Theory**. This theory was developed from the kinetic theory of gases in order to account for the factors that influence the rate of a reaction. The basic assumptions of this theory are:

(i) For a reaction to occur, the reacting particles must collide with each other.

(ii) A collision only results in the formation of products if a certain minimum energy is exceeded in the collision. Such a collision is called an effective collision.

When an effective collision occurs, bonds are broken and new bonds are formed. This brings about the formation of the products.

According to the collision theory, the rate of a reaction depends not only on the number of collisions per second that take place between the reacting particles, but also on the fraction of these collisions that are effective.

Suppose that one mole of oxygen gas and one mole of hydrogen gas are placed in a one litre container at room temperature and at a pressure of one atmosphere. It has been calculated that 1×10^{30} collisions take place every second between the molecules. If every one of these collisions were to lead to a chemical reaction, the reaction would be over in a fraction of a second. However, by experiment, this is not found to be the case. Chemists conclude that it is not the total number of collisions that will determine the rate of reaction but rather the number of **effective** collisions. An effective collision is one that leads to the formation of products.

Suppose you were to measure the walking speeds of people in the street. You would find that some were walking very slowly, some walking very fast and the majority would be walking at moderate speeds. Similarly, if you calculate the amount of energy possessed by the molecules in a sample of gas, it is found that most of the molecules have an average kinetic energy, some have a lower than average kinetic energy and some have a higher than average kinetic energy. This distribution of kinetic energies of the molecules of a gas at 15 °C is shown in Fig 16.31. (This type of graph is called the Maxwell-Boltzmann distribution.)

If the temperature of the gas is increased, the kinetic energies of the gas molecules will increase also due to the heat energy supplied. This will mean that a greater proportion of molecules will now have higher energies as shown in Fig 16.32.

This distribution of energy among the molecules in a sample of gas explains why it is that not all the collisions between molecules bring about a reaction. When molecules with low energies

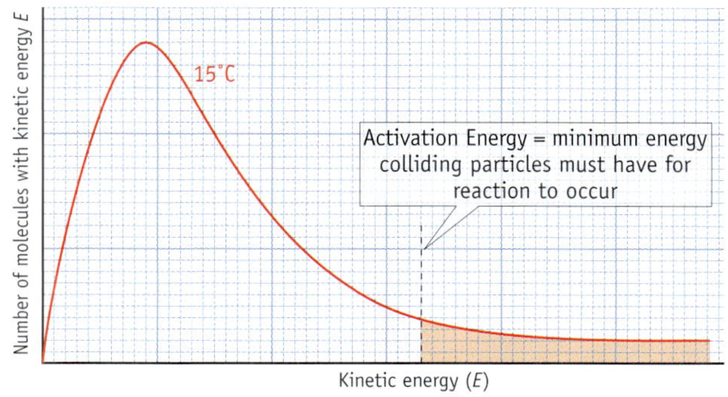

Fig 16.31
This graph shows the distribution of energies of the molecules in a sample of gas at a temperature of 15 °C.

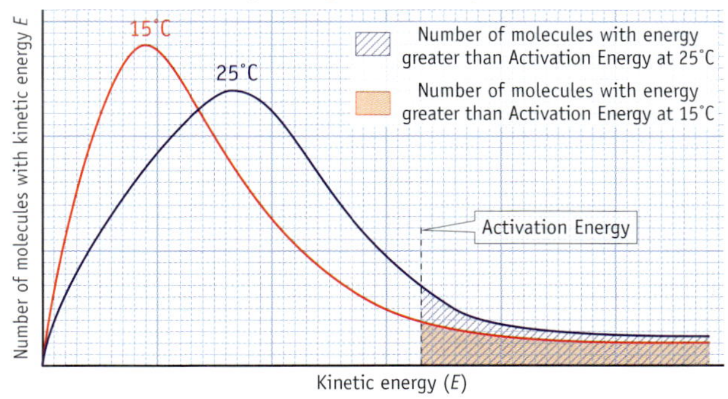

Fig 16.32
Increasing the temperature causes a greater number of molecules to have energy in excess of the Activation Energy.

collide, they simply bounce apart without any reaction taking place. It is only those molecules with high enough energies that form the products when they collide. A certain amount of energy is necessary for a chemical reaction to occur when molecules collide. This energy is called the Activation Energy and is defined as follows:

> The Activation Energy is the minimum energy which colliding particles must have for a reaction to occur.

Thus, the collision theory explains why reactions go faster at higher temperatures because more of the colliding molecules have the minimum Activation Energy needed to react.

The size of the Activation Energy depends on the nature of the reactants. Some reactions have a large Activation Energy. This means that these reactions are slow or may not occur at all, since only a small fraction of the particles have enough energy to supply the Activation Energy. In other cases, a greater fraction of the molecules have sufficient energy to supply the Activation Energy. This means that a greater fraction of the collisions are effective, i.e. these reactions are faster. Activation Energy is often represented as shown in FIG 16.33.

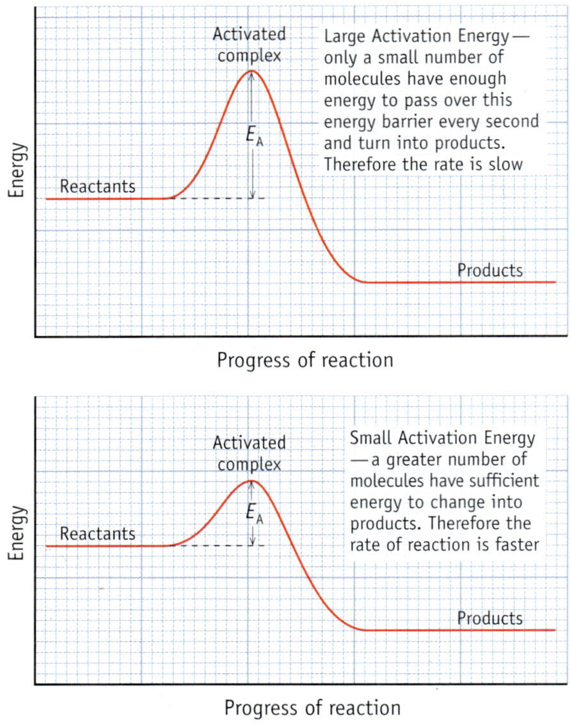

Fig 16.33
The size of the Activation Energy determines the rate of reaction.

This type of diagram is called a **Reaction Profile Diagram**. Note from the diagrams that the Activation Energy is the difference in energy between the average energy of the reactant molecules and the energy of the molecules in the activated complex. The activated complex is simply an intermediate substance formed by the collision of molecules with enough energy to react.

Reaction profile diagrams may be compared to a ball being rolled up a hill. If the ball has a lot of energy going up the hill, it may go over the hump and roll down the other side, i.e. a chemical reaction occurs (FIG 16.34). If the ball has only a small amount of energy, it will only go a little way up the hill before falling back down, i.e. no chemical reaction occurs.

We have already discussed how the Collision Theory explains how a rise in temperature increases the rate of reaction. This theory may also be used to explain why the other factors that we have studied in this chapter affect the rate of a reaction.

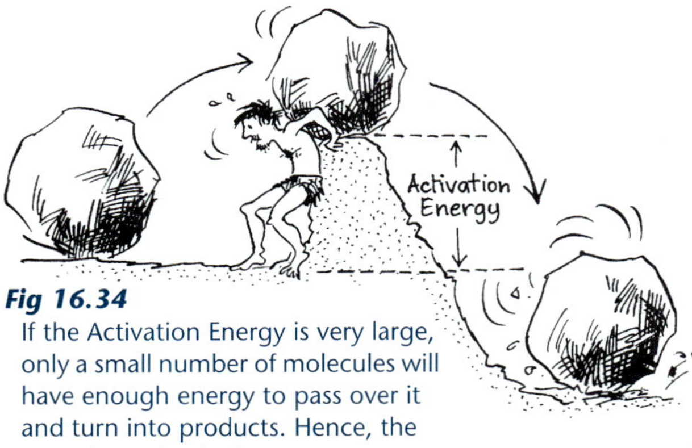

Fig 16.34
If the Activation Energy is very large, only a small number of molecules will have enough energy to pass over it and turn into products. Hence, the rate of reaction is slow.

(i) **Nature of reactants**: Reactions with covalent compounds involve bond breaking taking place during a collision. This breaking of bonds requires energy and hence these type of reactions have a much higher Activation Energy than reactions which merely involve oppositely charged ions colliding together in solution.

(ii) **Particle size**: The smaller the particle size, the more surface area is exposed, i.e. the greater the number of collisions that take place. Since the number of collisions that take place is greatly increased, the number of these collisions that are effective will also increase.

(iii) **Concentration**: If the concentration of the reactants is increased, the number of collisions will also be increased. Therefore, the number of effective collisions will be increased giving rise to an increased rate of reaction.

(iv) **Catalyst**: A catalyst works by providing an alternative reaction pathway with a lower activation energy (FIG 16.35). Therefore, more reactant molecules now possess the energy required for effective collisions to occur. Thus, the rate of reaction increases.

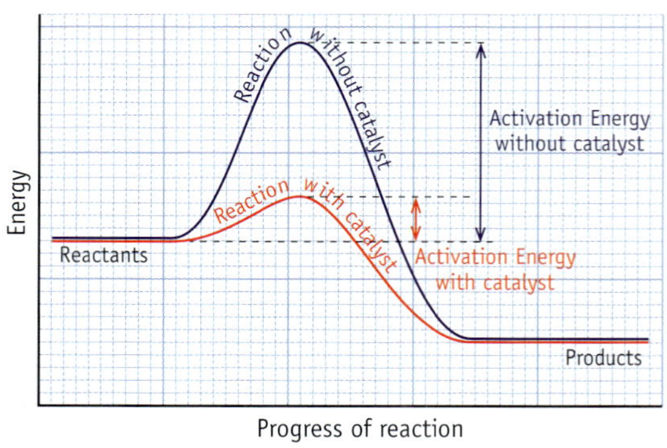

Fig 16.35
A catalyst reduces the Activation Energy for a reaction.

Example 16.2

For each of the following statements explain briefly why the rate of the reaction alters, stating the factor or factors involved:

(i) Silver nitrate crystals do not seem to react with sodium chloride crystals, but in solution an immediate white precipitate is formed.

(ii) If hydrogen and oxygen are mixed at room temperature no reaction can be detected, but if a spark is passed through the mixture a violent reaction occurs.

(iii) When hydrogen is being collected from the reaction of excess zinc with 2 M hydrochloric acid, it takes longer to fill a gas jar of a given size as the reaction proceeds. *[LCH]*

Solution

(i) The ions are not free to move in the solid state as they are bonded in fixed positions in the crystal lattice. However, when the crystals are dissolved in water the ions now have more freedom of movement. Therefore, there is a greater probability of collisions occurring to give the white precipitate of silver chloride.

(ii) At room temperature, the hydrogen and oxygen molecules are simply colliding with each other but there is not enough combined energy in the collision to reach the activation energy of the reaction. When a spark is introduced, some molecules get extra energy (heat energy) which gives them enough energy to overcome the activation energy barrier. In other words, the spark supplies the activation energy and a chemical reaction occurs. (When the reaction occurs it actually gives out heat. This heat causes more molecules to gain enough energy to overcome the activation energy barrier giving rise to a very fast reaction which is explosive.)

(iii) The rate of a reaction depends on the concentration of the reactants. As the reaction proceeds, the concentration of the reactants is decreasing. Therefore, the reaction slows down because there will be fewer collisions between the sulfuric acid and the zinc. Thus, hydrogen is given off more slowly, i.e. it takes longer to fill the gas jar.

Test Yourself:
Attempt questions 16.1 – 16.5 and W16.1 – W16.10

CHAPTER SUMMARY

- The rate of reaction is defined as the change in concentration per unit time of any one reactant or product.
- The instantaneous rate is determined by measuring the slope of the tangent to the curve at that instant.
- By experiment, it is found that the rate of a reaction depends on the nature of the reactants, the particle size, the concentration, the temperature and the presence of a catalyst.
- The effect of concentration and temperature on the rate of reaction is studied using the reaction between sodium thiosulfate and hydrochloric acid.
- A catalyst is a substance that alters the rate of a chemical reaction but is not consumed in the reaction.
- Enzymes are biological catalysts.
- There are three types of catalysis – homogeneous catalysis, heterogeneous catalysis and autocatalysis.

CHAPTER SUMMARY

- The Intermediate Formation Theory and the Surface Adsorption Theory are used to explain how catalysts work.
- A catalytic converter is a device that is placed in the exhaust system of a motor vehicle in order to change the harmful products from the engine into harmless products.
- The main assumptions of the Collision Theory are (i) for a reaction to occur, the reacting particles must collide with each other, and (ii) a collision only results in the formation of products if a certain minimum energy is exceeded in the collision.
- The Activation Energy is the minimum energy which colliding particles must have for a reaction to occur.

EXERCISES

16.1. (a) Define the term rate of reaction ....

(b) What is a catalyst? ....

(c) In coal mines there is a great danger that coal dust may catch fire and start an explosion. Why is coal dust more dangerous than coal? ....

(d) Hydrogen gas is evolved more quickly when zinc powder is reacted with hydrochloric acid than when granulated zinc is used. Explain this observation. ....

(e) Explain the term autocatalysis. Give an example of a reaction that could be used to demonstrate autocatalysis. ....

(f) Perishable foods stay fresher for longer when stored in a fridge than in a warm room. Explain this observation. ....

(g) Write down the balanced equation to describe the reaction between sodium thiosulfate and hydrochloric acid. ....

(h) Catalytic activity is associated particularly with transition metals and their compounds. Suggest a reason for this. ....

(i) Give one example of an enzyme and state the substance on which it acts. ....

(j) Give two reasons why raising the temperature usually increases the rate of a chemical reaction. .... *[LCH]*

16.2. Define rate of reaction.

Describe simple experiments, one in each case, to illustrate the following statements:

(i) Ionic reactions in solution are almost instantaneous.

(ii) Particle size can have a significant effect on reaction rate.

Excluding particle size and the ionic or covalent nature of the reactants, list three factors that can affect the rate of a chemical reaction. *[LCH]*

16.3. Hydrogen peroxide decomposes rapidly, at room temperature in the presence of certain catalysts, according to the equation:

$2H_2O_2 \rightarrow 2H_2O + O_2$

In one such experiment, 2 g of manganese dioxide were added to 20 cm^3 of a solution of hydrogen peroxide. The volume of oxygen liberated was measured at intervals, with the following results:

Time (secs.)	0	15	30	45	60	90	120	150	180	210	240
Vol. of O_2 (cm^3)	0	60	110	134	150	170	184	193	197	200	200

(i) Using graph paper, plot the volume of oxygen liberated against time.
(ii) After what time could the reaction be said to be complete?
(iii) What fraction of the hydrogen peroxide decomposed in the first minute?
(iv) Explain why the rate of the reaction gets slower as the reaction proceeds.
(v) If this reaction were carried out at a temperature higher than room temperature, would the curve lie above or below the curve you have drawn at room temperature? Explain fully the reason for your answer.
(vi) Explain why the decomposition of hydrogen peroxide at room temperature is very slow in the absence of a catalyst. *[LCH]*

16.4. What is a catalytic converter? Name any two catalysts normally found in a typical catalytic converter. What harmful compounds are usually found in exhaust fumes. Describe what happens when these harmful compounds are passed through a catalytic converter.
Only unleaded petrol should be used in cars fitted with catalytic converters. What is the reason for this?

16.5. What is meant by the term rate of reaction?
In an experiment, 25 cm^3 of 0.1 M sulfuric acid were added to excess granulated zinc in a reaction vessel. The hydrogen evolved in the reaction was collected and its volume (at room temperature and pressure) measured at regular intervals of time. The results obtained are shown in the following table:

Time (minutes)	0	1	2	3	4	5	6	7	8	9	10	11
Vol. H_2 (cm^3)	0	30	45	52.5	56.3	58.2	60	61	61.5	62	62	62

The balanced equation for the reaction is:

$$Zn + H_2SO_4 \rightarrow ZnSO_4 + H_2$$

(i) Draw a diagram of the apparatus you would use in this experiment.
(ii) Using the balanced equation for the reaction, calculate the maximum volume of hydrogen gas **at s.t.p.** which would be evolved in the above experiment.
(iii) Plot on graph paper the results in the above table and comment on the shape of the graph.
(iv) Find, using the graph, the rate of the reaction after two minutes (i.e. the instantaneous rate), expressing the rate in terms of cm^3 of hydrogen liberated per minute.
(v) What changes would you expect in the graph if (a) 25 cm^3 of 0.5 M sulfuric acid were used in the experiment, (b) if the same quantity of 0.1 M sulfuric acid had been used but the temperature was higher? *[LCH]*

17 Chemical Equilibrium

17.1 INTRODUCTION

You may have come across the word 'equilibrium' in your study of physics at Junior Certificate level. If you look up the meaning of 'equilibrium' in a dictionary you will find it explained using words like 'state of balance'. A metre stick which is suspended at its centre of gravity is said to be balanced or in equilibrium. This metre stick remains stationary or static. Thus, this type of equilibrium is often referred to as **static equilibrium**. In other words, the entire system is not moving.

Consider now the case of a man on a treadmill. Overall, there is no change in the position of the man. He is running forward at the same speed as the belt is moving in the opposite direction. The two opposing motions balance each other. This is an example of a type of equilibrium called **dynamic equilibrium** (FIG 17.1). The word dynamic means 'moving'.

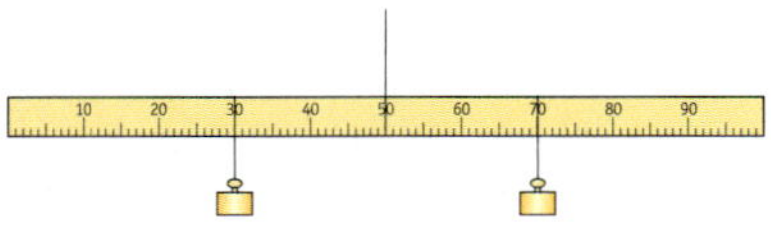

Static equilibrium

Dynamic equilibrium

Fig 17.1
Examples of static and dynamic equilibrium. The word equilibrium means 'state of balance'.

Another example of a dynamic equilibrium is walking down an escalator at the same speed as it is moving up. There is no overall change in your position because the two opposing motions are balanced.

In this chapter we will consider chemical reactions in which a state of dynamic equilibrium exists.

17.2 CHEMICAL EQUILIBRIUM

When magnesium ribbon is burned in air, a very bright light is given off and a white substance called magnesium oxide is formed. We may write this reaction as follows:

magnesium + oxygen → magnesium oxide

$$2Mg + O_2 \rightarrow 2MgO$$

It is very difficult to turn the magnesium oxide back into magnesium and oxygen. Chemists often say that the reaction 'goes to completion'. In other words, the reactants are almost completely converted to products.

By experiment it is found that there are many chemical reactions which do not 'go to completion'. For example, in 1908 a German chemist called Fritz Haber discovered that nitrogen gas and hydrogen gas reacted to form ammonia. (Ammonia is used in the manufacture of fertilisers, explosives, nylon, etc.)

nitrogen + hydrogen → ammonia

$$N_2 + 3H_2 \rightarrow 2NH_3$$

However, Haber discovered that the reaction did not go completely from left to right. He found that some of the ammonia which was formed, decomposed to give back the nitrogen and hydrogen. In other words, he found that the above reaction was reversible, i.e. it takes place both in the forward and backward directions. Such reactions are said to be equilibrium reactions and are indicated using the equilibrium arrow sign ($\rightleftharpoons$)

$$N_2 + 3H_2 \rightleftharpoons 2NH_3$$

Chemical reactions like the above are described as involving a dynamic equilibrium. When this state of equilibrium has been reached, it is found that the rate of the forward reaction is the same as the rate of the backward reaction. Therefore, there is no change in the overall amounts of either nitrogen or hydrogen or ammonia present in the container.

Chemical equilibrium is a state of dynamic balance where the rate of the forward reaction equals the rate of the reverse reaction.

Studying Fig 17.2 may help you to understand the concept of equilibrium. Note that initially the concentrations of nitrogen and hydrogen have fixed values depending on the number of moles of these gases placed in the sealed container. The initial concentration of ammonia is zero since the nitrogen and hydrogen have not yet reacted together. As the reaction proceeds, eventually at time **t** the rate of the forward reaction equals the rate of the backward reaction. At this stage, the concentrations of all three substances in the container become constant. It is called a **dynamic** equilibrium because the nitrogen and hydrogen are continually forming ammonia, whilst the ammonia is continually decomposing to form nitrogen and hydrogen. Since the rate of the forward reaction is equal to the rate of the reverse reaction, the two reactions balance out – hence the name **equilibrium**.

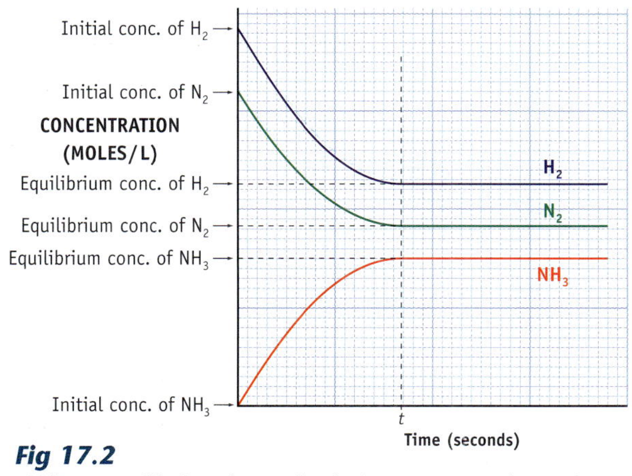

Fig 17.2

When equilibrium is reached, the concentrations of reactants and products become steady and no longer change with time.

17.3 LE CHATELIER'S PRINCIPLE

Many chemical processes involve equilibrium reactions. Therefore, for economic reasons it is important for chemists to try to shift the equilibrium as much as possible to the right, i.e. to maximise the amounts of products formed. A French chemist called Henri Le Chatelier (Fig 17.3) studied many equilibrium reactions. In 1888, he put forward a rule which allowed chemists to predict the direction taken by an equilibrium reaction if the conditions of the reaction were changed. This rule is know as Le Chatelier's Principle and may be stated as follows:

If a stress is applied to a system at equilibrium, the system readjusts to relieve the stress applied.

Fig 17.3

Henri Le Chatelier (1850 – 1936) was a very practical industrial chemist who worked as a mining engineer prior to his appointment in 1877 as professor of general chemistry at the School of Mines in Paris. He was particularly interested in finding ways to increase yields in industrial processes. This helped lead him to the discovery of the principle that bears his name.

In the above statement of Le Chatelier's Principle, the word 'stress' simply means a change in the conditions of the reaction, e.g. a change in the concentration of either reactants or products, a change in pressure, a change in temperature or adding a catalyst. In simple language, Le Chatelier's Principle tells us that if we change the conditions of a reaction that has reached a state of equilibrium, the equilibrium will be shifted in such a direction that the effect of the change will be minimised. When this has been achieved, the state of equilibrium will have been set up again.

To demonstrate Le Chatelier's Principle, consider the reaction in which two molecules of nitrogen dioxide, NO_2, join together to form a molecule of dinitrogen tetroxide, N_2O_4. Both of these substances are gases as indicated in the equation.

$$2NO_{2(g)} \underset{\text{endothermic}}{\overset{\text{exothermic}}{\rightleftharpoons}} N_2O_{4(g)}$$

nitrogen dioxide (dark brown) — dinitrogen tetroxide (colourless)

The forward reaction gives out heat (it is said to be **exothermic**) and the backward reaction takes in heat (it is said to be **endothermic**). Chemists also use the symbol ΔH to indicate whether a reaction gives out heat or takes in heat. For example, the above equation could also be written:

$$2NO_{2(g)} \rightleftharpoons N_2O_{4(g)} \quad \Delta H = -58.1 \text{ kJ mol}^{-1}$$

The negative sign indicates that the reaction gives out heat in the forward direction. (We shall be studying more about the ΔH convention in CHAPTER 21.)

The effect of change of temperature on the above reaction is illustrated in FIG 17.4.

Fig 17.4
An experiment to demonstrate Le Chatelier's Principle.

The three test-tubes contain an equilibrium mixture of nitrogen dioxide gas and dinitrogen tetroxide gas. The middle test-tube is at room temperature and contains a mixture of the two gases. Hence, the light brown colour.

The test-tube on the left is placed in ice. The lowering of the temperature constitutes a stress. In order to minimise the effect of the lowering of the temperature, the equilibrium is shifted in the direction which produces heat (Le Chatelier's Principle). Therefore, the forward reaction predominates (exothermic) and this results in the production of more dinitrogen tetroxide. Therefore, the brown colour becomes pale yellow.

$$2NO_{2(g)} \underset{\text{endothermic}}{\overset{\text{exothermic}}{\rightleftharpoons}} N_2O_{4(g)}$$

The test-tube on the right is placed in hot water. In order to absorb the stress of increased temperature, the reaction that absorbs heat will predominate, i.e. the equilibrium is shifted in the backward direction as predicted by Le Chatelier's Principle. Therefore, more nitrogen dioxide is produced. Hence, the contents of the test-tube turn dark brown.

$$2NO_{2(g)} \underset{\text{endothermic}}{\overset{\text{exothermic}}{\rightleftharpoons}} N_2O_{4(g)}$$

We now consider the effect of change of pressure on reactions involving gases. This is illustrated in FIG 17.5.

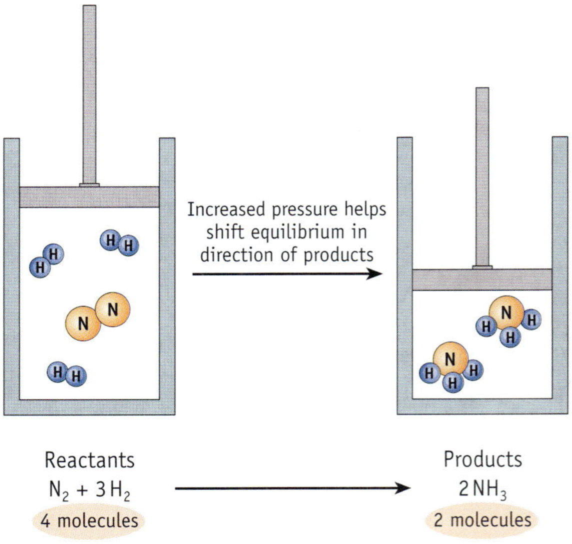

Fig 17.5
If the pressure is increased, the equilibrium is shifted to the side of the equation that has the smaller number of molecules. The smaller number of molecules occupy less volume and this helps to absorb the stress of increased pressure.

We can summarise the effect of pressure change as follows:

Le Chatelier's Principle predicts that in an all-gaseous reaction an increase in pressure will favour the reaction which takes place with reduction in volume, i.e. towards the side with the smaller number of molecules.

A change in the pressure on a system affects only an equilibrium reaction that has an unequal number of moles of gaseous reactants and products in the balanced equation.

Example 17.1

The following reaction occurs in a closed vessel at a certain temperature. The forward reaction is exothermic.

$PCl_{3(g)} + Cl_{2(g)} \rightleftharpoons PCl_{5(g)}$

What would be the effect on the concentration of PCl_5 in the equilibrium mixture of:
(i) introducing more PCl_3 into the mixture, (ii) increasing the pressure, (iii) increasing the temperature? *[LCH]*

Solution

(i) If more PCl_3 is introduced into the container, the equilibrium will shift to the right to absorb the increase. Therefore, the concentration of PCl_5 will increase.

(ii) Note that all of the substances in the above reaction are gaseous. There are two molecules on the left-hand side of the equation but only one on the right-hand side of the equation. If we increase the pressure, the equilibrium shifts to the side that has the smaller number of molecules (in accordance with Le Chatelier's Principle). Therefore, the equilibrium shifts to the right-hand side. Thus, the concentration of PCl_5 increases.

(iii) We are told that the forward reaction is exothermic. This implies that the backward reaction is endothermic.

$$PCl_{3(g)} + Cl_{2(g)} \underset{\text{endothermic}}{\overset{\text{exothermic}}{\rightleftharpoons}} PCl_{5(g)}$$

If the temperature is increased, the equilibrium will be shifted in the backward direction in order to absorb this heat. Therefore, the concentration of PCl_5 will decrease.

Note:

1. If there are the same number of moles on each side of the equation for an all-gaseous equilibrium reaction, then pressure changes will have no effect on the equilibrium.

2. We have already seen in Chapter 16 that a catalyst alters the rate of a chemical reaction. In the case of a reaction which has reached a state of dynamic equilibrium, a catalyst increases (or in some cases decreases) both the rate of the forward reaction and the reverse reaction by the same amount. In other words, the catalyst enables the system to reach a state of equilibrium in a shorter time. However, adding a catalyst to an equilibrium reaction does not affect the position of equilibrium in the reaction or the composition of the equilibrium mixture. Thus, a catalyst does not alter the yield of an equilibrium reaction. In short, a catalyst speeds up the rate at which equilibrium is reached but it does not change the position of equilibrium.

EXPERIMENT 17.1 MANDATORY STUDENT EXPERIMENT NO. 16

To investigate Le Chatelier's Principle using three named experiments.

Introduction

In this experiment, we shall carry out a number of experiments to illustrate Le Chatelier's Principle:

(A) $CoCl_4^{2-} + 6H_2O \rightleftharpoons Co(H_2O)_6^{2+} + 4Cl^-$

This experiment will be used to demonstrate the effects of both temperature changes and concentration changes on an equilibrium mixture.

(B) $Cr_2O_7^{2-} + H_2O \rightleftharpoons 2CrO_4^{2-} + 2H^+$

This experiment will be used to demonstrate the effects of concentration changes on an equilibrium mixture.

(C) $Fe^{3+} + CNS^- \rightleftharpoons Fe(CNS)^{2+}$

This experiment will also be used to demonstrate the effects of concentration changes on an equilibrium mixture.

PART A

We first consider the equilibrium reaction:

$CoCl_4^{2-} + 6H_2O \rightleftharpoons Co(H_2O)_6^{2+} + 4Cl^-$

Apparatus required

Boiling tubes, dropper, beaker (Pyrex).

Chemicals required

Cobalt(II) chloride, concentrated hydrochloric acid, deionised water.

Procedure

1. Dissolve some cobalt(II) chloride crystals in about 10 cm^3 of water in a boiling tube to give a dark red solution. The red colour is caused by the presence of the Co^{2+} ions - or more correctly, by hydrated Co^{2+} ions, $Co(H_2O)_6^{2+}$.
2. Using the fume hood and with the aid of a dropper, slowly add concentrated hydrochloric acid to the solution until it just turns blue. The following equilibrium is set up:

   $$\underset{\text{blue}}{CoCl_4^{2-}} + 6H_2O \rightleftharpoons \underset{\text{red}}{Co(H_2O)_6^{2+}} + 4Cl^-$$

   Adding the concentrated HCl causes the equilibrium to be shifted to the left-hand side. This is because adding the concentrated HCl increases the concentration of Cl^- ions and, in keeping with Le Chatelier's Principle, the concentration of these ions is decreased by the backward reaction taking place.
3. Add water to the boiling tube until the colour of the solution changes from blue to red. The equilibrium has now shifted to the right-hand side as the forward reaction absorbs the stress of the increased concentration of water.
4. We now study the effect of temperature on equilibrium.

   $$\underset{\text{blue}}{CoCl_4^{2-}} + 6H_2O \underset{\text{endothermic}}{\overset{\text{exothermic}}{\rightleftharpoons}} \underset{\text{red}}{Co(H_2O)_6^{2+}} + 4Cl^-$$

   Place the boiling tube in a beaker of hot water. Note that the colour changes from red to blue. This is in keeping with Le Chatelier's Principle, i.e. the endothermic reaction (reverse reaction) predominates in order to absorb the added heat.

PART B

We now consider the equilibrium reaction:

$Cr_2O_7^{2-} + H_2O \rightleftharpoons 2CrO_4^{2-} + 2H^+$

Apparatus required

Boiling tubes.

Chemicals required

Sodium dichromate, sodium hydroxide (2 M), dilute HCl (3 M), deionised water.

Procedure

1. Add a few crystals of sodium dichromate to about half a boiling tube of deionised water to give an orange colour. The following equilibrium is set up:
   $Cr_2O_7^{2-} + H_2O \rightleftharpoons 2CrO_4^{2-} + 2H^+$
   orange yellow
   Since the orange colour predominates, the equilibrium must lie on the left-hand side of the equation.
2. Carefully add some bench dilute sodium hydroxide until the orange colour changes to yellow. The sodium hydroxide removes the H^+ ions on the right-hand side of the equation giving rise to a stress. Therefore, in keeping with Le Chatelier's Principle, the forward reaction predominates to produce more H^+ ions.
3. Carefully add dilute hydrochloric acid until the yellow colour changes back to orange. The added hydrochloric acid creates an excess of H^+ ions which causes the equilibrium reaction to be shifted to the left in order to absorb this excess of H^+ ions.

PART C

We now consider the equilibrium reaction:

$Fe^{3+} + CNS^- \rightleftharpoons Fe(CNS)^{2+}$

Apparatus required

Boiling tubes.

Chemicals required

Iron(III) chloride (0.05 M), potassium thiocyanate (0.5 M), hydrochloric acid (3 M).

Procedure

1. Mix together about 5 cm^3 of solutions of iron(III) chloride and potassium thiocyanate in a boiling tube. Note the formation of the red complex.
   $FeCl_3 + CNS^- \rightleftharpoons Fe(CNS)^{2+} + 3Cl^-$
   yellow red
   Since the red complex is formed, the equilibrium must lie on the right-hand side of the above equation.
2. Add in some dilute hydrochloric acid until the red colour disappears. In keeping with Le Chatelier's Principle, the red colour disappears as the equilibrium is shifted to the left-hand side to absorb the Cl^- ions added.
3. Add in some iron(III) chloride (or potassium thiocyanate) solution. Note that the solution becomes red as the equilibrium is shifted to the right-hand side to absorb the substance added. This is in keeping with Le Chatelier's Principle.

17.4 INDUSTRIAL APPLICATIONS OF LE CHATELIER'S PRINCIPLE

In this section we consider how Le Chatelier's Principle is applied in the industrial manufacture of ammonia and sulfuric acid.

(A) Manufacture of Ammonia by the Haber Process

Ammonia is one of the most important chemicals manufactured by the chemical industry.

Ammonia is used to manufacture fertilisers, explosives, cleaning agents, etc.

Fig 17.6
Ammonia is manufactured in this plant by the Haber Process. Fritz Haber made use of Le Chatelier's Principle when selecting the right conditions for the manufacturing process.

Approximately 80% of all ammonia produced is used to manufacture fertilisers (Fig 17.6). We shall be making a detailed study of the manufacture of ammonia at the IFI plnant in Cobh, Co. Cork, in Chapter 25. However, for the present we need only concern ourselves with the application of Le Chatelier's Principle to the manufacturing process.

Ammonia is manufactured by reacting nitrogen (obtained from the air) with hydrogen (usually obtained from natural gas). The reaction may be summarised as follows:

$$\text{nitrogen} + \text{hydrogen} \xrightleftharpoons{\text{iron (catalyst)}} \text{ammonia}$$

$$N_2 + 3H_2 \xrightleftharpoons{Fe} 2NH_3 \qquad \Delta H = -92.4 \text{ kJ mol}^{-1}$$

The amount of ammonia produced depends on the temperature and the pressure inside the reaction vessel. It was a knowledge of Le Chatelier's Principle that helped Fritz Haber select the right conditions for the manufacture of ammonia. From the above equation, we see that there are four moles of reactants on the left-hand side but only two moles on the right-hand side (Fig 17.7). Therefore, from Le Chatelier's Principle, we know that a good yield of ammonia will be obtained at high pressure since at high pressure the equilibrium tends to be pushed to the right. By experiment, this is found to be true. However, a high-pressure plant is expensive both to build and to maintain. The improvement in yield at very high pressure is not worth the extra cost. Therefore, a compromise pressure of approximately **200 atmospheres** is used in the production of ammonia. In addition, safety problems would arise if the reaction were carried out at much higher pressures.

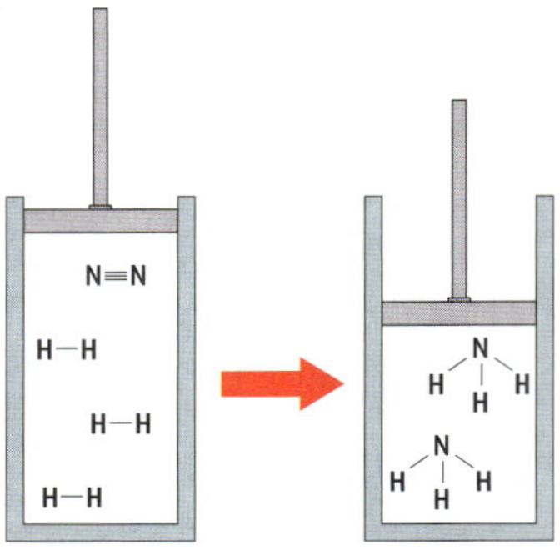

$$\text{nitrogen} + \text{hydrogen} \xrightleftharpoons{\text{iron (catalyst)}} \text{ammonia}$$

$$N_2 + 3H_2 \xrightleftharpoons{Fe_2O_3} 2NH_3 \qquad \Delta H = -92.4 \text{ kJ mol}^{-1}$$

Fig 17.7
An increase in pressure causes the equilibrium to be shifted from left to right.

The negative value for ΔH in the above equation tells us that the reaction is exothermic going from left to right. Therefore, if we lower the temperature in the reaction vessel, Le Chatelier's Principle tells us that the equilibrium will be shifted to the right, i.e. more ammonia will be produced. However, it is found that, if the temperature is too low, the rate of the reaction is too slow, i.e. the system takes a long time to come to equilibrium. Thus, at low temperature we succeed in getting a high yield of ammonia (the mixture of the three gases in the reaction vessel could consist of, say, about 90% ammonia) but the ammonia is being produced too slowly. Therefore, a compromise temperature of approximately **500 °C** is used in practice.

In short, Le Chatelier's Principle tells us that the best conditions to maximise the yield of ammonia in the Haber Process are **high pressure** and **low temperature**.

We shall see when we study the Haber Process in more detail that the process involves a method of removing the ammonia as it is formed (i.e. pushing the equilibrium to the right) and continually feeding in new reactants.

Note: In industrial processes, catalysts make a very valuable contribution to the economy of the process. For example, in the Haber Process, the percentage conversion of nitrogen and hydrogen to ammonia under the above conditions, at which a typical plant operates, is small (about 17% at the IFI plant in Cobh). However, using a catalyst enables the same percentage conversion to be achieved in a much shorter time, i.e. it brings the system to equilibrium more quickly. In addition, it allows the reaction to be carried out at a lower temperature than would be the case if no catalyst were present. This increases the productivity of the plant and hence makes the process more economical.

(B) Manufacture of Sulfuric Acid by the Contact Process

Sulfuric acid is a substance that is used in the manufacture of a wide range of materials. It is used in the manufacture of paints, detergents, fertilisers, plastics, fibres, car batteries, etc. In fact, the world consumption of sulfuric acid is about 156 million tonnes per year!

Sulfuric acid is manufactured by a method known as the **Contact Process**. The name arises from the fact that one of the stages in the manufacture of sulfuric acid involves passing sulfur dioxide gas (obtained by burning sulfur in air) and oxygen gas over a catalyst (Fig 17.8). Chemists found that there had to be very close contact between the two gases and the catalyst to ensure that the reaction took place satisfactorily.

$$\text{sulfur dioxide} + \text{oxygen} \xrightleftharpoons{\text{vanadium pentoxide}} \text{sulfur trioxide}$$

$$2SO_2 + O_2 \xrightleftharpoons{V_2O_5} 2SO_3 \quad \Delta H = -196 \text{ kJ mol}^{-1}$$

The sulfuric acid is made by reacting the sulfur trioxide formed in the above reaction with water. Platinum could also be used as a catalyst but it is found that the platinum is easily 'poisoned' by impurities in the reactants.

Fig 17.8
In this reactor vessel, sulfur dioxide and oxygen pass over vanadium pentoxide to form sulfur trioxide. This photograph was taken at the ICI sulfuric acid plant in Cheshire, England.

Since there are three molecules on the left-hand side of the above equation and only two molecules on the right-hand side, Le Chatelier's Principle suggests that the reaction should be carried out under high pressure. In practice, it is found that this is true – high pressure does favour the forward reaction. However, because high pressure plants are uneconomical (the improvement in yield is not worth the extra cost), a pressure just above **atmospheric pressure** is used in practice. Another reason is that high pressure tends to liquefy the sulfur dioxide.

Since the forward reaction is exothermic, Le Chatelier's Principle tells us that a high yield is obtained at low temperature. However, as in the case of the Haber Process, a low temperature would mean that the sulfur trioxide would be produced at a very slow rate. Therefore, a compromise temperature of **450 °C** is used as it is found that the catalyst works best at around this temperature.

Test Yourself:
Attempt questions 17.1 – 17.4 and W17.1 – W17.5.

17.5 THE EQUILIBRIUM CONSTANT

So far we have discussed equilibrium from a descriptive point of view. We have seen how various factors can affect the position of equilibrium and shift it to the left or to the right. When a system reaches a state of equilibrium, chemists and chemical engineers need to know the concentrations of reactants and products that exist in the equilibrium mixture. In some cases, it is possible to determine these concentrations by experiment. In other cases, it is necessary to determine these concentrations by performing calculations. In this section, we consider how chemists and chemical engineers perform calculations relating to systems which have reached a state of equilibrium.

By experiment, it has been discovered that, provided the temperature remains constant, there is a simple mathematical relationship between the concentrations of the reactants and products in an equilibrium mixture. This simple relationship is based on a law known as the **Law of Chemical Equilibrium** or the **Equilibrium Law**. Consider the reaction in which **a** moles of reactant A react with **b** moles of reactant B to form **c** moles of product C and **d** moles of product D at equilibrium.

$$aA + bB \rightleftharpoons cC + dD$$

By experiment, it has been found that the concentrations of all species at equilibrium are related by the expression:

$$K_c = \frac{[C]^c[D]^d}{[A]^a[B]^b}$$

equilibrium constant (↑ K_c) — concentration in moles/litre (↖ square brackets)

K_c is called the **equilibrium constant**. (The 'K' represents the constant and the subscript 'c' stands for concentration since the value of K depends on the concentrations of the reactants and products at equilibrium.) The square brackets represent concentration measured in moles per litre.

It is very important that you are capable of writing correct equilibrium constant expressions when given the balanced equation. Take note of the following:

1. The products are always written above the line (in the numerator) and the reactants are always written below the line (in the denominator).
2. The concentrations of the products are multiplied together. The concentrations of the reactants are multiplied together. The multiplication sign is usually left out.
3. The concentrations of all species are raised to the power of their coefficients in the balanced equation.

Thus, the equilibrium constant expression for the reaction:

$N_{2(g)} + 3H_{2(g)} \rightleftharpoons 2NH_{3(g)}$ is written as:

$$K_c = \frac{[NH_3]^2}{[N_2][H_2]^3}$$

Every equilibrium reaction has its own value for K_c at the particular temperature at which the reaction takes place. For example, at 458 °C the value of K_c for the reaction $H_2 + I_2 \rightleftharpoons 2HI$ is 46 approx. This constant value is always obtained no matter what concentrations of hydrogen and iodine are put into the container and allowed to reach equilibrium (FIG 17.9)

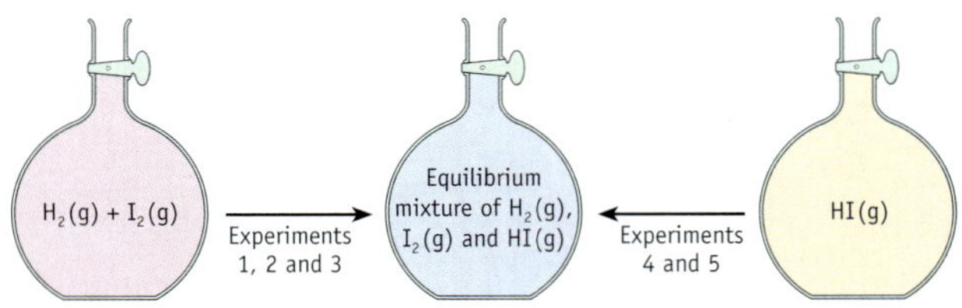

Fig 17.9
The equilibrium $H_2 + I_2 \rightleftharpoons 2HI$ can be set up starting either with a mixture of hydrogen and iodine, or with hydrogen iodide on its own, or with all three present in the container.

Alternatively, you could inject hydrogen iodide gas into the container and leave it stand until equilibrium is reached. The equilibrium concentrations of each of the three gases in the mixture are determined by experiment. The results from five experiments carried out are given in TABLE 17.1 on the following page. Note that a similar value for K_c is obtained in each case.

NOTE:

1. In the above example, the value of 46 for K_c tells us that the numerator is much greater than the denominator, i.e. the concentration of the product is much greater than that of the reactants. The value for K_c gives us an idea of how far the equilibrium is pushed to the right, i.e. how far a reaction goes to completion. The larger the value of K_c, the greater the extent to which the equilibrium is pushed towards products. A low value of K_c tells us that only a small fraction of the reactants have been converted to products.

Expt. No.	Initial concentrations (moles/litre)			Equilibrium concentrations (moles/litre)			
	$[H_2] \times 10^{-2}$	$[I_2] \times 10^{-2}$	$[HI] \times 10^{-2}$	$[H_2] \times 10^{-2}$	$[I_2] \times 10^{-2}$	$[HI] \times 10^{-2}$	$K_c = \frac{[HI]^2}{[H_2][I_2]}$
1	2.4	1.38	0	1.14	0.12	2.52	46.4
2	2.44	1.98	0	0.77	0.31	3.34	46.7
3	2.46	1.76	0	0.92	0.22	3.08	46.9
4	0	0	3.04	0.345	0.345	2.35	46.4
5	0	0	7.58	0.86	0.86	5.86	46.4

Table 17.1.
The results of five experiments carried out to determine the equilibrium constant for the reaction $H_2 + I_2 \rightleftharpoons 2HI$

In the preceding sections, we have used the term 'position of equilibrium' to describe equilibrium reactions. When we said that the position of equilibrium lies on the right-hand side of the equation, this means that the reaction has a large equilibrium constant. If the equilibrium lies on the left-hand side of the equation, the reaction has a small equilibrium constant.

2. Whenever a value of K_c for a particular reaction is quoted, the temperature is usually stated. This is because the value of K_c is temperature-dependent. (Changing the temperature shifts the equilibrium in a particular direction, this changes the equilibrium concentrations of the various species and hence the value of K_c changes.) For example, consider the reaction:

$$N_{2(g)} + 3H_{2(g)} \rightleftharpoons 2NH_{3(g)}$$

For this reaction, at 127 °C, K_c has a value of 4390. At 327 °C, K_c has a value of 4.03. At 527 °C, K_c has a value of 0.03.

3. The units of K_c depend on the reaction itself. For example, in the $H_2 + I_2 \rightleftharpoons 2HI$ reaction, the units of K_c can be found by proceeding as follows:

$$K_c = \frac{[HI]^2}{[H_2][I_2]}$$

$$= \frac{(\text{moles/L})^2}{(\text{moles/L})(\text{moles/L})}$$

Since the units cancel out, there are no units for K_c in this particular case.
However, for the reaction:
$N_{2(g)} + 3H_{2(g)} \rightleftharpoons 2NH_{3(g)}$

$$K_c = \frac{[NH_3]^2}{[N_2][H_2]^3}$$

$$K_c = \frac{(\text{moles/L})^2}{(\text{moles/L})(\text{moles/L})^3}$$

$$K_c = \frac{1}{(\text{moles/L})^2}$$

The units in this case are $(\text{moles/L})^{-2}$. In this course you need not worry about units for K_c and in all subsequent work the units of K_c will be ignored.

4. When stating a value for K_c it is essential to give the equation to which it refers. For example, we have already seen that at 458 °C, the value of K_c for the equilibrium:
$H_2 + I_2 \rightleftharpoons 2HI$ is 46 approx.

$$K_c = \frac{[HI]^2}{[H_2][I_2]} = 46$$

However, if we write the equilibrium reaction as $2HI \rightleftharpoons H_2 + I_2$, then the new K_c (represented as K_c') will be:

$$K_c' = \frac{[H_2][I_2]}{[HI]^2} = \frac{1}{46}$$

i.e.

$$K_c' = \frac{1}{K_c}$$

If the equilibrium reaction is written as:
$\frac{1}{2} H_2 + \frac{1}{2} I_2 \rightleftharpoons HI$
then a different value of K_c (represented as K_c'') will be obtained.

$$K_c'' = \frac{[HI]}{[H_2]^{\frac{1}{2}}[I_2]^{\frac{1}{2}}} = \sqrt{46}$$

5. While the value of the equilibrium constant gives us information about the extent of a reaction, it tells us nothing about the rate of a reaction. In other words, K_c tells us how **far** a reaction has gone but not how **fast** the reaction goes.

17.6 CALCULATIONS USING THE LAW OF CHEMICAL EQUILIBRIUM

Chemists and chemical engineers make great use of the Law of Chemical Equilibrium to deduce information about systems which have reached a state of equilibrium. The following examples demonstrate some of these uses.

A. To calculate the value of K_c

Example 17.2

The reaction:

$N_{2(g)} + 3H_{2(g)} \rightleftharpoons 2NH_{3(g)}$

was allowed to reach equilibrium at 450 °C in a one litre container. It was found that there were 13.6 moles N_2, 1.0 mole H_2 and 1.5 moles NH_3 in the container when equilibrium had been established. Calculate the value of K_c at this temperature.

Solution

To solve this problem simply write the equilibrium constant expression for the reaction and substitute in the appropriate values. Since we are told that the container is one litre, then the concentrations given are automatically in moles per litre.

$$K_c = \frac{[NH_3]^2}{[N_2][H_2]^3}$$

$$= \frac{(1.5)^2}{(13.6)(1.0)^3} = 0.165$$

Answer: $K_c = 0.165$

Example 17.3

A mixture of 27 g of ethanoic acid and 20.7 g of ethanol was allowed to come to equilibrium at a certain temperature. It was found by experiment that 9 g of ethanoic acid were present in the equilibrium mixture. The reaction is:

CH_3COOH	+ C_2H_5OH	$\rightleftharpoons$ $CH_3COOC_2H_5$	+ H_2O
ethanoic acid	ethanol	ethyl ethanoate	water

(i) Write the equilibrium constant expression (K_c) for the reaction.
(ii) State how you would experimentally determine the amount of ethanoic acid present in the equilibrium mixture.
(iii) Calculate the equilibrium constant (K_c). *[LCH]*

Solution

(i) $$K_c = \frac{[CH_3COOC_2H_5][H_2O]}{[CH_3COOH][C_2H_5OH]}$$

(ii) The amount of ethanoic acid could be determined by titrating it against a base whose concentration is known.

(iii) The first thing to note in the problem is that the quantities are given in grams. These must be converted to moles before we can begin to solve the problem.

Initially,

no. of moles of ethanoic acid = $\dfrac{\text{Mass}}{\text{Rel molecular mass}} = \dfrac{27}{60} = 0.45$

no. of moles of ethanol = $\dfrac{\text{Mass}}{\text{Rel molecular mass}} = \dfrac{20.7}{46} = 0.45$

At equilibrium,

no. of moles of ethanoic acid = $\dfrac{\text{Mass}}{\text{Rel molecular mass}} = \dfrac{9}{60} = 0.15$

In all problems of this sort, set out your work as follows:

	CH_3COOH +	C_2H_5OH ⇌	$CH_3COOC_2H_5$ +	H_2O
Initially	0.45	0.45	0	0
At equil.	0.15	?	?	?

The only difficult thing in these problems is to figure out the number of moles of each of the substances at equilibrium. We know from the problem that there are 0.15 moles of ethanoic acid at equilibrium. Therefore, there must also be 0.15 moles of ethanol at equilibrium since it is clear from the equation that ethanoic acid and ethanol always react in the ratio of 1:1. Thus, for every molecule of ethanoic acid that reacts to form products, a molecule of ethanol must also react to form products. Therefore, since we start with the same number of moles of ethanoic acid and ethanol, there must be the same number of moles of each at equilibrium.

Finally, we must figure out how many moles of ethyl ethanoate and water are formed. Since we started with 0.45 moles of ethanoic acid and there are only 0.15 moles at equilibrium, it must mean that 0.45 – 0.15 = 0.30 moles of ethanoic acid have reacted. For every mole of ethanoic acid that reacts, one mole of ethyl ethanoate is formed. Therefore, 0.3 moles of ethyl ethanoate are formed. Since the number of moles of water formed is the same as the number of moles of ethyl ethanoate formed, then the number of moles of water is also 0.3.

It is not necessary to go through the above detailed explanation in your examination. However, it is important that you work through it in your mind and proceed to lay out your work as follows:

	CH_3COOH +	C_2H_5OH ⇌	$CH_3COOC_2H_5$ +	H_2O
Initially	0.45	0.45	0	0
At Equil.	0.15	0.15	0.3	0.3
Conc. at equilib.	$\left[\dfrac{0.15}{V}\right]$	$\left[\dfrac{0.15}{V}\right]$	$\left[\dfrac{0.3}{V}\right]$	$\left[\dfrac{0.3}{V}\right]$

V is the volume of the container. We are not given it in the question as the V terms cancel out in the K_c expression.

$$K_c = \frac{[CH_3COOC_2H_5][H_2O]}{[CH_3COOH][C_2H_5OH]} = \frac{(0.3)(0.3)}{(0.15)(0.15)} = 4$$

Answer: $K_c = 4$.

Example 17.4

The thermal decomposition of phosphorus pentachloride is represented by the following equation:

$PCl_{5(g)} \rightleftharpoons PCl_{3(g)} + Cl_{2(g)}$

A mass of 20.85 g of phosphorus pentachloride was heated at 523 K in a closed eight litre vessel. When the reaction had come to equilibrium at that temperature, it was found that 80% of the pentachloride had dissociated.

(i) What is meant by chemical equilibrium? Why is it described as a dynamic process?

(ii) What would be the effect, if any, on the percentage decomposition of the phosphorus pentachloride if the reaction were carried out at a higher pressure.

(iii) Calculate K_c for the reaction. *[LCH]*

SOLUTION

(i) Chemical equilibrium is a state of dynamic balance where the rate of the forward reaction equals the rate of the reverse reaction.
It is described as a dynamic process because both the forward and reverse reactions are still taking place even though the overall reaction seems to have stopped.

(ii) Less PCl_5 would have dissociated since, according to Le Chatelier's Principle, at higher pressure the equilibrium is pushed to the left-hand side of the above equation due to the fact that there is a smaller number of moles on the left-hand side of the equation.

(iii) No. of moles of $PCl_5 = \frac{\text{Mass}}{\text{Rel molecular mass}} = \frac{20.85}{208.5} = 0.1$

Since 80% of PCl_5 dissociates, then the concentration of PCl_3 and Cl_2 will be 80% of 0.1 mole i.e. 0.08 mole.

	$PCl_{5(g)}$	$\rightleftharpoons$ $PCl_{3(g)}$	+ $Cl_{2(g)}$
Initially	0.1	0	0
At equil.	0.1 – 0.08 = 0.02	0.08	0.08
Conc. at equil. (V = 8L)	$\frac{0.02}{8}$ = [0.0025]	$\frac{0.08}{8}$ = [0.01]	$\frac{0.08}{8}$ = [0.01]

$$K_c = \frac{[PCl_3][Cl_2]}{[PCl_5]}$$

$$K_c = \frac{(0.01)(0.01)}{(0.0025)} = 0.04$$

Answer: $K_c = 0.04$

B. To determine the equilibrium concentrations given the value of K_c

Example 17.5

The value of K_c for the reaction: $H_2 + I_2 \rightleftharpoons 2HI$ is 50 at 450 °C. Calculate the final concentrations of each species if one mole of H_2 and one mole of I_2 come to equilibrium at 450 °C in a one litre flask.

Solution

	H_2 +	I_2 $\rightleftharpoons$	2HI
Initial conc.	[1]	[1]	[0]

To solve this problem, let x = number of moles of H_2 that must react in order to establish equilibrium. Each time one mole of H_2 reacts, one mole of I_2 also reacts. Therefore, x is also the number of moles of I_2 that reacts. From the balanced equation, if one mole of H_2 reacts, then 2 moles of HI must be formed. Therefore, if x moles of H_2 react, then 2x moles of HI must be formed.

	H_2	+	I_2	$\rightleftharpoons$	2HI
Initially	[1]		[1]		[0]
At equil.	1 – x		1 – x		2x
Conc. at equil. (V = 1)	[1 – x]		[1 – x]		[2x]

$$K_c = \frac{[HI]^2}{[H_2][I_2]}$$

$$= \frac{(2x)^2}{(1-x)(1-x)} = 50$$

Solving x = 0.78 or 1.39

(The easiest way to solve this equation is to take the square root of both sides.)

Chemically, 1.39 does not make sense since we started with only one mole of hydrogen. Therefore, we take the value of x = 0.78.

Therefore, equilibrium concentrations are:

$[H_2] = 1 - x = 1 - 0.78 = 0.22$ moles per litre

$[I_2] = 1 - x = 1 - 0.78 = 0.22$ moles per litre

$[HI] = 2x = 2 \times 0.78 = 1.56$ moles per litre

Note: In some cases it is not possible to simply take the square root of both sides or to factorise the equation. It is then necessary to use the following formula:

$$x = \frac{-b \pm \sqrt{b^2 - 4ac}}{2a}$$

for solving the quadratic equation.

It is very important that you have plenty of practice in solving quadratic equations. For this reason, it is strongly recommended that you work through the questions at the end of this chapter and in the Workbook.

A further worked example (Example 17.6) is given in the Workbook.

Test Yourself:
Attempt questions 17.5 – 17.8 and W17.6 – W17.12.

CHAPTER SUMMARY

- Chemical equilibrium is a state of dynamic balance where the rate of the forward reaction is equal to the rate of the reverse reaction.
- Le Chatelier's Principle states that if a stress is applied to a system at equilibrium, the system readjusts to relieve the stress applied.
- Le Chatelier's Principle predicts that in an all-gaseous reaction, an increase in pressure will favour the reaction which takes place with a reduction in volume, i.e. towards the side with the smaller number of molecules.
- Le Chatelier's Principle predicts that the best conditions to maximise the yield of ammonia in the Haber Process are **high pressure** and **low temperature**. In practice, the conditions are 200 atmospheres and 500 °C.
- Le Chatelier's Principle predicts that the best conditions to maximise the yield of sulfuric acid in the Contact Process are **high pressure** and **low temperature**. In practice, the conditions are atmospheric pressure and 450 °C.
- The Law of Chemical Equilibrium states that for:

  $aA + bB \rightleftharpoons cC + dD$

  $$K_c = \frac{[C]^c[D]^d}{[A]^a[B]^b}$$
- The Law of Chemical Equilibrium may be used in calculations involving systems at equilibrium.

EXERCISES

17.1. When cobalt(II) chloride crystals are dissolved in water and some concentrated hydrochloric acid is added, the following equilibrium is set up:
$CoCl_4^{2-} + 6H_2O \rightleftharpoons Co(H_2O)_6^{2+} + 4Cl^-$

(a) What colour is observed when the crystals are dissolved in water?

(b) What precautions should be taken when the concentrated hydrochloric acid is being added to the solution?

(c) What colour change is observed on adding the concentrated hydrochloric acid? Give a reason why this colour change is observed.

(d) Suggest one method by which the original colour could be restored.

(e) If the boiling tube containing the solution is heated, what colour would you observe? Explain your answer.

17.2. Potassium dichromate(VI) is an orange-red salt which is soluble in water giving an orange solution in which the following dynamic equilibrium is reached:
$Cr_2O_7^{2-} + H_2O \rightleftharpoons 2CrO_4^{2-} + 2H^+$

(i) Is the reaction shown a redox reaction? Explain your reasoning.

(ii) Explain why the solution turns yellow on the addition of sodium hydroxide solution.

(iii) Explain the visible and chemical effects of adding excess dilute hydrochloric acid to the alkaline solution in (ii). *[LCH]*

17.3. When bromine (a red liquid) is added to water, the bromine reacts with the OH^- ions in the water and the following equilibrium is established:

$Br_2 + 2OH^- \rightleftharpoons Br^- + BrO^- + H_2O$

All of the species on the right-hand side are colourless.

(a) Predict the direction in which the equilibrium will be shifted if some sodium hydroxide, NaOH, is added.

(b) What colour change will be observed on adding the sodium hydroxide?

(c) Suggest a way of reversing the colour change.

17.4. Sulfuric acid is manufactured on a large scale by the Contact Process. In this process sulfur dioxide and oxygen react to form sulfur trioxide. The reaction may be represented as:

$2SO_2 + O_2 \rightleftharpoons 2SO_3 \qquad \Delta H = -196 \text{ kJ mol}^{-1}$

The sulfur trioxide is then dissolved in water to make sulfuric acid.

(a) What does the symbol $\rightleftharpoons$ represent in the above equation?

(b) Explain the significance of the negative value for ΔH.

(c) The reaction is usually carried out in the presence of a catalyst called vanadium pentoxide. What is the purpose of the catalyst?

(d) Write a brief note on the conditions under which the above reaction is carried out.

17.5. Write the expression for K_c for each of the following reactions:

(a) $2CO_{(g)} + O_{2(g)} \rightleftharpoons 2CO_{2(g)}$

(b) $C_2H_{6(g)} \rightleftharpoons C_2H_{4(g)} + H_{2(g)}$

(c) $NO_{2(g)} \rightleftharpoons \frac{1}{2} N_2O_{4(g)}$

(d) $CO_{(g)} + H_2O_{(g)} \rightleftharpoons CO_{2(g)} + H_{2(g)}$

17.6. One mole of ethanoic acid and one mole of ethanol were mixed and allowed to reach equilibrium at 25 °C in a container.

$CH_3COOH + C_2H_5OH \rightleftharpoons CH_3COOC_2H_5 + H_2O$

When the mixture was analysed, it was found that 0.33 moles of unreacted ethanoic acid remained while the rest reacted to form ethyl ethanoate and water. Calculate the value of K_c for this reaction.

17.7. One mole of hydrogen iodide was injected into a one litre flask at 450 °C. The following equilibrium was set up:

$2HI \rightleftharpoons H_2 + I_2$

It was found that 0.80 moles of HI were formed at equilibrium. Calculate the value of K_c.

17.8. Bromine monochloride dissociated on heating according to the equation:

$2BrCl \rightleftharpoons Br_2 + Cl_2$

In an experiment 0.9 moles of BrCl were heated in a 5 litre vessel at a certain temperature until equilibrium was established. The amount of free chlorine in the mixture was found to be 0.4 moles.

(i) Write the equilibrium constant expression for this reaction.

(ii) Calculate the equilibrium constant for the reaction at this temperature.

(iii) Explain why it is not necessary to know the volume of the vessel in calculating the equilibrium constant.

(iv) Is the value of the equilibrium constant for the reaction when represented by the equation $BrCl \rightleftharpoons \frac{1}{2} Br_2 + \frac{1}{2} Cl_2$ the same as that calculated in (ii) above? Explain your answer.

[LCH]

18 pH and Indicators

18.1 INTRODUCTION

In CHAPTER 12 we learned about some of the properties of acids and bases. We saw that an acid could be described as a substance that donates protons. A base was described as a substance that accepts protons. In this chapter we will consider the strengths of acids and bases and the effect of adding acids and bases to water. We will then look at how chemists measure the levels of acidity in water and finally we will study how acid-base indicators work.

18.2 SELF-IONISATION OF WATER

Chemists have discovered that even pure water will conduct a tiny electric current. If the water is pure, how do we explain that there are ions present to carry the electric current? The presence of these ions is explained by the fact that water is a substance which dissociates slightly to form ions:

$$H_2O \rightleftharpoons H^+ + OH^-$$

The above occurrence is often referred to as the self-ionisation of water. (The self-ionisation of water may also be represented by the equation $2H_2O \rightleftharpoons H_3O^+ + OH^-$.) The concentration of H^+ and OH^- ions in pure water is extremely small, i.e. the equilibrium lies very much on the left-hand side of the equation. The equilibrium constant expression for the above reaction may be written as:

$$K_c = \frac{[H^+][OH^-]}{[H_2O]}$$

In the above expression, the value of the $[H_2O]$ term is easily calculated, i.e. the concentration of water in moles per litre. The density of water is 1 g/cm^3. Therefore, a litre of water (1000 cm^3) has a mass of 1000 g.

No. of moles of water in a litre of water =

$$\frac{\text{Mass}}{\text{Rel Molecular Mass}} = \frac{1000}{18} = 55.56 \text{ moles}$$

Because the degree of ionisation is so small, the term $[H_2O]$ may be taken as having a constant value. Thus, the above equation can be re-arranged to give:

$$K_c[H_2O] = [H^+][OH^-]$$

The two terms on the left-hand side of the above equation are constants. Therefore, these two constants can be replaced by a new constant indicated by the symbol K_w.

$$K_w = K_c[H_2O] = [H^+][OH^-]$$
$\uparrow$
ionic product of water

K_w is given a special name and, as it is calculated by multiplying together the concentrations (in moles per litre) of the H^+ and OH^- ions in water, it is called the **ionic product of water**.

The value of K_w has been measured experimentally and, at 25 °C, has been found to be 1×10^{-14}. (The units are $moles^2/litre^2$ but these need not concern us in this course.) As with all equilibrium constants, the value of K_w is temperature-dependent (TABLE 18.1).

Temp. (°C)	K_w
0	0.11×10^{-14}
10	0.30×10^{-14}
20	0.68×10^{-14}
25	1.01×10^{-14}
50	5.47×10^{-14}
100	51.3×10^{-14}

Table 18.1
The ionic product of water varies with temperature.

Let us now calculate the actual concentration of H^+ ions in pure water.

$K_w = [H^+][OH^-] = 1.0 \times 10^{-14}$

Since $[H^+] = [OH^-]$ then, $[H^+]^2 = 1.0 \times 10^{-14}$

Therefore, $[H^+] = \sqrt{1.0 \times 10^{-14}} = 1.0 \times 10^{-7}$

and $[OH^-] = 1.0 \times 10^{-7}$

In other words, for a **neutral solution** at 25 °C:

$[H^+] = [OH^-] = 1.0 \times 10^{-7}$

Now consider what happens when an acid (e.g. HCl) is added to the water. The acid increases the concentration of H^+ ions. However, since K_w is a constant, the value of K_w cannot change.

$K_w = [H^+][OH^-] = 1.0 \times 10^{-14}$

Therefore, if the value of $[H^+]$ increases (to say 10^{-5}), the value of $[OH^-]$ must decrease (to say 10^{-9}) so that the two values multiplied together will equal 1.0×10^{-14}.

In other words, for an **acidic solution**:

$[H^+] > [OH^-]$

e.g. $(10^{-5}) > (10^{-9})$

Now consider what happens when a base (e.g. NaOH) is added to water. The base increases the concentration of OH^- ions. However, since the value of K_w is a constant, then the value of $[H^+]$ must decrease in order to satisfy the equation:

$K_w = [H^+][OH^-] = 1.0 \times 10^{-14}$

Therefore, if the value of $[OH^-]$ increases (to say 10^{-6}), the value of $[H^+]$ must decrease (to say 10^{-8}) so that the two values multiplied together will equal 1.0×10^{-14}.

In other words, for a **basic solution**:

$[H^+] < [OH^-]$

e.g. $(10^{-8}) < (10^{-6})$

The relationship of $[H^+]$ to levels of acidity and basicity at 25 °C may be summarised as shown in FIG 18.1.

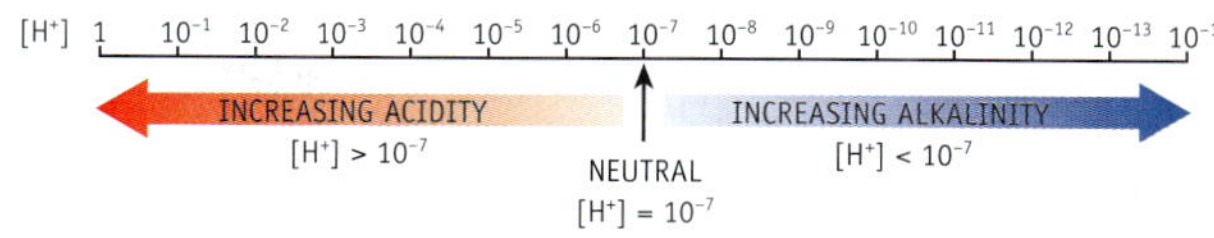

Fig 18.1
If $[H^+] > 10^{-7}$, the solution is acidic. If $[H^+] < 10^{-7}$, the solution is basic. If $[H^+] = 10^{-7}$, the solution is neutral. All concentrations are measured in moles per litre.

You may have discovered when reading the above that expressing concentrations of solutions in terms of hydrogen ion concentrations is a bit confusing! Chemists don't like having to use numbers like 10^{-7} to express the concentrations of hydrogen ions in solution. It was for this reason that a much simpler method for indicating acidity or basicity was developed. This method will be discussed in the next section.

Note: If you examine TABLE 18.1, you will observe that the value of K_w increases with increasing temperature. This must mean that the self-ionisation of water is endothermic since it is favoured by higher temperatures:

$H_2O \rightleftharpoons H^+ + OH^- \quad \Delta H = +$

However, pure water is neutral at all temperatures since there is the same number of H^+ and OH^- ions present in the water.

18.3 MEASUREMENT OF ACIDITY – THE pH SCALE

In the previous section, we have seen that expressing the concentration of a solution in terms of moles of H^+ ions per litre involves awkward numbers. In 1909, a Danish chemist called Søren Peter Sørensen (FIG 18.2) devised an easier method to express the degree of acidity of a solution. Sørensen was working for a Danish brewery and decided to introduce a less cumbersome method of reporting levels of acidity in beer. Instead of reporting that a solution contains 0.00001 (or 10^{-5}) moles of H^+ ions per litre, he referred to the solution as 'having a pH of 5'. To do

Fig 18.2
Sørensen introduced the idea of pH.

this, he had to get rid of the power in the expression 10^{-5} and also get rid of the negative sign in the power. Therefore, he defined pH as follows:

> $pH = -log_{10}[H^+]$ where the square brackets indicate concentration in moles per litre.

The above definition of pH may also be stated in words as follows: **The pH of a solution is the negative logarithm to the base 10 of the hydrogen ion concentration measured in moles per litre.**

The 'p' in pH comes from the German word 'potenz' meaning power and the 'H' in pH refers to the H^+ ion. Thus, pH may be roughly translated as 'power of hydrogen ions'.

The log to the base 10 in the above definition of pH ensures that the powers of ten do not appear in the pH scale. The negative sign in the above definition of pH ensures that all pH values are positive.

Example 18.1

A solution has a concentration of 10^{-4} moles of H^+ ions per litre. What is the pH of this solution?

Solution

$$pH = -log_{10}[H^+] = -log_{10}(10^{-4}) = -(-4) = +4$$

Answer: pH = 4

Studying Table 18.2 may help you to see the relationship between concentration in moles/litre of H^+ ions and pH.

Solution	$[H^+]$	pH
A	$0.001 = 1 \times 10^{-3}$	3
B	$0.000001 = 1 \times 10^{-6}$	6
C	$0.0000001 = 1 \times 10^{-7}$	7
D	$0.000000001 = 1 \times 10^{-9}$	9
E	$0.000000000001 = 1 \times 10^{-12}$	12

Table 18.2
Note that a pH value is simply the number in the power with the negative sign changed to a positive sign.

From Table 18.2, we see that solution A has the highest concentration of H^+ ions and the lowest pH. Solution E has the lowest concentration of H^+ ions and the highest pH. Solution C has the same concentration of H^+ ions as pure water. Thus, a pH of 7 indicates a neutral solution at 25 °C. The relationship between concentration in moles of H^+ ions per litre and pH is summarised in Fig 18.3. Note from Fig 18.3 that as the concentration of H^+ ions increases, the pH value decreases. Also, as the concentration of H^+ ions decreases, the pH value increases.

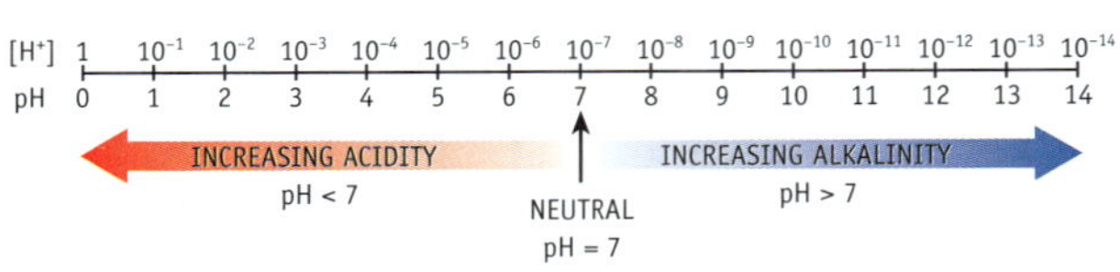

Fig 18.3
The concentration of H^+ ions (measured in moles per litre) in solution and the corresponding pH values.

Thus, at 25 °C we can say that:

> The pH of an acidic solution is less than 7.
> The pH of an alkaline solution is greater than 7.
> The pH of a neutral solution is equal to 7.

Note: In all subsequent calculations involving pH, it is assumed that the student has access to a scientific calculator. There are three operations that you need to be able to perform:

(a) Entering powers.

(b) Getting logs.

(c) Getting antilogs.

Consult the instruction booklet of your calculator to ensure that you know how to carry out these operations.

Example 18.2

Calculate the pH of a solution whose hydrogen ion concentration is 7.1×10^{-7} mol/L.

Solution

$$pH = -\log_{10}[H^+] = -\log_{10}(7.1 \times 10^{-7})$$
$$= 6.15$$

Answer: pH = 6.15

Example 18.3

Calculate the hydrogen ion concentration (in moles per litre) of a solution whose pH is 3.7.

Solution

$$pH = -\log_{10}[H^+] = 3.7$$
$$\log_{10}[H^+] = -3.7$$
$$[H^+] = \text{antilog}(-3.7)$$
$$= 1.995 \times 10^{-4}$$

Answer: $[H^+] = 1.995 \times 10^{-4}$ mol/L

It is possible to measure the pH of a substance by a number of methods. One method which you may have come across at Junior Certificate level was to use universal indicator paper, i.e. paper which turns different colours depending on the pH of the solution into which it is dipped (Fig 18.4). Universal indicator solution may also be used. The colour of the paper or solution is then matched against a chart of colours supplied with the indicator. We shall be studying indicators in more detail in Section 18.7. Another method of measuring pH is to use an instrument called a pH meter (Fig 18.5). This is simply an instrument that is sensitive to the concentration of H^+ ions in solution. One part of the pH meter (called the 'glass electrode') is placed in the solution whose pH is to be measured and the pH is then read off directly from a scale. pH meters are widely used, e.g. in hospitals to monitor changes of pH in blood and other body fluids, in industry to monitor the pH of sewage and effluents, in agriculture to monitor the pH of soil, etc.

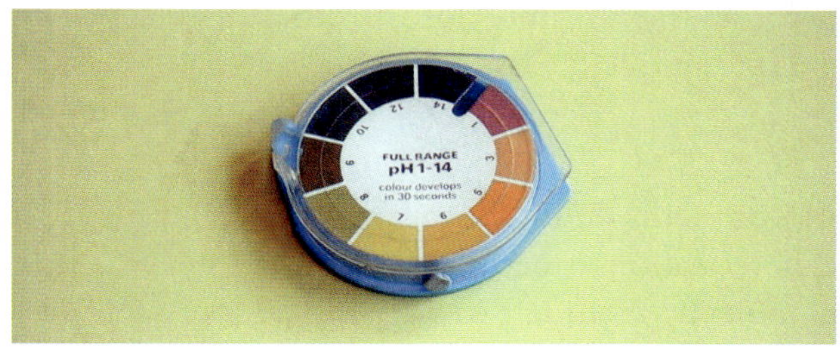

Fig 18.4
Universal indicator paper or solution may be used to determine the pH of a solution.

Fig 18.5
The pH of the yoghurt is being measured using a pH meter.

The pH values of common substances are shown in Fig 18.6.

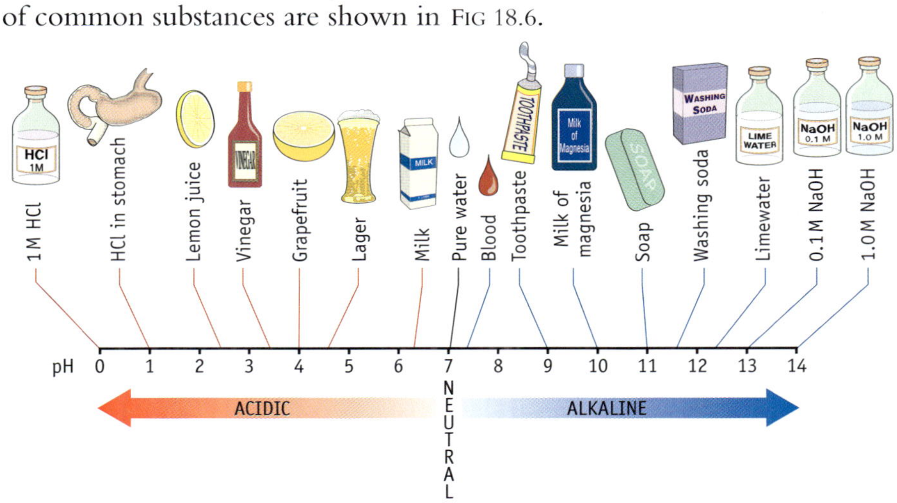

Fig 18.6
The approximate pH values of some common substances.

Test Yourself:
Attempt questions 18.1 – 18.3 and W18.1 – W18.3.

18.4 STRENGTHS OF ACIDS AND BASES

In CHAPTER 12 we defined an acid as a proton donor. However, not all acids donate protons to the same extent. **Strong acids** are those which have a great tendency to donate protons. Acids like HCl, H_2SO_4 and HNO_3 are classified as strong acids because they readily lose protons, i.e. they are almost completely dissociated in water.

$$HCl \rightarrow H^+ + Cl^-$$

Note that the equilibrium arrow $\rightleftharpoons$ is not used for this equation since the reaction goes almost completely from left to right. (By experiment it is found that there is about 99.996% dissociation.)

Do not get confused between the words **concentrated** and **strong**. As already discussed in CHAPTER 13, a concentrated solution is one that contains a large amount of solute in a given volume of solution. In chemistry, the word strong refers to the extent to which an acid can donate H^+ ions. Unfortunately, in everyday speech we use the word strong to mean concentrated. Instead of asking for a cup of strong tea you should really ask for a cup of concentrated tea!

However, not all acids are strong acids. Many acids are described as **weak acids**, i.e. they have a small tendency to donate protons. For example, when ethanoic acid (acetic acid) is added to water, it does not fully dissociate in water. In fact, only about 1.3% of the ethanoic acid molecules are dissociated in a 0.1 M solution and the following equilibrium is set up:

$$CH_3COOH \rightleftharpoons CH_3COO^- + H^+$$

In order to represent the strength of an acid, chemists use the symbol $\mathbf{K_a}$ which is known as the **acid dissociation constant**. For the general case of an acid HA undergoing dissociation:

$$HA \rightleftharpoons H^+ + A^-$$

The acid dissociation constant, K_a, is defined as:

$$K_a = \frac{[H^+][A^-]}{[HA]}$$

Sometimes, the formation of the hydronium ion, H_3O^+, is shown in the equilibrium reaction:

$$HA + H_2O \rightleftharpoons H_3O^+ + A^-$$

In this case, K_a is written as:

$$K_a = \frac{[H_3O^+][A^-]}{[HA]}$$

Note that $[H_2O]$ does not appear in the K_a expression. This is because K_a is defined purely in terms of the dissociation of the acid. (The $[H_2O]$ would be included if you were simply asked to write K_c for the reaction.)

Every acid has its own value of K_a which only varies with temperature. The value of K_a gives us an accurate measure of the extent to which an acid is dissociated. In other words, it tells us how strong the acid is. For example, for an acid like hydrochloric acid which is almost completely dissociated in water, the value of K_a is very large. This is because for a strong acid the concentrations of H^+ ions and A^- ions in solution are very large. On the other hand, for a weak acid like ethanoic acid, little dissociation occurs and therefore the concentrations of H^+ and A^- are small. Therefore, the value of K_a is small. The values of K_a for some acids are listed in TABLE 18.3.

Name of acid	Formula	K_a
Hydrochloric acid	HCl	1×10^7
Nitric acid	HNO_3	40
Methanoic acid	HCOOH	1.6×10^{-4}
Ethanoic acid	CH_3COOH	1.7×10^{-5}
Hydrocyanic acid	HCN	4.9×10^{-10}

Table 18.3

The dissociation constants of some acids at 25 °C. The value of K_a provides an accurate measure of the extent to which an acid dissociates, i.e. the strength of that acid.

Thus, strong acids have relatively large values of K_a and weak acids have small values of K_a. As a general rule, if the value of K_a is less than 10^{-2}, the acid is classified as a weak acid.

Strong bases and weak bases also exist. A **strong base** (e.g. NaOH, KOH, etc.) is one that accepts protons readily. A weak base is one that has little tendency to accept a proton, e.g. NH_3.

$$NH_3 + H_2O \rightleftharpoons NH_4^+ + OH^-$$

The value of K_b, the base dissociation constant, gives an indication of the degree of dissociation of a base, e.g. K_b for ammonia is only 1.74×10^{-5}, implying that ammonia is a weak base.

$$K_b = \frac{[NH_4^+][OH^-]}{[NH_3]}$$

Note: A weak acid like ethanoic acid has a strong conjugate base. This may be understood by considering the equilibrium:

$$CH_3COOH \rightleftharpoons CH_3COO^- + H^+$$

weak acid — strong conjugate base

Since the equilibrium lies on the left-hand side, this means that the CH_3COO^- ion has a great tendency to accept a proton and push the equilibrium to the left, i.e. the CH_3COO^- ion acts as a strong base. Similarly, a strong acid like HCl has a weak conjugate base (Cl^-) since the Cl^- ion has little tendency to accept a H^+ ion and push the equilibrium in the reverse direction.

The stronger the acid, the weaker is its conjugate base.
The weaker the acid, the stronger is its conjugate base.

Example 18.4

Define (a) acid, (b) conjugate pair, in terms of the Brønsted-Lowry theory

Sulfuric acid dissociates in water as follows: $H_2SO_4 + H_2O \rightleftharpoons H_3O^+ + HSO_4^-$

Why is (i) sulfuric acid considered to be a strong acid, (ii) the conjugate base of sulfuric acid described as a weak base?

Show, by means of an equation that the conjugate base of sulfuric acid can itself act as an acid. [LCH]

SOLUTION

(a) An acid is a proton donor.

(b) A conjugate acid-base pair is any pair consisting of an acid and a base which differ by one proton.

(i) Sulfuric acid is considered to be a strong acid because it is almost fully dissociated into ions in aqueous solution, i.e. it has a great tendency to donate protons.

(ii) The conjugate base of sulfuric acid, HSO_4^-, is described as a weak base because it has little tendency to accept a proton to form H_2SO_4. In other words, since H_2SO_4 is a strong acid, the equilibrium $H_2SO_4 + H_2O \rightleftharpoons H_3O^+ + HSO_4^-$ lies very much on the right-hand side.

The following equations show that the conjugate base of sulfuric acid can itself act as an acid:

$HSO_4^- \rightleftharpoons SO_4^{2-} + H^+$ or

$HSO_4^- + H_2O \rightleftharpoons SO_4^{2-} + H_3O^+$

(It is quite acceptable in this case to replace the $\rightleftharpoons$ sign with $\rightarrow$ since the equilibrium lies so far to the right.)

Measuring the pH of a solution does not allow us to distinguish between a strong acid and a weak acid. The pH of a solution depends only on the concentration of H^+ ions in that solution. For example, suppose the pH of a solution is found to be 5 (i.e. there are 1×10^{-5} moles of H^+ ions per litre). This solution could be obtained by making up a dilute solution of a strong acid or a concentrated solution of a weak acid (FIG 18.7). Think carefully about this!

Fig 18.7
Measurement of pH alone does not allow us to distinguish between a weak acid and a strong acid.

However, it is possible to work out the pH of a solution if we have certain information about that solution. In this course, you are required to be able to calculate the pH of solutions of (a) strong acids and strong bases and (b) weak acids and weak bases. We now consider these two types of calculations.

18.5 CALCULATING THE pH OF STRONG ACIDS AND STRONG BASES

Calculating the pH of strong acids and strong bases is very straightforward since strong acids and strong bases can be assumed to be fully dissociated in water. The following examples will illustrate the method used.

Example 18.5

Calculate the pH of a 0.01 M HCl solution.

SOLUTION

Since the acid is fully dissociated, the 0.01 mole of HCl will give 0.01 mole of H^+ ions in solution.

$$HCl \rightarrow H^+ + Cl^-$$
$$1 \text{ mole} \rightarrow 1 \text{ mole}$$
$$\therefore 0.01 \text{ mole} \rightarrow 0.01 \text{ mole}$$

$pH = -\log_{10}[H^+] = -\log_{10}(0.01) = 2$

Answer: pH = 2

Example 18.6

Calculate the pH of a 0.4 M H_2SO_4 solution.

SOLUTION

In this case, it is important to realise that one mole of H_2SO_4 gives two moles of H^+ ions in solution.

$$H_2SO_4 \rightarrow 2H^+ + SO_4^{2-}$$
$$1 \text{ mole} \rightarrow 2 \text{ moles}$$
$$\therefore 0.4 \text{ mole} \rightarrow 0.8 \text{ mole}$$

$pH = -\log_{10}[H^+] = -\log_{10}(0.8) = 0.097$

Answer: pH = 0.097

Example 18.7

Calculate the pH of a solution containing 5 g of sulfuric acid in 250 cm^3 of solution.

SOLUTION

First calculate the concentration of the acid in moles per litre.

5 g H_2SO_4 /250 cm^3 = 20 g H_2SO_4 /L

No. of moles of $H_2SO_4 = \frac{20}{98} = 0.2$

i.e. concentration of H_2SO_4 solution = 0.2 mol/L

$$H_2SO_4 \rightarrow 2H^+ + SO_4^{2-}$$
$$1 \text{ mole} \rightarrow 2 \text{ moles}$$
$$\therefore 0.2 \text{ mole} \rightarrow 0.4 \text{ mole}$$

$pH = -\log_{10}[H^+] = -\log_{10}(0.4) = 0.398$

Answer: pH = 0.398

A similar method is used to find the pH of a solution of a strong base. We know from our study of the ionic product of water that:

$[H^+][OH^-] = 10^{-14}$

Taking the negative log of each term:

$$pH + pOH = 14$$

Thus, pOH is defined in a similar fashion to pH:

$$pOH = -\log_{10}[OH^-]$$

To calculate the pH of a solution of a base, we first work out the pOH and subtract this value from 14.

pH = 14 - pOH

Example 18.8

Calculate the pH of a 0.15 M NaOH solution.

SOLUTION

$NaOH \rightarrow Na^+ + OH^-$

1 mole $\rightarrow$ 1 mole

$\therefore$ 0.15 mole $\rightarrow$ 0.15 mole

$pOH = -log_{10}[OH^-] = -log_{10}(0.15) = 0.82$

$pH = 14 - pOH = 14 - 0.82 = 13.18$

Answer: pH = 13.18

Example 18.9

Calculate the pH of a solution containing 9 g of potassium hydroxide, KOH, in 500 cm^3 of solution.

SOLUTION

9g KOH/500 cm^3 = 18 g KOH/ litre

No. of moles of KOH = $\frac{18}{56} = 0.32$

Therefore, concentration of KOH = 0.32 mol/L

$KOH \rightarrow K^+ + OH^-$

1 mole $\rightarrow$ 1 mole

$\therefore$ 0.32 mole $\rightarrow$ 0.32 mole

$pOH = -log_{10}[OH^-] = -log_{10}(0.32) = 0.49$

$pH = 14 - pOH = 14 - 0.49 = 13.51$

Answer: pH = 13.51

Example 18.10

Calculate the pH of a 0.01 M solution of calcium hydroxide, $Ca(OH)_2$. Assume that the calcium hydroxide is fully dissociated.

SOLUTION

$Ca(OH)_2 \rightarrow Ca^{2+} + 2OH^-$

1 mole $\rightarrow$ 2 moles

$\therefore$ 0.01 mole $\rightarrow$ 0.02 mole

$pOH = -log_{10}[OH^-] = -log_{10}(0.02) = 1.7$

$pH = 14 - pOH = 14 - 1.7 = 12.3$

Answer: pH = 12.3

Note:

1. The pH scale is usually taken between 0 and 14 even though pH values outside this range are possible. The pH of 1 M HCl is 0 and that of 1 M NaOH is 14. It is found that, once the concentration goes above 1 M, complete dissociation does not always occur, i.e. you cannot predict the actual number of H^+ or OH^- ions in solution. Thus, calculations of pH for solutions of concentration greater than 1 M are not accurate. For example, if you calculate the pH of 10 M HCl solution using the above method, you obtain a pH value of -1. Similarly, if you calculate the pH value of 10 M NaOH solution, you obtain a pH value of 15. However, these calculations are based on the assumption that full dissociation occurs and this is not the case with very concentrated solutions.

2. Since the pH scale is a logarithmic one, a difference in one pH unit corresponds to a ten-fold difference in the concentration. For example, the pH of 0.01 M HCl is 2 but that of 0.1 M HCl (ten times more concentrated) is 1.

Test Yourself:
Attempt questions 18.4 – 18.5 and W18.4 – W18.5.

18.6 CALCULATING THE pH OF WEAK ACIDS AND WEAK BASES

To calculate the pH of a solution of a weak acid or weak base, we must know the extent to which the acid or base dissociates. This depends on the value of the dissociation constant for the acid or base (K_a or K_b) and also on the concentration of the solution. To explain how the pH is calculated, we will first work out the pH of a solution from first principles and then use this to deduce a general expression for solving similar problems.

Example 18.11

Calculate the pH of a 0.1 M solution of ethanoic acid given that the value of K_a is 1.7×10^{-5} at 25 °C.

SOLUTION

Let x = number of moles of H^+ ions produced.

	CH_3COOH	$\rightleftharpoons$ CH_3COO^-	+ H^+
Initially	[0.1]	[0]	[0]
At equil.	[0.1 - x]	[x]	[x]

Now write the dissociation constant (K_a) expression for this reaction and substitute the above equilibrium concentrations into the expression.

$$K_a = \frac{[CH_3COO^-][H^+]}{[CH_3COOH]}$$

$$1.7 \times 10^{-5} = \frac{(x)(x)}{(0.1 - x)}$$

To make it easier to solve this equation, we make the assumption that 0.1 - x ≈ 0.1. This is a valid assumption since, for a weak acid, the amount of dissociation is small.

Thus, the above expression now simplifies to:

$$\frac{(x)(x)}{0.1} = 1.7 \times 10^{-5}$$

$$x^2 = 1.7 \times 10^{-5} \times 0.1$$

$$x = \sqrt{1.7 \times 10^{-5} \times 0.1} = 1.3 \times 10^{-3}$$

i.e. $[H^+] = 1.3 \times 10^{-3}$

$pH = -\log_{10}[H^+] = -\log_{10}(1.3 \times 10^{-3}) = 2.9$

Answer: pH = 2.9

Note: You could check the validity of the assumption made above that 0.1 - x ≈ 0.1 by solving the quadratic equation obtained without making the assumption. This would be a much longer and more tedious route and there would be only a very slight (negligible) difference between the two results.

In the above problem, we obtain a very simple expression at the end.

$$x = \sqrt{1.7 \times 10^{-5} \times 0.1} = 1.3 \times 10^{-3}$$

We can write a more general expression as:

$$[H^+] = \sqrt{K_a \times M_{acid}}$$

where M_{acid} is the concentration of the acid in moles per litre.

This is a very useful formula for finding the concentration of hydrogen ions (in moles per litre) in a solution of a weak acid when given the value of K_a and the concentration of the acid in moles per litre. Note that all concentrations are measured in moles per litre.

Similarly, it can be shown that the concentration of OH^- ions in a solution of a weak base is given by:

$$[OH^-] = \sqrt{K_b \times M_{base}}$$

where M_{base} is the concentration of the base in moles per litre.

The above formulas are well worth remembering as they greatly simplify the calculations involved in determining the pH values of weak acids and weak bases as shown in the following examples.

Example 18.12

Calculate the pH of a 0.1 M solution of methanoic acid (formic acid) given that K_a for this acid is 2.1×10^{-4}. *[LCH]*

SOLUTION

$$[H^+] = \sqrt{K_a \times M_{acid}}$$

$$= \sqrt{2.1 \times 10^{-4} \times 0.1}$$

$$= \sqrt{0.000021}$$

$$= 0.0046$$

$pH = -\log_{10}[H^+] = -\log_{10}(0.0046) = 2.34$

Answer: pH = 2.34

Example 18.13

Find the pH of an aqueous solution containing 1.48 g of propanoic acid (CH_3CH_2COOH) in 200 cm^3 of solution, given that the dissociation constant K_a for propanoic acid is 1.36×10^{-5}. *[LCH]*

SOLUTION

First calculate the concentration in mol/L.

Rel. molecular mass $CH_3CH_2COOH = 74$

1.48 g $CH_3CH_2COOH/200\ cm^3$

$= 1.48 \times 5$ g CH_3CH_2COOH/L

$= 7.4$ g CH_3CH_2COOH/L

No. of moles of $CH_3CH_2COOH = \frac{7.4}{74} = 0.1$

$\therefore M_{acid} = 0.1$ mol/L

$$[H^+] = \sqrt{K_a \times M_{acid}}$$

$$= \sqrt{1.36 \times 10^{-5} \times 0.1}$$

$$= \sqrt{1.36 \times 10^{-6}}$$

$$= 0.0012$$

$pH = -\log_{10}[H^+] = -\log_{10}(0.0012) = 2.92$

Answer: pH = 2.92

Example 18.14

The dissociation of the weak base, ammonia, in aqueous solution may be represented as:

$NH_3 + H_2O \rightleftharpoons NH_4^+ + OH^-$

Find the pH of a 0.01 M solution of ammonia in water given that its dissociation constant, K_b, is 1.8×10^{-5}. [LCH]

SOLUTION

$$[OH^-] = \sqrt{K_b \times M_{base}}$$

$$= \sqrt{1.8 \times 10^{-5} \times 0.01}$$

$$= \sqrt{1.8 \times 10^{-7}}$$

$$= 0.00042$$

$pOH = -\log_{10}[OH^-] = -\log_{10}(0.00042) = 3.38$

$pH + pOH = 14$

$pH = 14 - pOH = 14 - 3.38 = 10.62$

Answer: pH = 10.62

Test Yourself:
Attempt questions 18.6 – 18.11 and W18.6 – W18.15

18.7 ACID-BASE INDICATORS

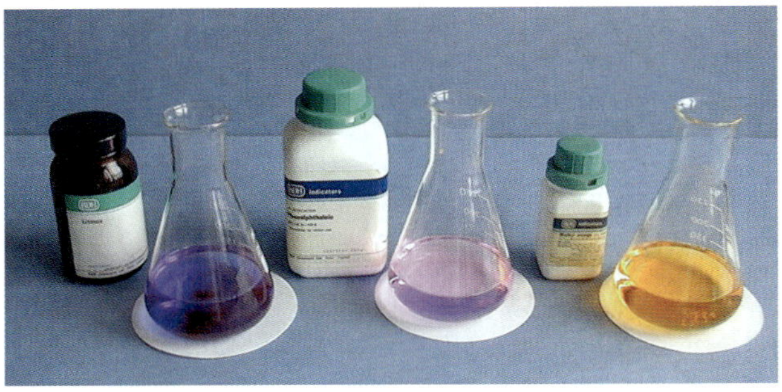

Fig 18.8
Some common acid-base indicators.

In Chapter 13 we saw that acid-base indicators were used to find the end point when carrying out a titration. We now consider the theory behind how these indicators work and discuss how we choose an indicator for a particular titration.

Acid-base indicators are substances that change colour according to the pH of the solution in which they are placed. Many natural dyes act as indicators, e.g. the dye in red cabbage, elderberry juice, blackberry juice, etc. Examples of indicators which you meet in the school laboratory are litmus, methyl orange and phenolphthalein (Fig 18.8). Litmus is obtained from plants called lichens that grow in warm climates on clifftops and trees. Acid-base indicators are often simply referred to as 'indicators'. It is found that most indicators are weak acids in which the colour of the undissociated molecule is different from that of the dissociated molecule. Suppose we represent the indicator litmus as HIn. Litmus has a red colour in an acidic solution and is blue in an alkaline solution.

$$\underset{\textbf{(acid)}}{\underset{\text{red}}{HIn}} \rightleftharpoons H^+ + \underset{\textbf{(base)}}{\underset{\text{blue}}{In^-}}$$

If H^+ ions from an acid are added to blue litmus, the blue litmus turns red. This is because, in accordance with Le Chatelier's Principle, the above equilibrium shifts from right to left in order to absorb the added H^+ ions. Therefore, the red colour of the HIn predominates, i.e. the blue litmus turns red.

If a base containing OH^- ions is added to red litmus, the OH^- ions react with the H^+ ions in the above equilibrium reaction to remove them to form water. In accordance with Le Chatelier's Principle, more H^+ ions must be formed and therefore the equilibrium is shifted to the right. Thus, the blue colour of the In^- predominates, i.e. the red litmus turns blue.

A small number of indicators may be considered to be weak bases that dissociate slightly as follows:

$$XOH \rightleftharpoons X^+ + OH^-$$

The undissociated XOH has a different colour to the dissociated X^+. Addition of acid removes some of the OH^- ions from the right-hand side and thus the equilibrium shifts to the right. Therefore, the colour of X^+ predominates. Addition of base increases the concentration of OH^- ions on the right-hand side and this shifts the equilibrium to the left-hand side. Hence, the colour of XOH predominates. As a general rule, the concentration of one coloured form of the indicator must be at least ten times that of the other for a change to be visible to the eye.

Recall that when an indicator is used in a titration there is usually very little of it used, e.g. only two or three drops of an indicator are added to 25 cm^3 of the solution in the conical flask. Therefore, the indicator makes a negligible contribution to the H^+ ion concentration.

The molecules of indicators tend to be rather complicated! Some examples of these molecules are shown in Fig 18.9. You do not have to remember the structure of any of these molecules!

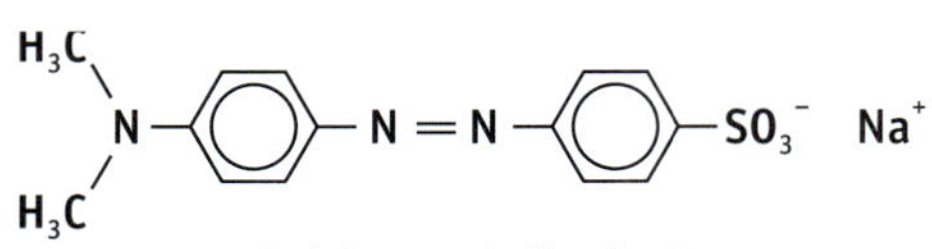

Methyl orange (yellow form)

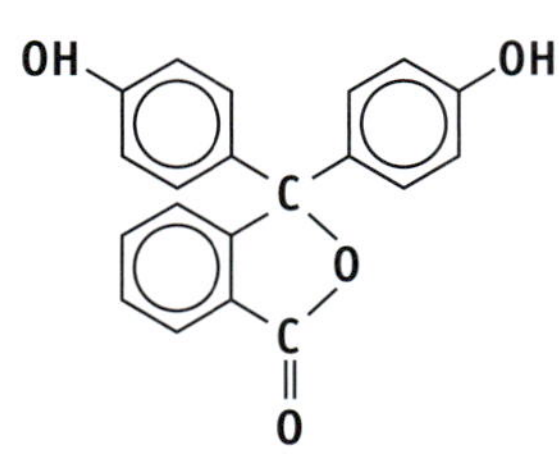

Phenolphthalein (colourless form)

Fig 18.9
The molecules of the indicators methyl orange and phenolphthalein are fairly complicated!

Due to the limits of sensitivity of the human eye, it is found that indicators need a change of about two pH units in order to show a distinct change from one colour to another. The pH range over which there is a clear change of colour for a particular indicator is called the **range** of the indicator. Each indicator has its own range and examples of the ranges of common indicators are given in TABLE 18.4.

Name of indicator	Approx. Range	Acid Colour (Lower pH)	Base Colour (Higher pH)
Methyl orange	3–5	red	yellow
Litmus	5–8	red	blue
Phenolphthalein	8–10	colourless	pink

Table 18.4
The range of some common indicators.

From TABLE 18.4, we can conclude that methyl orange has a distinct red colour below a pH of 3 and a distinct yellow colour above a pH of 5. Between a pH of 3 and 5, the indicator is neither distinctly red or yellow but an intermediate colour. Similarly, litmus has a definite red colour below a pH of 5 and a definite blue colour above a pH of 8. Also, phenolphthalein is colourless below a pH of 8 and has a definite pink colour above a pH of 10. It is recommended that TABLE 18.4 be learned off by heart!

It is clear that indicators change colour over widely differing ranges of pH values. The reason for this need not concern us in this course – it is related to the different dissociation constants of the various indicators.

Note: Universal indicator is a mixture of indicators that gives a gradual change in colour over the entire range of pH. A few drops of universal indicator are added to a solution whose pH is to be measured and the colour observed is compared with a colour chart supplied with the universal indicator. This gives an approximate value of the pH of the solution.

It is quite easy to follow the changes in pH as a titration is carried out. Such an experiment is called a **pH titration** and the experimental set up is shown in FIG 18.10. In this experiment, the pH meter measures the pH as the base is added from the burette to the acid (kept stirred with the magnetic stirrer) in the beaker. Note that the base is being added to an acid (the reverse of a normal titration) in order to allow us to start with a low pH value and work upwards. Note that a beaker is used instead of a conical flask as a beaker is more suitable as a container for the magnetic stirring pellet and also for the electrode of the pH meter.

Fig 18.10
A pH titration enables us to follow the changes in pH as the titration proceeds.

We now consider each of the four possible types of acid-base titrations.

(I) TITRATING A STRONG ACID AGAINST A STRONG BASE

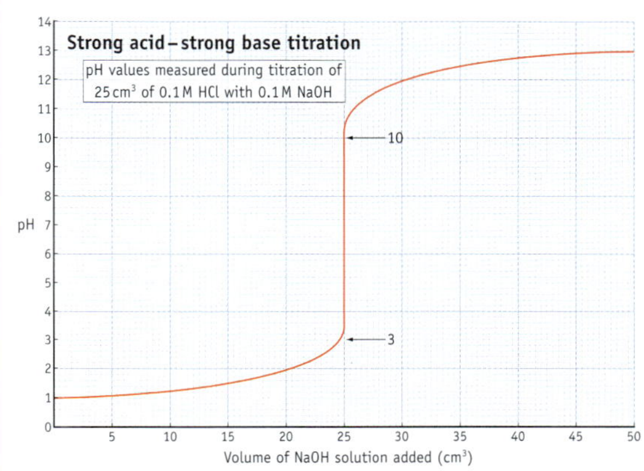

Fig 18.11
This graph shows the pH changes observed as a strong acid is titrated against a strong base. Litmus, methyl orange and phenolphthalein are all suitable indicators for this titration.

FIG 18.11 shows how the pH changes when titrating a strong acid against a strong base. In this titration, 0.1 M NaOH is slowly added to 25 cm^3 of 0.1 M HCl. Note that the curve is almost vertical from 3 – 10 approx., i.e. there is a sudden jump in the pH at the end point. This sudden increase in the pH is caused by the addition of just one extra drop of base at the end point.

What indicator would be suitable for this titration? Any indicator that has one colour at a pH of around 3 and another colour at a pH of around 10 will identify the end point accurately. In other words, we need an indicator which has one colour at the bottom of the vertical portion of the curve and another colour at the top of the vertical portion of the curve. Chemists often express this idea by saying that the end point must be 'sharp', i.e. a very small addition of acid or alkali (one drop) must cause a complete colour change in the indicator. Referring to TABLE 18.4, we see that all three indicators satisfy this criterion. Therefore, all three indicators would be suitable for carrying out a titration of a strong acid against a strong base. (Litmus is not as reliable as the other two indicators and hence is not usually used in accurate work.)

(II) TITRATING A STRONG ACID AGAINST A WEAK BASE

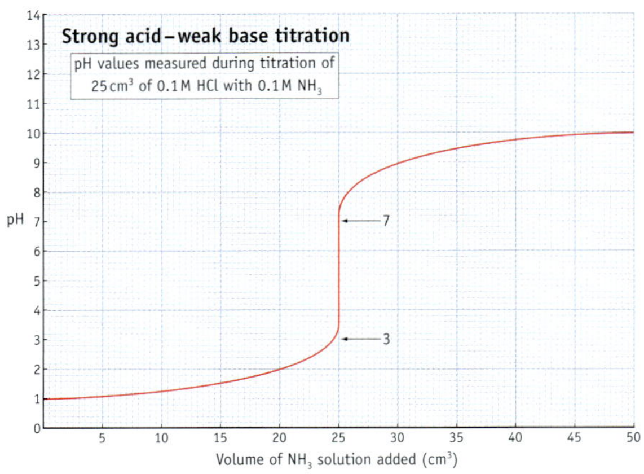

Fig 18.12
This graph shows the pH changes observed as a strong acid is titrated against a weak base. Methyl orange is a suitable indicator for this titration.

FIG 18.12 shows how the pH changes when titrating a strong acid against a weak base. In this titration, 0.1 M NH_3 solution is slowly added to 25 cm^3 of 0.1 M HCl. Note that the curve is almost vertical from 3 – 7 approx. Therefore, any indicator which has one colour at pH = 3 and another colour at pH = 7 will identify the end point accurately. Referring to TABLE 18.4, we see that the only one of the three indicators which would be suitable is methyl orange. (Litmus would not be suitable as it would not have completely changed colour until a pH of 8 had been reached. Phenolphthalein would not be suitable as its range is completely outside the 3 – 7 range. In fact, it would not begin to change colour until a pH of 8 was reached.)

(III) TITRATING A WEAK ACID AGAINST A STRONG BASE

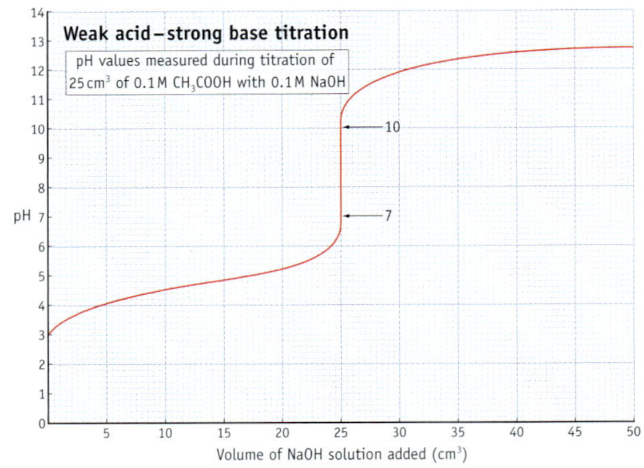

Fig 18.13
This graph shows the pH changes observed as a weak acid is titrated against a strong base. Phenolphthalein is a suitable indicator for this titration.

FIG 18.13 shows how the pH changes when titrating a weak acid against a strong base. In this titration, 0.1 M NaOH is slowly added to 25 cm^3 of 0.1 M CH_3COOH. Note that the curve is almost vertical from 7 – 10 approx. Therefore, we need an indicator which has one colour at pH = 7 and another colour at pH = 10 to identify the end point accurately. From TABLE 18.4, we see that phenolphthalein is a suitable indicator for this titration as it is colourless at a pH of 7 and pink at a pH of 10. None of the other two indicators would be suitable as they are outside the 7 – 10 range.

(IV) TITRATING A WEAK ACID AGAINST A WEAK BASE

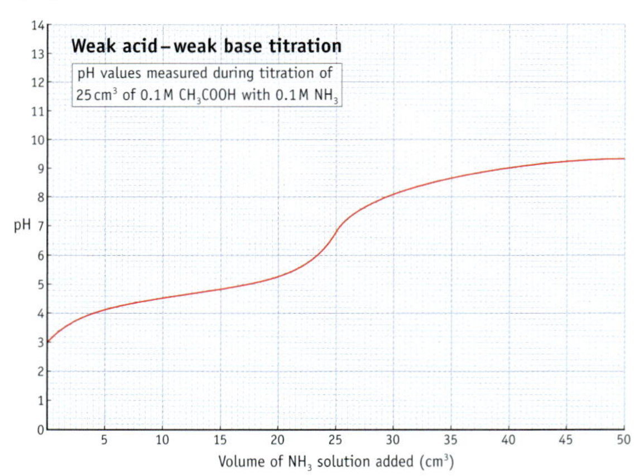

Fig 18.14
This graph shows the pH changes observed as a weak acid is titrated against a weak base. There is no suitable indicator for this titration as there is no sudden increase in the pH at any stage of the titration.

FIG 18.14 shows how the pH changes when titrating a weak acid against a weak base. In this titration,

0.1 M NH_3 is slowly added to 25 cm^3 of 0.1 M CH_3COOH. Note that the pH rises so gradually that there is no sudden jump in the pH value at any stage of the titration. Therefore, it is not possible to detect the end point using an indicator. This type of titration is not used to determine the concentration of a solution of a weak acid or a weak base. Instead, the weak acid is titrated against a strong base and the weak base is titrated against a strong acid.

Test Yourself:
Attempt questions 18.12 – 18.17 and W18.16 – W18.20

CHAPTER SUMMARY

- $K_w = [H^+][OH^-]$. K_w is called the ionic product of water.
- $pH = -\log_{10}[H^+]$ where the square brackets indicate concentration in moles per litre.
- At 25 °C the pH of an acidic solution is less than 7, the pH of an alkaline solution is greater than 7 and the pH of a neutral solution is equal to 7.
- $[H^+] = \sqrt{K_a \times M_{acid}}$ and $[OH^-] = \sqrt{K_a \times M_{base}}$ are useful expressions for calculations involving weak acids and weak bases.
- Acid-base indicators are substances that change colour according to the pH of the solution in which they are placed.

EXERCISES

18.1. (a) Why does pure water conduct a tiny electric current? ....

(b) Write down the symbol used to denote the ionic product of water ....

(c) Write down the concentration in moles per litre of H^+ ions in pure water at 25 °C...........

(d) A solution contains 10^{-9} moles of H^+ ions per litre. Is it acidic or basic?.............................

(e) A solution with a high concentration of H^+ ions has a .... pH value.

(f) Write down the pH of a solution which has 10^{-11} moles of H^+ ions per litre ....

18.2. Calculate the pH of the following solutions whose hydrogen ion concentrations are: (a) 10^{-7} mol/L, (b) 4.0×10^{-3} mol/L, (c) 2.62×10^{-8} mol/L, (d) 5.2×10^{-3} mol/L, (e) 8.7×10^{-11} mol/L.

18.3. Calculate the hydrogen ion concentration (in moles per litre) of the solutions whose pH values are as follows: (a) 5.65, (b) 11.15, (c) 1.09, (d) 7, (e) 9.65.

18.4. Calculate the pH of the following solutions: (a) 0.1 M HCl, (b) 0.01 M H_2SO_4, (c) 12.6 g HNO_3 in 500 cm^3 solution, (d) 5.48 g HCl in 500 cm^3 solution, (e) 5.8 g H_2SO_4 in 250 cm^3 solution.

18.5. Calculate the pH of the following solutions: (a) 0.95 M NaOH, (b) 5 M NaOH, (c) 1 g NaOH in 500 cm^3 solution, (d) 0.024 grams KOH per litre, (e) 0.37 g $Ca(OH)_2$ per 500 cm^3 solution.

18.6. (i) Define the pH of a solution. Explain what is meant by the statement that the dissociation constant of ethanoic acid in aqueous solution is 1.8×10^{-5}.

(ii) Find the approximate pH of a solution of 0.3 g ethanoic acid, CH_3COOH, in 50 cm^3 water given that the dissociation constant for ethanoic acid is 1.8×10^{-5}.

18.7. The dissociation constant (K_a) of a monobasic (monoprotic) acid at room temperature is 5.5×10^{-5}. Find the pH of a solution of this acid containing 5.0×10^{-2} mole in 100 cm^3 of the solution.

18.8. The dissociation constant of methanoic acid is 1.6×10^{-4} and the dissociation constant of ethanoic acid is 1.8×10^{-5}. Which is the stronger acid?
Calculate the $[H^+]$ in a 0.1 M solution of ethanoic acid given that $K_a = 1.8 \times 10^{-5}$.

18.9. The pH of a 1.0 M solution of ethanoic acid is 2.37. Calculate the hydrogen ion concentration and show that the approximate value of the dissociation constant is 1.8×10^{-5}.
What is the approximate value of the pH of a 0.05 M solution of ethanoic acid?

18.10. If methanoic acid is 1.3% dissociated in a 1.0 M aqueous solution, show clearly that the approximate value of the acid dissociation constant is 1.7×10^{-4}.
Calculate the pH of the 1.0 M solution of methanoic acid. *[LCH]*

18.11. The compound $C_2H_5NH_2$ acts as a base by accepting a proton. Calculate the pH of a solution containing 1.5 g of this compound in 250 cm^3 water. K_b for this base is 4.4×10^{-4}.

18.12. An indicator called bromothymol blue, represented as HIn, has a yellow colour in the associated form and a blue colour in the dissociated form.
$HIn \rightleftharpoons In^- + H^+$
Use Le Chatelier's Principle to predict the colour change that occurs when (a) acid and (b) base is added to the indicator.

18.13. Explain the terms strong and weak as applied to acids and bases. The dissociation of an indicator (HIn) which is a weak acid may be represented by:
$HIn \rightleftharpoons H^+ + In^-$
How is the degree of dissociation affected (i) by the addition of a strong acid, (ii) by the addition of a strong base? Explain your reasoning.
The following table shows the results of titrating 25.0 cm^3 volumes of 0.1 M ethanoic acid solution with a 0.1 M solution of sodium hydroxide using two different indicators.

Indicator	pH range	Volume of sodium hydroxide solution
Methyl Red	4.2 - 6.3	22.0 cm^3
Phenolphthalein	8.3 - 10.0	25.0 cm^3

(i) Which is the correct volume of sodium hydroxide solution?
(ii) Explain why the colour change coincided with the end point of the titration for one indicator but not for the other indicator. *[LCH]*

18.14. (i) The hydrogen ion concentration in pure water at 25 °C is 1.0×10^{-7} mol/L but at 57 °C it is about 3.0×10^{-7} mol/L. Does this mean that water is acidic at 57 °C? Explain your reasoning.
(ii) Methyl red is an indicator used in titrations of strong acids against weak bases. It changes from red to yellow in the pH range 4.2 to 6.3. If it can be regarded as a weak base represented as XOH explain how it acts as an indicator. *[LCH]*

18.15. No indicator will detect the end point when titrating 0.1 M CH_3COOH against 0.1 M NH_3. Explain this observation.

18.16. Discuss the principle involved in the use of indicators in acid-base titrations. Suggest suitable indicators for the following titrations: (i) strong acid-strong base, (ii) strong acid-weak base, (iii) weak acid-strong base.

18.17. Methanoic acid (formic acid) is a weak acid that reacts with sodium hydroxide according to the following equation:
$HCOOH + NaOH \rightarrow HCOONa + H_2O$
(i) What volume of 0.1 M methanoic acid would be needed to neutralise completely 25 cm^3 of 0.108 M sodium hydroxide solution?
(ii) Name a suitable indicator for this titration.
(iii) Assuming that the indicator you choose is a weak acid, explain briefly how it works.

Environmental Chemistry – Water

19

19.1 INTRODUCTION

Water is one of the most abundant and most important chemicals on Earth. About 80% of the Earth's surface is covered with water and it makes up about two-thirds of our body weight. Water is essential for life. We could survive for several weeks without food but we could not live for more than a few days without water. In fact, we need to take in about two litres of water every day in order to stay healthy.

Each day we use large quantities of water for various purposes – drinking, washing, flushing the toilet, watering the garden, etc. Industry uses vast quantities of water for cooling, manufacturing various products, generating electricity, etc. We take water very much for granted and yet it is one of the most important items that we require. When we turn on the tap we expect to have a plentiful supply of clean safe drinking water. However, countries that have a piped water system are in a minority. In many countries water has to be carried many miles from the nearest source to the home. In many under-developed countries, the lack of water contributes to the death of millions of people each year.

In this chapter we will study the importance of water in our lives. We will learn that water is an excellent solvent and we will study how the substances dissolved in water affect the properties of water. We will study how water is treated in a water treatment plant and also learn about water pollution. We will carry out some experiments on water to analyse its composition and, finally, we will study how sewage is treated.

19.2 HARDNESS OF WATER

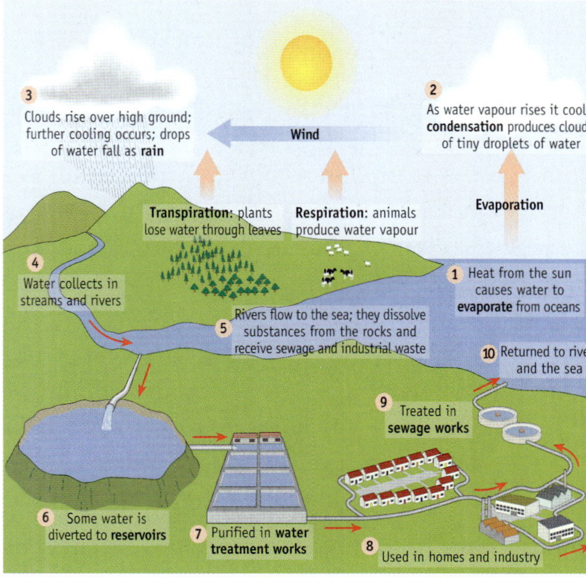

Fig 19.1
The water cycle represents the way that water circulates around the Earth.

The way water circulates around the Earth may be represented by the water cycle (Fig 19.1). From studying Fig 19.1, you will understand why pure water is very rare. Rainwater is fairly pure water but even rainwater contains dissolved gases, e.g. oxygen, nitrogen, carbon dioxide, etc. Since water has such a great ability to dissolve substances, it will dissolve many of the minerals found in the earth as it flows over the soil to form streams and rivers.

In some parts of Ireland the water is **hard**, i.e. it is difficult to form a lather with soap. Instead of forming a lather, it is found that the soap forms a grey precipitate – often referred to as an insoluble **scum**. When chemists analysed the water to find out the cause of this problem, they discovered that hard water contained definite quantities of Ca^{2+} or Mg^{2+} ions.

Hard water is water that will not easily form a lather with soap. Hardness in water is caused by the presence of Ca^{2+} or Mg^{2+} ions.

One of the most common substances in soap is a chemical called sodium stearate, $C_{17}H_{35}COONa$. When soap is added to hard water, the calcium or magnesium ions in the water react with the stearate ions to form a grey insoluble compound called calcium stearate (scum) which floats on the water.

calcium ion	+	stearate ions	→	calcium stearate (scum)
Ca^{2+}	+	$2C_{17}H_{35}COO^-$	→	$(C_{17}H_{35}COO)_2Ca\downarrow$

A lot of soap is wasted in trying to form a lather as a lather is only formed when all of the calcium ions have been precipitated in the scum. For this reason, the use of soap for washing clothes went into decline with the advent of modern detergents that are not affected by hard water. (Modern detergents like washing powders, washing-up liquids, shampoos, etc. do not contain soap. They are made from crude oil and are very good at removing oil and grease but are too severe to be used on our skin.)

Chemists have found that hardness in water may be classified as either **temporary** or **permanent**.

Temporary hardness is hardness that can be removed by boiling the water. To understand how boiling does this, we must consider how temporary hardness arises. Rainwater is slightly acidic due to the fact that it dissolves carbon dioxide as it falls through the air.

water	+	carbon dioxide	⇌	carbonic acid
H_2O	+	CO_2	⇌	H_2CO_3

Limestone (calcium carbonate) is insoluble in water but does react with carbonic acid. The acidic solution of carbonic acid reacts with the limestone in the soil to form calcium hydrogencarbonate.

limestone	+	carbonic acid	→	calcium hydrogencarbonate
$CaCO_3$	+	H_2CO_3	→	$Ca(HCO_3)_2$
insoluble				soluble

Thus, water that has temporary hardness is really a dilute solution of calcium hydrogencarbonate. The Ca^{2+} ions cause hardness. The HCO_3^- ions have no effect on hardness. However, if water containing these two ions is heated, a chemical reaction occurs which removes the Ca^{2+} ions from the water, i.e. the water is softened.

calcium hydrogencarbonate	→	calcium carbonate	+	carbon dioxide	+	water
$Ca(HCO_3)_2$	→	$CaCO_3\downarrow$	+	CO_2	+	H_2O
soluble		insoluble				

Note that this equation is easy to remember – just reverse the first equation to get the second! The overall effect is that the calcium ions that were dissolved in the water are now precipitated as insoluble calcium carbonate (limestone). This insoluble calcium carbonate is often referred to as 'fur' or 'scale' and tends to build up in kettles, boilers, hot water pipes, etc. (FIG 19.2). The presence of the carbonate ion in the 'fur' of the kettle may be verified by carrying out the same tests as in EXPERIMENT. 6.1.

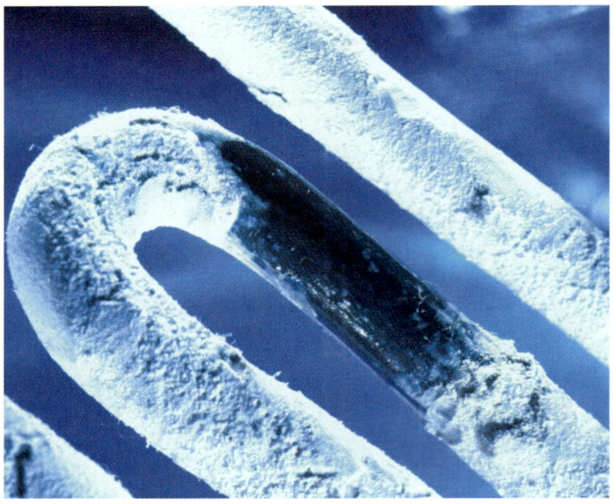

Fig 19.2
Temporary hardness causes the element of this kettle to be coated with a deposit of 'fur'. This 'fur' is limestone or calcium carbonate.

Note: We have already mentioned that hardness may be caused by the presence of Mg^{2+} as well as Ca^{2+} ions. The Mg^{2+} ions often come from dolomite rock, which contains a mixture of $MgCO_3$ and $CaCO_3$ and is often represented as $MgCO_3.CaCO_3$. In some cases, it is found that the hardness in the water cannot be removed by boiling the water. Such hardness is called **permanent hardness**. Permanent hardness is caused by the presence of calcium sulfate, $CaSO_4$, or magnesium sulfate,

$MgSO_4$, in the water. When water containing these substances is heated, no chemical reaction occurs to precipitate the Ca^{2+} or Mg^{2+} ions. Therefore, the hardness is not removed by boiling.

As already stated, temporary hardness is removed by boiling but there are a number of methods that remove both temporary and permanent hardness. Let us now consider these.

Methods of removing hardness from water

1. **Distillation**. This method involves boiling the water and then cooling the vapour. All dissolved solids and dissolved liquids are removed from the water by this method. However, it is not feasible to soften water on a large scale using distillation due to the expense involved in boiling the water.
2. **Using washing soda**. The chemical name for washing soda is hydrated sodium carbonate, which has the formula $Na_2CO_3.10H_2O$. Before the introduction of modern detergents, clothes were washed with soap, and crystals of this compound were added to the water to soften the water, i.e. to make it easier to form a lather and to avoid wasting soap. Washing soda softens the water because the carbonate ions in the washing soda react with the calcium ions in the water and removes them as insoluble calcium carbonate.

   $Ca^{2+} + CO_3^{2-} \rightarrow CaCO_3 \downarrow$

   The Na^+ ions from the sodium carbonate do not play any part in the reaction – they simply remain dissolved in the water.

   Washing soda is commonly used in bath salts to help a lather form more easily (Fig 19.3).

Fig 19.3
Bath salts are simply crystals of washing soda which have been coloured and to which perfume has been added.

3. **Ion exchange resin**. The easiest way of removing hardness from water is to pass the water through ion exchange resin. Modern ion exchange resins are man-made materials which exchange or 'swap' ions that cause hardness with ions that do not cause hardness. Houses that have a supply of hard water need to soften it by installing an ion exchanger like that shown in Fig 19.4.

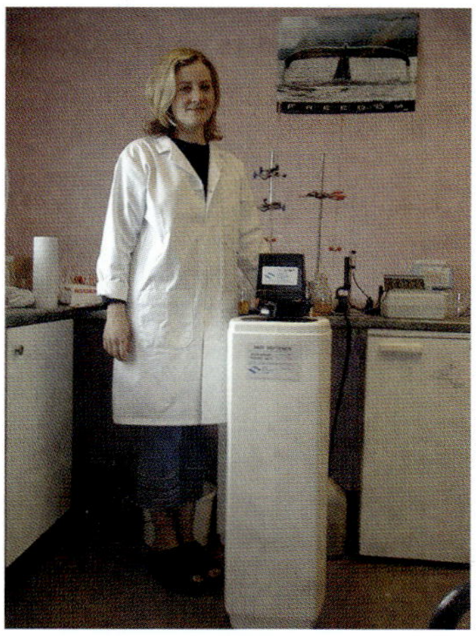

Fig 19.4
This type of ion exchanger is used in houses to remove the hardness from the water supply. It is packed with cation exchange resin that replaces Ca^{2+} and Mg^{2+} ions with Na^+ ions.

Studying Fig 19.5 will help you to understand how an ion exchanger works.

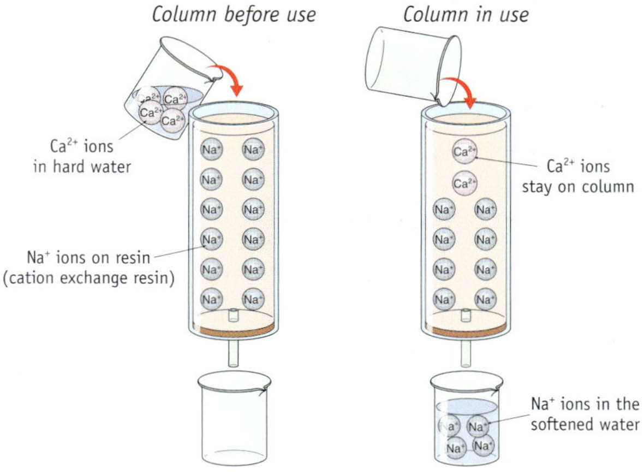

Fig 19.5
A cation exchange resin swaps all the positive ions in the water for Na^+ ions. Thus, the water is said to be softened.

Ion exchange units used to soften water contain **cation exchange resin**. A cation is another word for a positive ion. Thus, cation exchange resin exchanges the positive ions in the water (e.g. Ca^{2+} and Mg^{2+} ions) for Na^+ ions. This particular type of ion exchange resin is a fairly complicated molecule but may be regenerated as Na_2R. The exchanging or 'swapping' of the Na^+ ions for Ca^{2+} ions may be represented as:

$$Ca^{2+} + Na_2R \rightarrow CaR + 2Na^+$$

Thus, the Ca^{2+} ions remain behind in the resin. Eventually, the resin loses all of its Na^+ ions and it needs to be regenerated by passing a concentrated solution of sodium chloride through it. The Ca^{2+} and Mg^{2+} ions in the resin are replaced by Na^+ ions and the resin is ready for use again.

In some cases (e.g. for laboratory or medical purposes), it is necessary to have **all** of the ions removed from the water, i.e. **deionised water** is needed. Such water is produced by passing ordinary water through a water deioniser that contains a mixture of cation (positively charged ion) exchange resin and anion (negatively charged ion) exchange resin. The mixture is commonly referred to as **mixed bed resin**. The cation exchange resin removes all the positive ions and the anion exchange resin removes all the negative ions from the water (FIG 19.6).

Fig 19.6
The cation exchange resin replaces all the positive ions in water by H^+ ions. The anion exchange resin replaces all the negative ions in water by OH^- ions.

Suppose the water contains Na^+ ions and Cl^- ions. The cation exchange resin, represented as RH, replaces the Na^+ ions with H^+ ions as follows:

$$RH + Na^+ \rightarrow RNa + H^+$$

The Cl^- ions are replaced by the anion exchange resin (represented as ROH) as follows:

$$ROH + Cl^- \rightarrow RCl + OH^-$$

Thus, all the positive ions in the water are replaced by H^+ ions and all the negative ions in the water are replaced by OH^- ions (FIG 19.7). The H^+ and OH^- ions then combine together to form water.

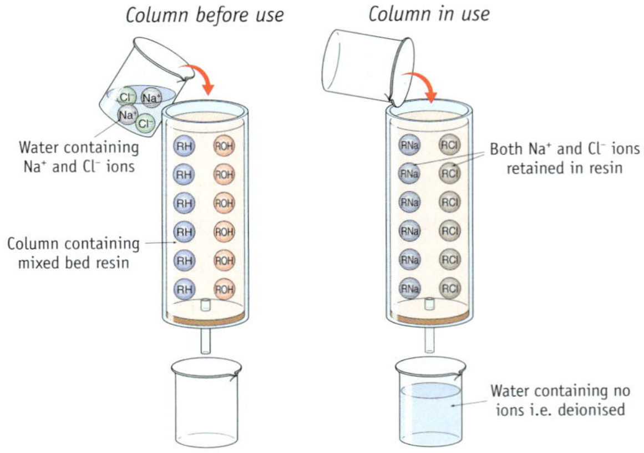

Fig 19.7
The mixture of cation exchange resin and anion exchange resin (mixed bed resin) swaps all positive ions for H^+ ions and all negative ions for OH^- ions. Thus, the water is said to be deionised, i.e. it contains no ions.

During World War II, small ion exchange resin columns were part of a soldier's survival kit. Can you think of a situation in which soldiers could use them?

Note: Since deionised water is so easily and cheaply produced, its use is far more widespread than distilled water. However, there is a slight difference between deionised water and distilled water. Distilled water is the purest form of water as all dissolved and suspended solids as well as dissolved gases have been removed from the water. However, deionised water normally contains dissolved gases as well as any non-ionic material (organic material) which may be dissolved in the water. For some specialised industries (e.g. the electronics industry) the water must not contain any organic materials as this affects the purity of the silicon chips being manufactured. Therefore, in recent years, ion exchange resins have been developed which remove organic material from water as well as ions. These specialised types of ion exchange resins are about twice the price of ordinary ion exchange resins.

The advantages and disadvantages of hard water are summarised in TABLE 19.1.

ADVANTAGES	DISADVANTAGES
1. Provides calcium for teeth and bones.	1. Blocks pipes, leaves scale on kettles and boilers.
2. Nicer taste	2. Wastes soap.
3. Good for brewing and tanning.	3. Produces scum.

Table 19.1.
Advantages and disadvantages of hard water

Test Yourself:
Attempt questions 19.1 – 19.2 and W19.1.

EXPERIMENT 19.1 MANDATORY STUDENT EXPERIMENT NO. 17

To determine the total hardness in a water sample.

Introduction

We have already learned that hardness in water is due to the presence of calcium and magnesium ions dissolved in the water. These ions come from soluble calcium or magnesium salts, e.g. $Ca(HCO_3)_2$, $CaSO_4$, $CaCl_2$, $MgCO_3$, $Mg(HCO_3)_2$, etc. Measuring the hardness of water involves measuring the concentration of calcium and magnesium ions in a sample of water. The sum of the concentration of calcium and magnesium ions in the water is called the **total hardness** of the water.

Total hardness = calcium hardness + magnesium hardness

For the sake of simplicity, it has been traditional to express the hardness of water in terms of mg per litre of $CaCO_3$ (p.p.m. $CaCO_3$). (In fact, as we have seen, calcium carbonate itself is not soluble in water.)
The way that hardness of water is classified is shown in TABLE 19.2.

Degree of Hardness	p.p.m. as $CaCO_3$
Soft	0 – 75
Moderately hard	75 – 150
Hard	150 – 300
Very hard	> 300

Table 19.2
Water may be classified according to the degree of hardness in it.

The concentration of calcium and magnesium ions is determined using a chemical called **ethylenediaminetetraacetic acid** commonly referred to as **edta** for short. A molecule of edta has the following structural formula:

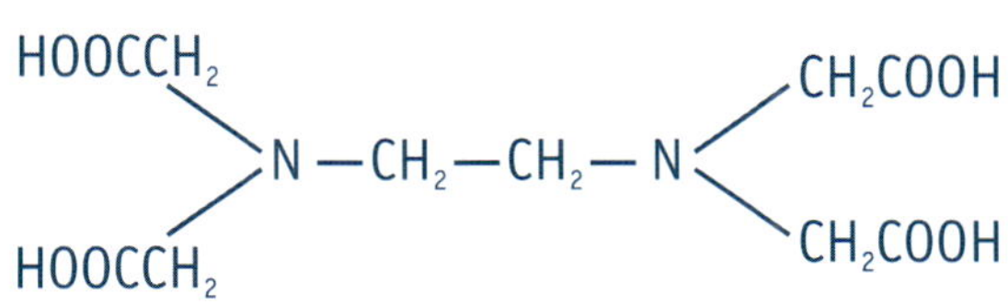

In the name **e**thylene**d**iamine**t**etra**a**cetic acid, the **ethylene** part of the name refers to the CH_2CH_2 unit in the molecule; the **diamine** part of the name refers to the two nitrogen atoms in the molecule; the **tetraacetic** part of the name refers to the four ethanoic acid (acetic acid) molecules in the structure.
In practice, it is the disodium salt of edta which is normally used in titrations. The disodium salt may be represented as Na_2H_2X. The edta works by wrapping itself around the Ca^{2+} and Mg^{2+} ions in the water to form what is called a complex. The reaction may be represented as:

$$\underset{\text{edta ion}}{H_2X^{2-}} + Ca^{2+} \rightarrow \underset{\text{edta complex}}{CaX^{2-}} + 2H^+$$

It is important that you remember that edta and Ca^{2+} ions react in the ratio of 1 : 1 as you will need to know this ratio in future calculations.

In the presence of an indicator called **Eriochrome Black T** (also called **Solochrome Black T**), calcium and magnesium ions have a wine red colour. However, if sufficient edta is added to form a complex with all of the calcium and magnesium ions, the colour changes to blue. Thus, the amount of total hardness in water is determined by titrating a sample of edta against a sample of water using Eriochrome Black T as the indicator. **At the end point, the indicator changes from wine red to blue**. This colour change tells us that all the Ca^{2+} and Mg^{2+} ions have been converted to the corresponding edta complexes. The end point can be quite difficult to detect, as the colour change is not as obvious as in the case of acid-base indicators. This type of titration is often referred to as a **complexometric titration**.

Note: In order for the indicator to work satisfactorily, it is necessary to keep the pH around 10. For this reason, a special type of solution called a **buffer solution** is added.

Apparatus required

Pipette (25 cm^3), burette, conical flask, wash bottle, retort stand, three beakers, white tile, funnel.

Chemicals required

Water sample (with some degree of hardness), edta solution (0.01 M), buffer solution, Eriochrome black T (Solochrome black T) indicator, deionised water.

Procedure

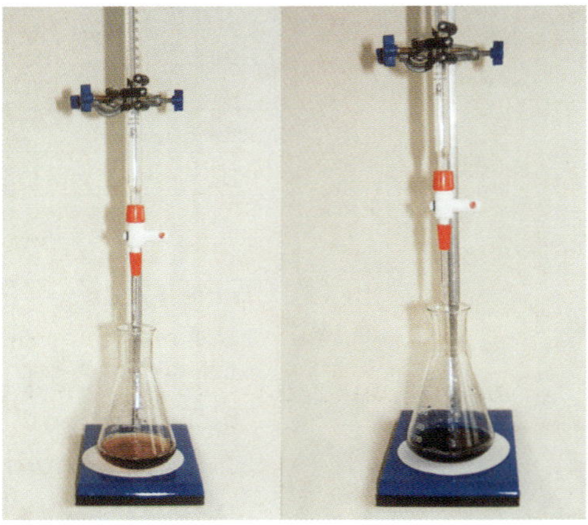

Fig 19.8
A sample of edta is being titrated against a sample of hard water using Eriochrome Black T as indicator. At the end point, the indicator changes from red to blue.

1. Wash out the pipette, burette and conical flask with deionised water.
2. Pour about 100 cm^3 of the edta solution into a clean, dry beaker. Also pour about 200 cm^3 of the hard water sample into another beaker. Label both beakers. Note that the edta solution is stored in a plastic container as edta reacts with the ions in glass if stored in glass for long periods.
3. Rinse out the pipette with some of the hard water sample and the burette with some of the edta solution.
4. Pipette 50 cm^3 (i.e. two 25 cm^3 samples) of the hard water sample into a conical flask (Fig 19.8).
5. Add 1 cm^3 of the buffer solution (pH = 10) and a small amount of the solid Eriochrome black T indicator to the conical flask. (The quantity of indicator taken should be about the amount that will fit on the tip of a spatula.)
6. Clamp the burette vertically on the retort stand.
7. Using a small funnel, fill the burette almost completely with the edta solution. Open the tap briefly and allow the liquid to run out into a beaker marked 'waste'. Remove the funnel from the burette. Adjust the meniscus of the liquid to the zero graduation mark. Ensure that the space below the tap is filled. Check that there are no air bubbles in the nozzle.
8. Place a white tile on the base of the retort stand and put the conical flask on the white tile. The colour of the solution in the conical flask should be wine red.
9. Titrate in the usual manner until the colour of the indicator changes from wine red to blue. Carry out one rough and two accurate titrations which should agree to within 0.1 cm^3.

Calculations

The method of calculating the total hardness of the water is shown in Example 19.1.

Example 19.1

A sample of water was analysed in order to determine the total amount of hardness in the water. 50 cm^3 of the water were placed in a conical flask and some buffer solution and Eriochrome black T indicator added. The solution was titrated against 0.01M edta solution. The average titration figure was 15.2 cm^3. Calculate the total amount of hardness in the water.

SOLUTION

The edta complexes with calcium and magnesium ions in the ratio of 1:1, i.e. in calculations involving edta the mole ratio is always 1:1. We consider all of the hardness in the water to be due to the presence of Ca^{2+} ions.

GIVEN:
$V_{Ca} = 50$
$M_{Ca} = ?$
$n_{Ca} = 1$
$V_{ed} = 15.2$
$M_{ed} = 0.01$
$n_{ed} = 1$

(Ca^{2+}) (edta)

$$\frac{V_{ca} \times M_{ca}}{n_{ca}} = \frac{V_{ed} \times M_{ed}}{n_{ed}}$$

$$\frac{50 \times M_{ca}}{1} = \frac{15.2 \times 0.01}{1}$$

$$M_{ca} = \frac{15.2 \times 0.01}{50}$$

$= 0.003$ moles/L $CaCO_3$
$= 0.003 \times 100$ g/L $CaCO_3$ (Rel molecular mass of $CaCO_3$ = 100)
$= 0.3$ g/L $CaCO_3$
$= 0.3 \times 1000$ mg/L $CaCO_3$
$= 300$ mg/L $CaCO_3$
$= 300$ p.p.m. $CaCO_3$

Answer: Total hardness of water = 300 p.p.m.

Test Yourself:
Attempt questions 19.3 – 19.5 and W19.2 – W19.5.

19.3 WATER TREATMENT

It has been estimated that each household in Ireland uses about 300 litres of water each day. The water supplied to each of our houses by the various councils and corporations around the country must fulfil certain criteria. It must be attractive looking (i.e. colourless), it must be odourless, it must be safe to drink (no active bacteria present) and it must be fluoridated. In order to ensure that the quality of domestic water supplies is of an adequate standard, it is necessary to carry out certain procedures on the water taken from a river or lake. Water treatment plants have been set up throughout the country. The various stages in the process of water treatment may be summarised as follows:

1. **Screening**. This involves passing the water (which is being taken from a river or lake) through a wire mesh to remove any floating debris like twigs, plastic bags, etc.
2. **Flocculation.** The word 'flocculation' means coagulation or precipitation. It refers to the fact that very small suspended solids in the water can be made to form larger particles (called 'flocs') by adding certain chemicals to the water. Such chemicals are called **flocculating agents**. The most commonly used flocculating agent is aluminium sulfate, $Al_2(SO_4)_3$. The addition of aluminium sulfate to the water causes the suspended solids to coagulate or stick together to form larger particles. These larger particles can be allowed to settle out in the settlement stage.

A flocculant is a chemical added to water to coagulate suspended particles.

3. **Settlement (Sedimentation).** In this stage, water is passed into settlement tanks. The water flows in at the bottom of these large tanks and rises up slowly to the surface where the clear water is collected in channels.

Fig 19.9
Settlement tanks at Inniscarra Water Treatment Plant, Co. Cork.

The principle used in the design of settlement tanks is that the water is allowed to rise up towards the surface very slowly (at a speed of less than 2 metres per hour) to allow maximum settlement to take place at the bottom of the tank. Over 90% of the particles are removed in the settlement stage. If these particles were not removed, the water would look cloudy.

4. **Filtration**. The water from the top of the settlement tanks is then passed through large beds of sand. These filter beds remove any remaining suspended solids. The sand acts just like a sieve or filter paper in removing the suspended solids. The sand is supported on a layer of gravel and the layers of sand are cleaned regularly.

   The water that comes out of the filter beds is now clear. However, it is not fit to drink as it may contain harmful bacteria. This problem is dealt with in the next stage.

5. **Chlorination**. Chlorine is added to the water in order to sterilise it, i.e. to get rid of any harmful micro-organisms in the water (FIG 19.10). The chlorine need only be present in small quantities (approximately 0.2 – 0.5 p.p.m.). The amount of chlorine gas added to the water is carefully controlled. If too much chlorine is added, it gives the water an unpleasant taste and smell.

Fig 19.10
These cylinders contain chlorine which is injected periodically into the water to kill bacteria.

The effectiveness of the chlorine treatment is continually monitored by a bacteriological examination of the water.

6. **Fluoridation**. By law, all councils and corporations are compelled to include fluoridation as part of the water treatment process. This involves adding small quantities (about 1 p.p.m.) of fluorine compounds to the water (FIG 19.11). Examples of fluorine compounds commonly added are sodium fluoride, NaF, and hexafluorosilicic acid, H_2SiF_6. Water is fluoridated because it has been shown that the presence of fluoride ions in water helps to reduce dental decay. It does this by strengthening the enamel of teeth. It is estimated that for every £1 spent on fluoridation, between £30–£50 are saved in treating tooth decay.

Fig 19.11
This unit measures the level of fluoridation of the water at Inniscarra Water Treatment Plant, Co. Cork.

7. **pH adjustment**. Some water authorities may find it necessary to alter the pH of the water before distributing it to homes and factories. For example, if the water supply is slightly acidic, it may cause corrosion of pipes. Therefore, the pH has to be raised above 7 and a substance like calcium hydroxide is commonly used for this. Alternatively, if the water is very hard, the water authority may soften it by adding in substances like sodium carbonate (a base). This could cause the pH to be too high and it could be lowered using dilute sulfuric acid or carbon dioxide. The optimum pH of water for distribution to homes and factories is in the range of 7 – 9.

EXPERIMENT 19.2 MANDATORY STUDENT EXPERIMENT NO. 18

To determine (a) the total suspended solids (in p.p.m) of a sample of water by filtration, (b) the total dissolved solids (in p.p.m.) of a sample of water by evaporation and (c) the pH of a sample of water.

Introduction

In part A of this experiment, we will determine the amount of suspended solids in a litre of water. This is done by filtering a known volume of water through a weighed filter paper. From the increase in weight of the filter paper, we can calculate the total amount of suspended solids in p.p.m.

In part B of the experiment, we will measure the amount of dissolved solids by taking the filtered water in part A and boiling off all of the water. From the weight of the residue obtained, we can calculate the total amount of dissolved solids in p.p.m.

The method of measuring the total suspended and total dissolved solids is summarised in FIG 19.12.

In Part C of the experiment, we will measure the pH of the tap water supplied in your locality.

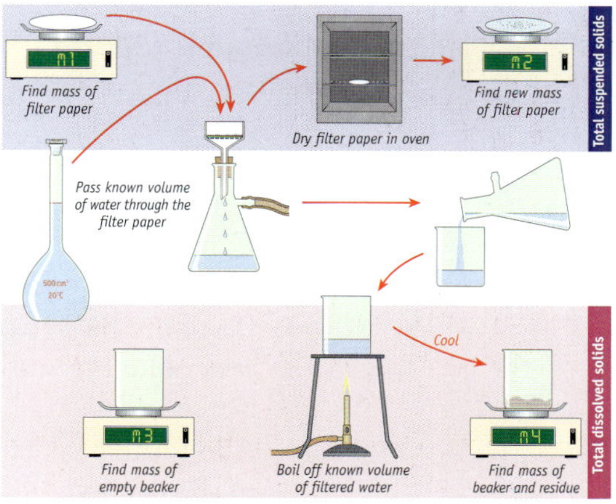

TOTAL SUSPENDED SOLIDS = M2 – M1 expressed in mg/L
TOTAL DISSOLVED SOLIDS = M4 – M3 expressed in mg/L

Fig 19.12
The steps involved in measuring the total suspended solids and total dissolved solids in a sample of water.

PART A: TO MEASURE THE TOTAL SUSPENDED SOLIDS BY FILTRATION

Apparatus required

Volumetric flask (500 cm^3), filter paper (fine pore), Buchner flask and funnel, vacuum pump, retort stand and clamp, large beaker (at least 1L), laboratory balance.

Chemicals required

Water sample (e.g. seawater, river water, pond water, etc.). Tap water is not suitable for this experiment due to the very small amount (if any) of suspended solids present.

Procedure

1. Fill the volumetric flask to the mark with the sample of water. To save time a 500 cm^3 volumetric flask is used in preference to a 1 L flask.
2. Find the mass of a filter paper.
3. Filter the known quantity of water through the filter paper. Retain the filtered water, as this will be required for part B of the experiment.

4. Dry the filter paper (allow it to air dry or use a warm oven or desiccator).
5. Find the new mass of the filter paper. By subtraction, find the mass of suspended solids in that particular volume of water. Express your results in mg/L, i.e. p.p.m.
6. Summarise your results in the form of a table as follows:

Mass of filter paper	$= m_1$ g
Mass of dry filter paper after filtration	$= m_2$ g
Mass of suspended solids	$= (m_2 - m_1)$ g/500 cm^3
	$= (m_2 - m_1) \times 2$ g/L
	$= (m_2 - m_1) \times 2 \times 1000$ mg/L
∴ Total suspended solids	= .... p.p.m.

Part B: To measure the total dissolved solids by evaporation

Apparatus required

Pyrex beaker (at least 1L), graduated cylinder (100 cm^3), hot plate or Bunsen burner or oven, laboratory balance.

Chemicals required

Filtered water.

Procedure

1. Find the mass of a clean, dry beaker.
2. Using a graduated cylinder, place a known quantity (e.g. 200 cm^3) of the filtered water from the previous experiment in the beaker.
3. Place the beaker on a hot plate, or over a Bunsen burner or in a hot oven and evaporate the water to dryness. Note that the dissolved solids remain in the beaker.
4. Allow the beaker to cool and find the new mass. By subtraction, find the mass of the dissolved solids in that particular volume of water. Express your results in mg/L, i.e. p.p.m.
5. Summarise your results in the form of a table as follows:

Mass of beaker	$= m_3$ g
Mass of beaker + dissolved solids	$= m_4$ g
Mass of dissolved solids	$= (m_4 - m_3)$ g/200 cm^3
	$= (m_4 - m_3) \times 5$ g/L
	$= (m_4 - m_3) \times 5 \times 1000$ mg/L
∴ Total dissolved solids	= .... p.p.m.

Part C: To measure the pH of a sample of tap water

The most accurate method of measuring the pH of tap water is to use a **pH meter**. Simply place the electrode of the pH meter (previously calibrated using a buffer solution) in a sample of the tap water and read off the pH. If a pH meter is not available, pH paper may be used.

Example 19.2

A student filtered 500 cm^3 of water and found that the mass of the filter paper after drying had increased by 0.68 g. 100 cm^3 of the filtered water were then evaporated to dryness in a beaker and it was found that the mass of the beaker had increased by 0.13 g.
What is the amount of suspended solids and the amount of dissolved solids in the water expressed as parts per million (p.p.m.)? *[LCH]*

SOLUTION

Total suspended solids $= 0.68\ g/500\ cm^3$
$= 0.68 \times 2\ g/L$
$= 0.68 \times 2 \times 1000\ mg/L$
$= 1360$ p.p.m.

Total dissolved solids $= 0.13\ g/100\ cm^3$
$= 0.13 \times 10\ g/L$
$= 0.13 \times 10 \times 1000\ mg/L$
$= 1300$ p.p.m.

Answer: Total suspended solids = 1360 p.p.m.
Total dissolved solids = 1300 p.p.m.

Test Yourself:
Attempt questions 19.6 – 19.7 and – W19.6.

19.4 WATER POLLUTION

DISSOLVED OXYGEN – THE BREATH OF LIFE UNDER WATER.

Dissolved oxygen in rivers, lakes and the sea is vital for the survival of fish and other forms of life. Since O_2 is a non-polar molecule, the solubility of oxygen in water (a polar solvent) is quite low and depends on the temperature of the water. For example, at 10 °C the solubility of oxygen is 11.3 mg/L and at 20 °C the solubility of oxygen drops to 9.2 mg/L. (The fact that solubility of gases in water drops with increase in temperature is observed when water is heated. As the water is heated, bubbles of air are seen to come out of solution.)

Now consider what happens when organic waste is discharged into a waterway. Examples of organic waste are domestic sewage, animal slurry, silage effluent from farms, effluents from food processing factories, milk, industrial waste, etc. When the organic waste is discharged into the water, the bacteria and other organisms that are naturally present in the water are provided with nutrients. These nutrients cause the organisms to multiply into large numbers. As the organisms break down the waste into compounds like carbon dioxide and water, they use up oxygen dissolved in the water.

organic matter + oxygen → carbon dioxide + water

If a considerable amount of organic waste is present, it is possible that the dissolved oxygen level in the water will be reduced so much that fish life in the water may be reduced or possibly killed off completely. For example, trout and salmon need an approximate concentration of 5 p.p.m of dissolved oxygen in order to survive. At this stage, the dissolved oxygen used by bacteria (breaking down the waste) cannot be replaced fast enough by more oxygen from the air dissolving in the water.

If the dissolved oxygen drops to zero concentration, anaerobic bacteria (bacteria which do not require oxygen) will take over and the river will become foul smelling due to the presence of hydrogen sulfide gas, H_2S.

The release of substances into the environment that damage the environment is called **pollution**. Let us now consider the method used by chemists to indicate the level of pollution in a sample of water.

In the early 1900s, the Royal Commission on Sewage Disposal introduced a test to measure the demand that water has for dissolved oxygen. This test is called the **Biochemical Oxygen Demand** test or, more simply, the **B.O.D.** test.

Biochemical oxygen demand is defined as:

- The amount of dissolved oxygen
- Consumed by biological action
- When a sample of water is kept at 20 °C
- In the dark
- For five days.

The B.O.D. test is carried out by completely filling two bottles with the water to be tested. The dissolved oxygen in one of the bottles is then measured (by using a dissolved oxygen meter or by a titration called the Winkler method – see EXPERIMENT 19.3). The second bottle is incubated at 20 °C in darkness for five days. The bottle is kept in darkness to prevent photosynthesis taking place, as this would increase the level of oxygen in the water. The purpose of the fixed temperature of 20 °C is because a fixed temperature is required for a valid comparison of different water samples – the amount of oxygen dissolved in water depends on temperature. Also, the rate at which oxygen is consumed depends on temperature.

When the five-day period has elapsed, the amount of dissolved oxygen remaining in the second bottle is measured. **The B.O.D. is the difference in the two dissolved oxygen levels as this is the amount of oxygen that has been used up during the test**. B.O.D. is measured in mg/L of oxygen. Examples of B.O.D. values are given in TABLE 19.3. Note that the higher the B.O.D. value, the more polluted the water. A high B.O.D. value means that there is a large amount of organic waste in the water and as this organic waste is acted on by bacteria, the level of oxygen decreases.

B.O.D. (mg/L)	Example
1–2	Clean water
20–40	Treated sewage
100	Polluted water (fish die)
300	Raw sewage
500	Brewery effluent
30,000	Pig slurry
54,000	Silage effluent

Table 19.3

Some examples of B.O.D. values.

Most of the fish kills in Ireland are caused by silage and slurry effluent being discharged into rivers and lakes (FIG 19.13).

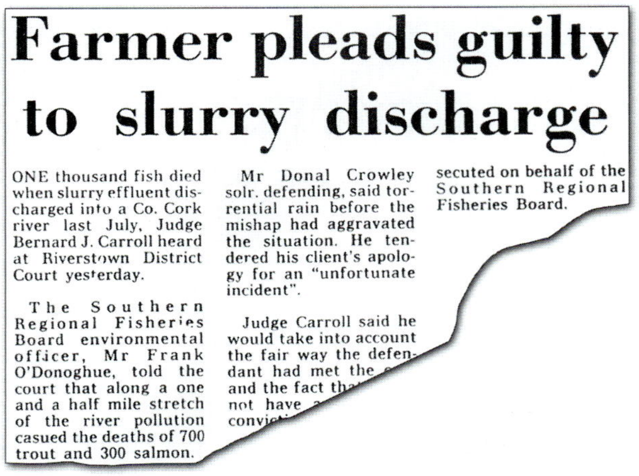

Farmer pleads guilty to slurry discharge

ONE thousand fish died when slurry effluent discharged into a Co. Cork river last July, Judge Bernard J. Carroll heard at Riverstown District Court yesterday.

The Southern Regional Fisheries Board environmental officer, Mr Frank O'Donoghue, told the court that along a one and a half mile stretch of the river pollution casued the deaths of 700 trout and 300 salmon.

Mr Donal Crowley solr. defending, said torrential rain before the mishap had aggravated the situation. He tendered his client's apology for an "unfortunate incident".

Judge Carroll said he would take into account the fair way the defendant had met the ... and the fact th... not have ... convi...

...secuted on behalf of the Southern Regional Fisheries Board.

Fig 19.13

Most fish kills in Ireland are caused by the discharge of slurry and silage effluent into rivers and lakes.

The B.O.D. test may be used to measure the level of pollution in effluents. However, because of the low solubility of oxygen in water (9.2 mg/L at 20 °C), effluents with B.O.D.s greater than 9 mg/L must be diluted by a fixed amount with well-oxygenated water. This is done in order to ensure that dissolved oxygen will be present throughout the five-day period of the test and that a measurable amount of oxygen (at least 2 p.p.m.) will be left after the five-day period. Fresh distilled water is used for diluting the sample and the mixture must be well shaken for a few minutes in order to oxygenate the water fully. If the effluent is not diluted, it is possible that all of the oxygen will be used up before the five days have elapsed.

Example 19.3

To find the biochemical oxygen demand (B.O.D.) of a sample of polluted river water, 25 cm^3 of the water were diluted to one litre with well-oxygenated pure water. Two bottles, A and B, were filled with the diluted water and their dissolved oxygen concentrations were determined. The analysis was carried out immediately for bottle A and five days later for bottle B. The results obtained were 12.8 p.p.m. and 8.2 p.p.m. respectively.

(i) Why was it necessary to dilute the polluted river water? What was the advantage in using well-oxygenated water for this purpose?

(ii) Under what conditions should bottle B have been kept for the five days before it was analysed? In the case of one of the conditions you have stated, explain why it is necessary.

(iii) What is the B.O.D. of the polluted river water? *[LCH]*

SOLUTION

(i) It is necessary to dilute the water because, if you used the undiluted water, there might not be enough oxygen present after five days to get an accurate result.
The advantage of using well-oxygenated water is that it will ensure a high oxygen level to start with so that there will be a measurable amount of oxygen left after five days.

(ii) The conditions are that the bottle should be kept in the dark at 20 °C. The bottle must be kept in the dark to prevent photosynthesis, as this would add oxygen to the water.
A constant temperature of 20 °C is required as the amount of dissolved oxygen depends on the temperature. Also, the rate at which oxygen is consumed depends on the temperature. (Alternatively, one could say that a constant temperature is required for a valid comparison of results from various samples of water.)

(iii) B.O.D. for sample = 12.8 p.p.m. – 8.2 p.p.m. = 4.6 p.p.m.
But 25 cm^3 of the water were diluted to 1 litre, i.e. a dilution factor of 40
$\therefore$ B.O.D. of undiluted water = 4.6×40 = 184 p.p.m.

Answer: 184 p.p.m.

Note that in this problem the dilution factor must be taken into account in order to obtain the B.O.D., i.e. the value obtained is multiplied by the dilution factor in order to obtain the B.O.D.

The dilution factor is based on the expected result. For example, a dilution factor of 20 is used for a B.O.D. range of 20 – 140 and a dilution factor of 50 is used for a B.O.D. range of 50 – 350.

When collecting a sample of water for a B.O.D. analysis, it is important that care is exercised to ensure that the sample is representative of the body of water being tested. It is also important to ensure that the sampling process itself does not alter the concentration of the oxygen in the sample. The following precautions should be observed when collecting water for B.O.D. analysis:

(i) Fill the bottle under the surface of the water to prevent atmospheric oxygen being trapped and, thus, giving an artificially high oxygen level.

(ii) Fill the bottle completely to ensure that no air is trapped between the top of the water and the stopper of the bottle.

(iii) Place the second bottle in the dark immediately to prevent photosynthesis.

Another way to reduce the amount of dissolved oxygen in water is to overload the water with plant nutrients, especially nitrate ions (NO_3^-) and phosphate ions (PO_4^{3-}). This 'over-enrichment' is called **eutrophication** (a Greek word meaning 'well nourished'). As the nitrate and phosphate levels rise, many plants and floating algae undergo population explosions. These algae are short lived and when they decay, a heavy oxygen demand is placed on the water. This causes the dissolved oxygen level to drop and many forms of animal life are killed. The 'algal bloom' covers much of the surface of the water with a green scum (FIG 19.14).

Fig 19.14
Algal bloom caused by eutrophication.

Eutrophication is the enrichment of water with nutrients, which leads to excessive growth of algae.

Eutrophication may occur naturally or artificially. Natural eutrophication occurs mainly in lakes and is caused by a gradual increase in nitrogen and phosphorus levels caused by sediments building up in the lake.

Artificial eutrophication is caused by a sudden increase in nutrients caused by artificial fertilisers being washed into rivers and lakes or by domestic sewage or waste from intensive farming entering these waterways.

There is some concern about the levels of nitrate ions in water. It is thought that high levels of nitrates may cause stomach cancer and also death in babies.

Water pollution is also caused by the release of toxic metal ions into the water, e.g. lead ions (Pb^{2+}), mercury ions (Hg^{2+}) and cadmium ions (Cd^{2+}). Due to the high relative atomic masses of the metals, they are commonly referred to as 'heavy metals'. Lead, mercury and cadmium are often classified as cumulative poisons, i.e. their concentrations tend to build up in the body upon continuous exposure. These metals can be found in rivers and lakes as a result of the discharge of industrial effluents or the dumping of batteries that contain these metals (in the absence of recycling). Lead in drinking water is a problem in very old houses containing lead plumbing. (The word *plumbing* comes from the Latin word *plumbum* for lead).

As an example of the problems associated with heavy metals, let us consider the case of mercury poisoning. Metallic mercury is dangerous when inhaled into the lungs. It is less dangerous when swallowed because most of the mercury passes out of the body in a few days. Inorganic mercury salts are dangerous at high levels and can cause damage to the intestine and to the kidneys. However, our bodies normally excrete these salts in the urine before they accumulate to high levels.

The best known case of mercury poisoning occurred in Minamata Bay, Japan in the late 1950s. Industrial waste containing mercury salts was discharged into Minamata bay. The mercury salts became incorporated into the food chain and high levels of mercury were found in fish that were consumed by the local population. The mercury caused birth defects and death, and the disease caused by mercury poisoning became known as *Minamata disease*. Nowadays, the mercury is recovered from the industrial waste before it is discharged into water. Heavy metal ions are generally removed from waste by precipitation, e.g. lead ions are reacted with dilute HCl and the lead is precipitated as lead chloride:

$$Pb^{2+} + 2Cl^- \rightarrow PbCl_2\downarrow$$

Examples of limits set by the EU for various chemical species in water are listed in TABLE 19.4.

SUBSTANCE	MAXIMUM ADMISSIBLE CONCENTRATION
Nitrates	50 mg/L
Phosphates	2.2 mg/L
Lead	50 mg/L
Chromium	50 mg/L
Cadmium	5 mg/L
Mercury	1 mg/L
Pesticides	0.5 mg/L

Table 19.4
Examples of EU limits for some chemical species.

EXPERIMENT 19.3 MANDATORY STUDENT EXPERIMENT NO. 19

To measure the amount of dissolved oxygen in a sample of water by means of a redox titration.

Introduction

In this experiment, we will measure the amount of dissolved oxygen in a sample of water using a method commonly known as the **Winkler method**. This experiment involves collecting a sample of water in a bottle. The bottle is then stoppered. When collecting the water, it is important that no air is present in the bottle as this would raise the oxygen level and give an incorrect result. Therefore, the bottle must be completely filled to the top.

Small quantities of concentrated manganese(II) sulfate solution, $MnSO_4$, and alkaline potassium iodide solution (a mixture of NaOH and KI) are then added to the water. The Mn^{2+} ions and the OH^- ions from the alkali react together to form a white precipitate of manganese(II) hydroxide as shown in the following equation:

$$Mn^{2+} + 2OH^- \rightarrow Mn(OH)_2\downarrow \qquad \text{(a)}$$

manganese(II) hydroxide
(white precipitate)

This white precipitate then reacts with the dissolved oxygen in the water to form a brown precipitate of manganese(III) hydroxide.

$$4Mn(OH)_2 + O_2 + 2H_2O \rightarrow 4Mn(OH)_3\downarrow \qquad \text{(b)}$$

manganese(III) hydroxide
(brown precipitate)

If the white precipitate does not turn brown, this indicates that there is no dissolved oxygen in the water.

Note: There are a number of ways of representing the formula of this brown precipitate, e.g. $Mn(OH)_3$, Mn_2O_3, $MnO(OH)_2$, etc. These different formulas are due to the fact that in solution, the composition of the brown precipitate may be variable. Its exact composition need not concern us in this course.

A small quantity of concentrated sulfuric acid is then added to the bottle. The sulfuric acid causes the iodide ions (from the alkaline KI) to react with the manganese(III) hydroxide to liberate iodine.

$$2Mn(OH)_3 + 2I^- + 6H^+ \rightarrow I_2 + 6H_2O + 2Mn^{2+} \qquad \text{(c)}$$

The brown precipitate dissolves and the red/brown colour of iodine is now observed in the bottle.

The amount of iodine is determined by titrating it against standard sodium thiosulfate solution.

$$I_2 + 2S_2O_3^{2-} \rightarrow 2I^- + S_4O_6^{2-} \qquad \text{(d)}$$

Knowing the amount of iodine liberated, it is quite easy to calculate the amount of dissolved oxygen in the water. The method of calculation is shown in EXAMPLE 19.4.

Note: The Winkler method is usually not used for water that has been chlorinated. This is because chlorine, an oxidising agent, would react with the iodide ions of the potassium iodide to liberate iodine.

$$Cl_2 + 2I^- \rightarrow 2Cl^- + I_2$$

This 'extra' iodine coming from the above reaction would give a titration figure higher than the true value. (The concentration of chlorine in tap water is very small - only about 0.2 – 0.5 p.p.m. Therefore, this would introduce only a small error into the measurement of dissolved oxygen.)

Apparatus required

Pipette (25 cm^3), pipette filler, burette, conical flask, wash bottle, retort stand, three beakers, white tile, funnel, graduated cylinder, stirring rod, volumetric flask, bottle of about 250 cm^3 capacity (clear glass with stopper), plastic disposable droppers (graduated).

Chemicals required

Manganese(II) sulfate solution (2 M approx.), alkaline potassium iodide solution (125 g NaOH and 37.5 g KI per 250 cm^3 solution), sodium thiosulfate solution (0.02 M), starch indicator, concentrated sulfuric acid.

Procedure

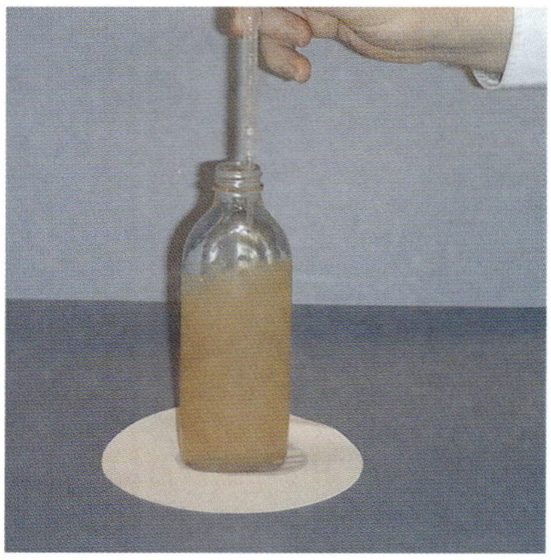

Fig 19.15
A brown precipitate is formed on adding the manganese(II) sulfate and alkaline potassium iodide solutions to the water.

1. Place some water into the bottle, shake the bottle a few times and empty out the water. Rinsing out the bottle with water removes any impurities, wets the inside of the bottle and helps avoid trapped air bubbles.
2. Submerge the bottle completely in the water to be analysed and fill the bottle right to the top.
3. Use a graduated dropper, with the end of the dropper under the surface of the water in the bottle, to add 1 cm^3 of the manganese (II) sulfate solution to the water.
4. Use another graduated dropper, with the end of the dropper under the surface of the water, to add 1 cm^3 of the alkaline potassium iodide solution. Since both solutions are quite dense, note that these solutions sink to the bottom of the bottle. Note the brown precipitate being formed in the bottle (Fig 19.15).
5. Stopper the bottle. There will be a slight overflow of water.
6. Shake the bottle vigorously for about half a minute. Allow the brown precipitate to settle for a few minutes until there is at least a 5 cm depth of clear liquid below the stopper.
7. Using a graduated dropper, carefully add 1 cm^3 of concentrated sulfuric acid to the bottle.
8. Re-stopper the bottle carefully to avoid dissolving any oxygen from the air. Shake the bottle to redissolve the precipitate. If the precipitate is difficult to dissolve, carefully add a few more drops of concentrated sulfuric acid. At this stage, the solution should be a red/brown colour due to the liberated iodine.
9. Now perform two titrations to determine the concentration of iodine in the water. Using a pipette, place a 100 cm^3 sample from the bottle into a conical flask.
10. Fill the burette to the 0 cm^3 mark with the sodium thiosulfate solution and titrate in the usual manner until a pale straw colour is obtained.
11. Add a few drops of starch solution and continue titrating until the blue/black colour disappears.

12. Repeat the procedure with a second sample from the bottle. Since it will not be possible to carry out a third titration (due to the capacity of the bottle) the second titration figure obtained must be as accurate as possible.
13. Calculate the concentration of oxygen in the water expressing your results in p.p.m. The method of calculating this is shown in EXAMPLE 19.4.

Example 19.4

In a Winkler determination of dissolved oxygen in water, 100 cm^3 samples of water were analysed. The iodine liberated in the experiment was titrated against 0.02 M sodium thiosulfate solution. The average titration figure was 5.7 cm^3. Calculate the amount of dissolved oxygen in the water in p.p.m.

SOLUTION

In solving this problem we treat the water as a solution of oxygen. On page 277, we discussed the following equations:

$$4Mn(OH)_2 + O_2 + 2H_2O \rightarrow 4Mn(OH)_3\downarrow \quad \text{(b)}$$

$$2Mn(OH)_3 + 2I^- + 6H^+ \rightarrow I_2 + 6H_2O + 2Mn^{2+} \quad \text{(c)}$$

$$I_2 + 2S_2O_3^{2-} \rightarrow 2I^- + S_4O_6^{2-} \quad \text{(d)}$$

From equation (b), we see that one mole of oxygen produces four moles of $Mn(OH)_3$. From equation (c), we see that four moles of $Mn(OH)_3$ would produce two moles of I_2. From equation (d), we see that two moles of I_2 would react with four moles of $S_2O_3^{2-}$ ions.

Therefore, overall the ratio between O_2 and $S_2O_3^{2-}$ is 1: 4. This ratio is used for calculations.

This means that $n_o = 1$ and $n_r = 4$, i.e. $1\ O_2 \equiv 4\ Mn(OH)_3 \equiv 2\ I_2 \equiv 4\ S_2O_3^{2-}$

GIVEN:

$V_o = 100$

$M_o = ?$

$n_o = 1$

$V_r = 5.7$

$M_{red} = 0.02$

$n_r = 4$

$$\underset{(O_2)}{\frac{V_o \times M_o}{n_o}} = \underset{(S_2O_3^{2-})}{\frac{V_r \times M_{red}}{n_r}}$$

$$\frac{100 \times M_o}{1} = \frac{5.7 \times 0.02}{4}$$

$$M_o = \frac{5.7 \times 0.02}{4 \times 100}$$

$= 0.000285$ moles/L

$= 0.000285 \times 32$ g/L (Rel molecular mass $O_2 = 32$)

$= 0.00912$ g/L

$= 0.00912 \times 1000$ mg/L

$= 9.12$ mg/L

$= 9.12$ p.p.m.

Answer: Dissolved oxygen = 9.12 p.p.m.

Example 19.5

A sample of water was analysed in order to determine its B.O.D. A 100 cm^3 sample of the water was diluted to two litres with well-oxygenated water. 100 cm^3 of this diluted solution were analysed by the Winkler method and 7.15 cm^3 of 0.025 M sodium thiosulfate solution were required in the titration.

A second 100 cm^3 sample was stored under the appropriate conditions for five days and then analysed for dissolved oxygen. The liberated iodine required 3.2 cm^3 of 0.025 M sodium thiosulfate solution. Calculate the B.O.D. of the water.

SOLUTION

First calculate the amount of dissolved oxygen in the water at the beginning of the five-day period.

GIVEN:
$V_o = 100$
$M_o = ?$
$n_o = 1$
$V_r = 7.15$
$M_{red} = 0.025$
$n_r = 4$

$$(O_2) \qquad (S_2O_3^{2-})$$

$$\frac{V_o \times M_o}{n_o} = \frac{V_r \times M_{red}}{n_r}$$

$$\frac{100 \times M_o}{1} = \frac{7.15 \times 0.025}{4}$$

$$M_o = \frac{7.15 \times 0.025}{4 \times 100}$$

$= 0.000447$ moles/L

$= 0.000447 \times 32$ g/L (Rel molecular mass $O_2 = 32$)

$= 0.0143$ g/L

$= 0.0143 \times 1000$ mg/L

$= 14.3$ mg/L

$= 14.3$ p.p.m.

Now calculate the amount of dissolved oxygen in the water at the end of the five-day period.

GIVEN:
$V_o = 100$
$M_o = ?$
$n_o = 1$
$V_r = 3.2$
$M_{red} = 0.025$
$n_r = 4$

$$(O_2) \qquad (S_2O_3^{2-})$$

$$\frac{V_o \times M_o}{n_o} = \frac{V_r \times M_{red}}{n_r}$$

$$\frac{100 \times M_o}{1} = \frac{3.2 \times 0.025}{4}$$

$$M_o = \frac{3.2 \times 0.025}{4 \times 100}$$

$= 0.0002$ moles/L

$= 0.0002 \times 32$ g/L

$= 0.0064$ g/L

$= 0.0064 \times 1000$ mg/L

$= 6.4$ mg/L

$= 6.4$ p.p.m.

B.O.D. of diluted sample = 14.3 - 6.4 = 7.9 p.p.m.
But since the original water sample was diluted 20 times.
⇒ B.O.D. of original water sample = 7.9 × 20 = 158 p.p.m.

Answer: B.O.D. = 158 p.p.m.

Test Yourself:
Attempt questions 19.8 – 19.10 and W19.7 – W19.8.

19.5 SEWAGE TREATMENT

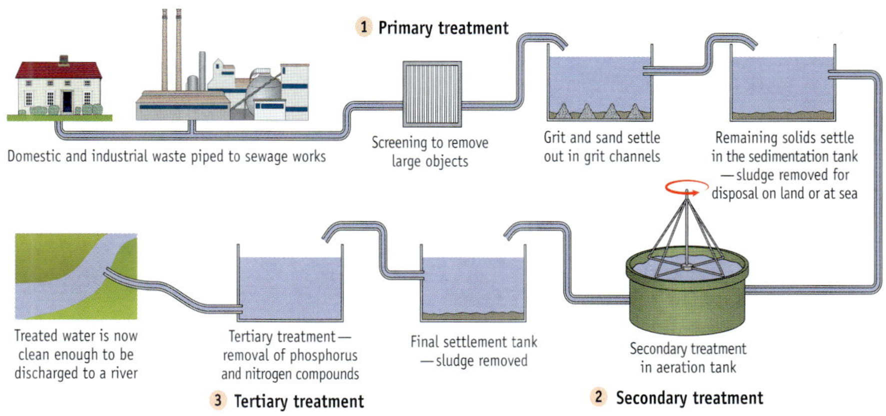

Fig 19.16
The stages involved in the treatment of sewage in a sewage treatment plant.

Sewage is the term used to describe the material delivered to waste treatment plants of towns and cities. The used water from homes, factories, businesses, etc. is discharged into sewers that take it to a sewage treatment plant. Typical raw sewage of towns and cities contains a wide variety of different substances, e.g. domestic waste, some industrial waste, discharges from commercial establishments, storm water from drains, etc.

Sewage may consist of organic waste (e.g. domestic waste, or oils and grease from industrial waste), inorganic waste (e.g. nitrates and phosphates from domestic and industrial waste) and gases like hydrogen sulfide and methane (formed when organic waste decomposes).

If sewage were simply dumped into the local river or lake, there would be such a high demand on the dissolved oxygen that the water would become very polluted. Thus, the essential purpose of a sewage treatment plant is to reduce the B.O.D. of the sewage before it is discharged into the local river or lake. The process for the treatment of sewage may be divided into three stages – primary treatment, secondary treatment and tertiary treatment. These stages are summarised in FIG 19.16.

PRIMARY TREATMENT

Primary treatment is essentially a mechanical process involving screening and settlement. Solids that are floating in the sewage are removed by passing the sewage through steel bars. This process is called **screening**. It is important to remove items like twigs, leaves, plastic bags, etc. as these could interfere with the equipment in the rest of the treatment plant. The sewage then passes through grit channels in which items like pebbles and grit settle and are removed periodically.

Many materials in sewage are present as suspended particles that are carried along in flowing liquids. However, these suspended particles will settle out when the flow of liquid stops. Therefore, the sewage is allowed to flow into large settling tanks where it remains for a few hours. The solids settle to the bottom of these tanks and form a sludge. This process is called **settlement** or **sedimentation**.

The solids that settle at the bottom of the tanks are removed at regular intervals for disposal on land or at sea. The liquid on top (the supernatant liquid) is then passed on to receive secondary treatment.

Primary treatment removes about one half of the suspended solids and removes about one third of the B.O.D. in the sewage.

SECONDARY TREATMENT

Secondary treatment is a biological process in which the levels of suspended and dissolved organic materials are reduced. One of the most common methods of bringing about this bacterial breakdown of the sewage involves using the **Activated Sludge Process**. This consists of an aeration tank followed by a settling tank. The sewage is pumped into a large aeration tank where the sewage provides nutrients for a large number of growing micro-organisms (called activated sludge). These micro-organisms digest the sewage in the same way that occurs if the sewage is discharged into a river. However, in this case, the breakdown of sewage occurs within the treatment plant. The liquid in the aeration tank is kept aerated by a mechanical rotor that continually churns the liquid and sludge with the air. In some aeration tanks, compressed air is continually bubbled into the tank. Thus, the aeration tank prevents the concentration of dissolved oxygen in the liquid from dropping too low (FIG 19.17).

Fig 19.17
Photograph of aeration tank in Ballincollig sewage treatment plant, Co. Cork.

From the aeration tank, the sewage flows into a settling tank. Some of the sludge is removed and some is recycled back to the aeration tank to come in contact with fresh sewage. (The sludge removed is rich in nutrients and may be used as fertiliser. Alternatively, the sludge could be stored in sealed tanks where anaerobic bacteria convert it to methane. This methane is then used as a fuel.) Thus, the micro-organisms are continually decomposing the organic waste in the sewage into compounds like carbon dioxide, nitrates, etc.

Although the Activated Sludge Process is the system being installed most commonly, some older sewage treatment plants, instead of using aeration tanks, spray the sewage (after primary treatment) over circular beds of rocks using rotating sprayers. The sewage percolates down through the rocks and a slime of micro-organisms grows on the surface of the rocks. Thus, the micro-organisms use the nutrients from the sewage, and oxygen from the air, to break down the sewage.

After secondary treatment, about 95% of the B.O.D. of the original sewage is removed. The effluent may then be discharged into a nearby waterway. However, the effluent may still contain compounds of phosphorus and nitrogen. These are removed in tertiary treatment.

Note: There may not be a sewage treatment plant near you if you are living near the sea. The sewage may possibly be discharged well out into the sea where the large dilution factor decreases the chance of serious pollution. Many of the micro-organisms are killed by the salt water.

TERTIARY TREATMENT

Tertiary treatment involves the removal of phosphorus and nitrogen compounds from effluents. Phosphates come from household detergents (washing powders and washing-up liquids) and nitrates come from organic materials in sewage. As already discussed, nitrates and phosphates cause pollution by eutrophication.

Phosphates are removed by precipitation, e.g. by adding aluminium sulfate. This precipitates the phosphates as aluminium phosphate. Phosphates may also be removed by adding iron(III) chloride. This removes the phosphates as iron(III) phosphate. Lime may also be used to precipitate phosphates. These insoluble phosphate compounds are allowed to settle out before the final effluent is discharged.

Removal of nitrogen compounds from effluent can be difficult. Nitrogen may be present in the form of ammonia, nitrite compounds, nitrate compounds or organic compounds containing nitrogen. The removal of all of these nitrogen compounds can be quite difficult and expensive.

Tertiary treatment plants are installed in places like Mullingar (to prevent eutrophication in Lough Ennell) and at Killarney (to prevent eutrophication in the local lakes).

Note: Do not get confused between water treatment (for domestic use) and sewage treatment (before discharging into local waterways) when answering examination questions!

19.6 INSTRUMENTAL METHODS OF WATER ANALYSIS

As well as the methods of analysing water already discussed, a number of other methods using laboratory instruments are used when analysing water. Common instrumental methods of analysing water will now be discussed:

(I) pH METER

In chapter 18 we saw that the pH of a solution may be measured using a pH meter. pH meters are used in laboratories attached to water treatment plants to monitor the pH of water taken from rivers and lakes (FIG 19.18). At Inniscarra water treatment plant, water is analysed from all over County Cork. Water coming from limestone areas tends to be basic with a pH of about 7.8. Water coming from sandstone areas tends to be acidic and could have a pH as low as 5.8. The pH is adjusted in the water treatment plant so that the final pH of the water before distribution is in the range 7 – 9.

Fig 19.18
This laboratory technician is measuring the pH of water samples using a pH meter.

(II) ATOMIC ABSORPTION SPECTROMETRY

Fig 19.19
This laboratory technician is using an atomic absorption spectrometer to measure the concentrations of various metals in water.

We have already discussed atomic absorption spectrometry in CHAPTER 3. We have seen that it is a very useful method for analysing for the presence of certain elements since each element has its own unique absorption spectrum. Atomic absorption is often used to detect and measure the concentrations of heavy metals in water, e.g. lead, cadmium, mercury, etc. The atomic absorption spectrometer shown in FIG 19.19 is being used to measure the concentration of various metals in a water sample.

(III) COLORIMETRY

Colorimetry is a technique whereby white light is passed through a coloured solution and the colour of the solution is then compared with the colour of solutions of known concentrations of that substance. The technique is illustrated in EXPERIMENT 19.4.

EXPERIMENT 19.4 MANDATORY STUDENT EXPERIMENT NO. 20

To estimate the concentration of free chlorine in swimming-pool water or bleach using (a) a comparator or (b) a colorimeter.

Introduction

Chlorine is added to swimming-pool water to prevent water-borne diseases. When chlorine is added to water, it reacts with the water to form hypochlorous acid and hydrochloric acid.

$$Cl_2 + H_2O \rightleftharpoons \underset{\text{hypochlorous acid}}{HOCl} + HCl$$

The hypochlorous acid is a weak acid and dissociates into hydrogen ions and hypochlorite ions.

$$HOCl \rightleftharpoons H^+ + \underset{\text{hypochlorite ion}}{OCl^-}$$

(In fact, the hypochlorite ions have little power to kill germs. Therefore, pool managers have to keep the pH of the pool within certain pH limits to maintain a high proportion of undissociated hypochlorous acid.) Chlorine that exists in water as hypochlorous acid and hypochlorite ion is referred to as **free chlorine**. This free chlorine is a very efficient disinfectant. Chlorine may also exist combined with ammonia or organic nitrogen compounds in water, e.g. NH_2Cl, $NHCl_2$, etc. This is referred to as **combined chlorine**. (The main sources of the nitrogen compounds in water are sweat and urine.)

In the past, chlorine was added to water from chlorine in gas cylinders. However, nowadays chlorine is not used. Instead a compound which produces hypochlorous acid is added to the water, e.g. calcium hypochlorite, $Ca(OCl)_2$. Find out what your local swimming pool uses as the source of hypochlorous acid.

In this experiment, we will use colorimetry to measure the amount of free chlorine in swimming-pool water. If the level of free chlorine is too low, bacteria may survive in the water and infect swimmers. If the level of free chlorine is too high, compounds like NCl_3 will be formed and these compounds cause a lot of irritation to swimmers' eyes. To avoid problems like the above, the level of free chlorine in swimming pools must be kept in the range 1 – 5 ppm.

To estimate the amount of chlorine in swimming-pool water, one could use a simple test kit (e.g. the *Hach* test kit) or a more expensive type of instrument called a colorimeter. We now describe both of these methods. You are required to know only **one** of these methods for your Leaving Certificate examination. If swimming-pool water is not available, this experiment may be performed with a very dilute solution of bleach.

Method A: The comparator method (using a Hach test kit).

In this experiment, a chemical called DPD is used to help us measure the amount of chlorine present in swimming-pool water. You need not worry about what this chemical is (DPD stands for **d**iethyl-**p**henylene**d**iamine) or how it reacts with chlorine. Just be aware of the fact that when this chemical and other chemicals are added to water containing chlorine, they react with the chlorine in the water to give a pink-coloured solution. The more chlorine that is present in the solution, the pinker the colour. The intensity of the pink colour is then compared (hence the name **comparator method**) against a coloured disc with different shades of pink. This coloured disc has already been calibrated using solutions containing known quantities of chlorine. Hence, by visually comparing the pink colour of the swimming-pool water with the pink colour on the disc, the concentration of chlorine in the water may be estimated.

Apparatus required

Hach test kit (0 – 35 mg/L, model CN-66).

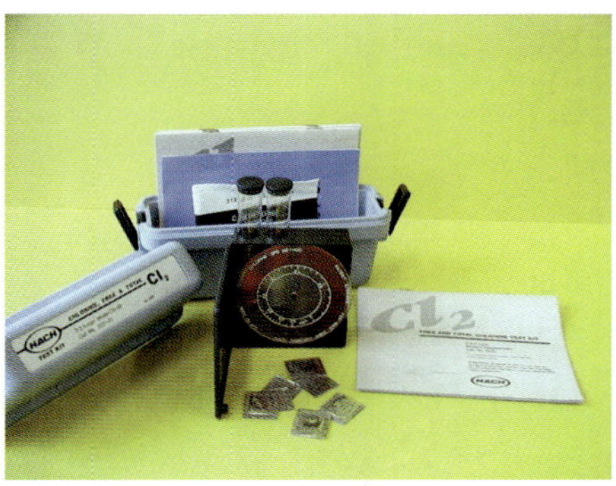

Fig 19.20
The Hach test kit.

Chemicals required

Deionised water, fresh sample of swimming-pool water (or dilute solution of bleach) and chemicals supplied with test kit.

Procedure

1. Take one of the colour viewing tubes (similar to small test-tubes) and place 5 cm^3 of deionised water in it. The level of the water will now be just below the 'frosted' part of the glass in the tube. Place the stopper on the tube. This is called the reference tube and its purpose is simply to indicate to us how much light will be absorbed by this depth of water.
2. Open the black-colour comparator box and place the tube with the clear water in the top left-hand opening (Fig 19.20). Note that the level of water is just above the circular opening through which the light passes.
3. Place 5 cm^3 of the swimming-pool water in the other viewing tube.
4. Open one of the small sachets marked *chlorine free DPD* using a scissors (or tear gently along the top) and add the contents of the sachet to the sample of swimming-pool water. The sachet contains a number of chemicals, the main one being DPD. Put the stopper on the tube and swirl gently to dissolve the solid. Note the pink colour of the solution.
5. Place the second sample tube in the right top opening of the comparator box.
6. Remove the coloured disc from its plastic protective cover and place it in the comparator box. Close the lid of the comparator box.
7. Hold the comparator box up to a source of white light, e.g. an ordinary lamp, window, etc. and view the two openings in the front of the box.
8. Rotate the disc (it protrudes slightly at the right-hand side of the box) until the colours in each opening match, i.e. the two colours are the same.
9. Immediately read the number that is visible through the opening at the bottom of the comparator box. This gives the value of chlorine dissolved in the water in units of mg/L (p.p.m.).
10. When finished rinse out both tubes with deionised water and return the coloured disc to its envelope. (This prevents the colours on the disc from fading.)

Method B: Using a colorimeter

This method uses an instrument called a colorimeter to measure the amount of chlorine in the sample of swimming-pool water. In this instrument, a narrow beam of light is passed through the solution being studied (Fig 19.21). A filter is used to select a particular wavelength of light as this improves the accuracy of the result. The light then falls on a photocell. A photocell is simply a device that converts light energy into an electrical signal. The current flowing is then registered on a meter. (In modern colorimeters, this is a digital display.) The colorimeter is calibrated to show the amount of light absorbed or transmitted by the sample being studied. It has been proved that the fraction of light absorbed depends on the concentration of the coloured substance in the solution being tested. Thus, it is possible to calculate the concentration of the solution.

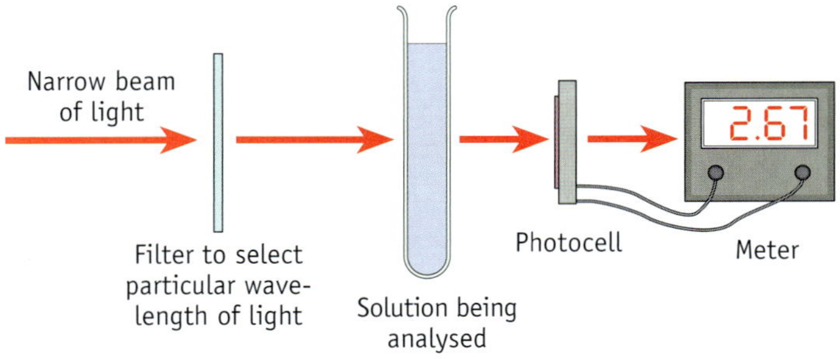

Fig 19.21
Colorimetry may be used to estimate the concentration of a coloured substance in solution.

Apparatus required

Colorimeter.

Chemicals required

Deionised water, calcium hypochlorite, fresh sample of swimming-pool water (or dilute solution of bleach), DPD number 1 tablet (for 'free chlorine').

Fig 19.22
This colorimeter is being used to measure the amount of chlorine in swimming-pool water.

Procedure

The instructions described here are for the Pasco CI-6747 colorimeter (Fig 19.22).

1. Dissolve 0.01 g calcium hypochlorite in deionised water and make the solution up to one litre. This gives a stock solution of 5 ppm free chlorine. This stock solution is now diluted with the aid of a graduated pipette to give a series of standard solutions as follows:

   10 cm^3 stock solution + 0 cm^3 deionised water gives a solution of 5 ppm free chlorine
   8 cm^3 stock solution + 2 cm^3 deionised water gives a solution of 4 ppm free chlorine
   6 cm^3 stock solution + 4 cm^3 deionised water gives a solution of 3 ppm free chlorine
   4 cm^3 stock solution + 6 cm^3 deionised water gives a solution of 2 ppm free chlorine
   2 cm^3 stock solution + 8 cm^3 deionised water gives a solution of 1 ppm free chlorine
   Each of the stock solutions may be made up in an ordinary laboratory sample bottle (with a lid).

2. Add a DPD number 1 tablet ('free chlorine') to each of the above solutions and allow time for the tablet to dissolve in each solution. If free chlorine is present, a pink colour will be observed in each solution.

3. When the tablets have dissolved, fill a cuvette to about 75% of its capacity with a sample of each solution. If air bubbles are observed in the solution, gently tap the side of the cuvette to dislodge the bubbles. Using a piece of tissue paper, wipe the sides of the cuvette clean. Using a marker, write the concentration of each solution on the lid of each cuvette.
4. Calibrate the colorimeter as follows:
   (a) Press the blue button marked ON/OFF to turn on the colorimeter. A message will be observed 'please calibrate'.
   (b) Press the SELECT and START/STOP buttons at the same time.
   (c) Insert the cuvette filled with deionised water making sure that the clear sides of the cuvette are in the line of the light beam (left to right).
   (d) Slide the door shut and press the SELECT button. The instrument is now calibrating itself at various wavelengths. When this is completed the message 'cal done' appears.
5. Press the SELECT button a number of times until 'green 565 nm' appears on the screen. You have now selected light of this particular wavelength to pass through the various solutions.
6. Insert the first solution and press START. Allow the digital display to stabilise and note the % transmittance reading. Enter this reading in a table like that shown below.

CONCENTRATION (ppm)	% TRANSMITTANCE	ABSORBANCE
1		
2		
3		
4		
5		

7. Remove the cuvette and repeat the procedure for the remaining four solutions. The absorbance value for each solution is calculated from the % transmittance using the relationship:
   Absorbance = 2 – log(% transmittance)
8. Plot a graph of absorbance (on the y-axis) against concentration (on the x-axis). A graph similar to that shown in FIG 19.23 will be obtained.

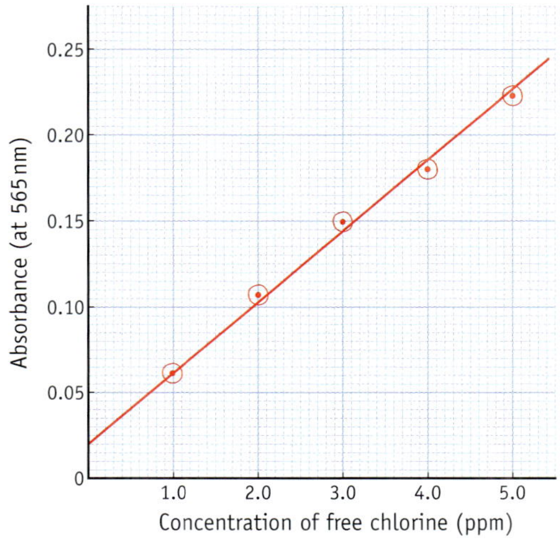

Fig 19.23
This graph of concentration of free chlorine vs absorbance may be used to calculate the concentration of chlorine in the sample being tested.

9. Insert a sample of the swimming-pool water into the colorimeter and note the percentage transmittance. Convert this to absorbance and read off the corresponding concentration from the graph. You now know the concentration of free chlorine in the water.

Note: This colorimeter may also be used in conjunction with a computer and the *DataStudio* software from Pasco Scientific. In this case, the absorbance is calculated automatically by the computer and the graph is also automatically drawn.

As well as analysing water for the presence of chlorine, colorimeters may also be used to analyse for the presence of a number of other substances in water, e.g. nitrates and phosphates (from fertilisers). (FIG 19.24)

Fig 19.24
This colorimeter is being used to test for the presence of nitrates in water

Test Yourself:
Attempt questions 19.11 – 19.12 and W19.9.

CHAPTER SUMMARY

- Hardness in water is caused by the presence of Ca^{2+} or Mg^{2+} ions.
- Temporary hardness may be removed by boiling the water.
- Permanent hardness is removed by ion exchange.
- The total hardness in a sample of water may be determined using edta.
- There are seven stages involved in the treatment of water.
- B.O.D. stands for Biochemical Oxygen Demand.
- Eutrophication is the enrichment of water with nutrients, which leads to excessive growth of algae.
- The Winkler method is used to measure the amount of dissolved oxygen in a sample of water.
- There are three stages involved in the treatment of sewage.
- Atomic absorption spectrometry is used to test for the presence of heavy metals in water.
- The concentration of chlorine in water may be determined using a comparator method or a colorimeter.

EXERCISES

19.1. (a) The way in which water circulates around the earth is represented by the ....
(b) Hard water is water which ....
(c) Why is the word 'temporary' used to describe a certain type of hardness found in water?
....
(d) Temporary hardness is found in water in limestone districts. Show, by means of an equation, how the hardness gets into the water. .... *[LCH]*
(e) What causes permanent hardness in water? .... *[LCH]*
(f) Give one advantage and one disadvantage of hard water. ....
(g) Why is washing soda rarely used nowadays for washing clothes? ....

19.2. What is meant by hardness of water? Distinguish clearly between temporary and permanent hardness. Which calcium compound is a cause of temporary hardness? How does this compound get into the water?
A whitish deposit ('fur' or 'scale') is often found on the insides of kettles in hard-water districts. Write an equation for the reaction involved in producing this deposit.

19.3. A 100 cm^3 sample of water was analysed in order to find the hardness of the water sample. On titrating the sample with 0.025 M edta, it was discovered that 25 cm^3 of the solution was required for complete reaction. Calculate the hardness of the water expressing your result as p.p.m. of $CaCO_3$.

19.4. From a bucket of spring water, containing dissolved calcium chloride and calcium hydrogencarbonate, 500 cm^3 of the water were accurately measured into a beaker and then boiled for about 30 minutes. When cool, the water from the beaker was filtered into a 500 cm^3 volumetric flask. Deionised water was then poured through the filter paper into the flask and finally the flask was made up carefully to the mark, again using deionised water. The water from the volumetric flask was then titrated in 100 cm^3 volumes with a 0.01 M solution of edta. The mean titration reading was 6.0 cm^3. When the measurement was repeated using 100 cm^3 volumes of the unboiled water from the bucket, the mean titration reading was 19.0 cm^3.

(i) Give the full name of edta. Suggest a suitable container for storing edta solutions.
(ii) Name the indicator you would use for the titration and state the colour change at the end point.
(iii) In order to ensure an accurate end point, a little of another solution must be added to the water. What type of solution is it and what purpose does it serve?
(iv) Write an equation for the reaction that took place when the water was boiled.
(v) Explain why it was necessary (a) to bring the volume back to 500 cm^3, (b) to use deionised water for this purpose. Why was deionised water poured through the filter paper into the volumetric flask?
(vi) Using H_2X^{2-} to represent the edta anion, the titration reaction may be represented as:
$H_2X^{2-} + Ca^{2+} \rightarrow CaX^{2-} + 2H^+$
Calculate the total hardness of the water in the bucket expressed as parts per million (p.p.m.) of calcium carbonate. Calculate also the hardness that was due to the presence of the calcium hydrogencarbonate. *[LCH]*

19.5. What is hard water?
A bucket of water from a stream was divided into three portions, A, B and C. A was left untreated, B was boiled and allowed to cool, and C was passed through a column of deionising resin. The hardness in the three portions was then determined by a titration experiment and the results (expressed in parts per million of calcium carbonate) are given in the following table:

Portion	A	B	C
Hardness	250	110	5

(i) State (a) the standard solution, (b) the indicator, used in the titration. What was the colour change at the end point?
(ii) Give the names or formulas of two compounds which could have been responsible for the hardness of the water in A.
(iii) Write an equation for a reaction that could explain the drop in hardness from A to B.
(iv) Explain why there was virtually no hardness left in C. *[LCH]*

19.6. A student filtered 250 cm^3 of river water and found that the mass of the filter paper after drying had increased by 0.52 g. 100 cm^3 of the filtered water were then evaporated to dryness in a beaker and it was found that the mass of the beaker had increased by 0.34 g.
Calculate the amount of suspended solids and the amount of dissolved solids in the water expressed as parts per million.

19.7. A sample of water was analysed, in order to determine the amount of dissolved oxygen in the water, using the Winkler method. The main reactions may be represented as follows:
$4Mn(OH)_2 + O_2 \rightarrow 2Mn_2O_3 + 4H_2O$
$Mn_2O_3 + 2I^- + 6H^+ \rightarrow 2Mn^{2+} + I_2 + 3H_2O$
The volume of water taken was 100 cm^3 and this required 8.8 cm^3 of 0.02 M sodium thiosulfate solution. Calculate the amount of dissolved oxygen in the water in p.p.m.

19.8. To determine the B.O.D. (Biochemical Oxygen Demand) of lake water, a sample was well aerated by shaking it thoroughly and then divided into two portions. One was immediately analysed for dissolved oxygen by the Winkler method; the other was kept for a period of time under certain conditions and then analysed for dissolved oxygen by the same method.

(i) Why was the sample well aerated?

(ii) In carrying out the analyses, the addition of two reagents to the water resulted in the formation of a brown precipitate. What were the two reagents?

(iii) For how long and under what conditions was the second portion kept before analysis? How was the B.O.D. calculated from the results of the two analyses?

19.9. A laboratory technician analysed a sample of river water, in order to determine the amount of dissolved oxygen in the water, using the Winkler method. The main reactions may be represented as follows:
$4Mn(OH)_2 + O_2 \rightarrow 2Mn_2O_3 + 4H_2O$
$Mn_2O_3 + 2I^- + 6H^+ \rightarrow 2Mn^{2+} + I_2 + 3H_2O$
The volume of water taken was 150 cm^3 and the liberated iodine required 21.7 cm^3 of 0.01 M sodium thiosulfate solution. Calculate the amount of dissolved oxygen in the water in p.p.m.

19.10. A sample of polluted water was analysed, in order to determine the amount of dissolved oxygen in the water, using the Winkler method. The main reactions may be represented as follows:
$4Mn(OH)_2 + O_2 \rightarrow 2Mn_2O_3 + 4H_2O$
$Mn_2O_3 + 2I^- + 6H^+ \rightarrow 2Mn^{2+} + I_2 + 3H_2O$
A 50 cm^3 sample of the water was diluted to one litre with pure well-oxygenated water.
300 cm^3 of this diluted solution were analysed by the Winkler method and 14.7 cm^3 of 0.02 M sodium thiosulfate were required.
A second 300 cm^3 sample was stored in the dark at 20 °C for five days and then analysed for dissolved oxygen. The liberated iodine required 4.8 cm^3 of 0.02 M sodium thiosulfate.
Calculate the amount of dissolved oxygen in each sample of water. Express your answers in p.p.m. Hence find the B.O.D. of the water.

19.11. (a) Name the three stages involved in the treatment of sewage and give a brief account of each stage.

(b) What is meant by the term *heavy metal* as used by chemists? Give one example of such a metal. Why is pollution by heavy metals a cause of such concern? Give one reason why these heavy metals may be found in rivers and lakes. How are heavy metal ions usually removed from water?

19.12. A student was asked to determine the concentration of chlorine in a sample of water taken from a swimming pool using either a comparator method or a colorimeter.

(i) Why is the method called a comparator method?

(ii) The contents of a sachet were added to a sample of the water. The active ingredient in the sachet is usually referred to in terms of a three-letter abbreviation. What is this abbreviation?

(iii) What colour does the water turn when the solid chemical is added to the water?

(iv) Describe the subsequent steps taken to measure the concentration of chlorine in the water.

Electrochemistry I

20

20.1 INTRODUCTION

On 6 October 1807, a young English chemist called Humphry Davy was working in the laboratory of the Royal Institution in London. Davy was only 28 years of age and was performing an experiment which was to open up a new area of chemistry and lead to the discovery of a number of new elements. Davy melted some potassium hydroxide, KOH, and decided to see what would happen if he passed electricity through the molten potassium hydroxide. To his delight, he saw tiny globules of a silvery metal break through the surface and burst into lilac flames. It is reported that he was as excited as a child and danced around the room in delight (FIG 20.1). Davy had just discovered the element potassium! A few days later, he discovered sodium by passing electricity through molten sodium hydroxide.

Fig 20.1
Humphry Davy discovered potassium by passing an electric current through molten potassium hydroxide.

Davy was carrying out research into a new branch of chemistry that is called **electrochemistry**. Electrochemistry may be divided into two areas: the use of an electric current to bring about chemical reactions (electrolysis) and the use of chemical reactions to produce an electric current (i.e. cells or batteries). Let us now consider each of these two areas of electrochemistry.

20.2 ELECTROLYSIS

We have seen above that Davy passed an electric current through some molten salts and brought about some chemical reactions. These reactions resulted in potassium and sodium being formed. This technique is called electrolysis.

> Electrolysis is the use of electricity to bring about a chemical reaction.

The term electrolysis was introduced into the language of chemistry by Davy's assistant, Michael Faraday, an even more brilliant chemist than Davy himself. (Faraday was said to be Davy's greatest discovery!) The word electrolysis comes from two Greek words *electro* and *lysis* which roughly translated mean 'splitting up using electricity'. The substance through which the electric current is passed is called the **electrolyte**. An electrolyte is a compound which, when made molten or dissolved in water will conduct an electric current. We will see later that the conduction of electricity through an electrolyte is due to the presence of ions. The two rods that dip into the electrolyte and make electrical contact with it are called the **electrodes** (FIG 20.2). In most cases, the electrodes do not react with the electrolyte into which they are dipping, i.e. the electrodes merely carry the current in and out of the electrolyte. Such electrodes are called **inert electrodes**. The most common inert electrodes are graphite and platinum. In some cases, the electrodes react with the electrolyte. Such electrodes are called **active electrodes,** e.g. copper, iron, etc.

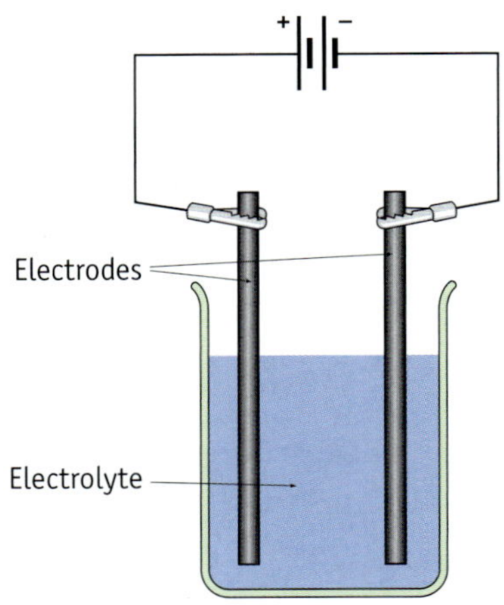

Fig 20.2
Two electrodes are used to make electrical contact with the electrolyte. Electrodes are often made of carbon or platinum since these are unreactive and do not react with the electrolyte.

In this chapter we will consider what happens when an electric current is passed through certain solutions. In the second part of Option 2 (CHAPTER 28) we will consider what happens when an electric current is passed through molten (melted) substances.

EXPERIMENT 20.1

To investigate what happens when an electric current is passed through a solution of potassium iodide using inert electrodes.

Electrolysis reactions may be easily demonstrated on the overhead projector. Some bench dilute potassium iodide solution is placed in a petri dish. Two lengths of platinum wire are used as electrodes. The electrodes are connected to a power supply unit (0 – 20 volts, DC) and the voltage adjusted to about 12 volts. To help us understand the reactions brought about by electrolysis, a few drops of phenolphthalein indicator are added to the petri dish before the current is switched on. When the current is switched on, we observe a pink colour appearing at the negative electrode and a brown colour appearing at the positive electrode (FIG 20.3).

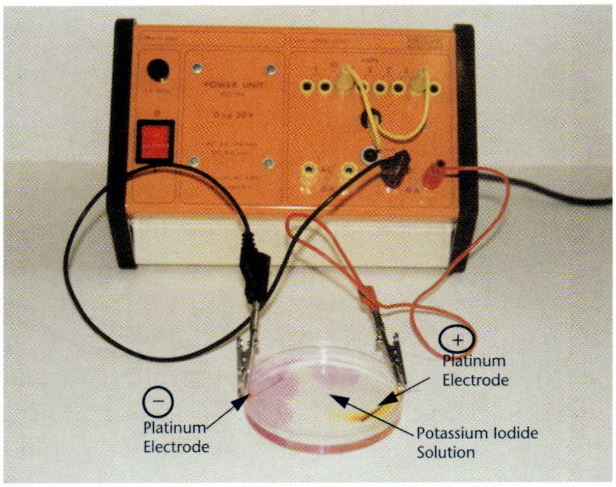

Fig 20.3
As electrolysis takes place, a pink colour appears at the negative electrode and a brown colour appears at the positive electrode.

On analysing the contents of the petri dish, it is found that there is a build up of iodine around the positive electrode – hence the brown colour. Chemists explain this by saying that the negative iodide ions in the solution are attracted to the positive electrode ('unlike charges attract'). At the positive electrode, each iodide ion loses an electron to form an iodine atom. The iodine atoms then join together to form iodine molecules.

Positive electrode: $2I^- \rightarrow I_2 + 2e^-$

↑ loss of electrons (oxidation)

At the negative electrode, bubbles of gas are seen to be evolved. On analysis, the gas is found to be hydrogen. The fact that the phenolphthalein turns pink implies that the solution is turning alkaline (OH^- ions present). Phenolphthalein is pink in a base and colourless in an acid. These facts lead chemists to conclude that the following reaction occurs at the negative electrode:

Negative electrode: $2H_2O + 2e^- \rightarrow H_2 + 2OH^-$

↑ gain of electrons (reduction)

Thus, what is actually happening at the negative electrode is that water molecules are forced to accept electrons from the negative electrode. These extra electrons cause the molecules to be destabilised and they break up to form hydrogen gas and hydroxide ions.

From the above experiment, we see that what is essentially happening in any electrolysis reaction is that chemical reactions occur at each electrode. The battery 'pumps' electrons on to the negative electrode where they are **gained** by some species, i.e. a **reduction** reaction occurs at the negative electrode. At the positive electrode, a chemical reaction occurs which involves a **loss** of electrons, i.e. an **oxidation** reaction occurs. If you remember that **O**xidation occurs at the **P**ositive electrode and **R**eduction occurs at the **N**egative electrode (OPRN), it will help you to remember the electrolysis reactions on your course.

Since electrons are being removed from one electrode and given up at the other electrode, there is a completed circuit and the bulb lights (FIG 20.4).

Note: You might be inclined to think that the reaction at the negative electrode should involve positive potassium ions gaining electrons to form potassium atoms. However, this does not occur as it requires a huge amount of energy for the reaction $K^+ + e^- \rightarrow K$ to take place. Potassium is such a reactive metal that it is far more stable in the form of K^+ ions. (However, it is possible to get the reaction to occur under high voltages and if no water is present. This is covered in more detail in CHAPTER 28.) Since the potassium ions play no part in the reaction, such ions are often referred to as 'spectator ions'.

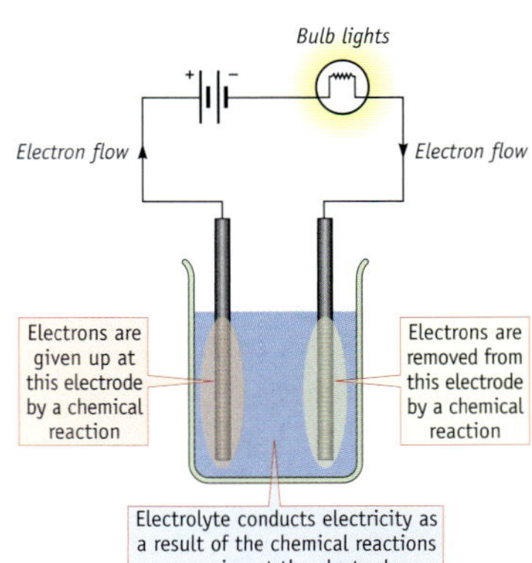

Fig 20.4
Chemical reactions cause electrons to be removed from the negative electrode and given up at the positive electrode. Thus, the electrolyte conducts electricity and the bulb lights since there is a completed circuit.

EXPERIMENT 20.2

To investigate the electrolysis of acidified water using inert electrodes

In this experiment, an electric current is passed through water to which some dilute sulfuric acid has been added. The sulfuric acid is added to assist the water to conduct electricity. A convenient apparatus in which the experiment can be carried out is the Hofmann voltameter or an apparatus that uses inverted test-tubes, (FIG 20.5). (A voltameter is an apparatus used to study the conduction of electricity through an electrolyte. Do not confuse this name with the name voltmeter, i.e. an instrument used to measure voltage.)

On passing the electricity through the water, it is found that the water has been split up into two gases. Oxygen is evolved at the positive electrode and hydrogen is evolved at the negative electrode. The oxygen may be tested for by checking to see if a glowing splint is rekindled. The hydrogen gas may be collected in an inverted test-tube. When ignited, it gives a characteristic 'pop'. The ratio of hydrogen gas to oxygen gas is 2 : 1 as predicted by the formula H_2O. The following reaction occurs at the negative electrode:

Negative electrode: $2H^+ + 2e^- \rightarrow H_2$

i.e. the H^+ ions from the sulfuric acid are attracted to the negative electrode. At the negative electrode, each H^+ ion accepts an electron to form a H atom (reduction). Two hydrogen atoms then combine to form a hydrogen molecule.

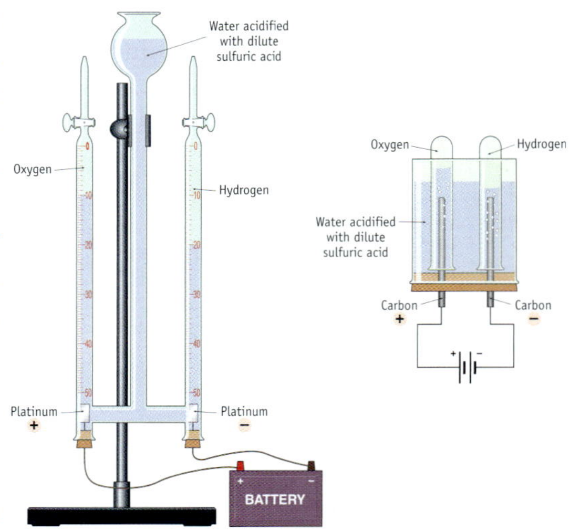

Fig 20.5
Electrolysis of water may be carried out in a Hofmann voltameter or using inverted test-tubes.

The following reaction occurs at the positive electrode:

Positive electrode: $H_2O \rightarrow 2H^+ + \frac{1}{2}O_2 + 2e^-$

i.e. electrons are removed from the water molecules around the positive electrode (oxidation). This destabilises the water molecule and causes it to break down as shown above. Thus, oxygen gas is evolved at the positive electrode.

Note that two H^+ ions are removed at the negative electrode and that two H^+ ions are produced at the positive electrode. Thus, the amount of sulfuric acid in the voltameter remains constant and the overall result is that water is broken down into hydrogen gas and oxygen gas.

Overall reaction: $H_2O \rightarrow H_2 + \frac{1}{2}O_2$

Note: You may wonder why the SO_4^{2-} ions from the sulfuric acid do not lose their two negative charges by a chemical reaction at the positive electrode. This does not happen because the SO_4^{2-} ions are very stable and cannot be forced to react at this low voltage. (If we had used hydrochloric acid, the Cl^- ions in solution would lose their electrons at the positive electrode. This would interfere with the electrolysis of water and would cause chlorine gas to be evolved.)

EXPERIMENT 20.3

To investigate the electrolysis of sodium sulfate solution using inert electrodes.

This electrolysis reaction emphasises the fact that neutral molecules may react at the electrodes during electrolysis. An electric current is passed through a dilute solution of sodium sulfate using platinum electrodes. A few drops of universal indicator solution are added to the sodium sulfate solution. (Universal indicator is red in acidic solution, blue in basic solution and green in neutral solution.) As the electrolysis proceeds, the green universal indicator turns red at the positive electrode and blue at the negative electrode. The sponge placed in the centre of the petri dish helps to keep the colours clearly separated (FIG 20.6).

Fig 20.6
The electrolysis of sodium sulfate.

How do we explain our observations? The fact that the universal indicator solution turns red at the positive electrode implies that H^+ ions are being formed. Bubbles of gas are seen to be evolved at the positive electrode. On analysis, this gas is found to be oxygen. All of these facts may be explained by the following reaction:

Positive electrode: $2H_2O \rightarrow O_2 + 4H^+ + 4e^-$

Thus, four electrons are lost (oxidation) from two water molecules which are destabilised and break down to give an oxygen molecule and four hydrogen ions.

How do we explain what happens at the negative electrode? The fact that the universal indicator solution turns blue at the negative electrode implies that OH^- ions are being formed. Bubbles of gas are seen to be evolved at this electrode. On analysis, this gas is found to be hydrogen. All of these facts may be explained by the following reaction:

Negative electrode: $2H_2O + 2e^- \rightarrow H_2 + 2OH^-$

Thus, two electrons are gained (reduction) by two water molecules which are destabilised and break down to give a hydrogen molecule and two hydroxide ions.

Note: The Na^+ ions and the SO_4^{2-} ions of the sodium sulfate do not take part in the electrolysis reactions. This is because they are both very stable ions and cannot be forced to react at this low voltage.

EXPERIMENT 20.4

To investigate the electrolysis of copper sulfate solution using copper electrodes.

So far we have studied electrolysis reactions using inert electrodes, i.e. the electrodes merely carry the current in and out of the electrolyte but do not take any part in the reaction. We now consider the electrolysis of copper sulfate solution using electrodes made of copper (FIG 20.7). Such electrodes are often called **active electrodes** because they play a part in electrolysis reactions.

As the electrolysis proceeds, the copper plate connected to the negative end of the battery gains weight and a fresh coating of copper metal is deposited on the plate. This is explained by saying that the positive Cu^{2+} ions from the copper sulfate solution are attracted towards the negative electrode. The following reaction takes place at the negative electrode:

Negative electrode: $Cu^{2+} + 2e^- \rightarrow Cu\downarrow$

In other words, each copper ion accepts two electrons to form a copper atom. Thus, copper metal is plated on to the negative electrode.

The SO_4^{2-} ions in solution are attracted towards the positive electrode but because these ions are so stable, they do not react. Instead, it is found that the following reaction occurs:

Positive electrode: $Cu \rightarrow Cu^{2+} + 2e^-$

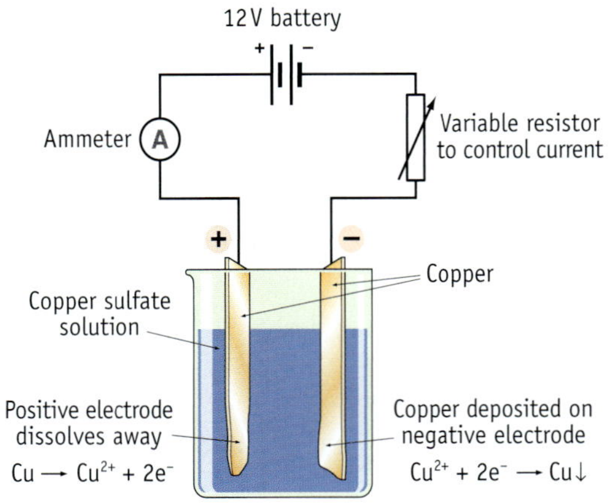

Fig 20.7
When copper sulfate is electrolysed using copper electrodes, the mass of the positive electrode decreases and the mass of the negative electrode increases.

In other words, copper atoms in the positive electrode lose two electrons and go into solution as Cu^{2+} ions. The 'driving force' for this reaction is the need to replace the Cu^{2+} ions being removed from solution at the negative electrode. Thus, the positive electrode is slowly eaten away and copper is continually coated on to the negative electrode.

The electrode at which oxidation occurs is called the **anode**. The electrode at which reduction occurs is called the **cathode**. An important point to remember is that 'the anode is eaten away'.

The electrolysis of copper sulfate solution using copper electrodes is used to purify copper for the electronics industry. This industry needs very pure copper because impurities reduce conductivity. The impure copper is made the positive electrode. A thin sheet of very pure copper is made the negative electrode. As a current is passed through the cell, pure copper is plated on to the negative electrode. The impurities fall to the bottom of the cell (FIG 20.8).

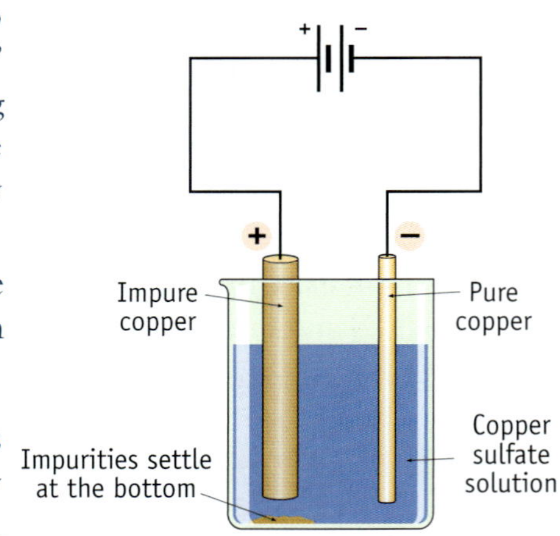

Fig 20.8
Impure copper is purified for the electronics industry by electrolysis.

EXPERIMENT 20.5

To demonstrate the movement of ions under the influence of an electric field.

Solid sodium chloride will not conduct an electric current because all of the ions are in fixed positions and cannot move. However, if the sodium chloride is dissolved in water or is melted, the ions are now free to move. This means that the ions can migrate to the electrodes where oxidation or reduction reactions take place, i.e. electrolysis can take place and the electrolyte conducts an electric current.

Thus, the reactions that take place during electrolysis may be explained in terms of ions moving towards the electrodes. To demonstrate this movement, one could use ions of different colour. For example, a green coloured solution is made up of blue Cu^{2+} ions (from copper sulfate) and yellow CrO_4^{2-} ions (from potassium chromate) dissolved in concentrated ammonia solution. A drop of the solution is placed on a piece of moist filter paper wrapped around a glass slide as a support. The filter paper is connected to a 20-volt supply as shown in Fig 20.9. Note that the blue Cu^{2+} ions (positive) move towards the negative electrode and the yellow CrO_4^{2-} (negative) ions move towards the positive electrode.

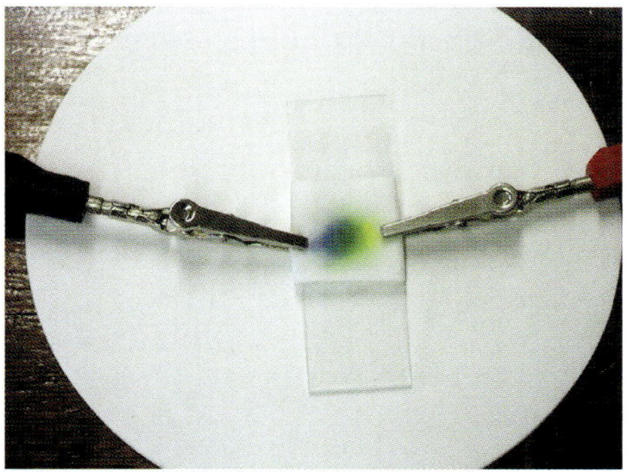

Fig 20.9
The blue Cu^{2+} ions (positive) move towards the negative electrode.
The yellow CrO_4^{2-} ions (negative) move toward the positive electrode.

20.3 ELECTROPLATING

Some very beautiful metals like silver and gold are very expensive. It is unlikely that objects of jewellery that you might own are made of pure silver or pure gold. Such items are usually made of a less expensive metal and are then **plated** with silver or gold. It is important that the layer of silver or gold should stick well to the surface. Also, the coating of the metal should be very thin to keep the cost of the item low. Chemists have found that electrolysis is an ideal way to plate an expensive metal on to a less expensive metal. The technique is called **electroplating**.

Electroplating is a process where electrolysis is used to put a layer of one metal on the surface of another.

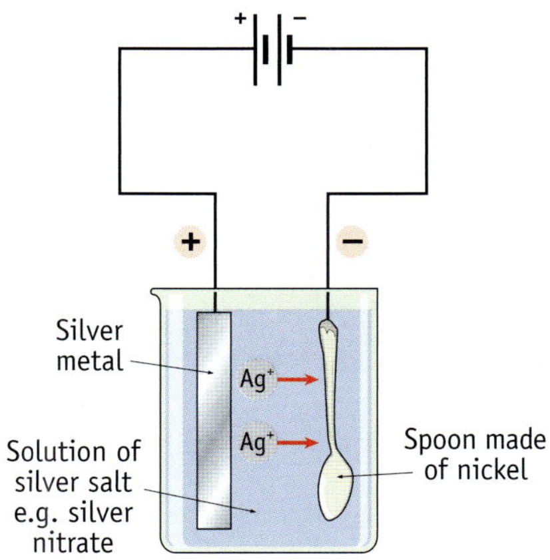

Fig 20.10
The spoon is being electroplated with silver metal.

To understand how electroplating works consider Fig 20.10 where the spoon is being coated with silver. To electroplate any object, the following conditions must be fulfilled:

1. The object to be plated must be connected to the negative terminal of the battery, i.e. the spoon is made the negative electrode.
2. The electrolyte must be a salt of the metal that is being plated on to the object, i.e. a solution of a salt like silver nitrate is the electrolyte.
3. The positive electrode must be the same metal as that being plated out, i.e. a bar of silver metal is used as the positive electrode.

The reactions that occur are as follows:

Negative electrode: $Ag^+ + e^- \rightarrow Ag\downarrow$

This is explained by saying that the positive Ag^+ ions from the silver nitrate solution are attracted towards the negative electrode. In other words, each silver ion accepts an electron to form a silver atom. Thus, silver metal is plated on to the negative electrode.

Positive electrode: $Ag \rightarrow Ag^+ + e^-$

In other words, the silver metal of the positive electrode dissolves in the electrolyte to form silver ions. This replaces the silver ions removed from the solution at the negative electrode. Thus, the positive electrode is slowly eaten away.

Some silver plated objects may have the letters EPNS printed on them. This stands for Electro Plated Nickel Silver, i.e. silver has been plated on to nickel (by electrolysis) (FIG 20.11).

Fig 20.11
This cutlery has been plated with silver by electrolysis.

Electroplating is not only used to make an object appear more attractive, it is also used to protect an object against corrosion. For example, bath fittings (taps, towel rails, etc.) are made of steel which has been coated with chromium. Also, the handlebars of your bicycle are probably made of steel coated with chromium. (In fact, chemists find that chromium does not stick to steel very well during electroplating. The steel is first plated with nickel which forms a very strong deposit on the steel. The steel is then plated with chromium.) Steel may be plated with nickel using a salt like nickel sulfate, $NiSO_4$ (FIG 20.12).

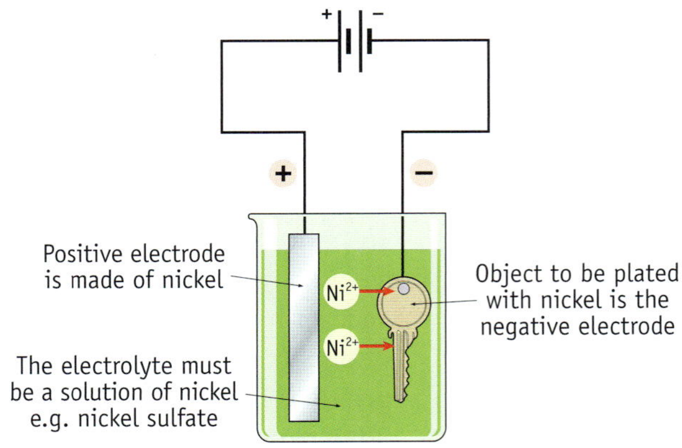

Fig 20.12
Electrolysis is being used to coat this key with nickel.

Silver, chromium, nickel and zinc are the metals most commonly electroplated on to cheaper metals.

Test Yourself:
Attempt questions 20.1 – 20.5 and W20.1 – W20.3

20.4 THE ELECTROCHEMICAL SERIES

In our study of electrochemistry so far, we have seen how an electric current may be used to bring about a chemical reaction. However, it is also possible to do the reverse, i.e. to produce an electric current using a chemical reaction. In 1771, an Italian scientist called Luigi Galvani discovered that the muscles in a dissected frog's leg twitched when touched by two different metals. Some years later, another Italian scientist called Alessandro Volta showed that electricity is produced whenever two different metals are immersed in a conducting solution. For example, if a bar of zinc is placed in a solution of zinc ions and a bar of copper is placed in a solution of copper ions as shown in FIG 20.13, it is observed that a current flows. A cell in which a chemical reaction results in the production of an electric current is called a **galvanic cell** or **voltaic cell**.

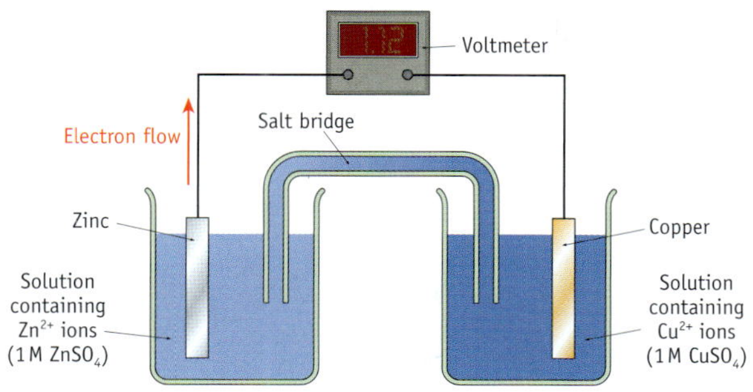

Fig 20.13
An electric current is generated by this cell.

In the above cell, a voltage of about 1.1 volts registers on the voltmeter. How does the current arise? At the zinc electrode, there is a tendency for zinc atoms to lose electrons and form Zn^{2+} ions, i.e. an oxidation reaction takes place:

$Zn \rightarrow Zn^{2+} + 2e^-$

These electrons flow around the wire to the copper electrode. At the copper electrode, copper ions accept electrons to become copper atoms, i.e. a reduction reaction takes place:

$Cu^{2+} + 2e^- \rightarrow Cu\downarrow$

(The salt bridge in the cell helps to maintain a balance of ions in each part of the cell. The salt bridge can simply consist of a piece of filter paper that has been soaked in a substance like potassium chloride or potassium nitrate solution.)

We have seen that the voltage generated by the zinc-copper cell is 1.1 volts. If we could assign a particular fraction of this voltage to each individual half-cell, we would then have a measure of the tendency for each of the above two reactions to occur. However, you cannot measure the voltage of a single half-cell because current will only flow when **two** half-cells are connected together. Chemists overcome this problem by choosing a particular half-cell as the standard with which to compare all half-cells. The half-cell chosen as the reference electrode is the hydrogen half-cell. This consists simply of a piece of platinum immersed in a 1 M solution of H^+ ions. Hydrogen gas (at a pressure of 1 atm) is then bubbled over the platinum electrode.

Some metals have a greater tendency to lose their outer electrons than other metals, e.g. if zinc is replaced by magnesium in the cell in FIG 20.13, a much higher voltage is observed. This is because magnesium has a far greater tendency to lose electrons than zinc. Chemists use the term **electrode potential** to describe the tendency of a metal to lose electrons, i.e. the tendency for a metal to be oxidised. Standard electrode potential values are obtained by measuring the voltage when the half- cell is connected to a hydrogen electrode as shown in FIG 20.14.

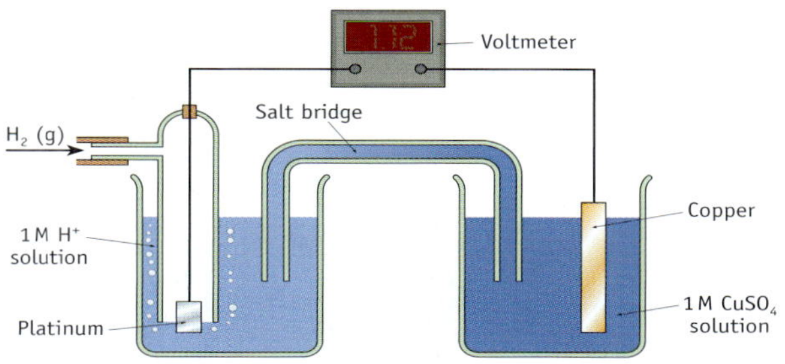

Fig 20.14
This cell is used to measure the standard electrode potential of copper.

Each element in turn may be connected to the hydrogen electrode and the standard electrode potential of the element measured. It is then possible to draw up a table of the elements in order of their standard electrode potentials. Such a table is called the **electrochemical series**.

The electrochemical series is a list of elements in order of their standard electrode potentials.

Part of the electrochemical series is shown in TABLE 20.1. Elements near the top of the table will have a great tendency to lose electrons to form positive ions. Hence, these elements tend to be very reactive. Elements near the bottom of the table have a small tendency to lose electrons and are therefore very unreactive.

Element	
Potassium	Metals near the top of the series have the greatest tendency to form positive ions in solution, i.e. they are very reactive.
Calcium	
Sodium	
Magnesium	
Aluminium	
Zinc	
Iron	REACTIVITY DECREASING ↓
Lead	
Hydrogen	
Copper	
Mercury	
Silver	Metals near the bottom of the series tend to be very unreactive.
Gold	

Table 20.1.
Part of the electrochemical series.

Note: You may have noticed that the Electrochemical Series is very similar to the Activity Series, part of which you would have studied in your Junior Certificate course. However, the order of elements is not identical in both cases. This is because the Activity Series is based on chemical reactivity while the Electrochemical Series is based on values of standard electrode potential. This causes some elements (e.g. calcium) to have different positions in each list. (The mnemonic ***Pretty Sally Could Marry A Zulu In Lovely Honolulu Causing Many Strange Gazes*** still holds – except for calcium!)

One of the great uses of the electrochemical series is that it helps us to predict the results of displacement reactions in solution. It is found that a metal will displace another metal that is below it in the electrochemical series from a solution of its ions. For example, if a piece of magnesium metal is dipped into a solution of copper ions, the magnesium metal becomes coated with copper. This is because the following reaction occurs:

$Mg + Cu^{2+} \rightarrow Mg^{2+} + Cu\downarrow$

Magnesium is above copper in the electrochemical series. Therefore, magnesium has a greater tendency than copper to lose electrons, i.e. the magnesium forces the Cu^{2+} ions to accept electrons. Thus, the copper is displaced out of solution as copper metal.

Another example of a displacement reaction is when steel wool (iron) is placed in a boiling tube containing some Cu^{2+} ions. Since iron is above copper in the electrochemical series, the Cu^{2+} ions are displaced out of solution as copper metal (FIG 20.15).

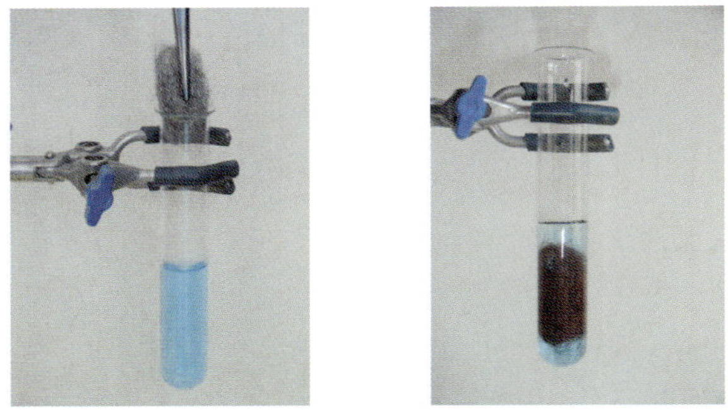

Fig 20.15
The iron in the steel wool displaces the Cu^{2+} ions out of solution as Cu metal.

One example of an application of this displacement reaction is the use of scrap iron to extract copper metal from a solution of copper salts:

$Fe + Cu^{2+} \rightarrow Fe^{2+} + Cu\downarrow$

Thus, as a general rule, an element higher up in the electrochemical series will displace one lower down in the series from a solution of its ions.

If you study Option 2, you will be meeting the electrochemical series again in CHAPTER 28 in the accompanying workbook. You will see how a knowledge of this series is of great help in understanding how metals are extracted from their ores.

Test Yourself:
Attempt questions 20.6 – 20.7 and W20.4 – W20.5.

CHAPTER SUMMARY

- Electrolysis is the use of electricity to bring about a chemical reaction.
- An electrolyte is a substance that conducts electricity as a result of the presence of ions.
- Electrodes may be either active or inert.
- When a solution of potassium iodide is electrolysed using inert electrodes, iodine is formed at the positive electrode and hydroxide ions are formed at the negative electrode.
- When acidified water is electrolysed using inert electrodes, oxygen gas is evolved at the positive electrode and hydrogen gas is evolved at the negative electrode.
- When a solution of sodium sulfate is electrolysed using inert electrodes, oxygen gas is evolved at the positive electrode and hydrogen gas is evolved at the negative electrode.
- When a solution of copper sulfate is electrolysed using copper electrodes, the positive electrode is eaten away and fresh copper is plated on to the negative electrode.
- Electroplating is a process where electrolysis is used to put a layer of one metal on the surface of another.
- The electrochemical series is a list of elements in order of their standard electrode potentials.

EXERCISES

20.1. (a) The chemist who used electrolysis to discover potassium and sodium was ....

(b) What is meant by the term 'inert' when applied to electrodes. ....

(c) When a solution of potassium iodide is electrolysed, at which electrode does a brown colour appear? ....

(d) Chromium plating is often used on door handles. What advantage has plating over painting? ....

(e) What is the meaning of the letters EPNS? ....

(f) Crystals of sodium chloride will not conduct electricity. However, sodium chloride solution will conduct electricity. Explain ....

20.2. Explain the meaning of the terms electrolysis, electrolyte, electrode.
A solution of potassium iodide to which a few drops of phenolphthalein indicator have been added is electrolysed using inert electrodes.

(a) Name a suitable material for use as electrodes.

(b) Describe what you observe as the electrolysis proceeds.

(c) For each electrode, write an ionic equation to describe the reaction happening at each electrode.

(d) In the case of the reactions described in (c), state whether each reaction is an oxidation reaction or a reduction reaction.

20.3. A student was asked to electrolyse some sodium sulfate (to which a few drops of universal indicator had been added) using inert electrodes.

(a) Draw a labelled diagram of the apparatus and the circuit.

(b) Describe what is observed in the container before electrolysis takes place and while electrolysis is taking place.

(c) Write an equation to describe the reaction taking place at the positive electrode and at the negative electrode.

(d) What is the purpose of adding the universal indicator.

(e) At which electrode does the reduction reaction take place?

20.4. Draw a labelled diagram of an apparatus and circuit that could be used to demonstrate the movement of ions under the influence of an electric field. Describe what is observed as the reaction proceeds.

20.5. A jeweller wishes to electroplate a nickel spoon with silver. Draw a labelled diagram of the apparatus and circuit. What electrolyte could be used? Why is silver plating so popular?

20.6. What is a galvanic cell? Draw a labelled diagram of any such cell. Write a brief note on the electrochemical series and explain how metals are placed in their correct order in the electrochemical series. Give one use of the electrochemical series.

20.7. Explain, with the aid of an ionic equation, what is happening in each of the following:

(a) A piece of zinc metal is added to a solution of copper sulfate and becomes coated with a brown metal and the blue colour of the solution slowly disappears.

(b) A length of copper wire is placed in a solution of silver nitrate. Shiny crystals are seen to form on the copper wire and the solution turns a blue colour.

21

Fuels and Heats of Reaction

21.1 INTRODUCTION

In this chapter and in the next two chapters we will be studying a branch of chemistry called organic chemistry. Originally, the word organic meant 'derived from plant or animal sources'. Substances like sugar, natural gas (methane), vegetable dyes, glycerol, etc. that were isolated from biological sources were called organic compounds. Substances that were isolated from mineral sources were called inorganic compounds. However, in 1828 the German chemist Friedrich Wöhler discovered that urea (a compound previously obtained only from urine) could be synthesised in the laboratory. This led to a redefinition of organic chemistry. **Organic chemistry is now defined as the study of the compounds of carbon**. (A few simple compounds containing carbon like carbon dioxide, carbon monoxide and carbonate salts are classified as inorganic compounds.)

Carbon is unique among the elements in that it can form stable chains and rings of carbon atoms. This means that there are millions of organic compounds. Thousands of new organic compounds are synthesised every year. Organic chemistry is the chemistry of pharmaceuticals, detergents, synthetic fibres, dyes, paints, fuels, explosives, adhesives, etc.

Even though there are millions of organic compounds, this does not mean that the study of organic chemistry is very difficult. In fact, the study of organic chemistry is very systematic and interesting. We begin our study by looking at a very important group of organic chemicals – hydrocarbons.

21.2 HYDROCARBONS

A **hydrocarbon** is a compound that contains only hydrogen and carbon. Common sources of hydrocarbons are coal, natural gas (methane) and petroleum (crude oil). These hydrocarbons have been formed in the earth over millions of years. It is possible to make methane quite readily using material of plant or animal origin. For example, some farmers allow manure to decay under anaerobic conditions (i.e. in the absence of air) in large tanks (FIG 21.1). Bacteria feed on the manure causing it to decay and methane is formed. Methane is a very good fuel and is burned to provide heating for the farm. The solid that remains in the tank forms an odourless compost that is used as a fertiliser. Methane forms an explosive mixture with air and this has led to many accidents because it is produced naturally in slurry pits, coal mines and refuse dumps. (Humphry Davy invented a special lamp for coal miners in 1815 to help prevent explosions caused by methane coming in contact with the flame of a candle.) Methane is a gas that is contributing to the greenhouse effect. Those of you who study OPTION 1 will learn more about this topic in CHAPTER 26.

Fig 21.1
Manure is allowed to decay in this tank and form methane. This type of tank is commonly called a 'biomass digester'. (Biomass is material of plant or animal origin.)

There are three main classes of hydrocarbons on your course. These are called **alkanes, alkenes** and **alkynes**. This classification is based upon the type of carbon-carbon bond in the molecules.

21.2.1 THE ALKANES

These compounds are a family of hydrocarbons in which all the atoms are linked by single bonds, i.e. there are no double or triple bonds in the molecules. Thus, the alkanes are said to be **saturated** compounds. The names and formulas of the first ten members of the alkane family are shown in TABLE 21.1.

Name	Molecular Formula
Methane	CH_4
Ethane	C_2H_6
Propane	C_3H_8
Butane	C_4H_{10}
Pentane	C_5H_{12}
Hexane	C_6H_{14}
Heptane	C_7H_{16}
Octane	C_8H_{18}
Nonane	C_9H_{20}
Decane	$C_{10}H_{22}$

Table 21.1
The first ten members of the alkanes.

The structural formulas of the first four alkanes are shown in FIG 21.2. The structural formula* of a compound shows how the atoms of the compound are arranged relative to one another. The full structural formula shows all the atoms and all the bonds in a compound. Note that the structural formulas do not show the actual shapes of the molecules, i.e. the molecules are drawn as 'flat' structures. In reality, each carbon atom has a tetrahedral structure as shown in the molecular models (FIG. 21.2). Studying the molecular models in your school laboratory will help you to understand this (FIG 21.3).

Name	Formula	Structural formula	Molecular model
Methane	CH_4	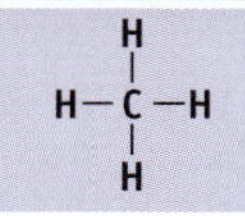	
Ethane	C_2H_6	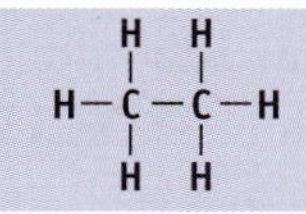	
Propane	C_3H_8	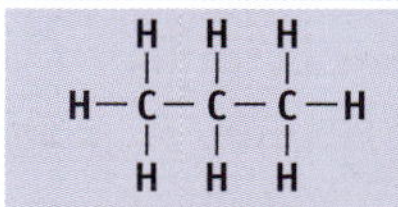	
Butane	C_4H_{10}	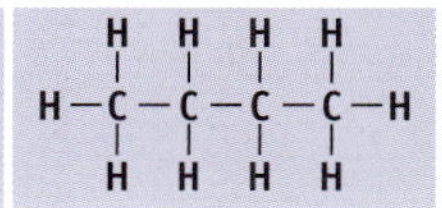	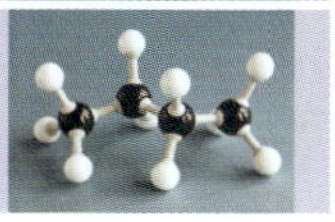

Fig 21.2
The first four membes of the alkanes.

Fig 21.3
A molecular model of octane.

Note that the name of each alkane ends in ***–ane***. The first part of the name (except for the first four) is derived from the Latin or Greek word for the number of carbon atoms in the molecules: *pent* = five, *hex* = six, *hept* = seven, *oct* = eight, non = nine, dec = ten.

The old name for the alkanes was the 'paraffins' (from the Latin *parum affinis* meaning little reactivity) because these compounds did not readily react with other compounds. Hence, many of the alkanes are found naturally occurring and they all have similar methods of preparation. In addition, all of the alkanes have similar chemical properties.

There is no need to prepare an alkane since you have one 'on tap' in your science laboratory. If your town has natural gas, this is almost pure methane. If you are using bottled gas, this is a mixture of propane and butane.

The first 4 members of the alkanes are gases, the next 12 are liquids (i.e. up to the C-16 compound) and the higher members are waxy solids – $C_{31}H_{64}$ is found in beeswax and $C_{29}H_{60}$ is found in cabbage

* The term structural formula may also be interpreted to mean a condensed structural formula, e.g. $CH_3CH_2CH_3$ for propane. However, in this textbook, the term structural formula is taken to mean the full structural formula as show in FIG 21.2.

leaves. This change in physical properties (boiling points, melting points, density, etc) is often referred to as a 'gradation in physical properties'.

You may notice from Table 21.1 that the formulas of the alkanes follow a definite pattern. The general formula for all alkanes is $\mathbf{C_nH_{2n+2}}$ where n is the number of carbon atoms. Thus, each member of the family of alkanes differs from the preceding member by a CH_2 unit. A family of compounds, like the alkanes, which have a lot of similar properties, is referred to as a homologous series.

> A homologous series is defined as:
> - A series of compounds of uniform chemical type
> - Showing gradations in physical properties
> - Having a general formula for its members
> - Each member having a similar method of preparation
> - Each member differing from the previous member by a (CH_2) unit.

In this course we will be meeting many other examples of homologous series.

Isomers

Take four carbon 'atoms' and ten hydrogen 'atoms' out of the box of molecular models in your school laboratory. Now build a model of the C_4H_{10} molecule. Is it possible to build more than one structure? If you 'play around' with the molecular model, you will find that there are two possible structures (Fig 21.4).

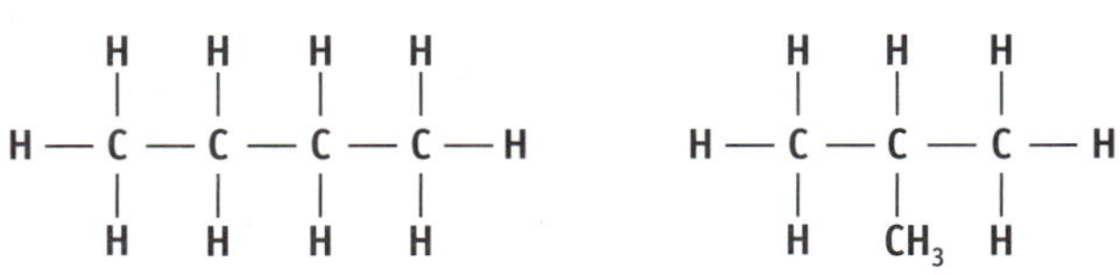

(a) Butane (b) Methylpropane

Fig 21.4
The two isomers of C_4H_{10}.

You will notice that compound (a), butane, has a 'straight' or unbranched chain of carbon atoms. However, compound (b), called methylpropane, has a CH_3 group attached to the middle carbon atom in the chain. (For neatness in drawing, we have not drawn out the full structure of this CH_3 group. This is normal practice in drawing the structural formulas of organic compounds.)

Butane and methylpropane are two different compounds with different physical properties, e.g. different boiling points. However, they both have the same molecular formula, C_4H_{10}. These compounds are called **structural isomers**.

> Structural isomers are compounds with the same molecular formula but different structural formulas.

How many isomers exist for C_5H_{12}? If you perform the same exercise as above, you will find there are three isomers. As an exercise, build models of these three isomers and draw their structural formulas. (For your Leaving Certificate examination, you only need to know the structures of the isomers of C_4H_{10} and C_5H_{12}. Obviously, all of the alkanes above C_5H_{12} have isomers also, but you need not go beyond the isomers of C_5H_{12}.)

Naming of Alkanes

The rules for naming organic compounds have been drawn up by the International Union of Pure and Applied Chemistry (IUPAC). The IUPAC names are more commonly referred to as the **systematic names** since the names are assigned according to a system of rules. Although the systematic names are now widely used, some compounds are still commonly referred to by their 'old' or traditional name. We will give examples of these as we come across them.

We will demonstrate the naming of alkanes using some worked examples. Side chains (also called substituent groups or branches) are simply groups of atoms attached to the main carbon chain. In alkanes, these are usually **alkyl groups**. An alkyl group is simply an alkane molecule from which a hydrogen atom has been removed, e.g. a CH_3 group is called a **methyl group**, a C_2H_5 group is called an **ethyl group**, a C_3H_7 group is called a **propyl group**, etc.

If two identical alkyl groups are present in a molecule, this is indicated using the prefix **di-**, e.g. dimethyl indicates two methyl groups, diethyl indicates two ethyl groups, etc.

Example 21.1

Give the systematic name of the compound whose structural formula is shown in FIG 21.5.

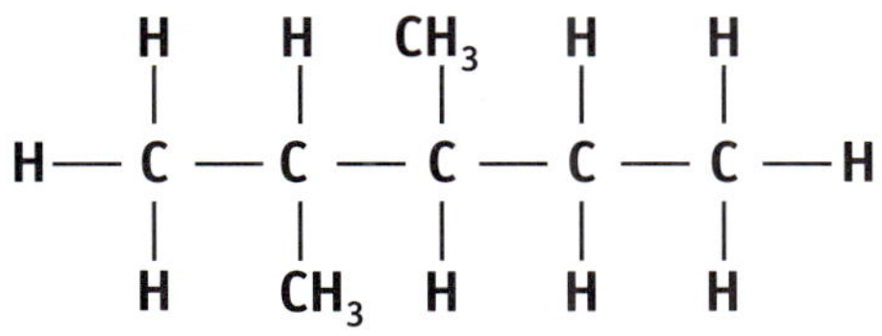

Fig 21.5

STEP 1

Identify the longest continuous chain of carbon atoms. This is the parent alkane. The compound is named as a derivative of this. Since there are five carbon atoms in the chain, the word **pent-** will be part of the name of this compound.

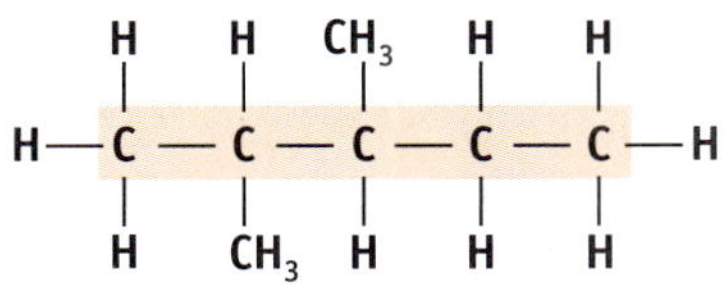

5 carbon atoms ⇒ **pent-**

Since this is a saturated hydrocarbon (alkane), the name will end in **-ane**.

5 C atoms ⇒ pentane

STEP 2

Number the carbon chain starting from the end that gives the substituents the lowest possible numbers.

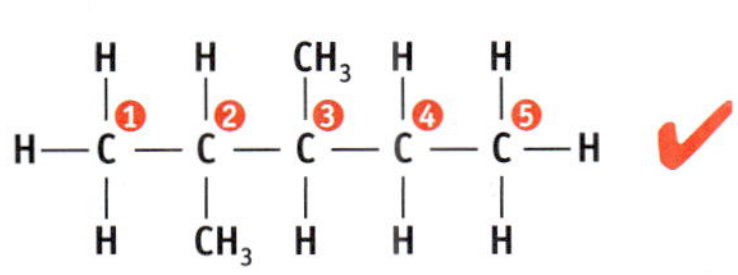

Methyl groups on C-2 and C-3

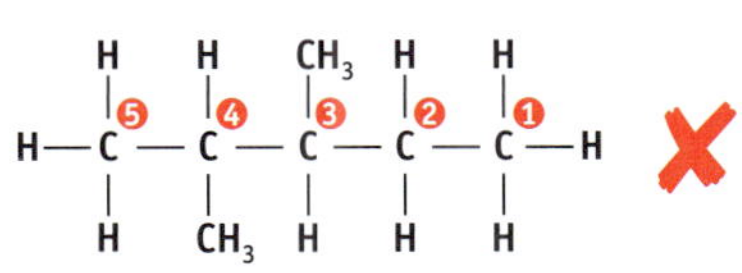

Methyl groups on C-3 and C-4

STEP 3

Indicate the type and position of substituents.

The name of the compound is **2,3-dimethylpentane.** (Not 3,4-dimethylpentane)

Some further examples of the naming of alkanes are shown in FIG 21.6. Examples (a) and (b) are two of the isomers of C_5H_{12} referred to when discussing isomers.

In example (c) note that substituents are named in alphabetical order, i.e. ethyl before methyl. Alphabetical order is always followed regardless of the numbering. Note also that the chain may not necessarily be drawn as a straight line. If you examine molecular models of these compounds, you will see that all the carbon atoms are tetrahedral. The structural formulas drawn on paper simply represent how the atoms are bonded to each other. They do not represent the actual shapes of the molecules.

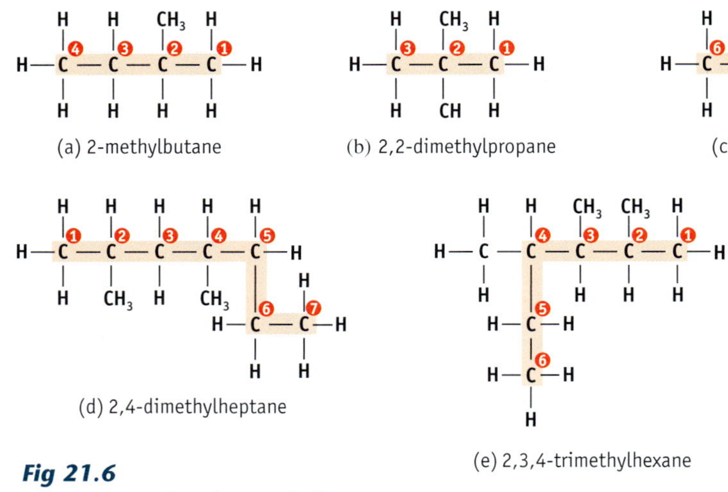

Fig 21.6
Some examples of named alkanes.

In your Leaving Certificate examination, the structural formulas may not be drawn out fully for you. Instead, the condensed structural formula could be given. For example, in FIG 21.6, example (a) could be simply written as $CH_3CH_2CH(CH_3)CH_3$. You would first have to draw out the full structural formula and proceed to name it as outlined in EXAMPLE 21.1 above.

A number of structures of alkanes are specifically mentioned on the syllabus due to their importance as fuels. These are shown in FIG 21.7. Compound (d), 2,2,4-trimethylpentane is an isomer of octane and its traditional name was iso-octane. Compound (e), cyclohexane, is the first hydrocarbon we have met in which the carbon atoms are arranged in the form of a ring. Build a model of this compound using the molecular models in your school laboratory. Note that the bond angles are the normal tetrahedral bond angles of 109°. Move the bonds around to get different overall shapes of the molecule. If you are really good you should be able to get a 'boat' and 'chair' shape!

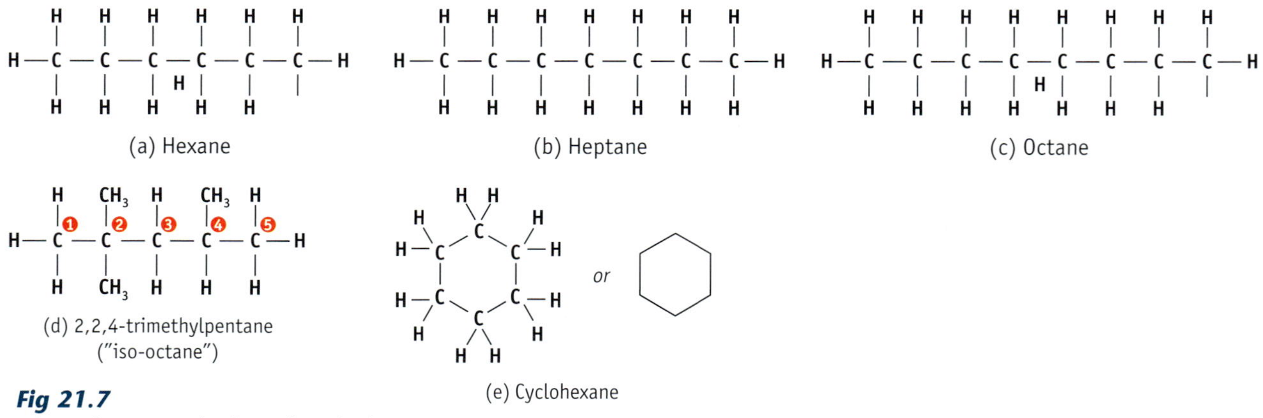

Fig 21.7
Some important hydrocarbon fuels.

Test Yourself:
Attempt questions 21.1 and W21.1.

Example 21.2

Draw the structural formula of 3-ethyl-2,5-dimethylhexane.

STEP 1

Examine the name and pick out the parent alkane.

'hexane' ⇒ six carbon atoms in the longest carbon chain.

STEP 2

Draw the six carbon atoms in the longest chain and number them. It does not matter whether you number them from left to right or right to left – you will end up with the same structural formula in both cases. (Think about this!) If we number them from right to left we get the chain shown in FIG 21.8(a).

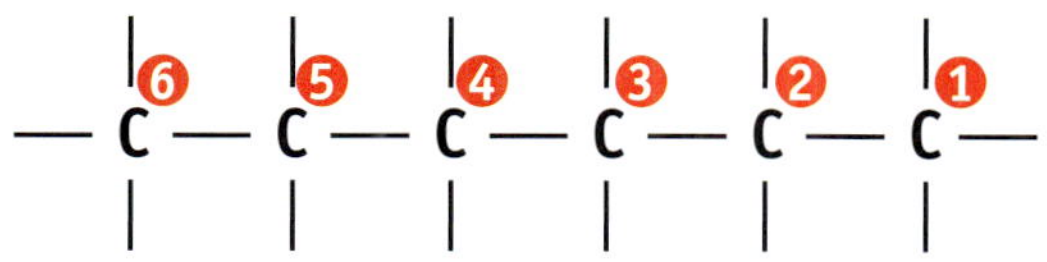

Fig 21.8(a)

STEP 3

Add in the side chains in the correct positions FIG 21.8(b):

'3-ethyl' ⇒ ethyl group on carbon atom 3

'2,5-dimethyl' ⇒ two methyl groups, one on carbon atom 2, one on carbon atom 5.
Students often ask whether the substituents should be drawn 'above' or 'below' the chain of carbon atoms. The answer is that it does not matter since, at room temperature, complete rotation about single bonds is continuously taking place.

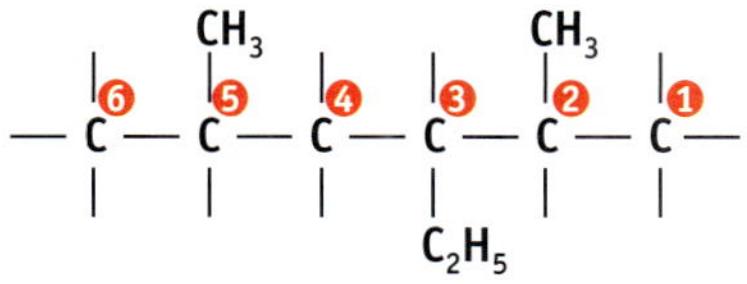

Fig 21.8(b)

STEP 4

Fill in the remaining bonds with hydrogen atoms, FIG 21.8(c).

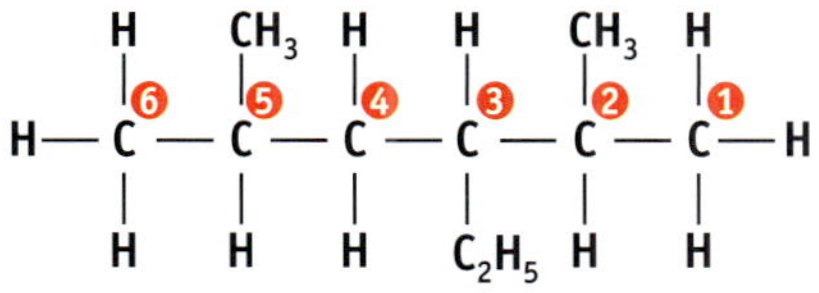

Fig 21.8(c)

Test Yourself:
Attempt questions 21.2 and W21.2.

21.2.2 THE ALKENES

The alkenes are the second homologous series of hydrocarbons that we shall be studying in this course. Alkenes contain a carbon-carbon double bond (>C=C<). Alkenes are said to be **unsaturated**. You may have seen margarine advertised as 'high in polyunsaturates'. This means that the various substances in margarine have a lot of C = C double bonds.

Alkenes are named from the corresponding alkanes by changing the ending **–ane** to **–ene**. Thus, the alkene family consists of ethene, propene, butene, pentene, hexene, etc. Note that the first member of the alkenes contains **two** carbon atoms. The first three members of the series are gases, the next eleven members are liquids and the higher alkenes are waxy solids at room temperature.

In this course, we need only be concerned with the structures of the first three alkenes, i.e. ethene, propene and butene. These are given in Fig 21.9. Build a model of the ethene molecule and note that it is a flat molecule with all bond angles equal to 120°.

From the pattern of formulas in Fig 21.9, you may observe that the general formula for alkenes is $\mathbf{C_nH_{2n}}$. You will also observe that there are three isomers of C_4H_8: but-1-ene (or 1-butene), but-2-ene (or 2-butene) and methylpropene. Let us now consider the origin of these names.

Name	Formula	Structural formula	Molecular model
Ethene (ethylene)	C_2H_4	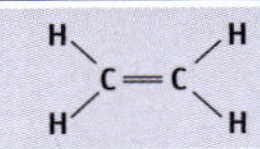	
Propene	C_3H_6	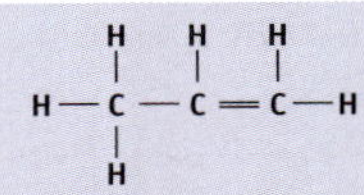	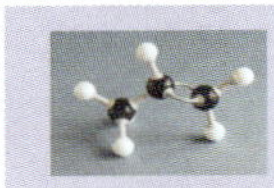
Butene (isomerism begins)	C_4H_8	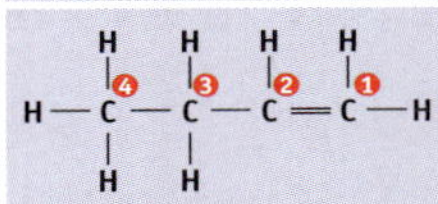 But-1-ene (1-butene)	
		or	
		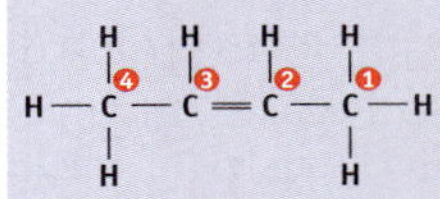 But-2-ene (2-butene)	
		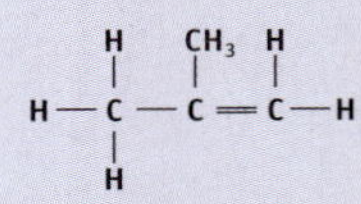 Methylpropene	

Fig 21.9
The members of the alkenes up to C-4.

NAMING OF ALKENES

The IUPAC rules for naming alkenes are very similar to those for naming alkanes. The rules are illustrated in Example 21.3.

Example 21.3

Name the compound whose structural formula is given in Fig 21.10 (a).

Step 1

Select the longest chain of carbon atoms containing the C = C double bond. The compound is named after the parent alkane by changing the –ane ending to –ene.

Six carbon atoms ⇒ hex-

C = C double bond ⇒ hexene

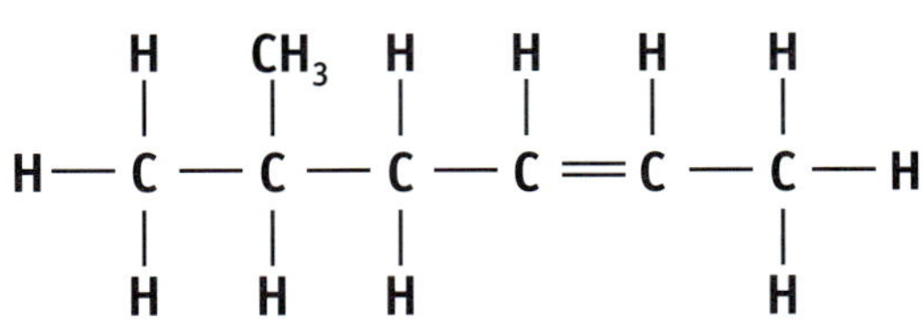

Fig 21.10(a)

Step 2

Number the chain starting at the end nearest to the C = C double bond. The position of the C = C double bond is indicated by the number of the carbon atom at which the double bond starts. In this example, the double bond starts at carbon atom 2 (Fig 21.10(b)).

C-2 ⇒ **hex-2-ene** (or **2-hexene**)

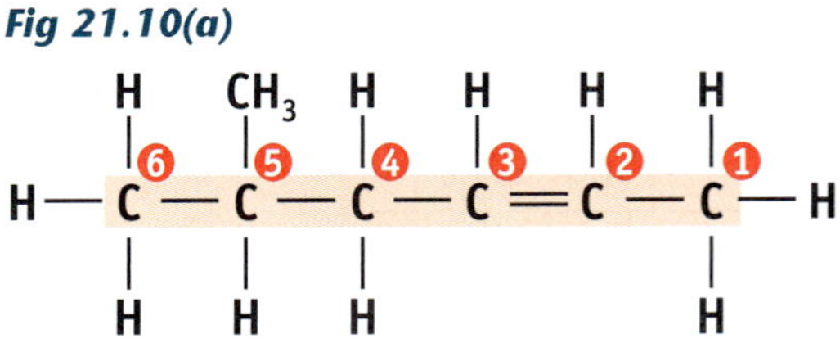

Fig 21.10(b)

Step 3

Indicate the type and position of substituents.

Methyl group on C-5 ⇒ **5-methyl-hex-2-ene** or **5-methyl-2-hexene**

Note: Since the hex-2-ene format is used more widely than the 2-hexene format, we will use the hex-2-ene format in this book.

> **Test Yourself:**
> Attempt questions 21.3 and W21.3.

Example 21.4

Draw the structural formula of 3,3-dimethyl-but-1-ene

If we examine the name carefully, the structure can easily be drawn.

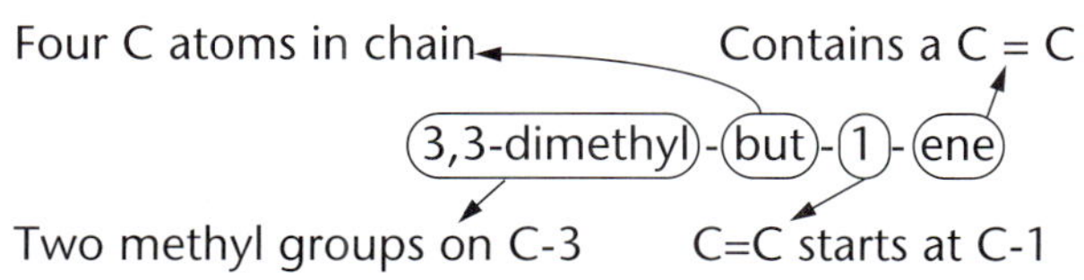

Hence, the structure is as shown in Fig 21.11.

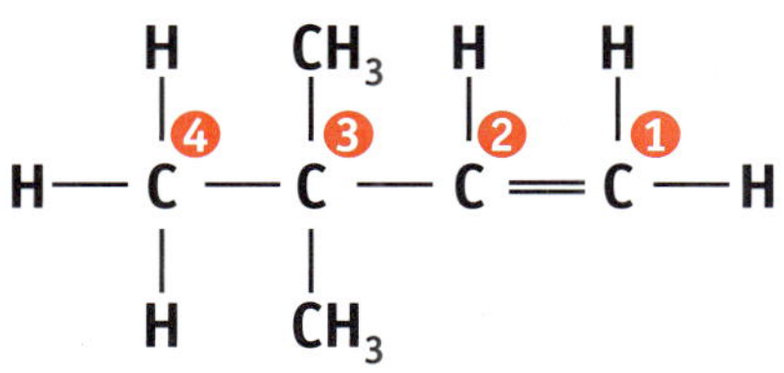

3,3-dimethyl-but-1-ene

Fig 21.11

Note: As in the case of cyclohexane, it is also possible to have the carbon atoms arranged in the form of a ring with a C = C between two of the carbon atoms, Fig 21.12. This compound is called cyclohexene.

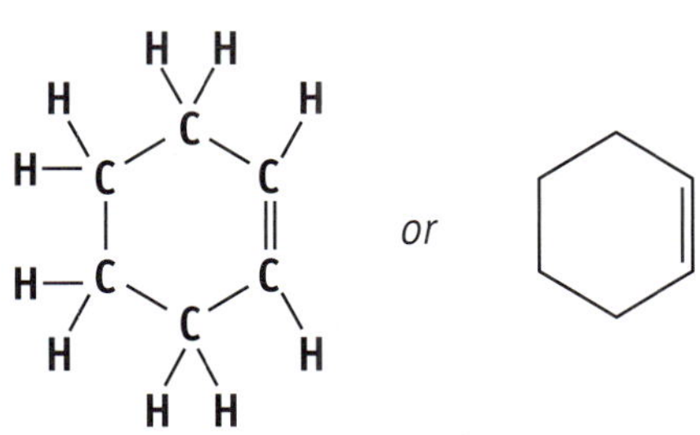

Cyclohexene

Fig 21.12

> **Test Yourself:**
> Attempt questions 21.4 and W21.4.

EXPERIMENT 21.1 MANDATORY STUDENT EXPERIMENT NO. 21

To prepare ethene and examine its properties.

Introduction

Ethene is often referred to as 'industry's most important organic chemical'. The reason for this is that ethene is the starting material from which many other substances (particularly plastics) are made, e.g. polythene, Terylene, synthetic rubber, polystyrene, PVC, etc. In this experiment, we will prepare a sample of ethene gas by dehydrating a substance called ethanol, i.e. using a chemical reaction to remove the elements of water from ethanol (FIG 21.13). (Ethanol is the drug found in alcoholic drinks, CHAPTER 22.)

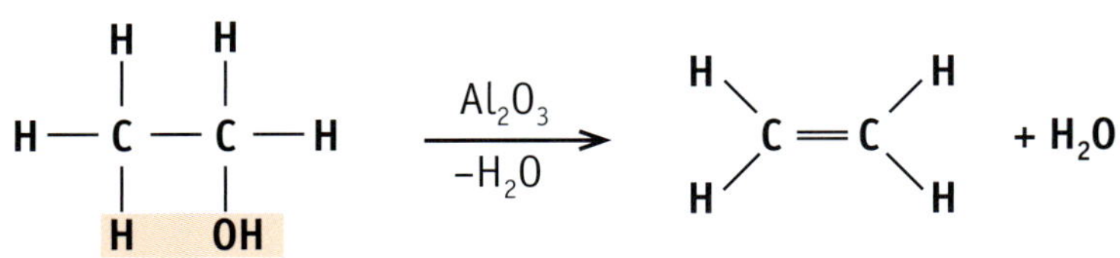

Fig 21.13

A white powder called aluminium oxide, Al_2O_3, is used in the reaction. This substance acts as a dehydrating agent and also as a catalyst for the reaction.

Apparatus required

Boiling tube, one-hole rubber stopper, delivery tube, trough, test-tube rack, five test-tubes, five corks or rubber stoppers to fit test-tubes, retort stand, spatula, Bunsen burner, wax taper, glass rod.

Chemicals required

Ethanol, aluminium oxide, limewater, glass wool, bromine water, acidified potassium permanganate solution (very dilute – pale purple).

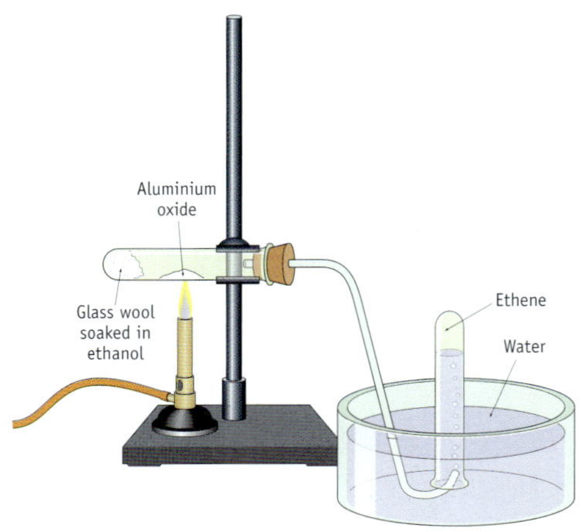

Fig 21.14
Preparation of ethene.

CAUTION: Beware of suckback (see step 8 below).

Procedure

1. Pour some ethanol into a boiling tube to a depth of about 2 – 3 cm.
2. Add some glass wool to soak up the ethanol. Use a glass rod to push the glass wool down the boiling tube. The glass wool is used to hold the ethanol in place while the reaction to produce ethene is occurring.
3. Using a retort stand, clamp the boiling tube near its mouth and in a horizontal position (FIG 21.14).
4. Using a spatula, place a small heap of aluminium oxide about half-way along the boiling tube.
5. Complete the arrangement of the apparatus as shown in FIG 21.14.
6. Light the Bunsen burner, adjust it to give a blue flame and gently heat the aluminium oxide. As the aluminium oxide becomes hot, the heat reaches the ethanol at the end of the tube. The ethanol changes into a vapour, passes over the hot aluminium oxide, and is dehydrated to form ethene. If the ethene is bubbling off too slowly, move the Bunsen burner nearer the ethanol for a brief moment.

7. Allow the bubbles to escape for a short while, as these are mainly bubbles of air displaced from the apparatus. Collect five test-tubes of the gas – put a stopper on each tube as it is filled.
8. When all the test-tubes have been filled, loosen the boss head on the retort stand and raise the apparatus up so that the delivery tube is no longer dipping into the water. Then turn off the Bunsen burner. If the delivery tube were simply left in the water trough, a suck back of water would occur since the boiling tube begins to cool as soon as the Bunsen burner is turned off. The cold water could cause the hot boiling tube to crack.
9. Note the **physical properties** of ethene. It is a colourless gas with a sweetish smell. It is insoluble in water but does dissolve in organic solvents, e.g. cyclohexane, chloroform, etc. ('like dissolves like'). If you were to test the solubility of methane in the laboratory, the same result would be obtained. The fact that the gas can be collected over water shows clearly that it is insoluble in water. Otherwise, the gas would simply dissolve in the water instead of bubbling through it.
10. **Combustion**. Remove the stopper from one of the test-tubes and, using a lighted wax taper, apply a light to the mouth of the test-tube. Note that the gas burns with a luminous flame. Allow the gas to burn. When it has stopped burning, add some limewater to the test-tube, stopper it and shake the test-tube a few times. The limewater turns milky showing that carbon dioxide is formed when ethene burns in air.
    The following reaction occurs when ethene burns in air:
    $C_2H_4 + 3O_2 \rightarrow 2CO_2 + 2H_2O$
    As a general rule, all organic compounds burn in air to form carbon dioxide and water.
11. **Addition of bromine**. Alkenes are more reactive than alkanes because of the presence of the C = C double bond in alkenes. For example, bromine adds across the C = C to form a colourless oily compound called 1,2-dibromoethane (Fig 21.15).

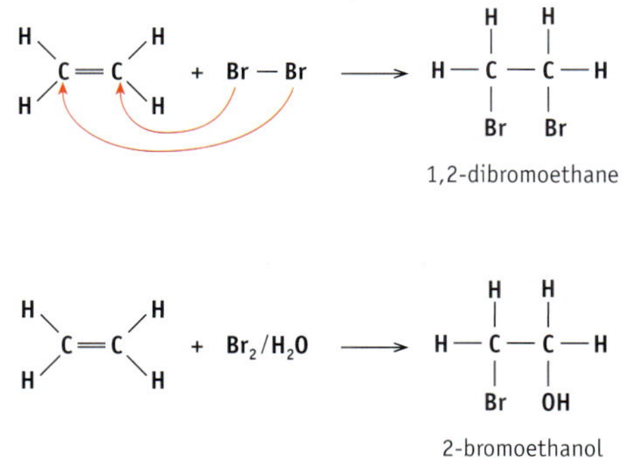

Fig 21.15
Addition of bromine and bromine water to ethene.

Similarly, if a solution of bromine in water (commonly referred to as 'bromine water') is added to some ethene, the yellowish/red colour disappears and a new colourless compound called 2-bromoethanol is formed (Fig 21.15). (You do not have to remember the name or formula of this product.) The decolorisation of bromine is the standard test for unsaturation in a compound. What would you observe if bromine water were added to a sample of margarine?

Great care should be taken when handling bromine water. Add about one-tenth of a test-tube of bromine water to a test-tube of ethene. Replace the stopper and shake the test-tube a few times. Note that the yellowish/red colour disappears.

12. **Addition of acidified $KMnO_4$.** Potassium manganate(VII) (potassium permanganate) is also used to test for the presence of a C = C double bond. Add about one-tenth of a test-tube of very dilute acidified $KMnO_4$ to a test-tube of ethene and shake the test-tube a few times. The disappearance of the purple colour to give a colourless solution is evidence for the presence of a C = C double bond. (The compound formed is one of the main constituents of antifreeze but you are not required to know the equation of this reaction for the Leaving Certificate examination.)
Note that, since ethene contains a C = C double bond, it is much more reactive than any of the alkanes.
Ethene is produced naturally by fruits as they ripen. Hence, you should never put a bowl of fruit next to flowers. The ripening fruit produces ethene that speeds up the 'ripening' of the flowers causing the petals to fall off. Ethene generators are produced commercially to speed up the ripening of fruit (FIG 21.16).

Fig 21.16
This apparatus is used to generate ethene to speed up the ripening of fruit like bananas. It produces ethene by the dehydration of ethanol over a heated catalyst.

21.2.3 THE ALKYNES

Name	Formula	Structural formula	Molecular model
Ethyne	C_2H_2	H–C≡C–H	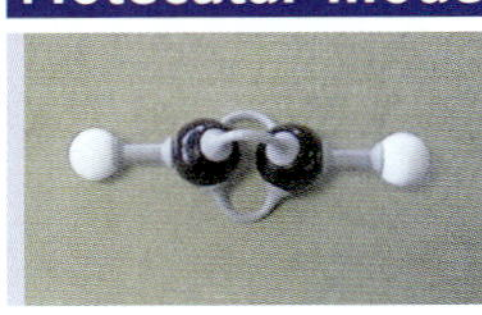

Fig 21.17
Ethyne.

The alkynes contain a carbon – carbon triple bond (-C ≡ C-), i.e. they are highly **unsaturated**. They form a homologous series and are named from the corresponding alkanes by changing the ending **–ane** to **–yne**. The general formula for alkynes is $\mathbf{C_nH_{2n-2}}$. The only alkyne on our course is ethyne (acetylene), C_2H_2 (FIG 21.17). This is the gas that is used in oxyacetylene welding and cutting. These two techniques require a cylinder of oxygen and a cylinder of ethyne (acetylene). It is used for this purpose because when ethyne is burned in controlled amounts of oxygen, temperatures in excess of 3000 °C may be obtained (FIG 21.18).
We will learn about the properties of ethyne in the next experiment.

Fig 21.18
Ethyne (acetylene) is used in oxyacetylene welding and cutting.

EXPERIMENT 21.2 MANDATORY STUDENT EXPERIMENT NO. 22

To prepare ethyne and examine its properties.

Introduction

Ethyne is prepared by the reaction of water with a substance called calcium carbide, CaC_2. (Calcium carbide is also called calcium dicarbide.) Before the invention of the battery, this substance was used in lamps where water dripped on to calcium carbide and the ethyne given off was lit at a jet. These types of lamps were called carbide lamps and were used on bicycles. Cave explorers still make use of this type of lamp. Calcium carbide cannot be obtained in a pure state – it is usually contaminated with calcium sulfide, CaS, calcium phosphide, Ca_3P_2 and calcium nitride, Ca_3N_2. (This is as a result of the process by which it is manufactured.) As a result, the ethyne obtained from it is usually contaminated with gases like hydrogen sulfide (H_2S), phosphine (PH_3) and ammonia (NH_3). These gases are removed by bubbling the gas through acidified copper(II) sulfate.

Apparatus required

Buchner flask, one-hole rubber stopper, dropping funnel, gas-washing bottle, spatula, delivery tube, trough, one gas jar, one gas-jar cover, test-tube rack, four test-tubes, four corks or rubber stoppers to fit test-tubes, retort stand, bee-hive shelf, wax taper.

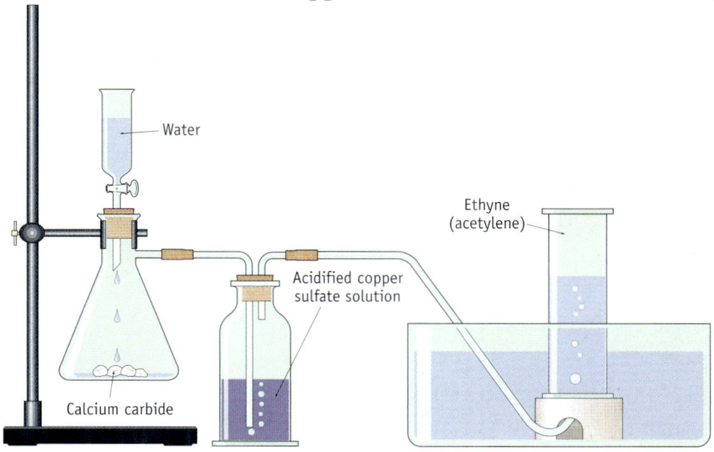

Fig 21.19
Preparation of ethyne.

Chemicals required

Calcium carbide, acidified copper(II) sulfate solution, limewater, bromine water, acidified potassium permanganate solution.

Procedure

1. Set up the apparatus as shown in FIG 21.19. Do not touch the calcium carbide with your fingers as moisture from your hands could start the reaction. Use a spatula when transferring the calcium carbide to the Buchner flask. Use only a small amount of calcium carbide to avoid problems with disposal later on. (You can always add more calcium carbide, if needed, at a later stage.)

NO NAKED FLAME SHOULD BE BROUGHT ANYWHERE NEAR THE GAS PREPARATION APPARATUS. ETHYNE GAS FORMS AN EXPLOSIVE MIXTURE WITH AIR. TESTS ON THE GAS SHOULD BE PERFORMED WELL AWAY FROM THE APPARATUS.

2. Double check that you have connected the gas-washing bottle correctly.
3. Using the dropping funnel, add water slowly to the calcium carbide to maintain an even (not too vigorous) evolution of ethyne. The following is the equation for the reaction which occurs:
   $CaC_2 + 2H_2O \rightarrow Ca(OH)_2 + C_2H_2$
   Note the formation of the white calcium hydroxide inside the flask. Place your hand on the outside of the flask. What do you notice?
4. Collect a few test-tubes of the gas first and, using a lighted taper, test for the presence of the gas. The first few test-tubes will be simply displaced air. The next few will be a mixture of ethyne and air and may burn with a slight 'pop'. When the gas burns quietly with a smoky flame, collect four test-tubes and a gas jar of the gas.
5. When the test-tubes and gas jar have been filled, dismantle the apparatus and dispose of the contents of the Buchner flask as directed by your teacher. Since we are going to be using a lighted taper in one of the tests, it is important that no more ethyne is being generated. Remember that ethyne forms an explosive mixture with air!
6. Note the **physical properties** of ethyne. It is a colourless gas with a sweetish type of smell. (If impurities are still present, it will not smell very nice!) Like methane and ethene, it is insoluble in water but soluble in organic solvents. It dissolves easily in a solvent called propanone (acetone). (This solvent is commonly used as nail varnish remover.) Because of the explosive nature of ethyne, it is normally stored in steel cylinders as a solution in acetone.
7. **Combustion**. This experiment is really for the thrill seekers! If your teacher allows it, bring the gas jar of ethyne to the fume cupboard, remove the gas jar cover and apply a lighted taper to the mouth of the gas jar. Note that it burns with a luminous smoky flame and a great deal of soot is formed. The soot is formed as a result of unburnt carbon. To burn the gas jar of ethyne completely to carbon dioxide and water, you would need two and a half gas jars of pure oxygen. Note the ratio in the balanced equation.
   $2C_2H_2 + 5O_2 \rightarrow 4CO_2 + 2H_2O$
   You could also burn a test-tube of ethyne but it's not as much fun as burning the gas jar of ethyne! Don't forget to wash out the gas jar when the experiment is finished.
8. **Addition of Bromine**. As in the case of ethene, bromine can be used to test for unsaturation in ethyne. Add about one-tenth of a test-tube of bromine water to a test-tube of ethyne. Replace the stopper and shake the test-tube a few times. Note that the yellowish/red colour disappears. The very reactive $C \equiv C$ triple bond reacts readily with the bromine. You are not required to know the equation of the reaction or the names of the products for this course. Hence, it will not be discussed further.
9. **Addition of acidified $KMnO_4$**. As in the case of ethene, potassium manganate(VII) (potassium permanganate) can be used to test for unsaturation. Add about one-tenth of a test-tube of potassium permanganate to a test-tube of ethyne. Replace the stopper and shake the test-tube a few times. Note that the purple colour disappears and a colourless solution is obtained. The very reactive $C \equiv C$ triple bond reacts readily with the potassium permanganate. You are not required to know the equation of this reaction or the names of the products for this course. Hence, it will not be discussed further.

21.2.4 AROMATIC HYDROCARBONS

So far we have been studying hydrocarbons in which the carbon atoms are arranged in open chains (branched or unbranched). We have also looked at a small number of compounds (cyclohexane and cyclohexene) in which the carbon atoms are arranged in rings. The term **aliphatic** is used to describe these compounds (Greek *aliphos* = fat) as many of these compounds were originally derived from fat tissue.

> An aliphatic compound is an organic compound that consists of open chains of carbon atoms and closed chain compounds (rings) that resemble them in chemical properties.

There are also many organic compounds which contain a benzene ring, C_6H_6 (FIG 21.20). Compounds that contain a benzene ring in their molecules are commonly called aromatic compounds because some of the early examples discovered had pleasant smells.

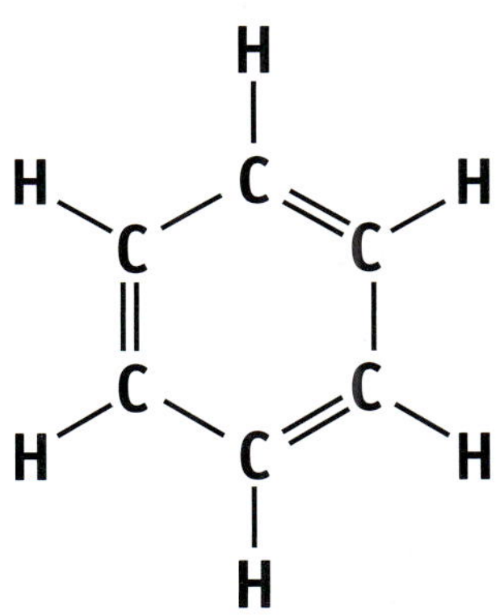

Fig 21.20
A molecule of benzene, C_6H_6.

> (c) Aromatic compounds are compounds that contain a benzene ring structure in their molecules.

Benzene was discovered by Michael Faraday in 1825. When its formula was determined as C_6H_6, it was realised that this compound was highly unsaturated. Therefore, it was predicted that benzene would be highly reactive due to the presence of the expected three double bonds. In fact, benzene was found to be a very unreactive compound, e.g. it did not decolorise bromine water or acidified potassium permanganate as does ethene. If the C = C double bond in ethene is so reactive and takes part so readily in addition reactions (CHAPTER 23), why is benzene so unreactive with regard to addition reactions?

Although the structure of benzene shown in FIG 21.20 satisfies the valencies of carbon and hydrogen, there is a problem about the alternating double and single bonds between the carbon atoms. When the bond lengths were measured (using X-rays), it was found that all carbon-carbon bond lengths were the same. A single bond is longer than a double bond. (The bond length of a C – C single bond is 0.154 nm where 1 nm = 1 × 10^{-9} m. The bond length of a C = C double bond is 0.134 nm.) The actual C – C bond lengths in benzene were measured as 0.139 nm, i.e. between that of a double bond and a single bond. (You don't have to remember these numbers!) We can imagine the six valence electrons, one from each carbon atom, belonging to the whole molecule, instead of being localised in three double bonds. This gives the benzene molecule extra stability and accounts for its unreactivity with regard to addition reactions.

Thus, the correct way to represent a molecule of benzene is as a hexagon with a circle inside it, FIG 21.21. Note that, for convenience, the carbon and hydrogen atoms are not written on the diagram.

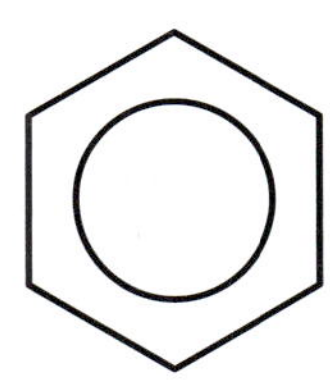

Benzene, C_6H_6

Fig 21.21
The correct way to represent a molecule of benzene.

Build a model of the benzene molecule using the molecular model kit in your school laboratory. Note that the molecule is a flat molecule with all bond angles equal to 120°. Contrast this shape with that of cyclohexane which is also represented as a hexagon but with no circle. A common error on Leaving Certificate examination scripts is for students to leave out the circle when indicating the structure of an aromatic compound. This is a major mistake! Refer back again to FIG 21.7(e) if you are not clear on the difference.

The only aromatic hydrocarbons, apart from benzene, whose structures you need to know for this course are methylbenzene and ethylbenzene (FIG 21.22). Note that the substituent group replaces one of the hydrogen atoms. The traditional name for methylbenzene is toluene. (You may have heard of the explosive TNT. These letters stand for trinitrotoluene. TNT is made from toluene. We will not be preparing TNT in this course!)

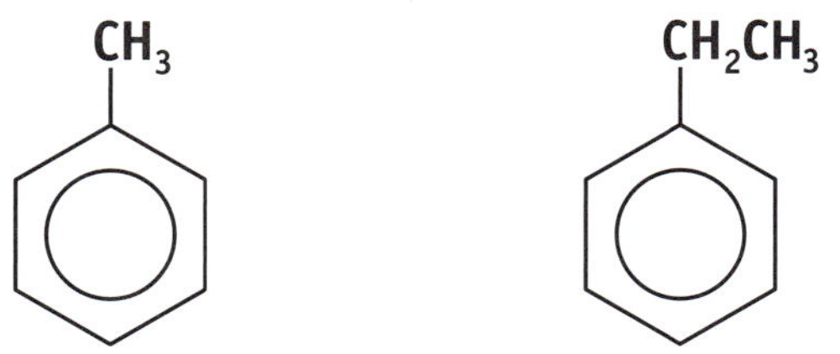

Methylbenzene, C_7H_8 (toluene)　　Ethylbenzene, C_8H_{10}

Fig 21.22
Methylbenzene and ethylbenzene are examples of aromatic hydrocarbons.

Benzene is a highly toxic and carcinogenic material. Methylbenzene is the recommended substitute for it. This is quite safe for small-scale occasional use in laboratories. Examine some methylbenzene in the school laboratory. Note that it is a liquid at room temperature. Add a small quantity to a test-tube containing water and note that methylbenzene is insoluble in water. However, it will dissolve in organic solvents like cyclohexane, propanone, etc. Methylbenzene itself is a good organic solvent. Add a few crystals of iodine (non-polar) to a test-tube containing some water and a test-tube containing some methylbenzene. What do you observe?

> **Test Yourself:**
> Attempt questions 21.5 – 21.8 and W21.5 – W21.7.

21.3 OIL REFINING AND ITS PRODUCTS

21.3.1 FRACTIONAL DISTILLATION OF CRUDE OIL

Crude oil (also called petroleum) is one of the most valuable and useful materials found in the earth. We depend on crude oil and its products to provide us with fuels for transport, heating and for the manufacture of a wide range of chemicals that we use in our everyday lives. You are probably aware that crude oil is found in large quantities in places like Iran, USA, Nigeria, Russia, the North Sea, etc.

Crude oil is commonly referred to as a fossil fuel. This is because it was formed from the bodies of tiny sea creatures that died millions of years ago. As the layers of mud and silt grew thicker over the years, decay caused by bacteria slowly turned the bodies of these creatures into crude oil and natural gas.

Crude oil as it is pumped from under the ground or from under the sea is of little use in itself. In fact, crude oil does not even burn very easily. It looks like a thick, black liquid and has an unpleasant smell. In order to separate it into a number of useful mixtures, a process called **fractional distillation** is used. This simply involves heating the crude oil and separating the various mixtures on the basis of their boiling points. Each one of these useful mixtures is called a **fraction** – hence the name fractional distillation. The various fractions obtained from crude oil and the approximate temperatures at which these fractions boil off are summarised in FIG 21.23. Crude oil from various sources contains different proportions of the various fractions. When oil refineries purchase crude oil, the crude oil is first analysed by the refinery chemists to determine its quality.

The crude oil is then heated in a furnace and enters the fractionating column as a partially vaporised mixture. The fractionating column in Whitegate Oil Refinery in Cork is shown in FIG 21.24. This is simply a tall cylindrical tower about 50 metres high containing a series of 'trays' to help collect the condensed liquids. The fractionating column is maintained at a high temperature at the bottom, with the temperature decreasing up along the column. Therefore, the larger hydrocarbons that have higher boiling points turn to liquids nearer the bottom of the column. However, the smaller hydrocarbons stay as gases and rise up along the column. The different fractions condense at different levels in the fractionating column. You are not required to remember the various temperatures at which the fractions are collected (FIG 21.23). However, you must know the uses to which the various fractions are put.

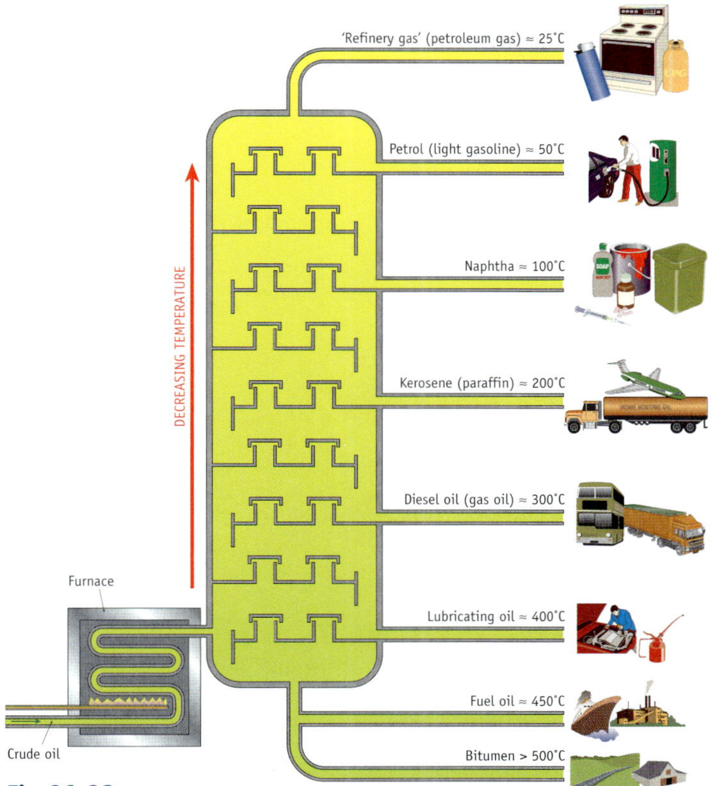

Fig 21.23
Fractional distillation of crude oil.

Fig 21.24
The fractionating column in Whitegate Oil Refinery.

At the top of the column, hydrocarbons like methane, ethane, propane and butane are gases at 25 °C. These come off as gases that are commonly referred to as '**refinery gas**' or '**petroleum gas**' Table 21.2. Some of this is used by the refinery as a source of fuel but most of it is bottled for sale as domestic gas. Since these gases are odourless, sulfur compounds called mercaptans are usually added to give an unpleasant smell in case of gas leaks. Bottled gas is mainly a mixture of propane and butane. These gases are easily liquefied under pressure – hence they are often referred to as LPG (liquid petroleum gas).

The next fraction is **petrol** (also called **light gasoline**) which is used as a motor fuel. Petrol or light gasoline is a mixture of at least 100 different compounds, mostly hydrocarbons. These hydrocarbons are mainly C_5 – C_{10} alkanes. Their boiling points vary over a range of 40 – 75 °C approx.

Fraction	Length of Carbon Chain
Refinery gas	$C_1 - C_4$
Petrol (light gasoline)	$C_5 - C_{10}$
Naphtha	$C_7 - C_{10}$
Kerosene (paraffin)	$C_{10} - C_{14}$
Diesel oil (gas oil)	$C_{14} - C_{19}$
Lubricating oil	$C_{19} - C_{35}$
Fuel Oil	$C_{30} - C_{40}$
Bitumen	$>C_{35}$

Table 21.2

Naphtha is a fraction that is very useful to the petrochemical industry. (The petrochemical industry is that part of the chemical industry based on crude oil.) It is the source of a huge number of very useful chemicals, e.g. medicines, plastics, synthetic fibres, detergents, solvents, etc. Naphtha consists of $C_7 - C_{10}$ compounds. The compounds in naphtha have boiling points in the range 75 – 150 °C approx.

Kerosene or **paraffin** is used as aviation fuel and also in certain stoves ('paraffin stoves') and lamps ('paraffin oil lamps'). It consists of $C_{10} - C_{14}$ compounds with boiling points in the range 150 – 240 °C approx. Note that as the number of carbon atoms increases, so also the boiling point increases. This is because the larger the molecule (i.e. the greater the relative molecular mass), the more energy is required to turn it into the gaseous state.

Diesel oil or **gas oil** is more difficult to vaporise than either petrol or kerosene. A diesel engine has a different design to that of a petrol engine in order to help the diesel to burn. Diesel is used as a fuel in trucks, buses, trains and some cars. Diesel consists of $C_{14} - C_{19}$ compounds with boiling points in the range 220 – 350 °C approx.

Lubricating oil is a viscous liquid. It consists of $C_{19} - C_{35}$ compounds with boiling points in the range 350 – 500 °C approx. Because of the high boiling point range, lubricating oil cannot be used as a fuel since it cannot be vaporised easily. Instead, it is used mainly as a lubricant to reduce the wear of engines.

Fuel oil consists of $C_{30} - C_{40}$ compounds with boiling points in the same range as lubricating oil, 350 – 500 °C approx. However, it is possible to use these compounds as a fuel by spraying them into burners as a fine mist of small droplets. Hence, fuel oil is used in ships, power stations, heating plants, etc.

Bitumen consists of compounds with a large number of carbon atoms ($> C_{35}$). Its boiling point is above 500 °C – too high to be used as a fuel. If you have ever seen a road being surfaced, bitumen is the black tar that is used. It has to be applied hot so that it can be spread. It is also used for waterproofing and for roofing.

Lubricating oil, fuel oil and bitumen are often referred to as **residue fractions**. This is because these are the fractions that are left over when the more volatile fractions boil off.

21.3.2 OCTANE NUMBER

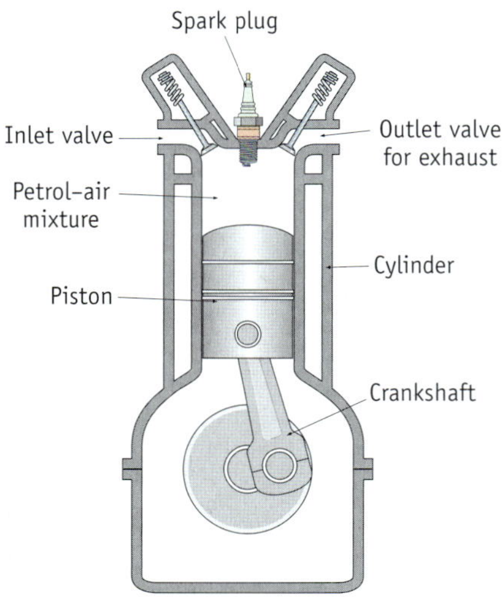

Fig 21.25
In the internal combustion engine, a mixture of petrol and air is drawn into the cylinder. The mixture is compressed and ignited by a spark. The gases produced by the explosion drive the piston down. This rotates the crankshaft causing the wheels to turn.

The petrol obtained directly from the fractionating column in an oil refinery would not be a very efficient fuel in a car. In the internal combustion engine of a motor car, the power to turn the wheels is produced in the cylinders by the explosive reaction between petrol and air (FIG 21.25). The smooth running of the engine depends on the explosion occurring in the cylinder at exactly the right moment. If the explosion occurs too early, the pistons vibrate and a metallic noise is heard from the engine. The early explosion occurs due to the petrol-air mixture exploding as it is compressed instead of the explosion being caused by the spark. Engineers refer to this occurrence as **knocking**. In addition to the noise, the engine also loses power and the inside of the cylinders can be damaged.

The early exploding of the petrol-air mixture is referred to as **auto-ignition**. When the problem of auto-ignition was investigated, it was discovered that straight chain alkanes like heptane, octane and nonane ignited very easily and exploded too soon in the cylinder causing knocking. On the other hand, branched-chain alkanes like 2,2,4-trimethylpentane (iso-octane) did not tend to undergo auto-ignition. Therefore, petrol mixtures that contain large amounts of these branched-chain alkanes are more efficient than those that contain straight-chain molecules. To indicate the efficiency of a particular type of petrol (i.e. its lack of tendency to auto-ignite), a number called the **octane number** is assigned to it.

> The octane number of a fuel is a measure of the tendency of the fuel to resist knocking.

Since **2,2,4-trimethylpentane (iso-octane)** is a very efficient fuel (low tendency to auto-ignite), this compound is assigned an **octane number of 100**. (The term *octane number* comes from iso-octane). On the other hand, heptane is not an efficient fuel (auto-ignites readily) and is assigned an octane number of 0. The octane number of a fuel can be measured by comparing its performance with known mixtures of heptane and 2,2,4-trimethylpentane in a test engine (FIG 21.26). Octane numbers can also be measured by detailed chemical analysis of the fuel. Good quality petrol could have an octane number of 97. This means that it is as efficient a fuel as a mixture of 97% 2,2,4-trimethylpentane and 3% heptane.

Fig 21.26
The octane number of the fuel is being measured in this test engine in Whitegate Oil Refinery, Co. Cork. If it has the same performance as a mixture of 90% 2,2,4-trimethylpentane (iso-octane) and 10% heptane, then its octane number is 90.

Research has shown that the **shorter the alkane chain, the higher the octane number**, e.g. the octane number of heptane is 0, the octane number of hexane is 25, that of pentane is 62 and that of butane is 94. The second factor that affects the octane number of the alkane is the degree of branching. The **more branched the chain, the higher the octane number** (FIG 21.27).

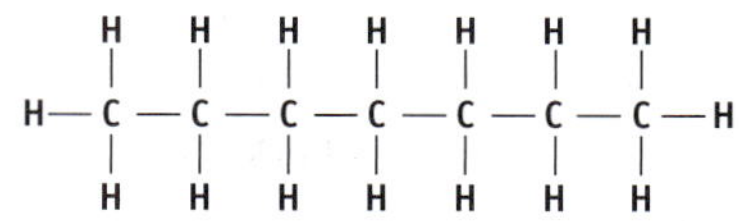

Heptane, C_7H_{16}
Octane number = 0

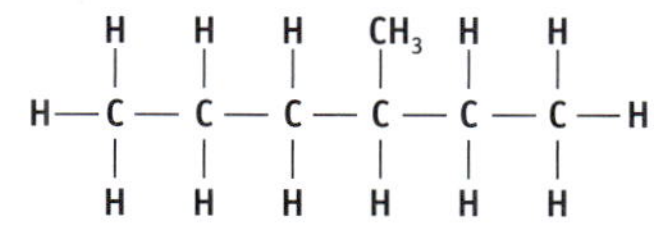

3-methylhexane, C_7H_{16}
Octane number = 65

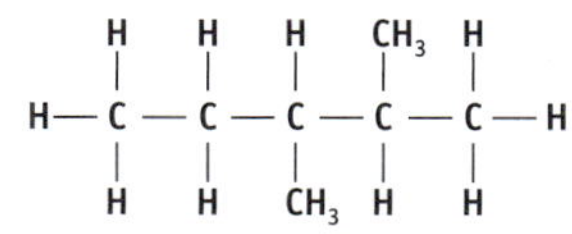

2,3-dimethylpentane, C_7H_{16}
Octane number = 91

Fig 21.27
The more branched the chain, the higher the octane number.

The third factor affecting octane number is whether the alkane chain is a straight chain or a cyclic structure. **Cyclic compounds have a higher octane number than straight-chain compounds**, e.g. hexane has an octane number of 25 but cyclohexane has an octane number of 83 (FIG 21.28).

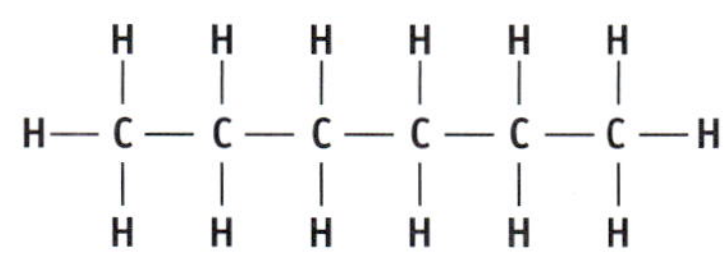

Hexane
Octane number = 25

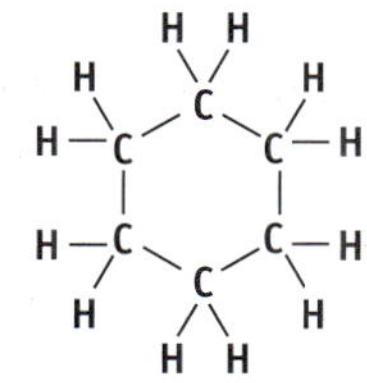

Cyclohexane
Octane number = 83

Fig 21.28
Cyclic compounds have a higher octane number than straight-chain compounds.

21.3.3 MAKING PETROL

In the 1920s, it was discovered that adding small amounts of a lead compound called tetraethyl lead, $Pb(C_2H_5)_4$, helped reduce the amount of knocking. It is called an anti-knock additive. However, awareness of the potential damage to health caused by lead pollution from car exhausts, particularly to city dwellers, has resulted in the phasing out of leaded petrol in Ireland since 2000 and its replacement with unleaded petrol. Lead compounds in the exhaust fumes are not only toxic, they also poison the metal catalysts in the catalytic converter of the car. (Hence, the catalytic converter can no longer reduce the levels of other pollutants in exhaust fumes.)

Adding lead compounds to petrol was an easy solution to the problem of knocking since it meant that the petrol did not have to contain compounds with high octane numbers. However, with the phasing out of lead additives, chemists in oil refineries had to design ways to increase the octane number of petrol. ('Straight run' petrol from the fractionating column has a typical octane number of only about 70. This would be a very inferior type of fuel.) Four ways are now commonly used to increase the octane number of petrol:

(i) Isomerisation

(ii) Catalytic cracking

(iii) Reforming (Dehydrocyclisation)

(iv) Adding oxygenates

(i) Isomerisation

Isomerisation, as the name suggests, involves changing straight-chain alkanes into their isomers. It is an important way of increasing the octane number of petrol. The alkanes are heated in the presence of a suitable catalyst and this causes the chains to break. The chain fragments are then allowed to join together again and when they do so, the final product is more likely to be a branched-chain alkane than a straight-chain alkane. Isomerisation is carried out in oil refineries on a large scale with pentane and hexane. These two compounds are found in the petrol fraction obtained from the distillation of crude oil. However, their octane numbers are too low: pentane has an octane number of 62 and hexane an octane number of only 25.

One example of an isomerisation reaction that happens with pentane is shown in FIG 21.29.

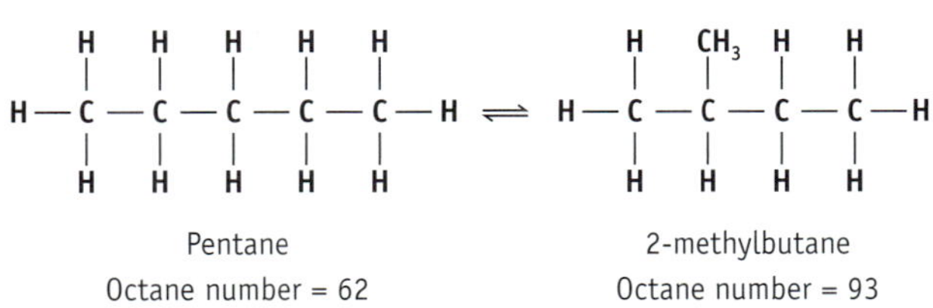

Fig 21.29
The isomerisation reaction changes pentane into one of its branched isomers.

Note that the isomerisation reaction does not go to completion but rather a state of equilibrium is reached. The mixture formed consists of straight-chain and branched isomers. In an oil refinery, the straight-chain and branched isomers are separated from each other. The straight-chain alkanes are then recycled over a suitable catalyst to convert them to branched isomers. Thus, the octane number of the petrol is increased. (You are not required to know the conditions under which the isomerisation reactions are carried out. Hence, these conditions are not given.)

(ii) Catalytic cracking

Of the various fractions obtained from crude oil, petrol is the one in greatest demand. You would probably have guessed that this is due to the large number of cars we see on our roads. In order to satisfy this demand and to avoid being left with a large surplus of the heavier fractions, chemists discovered a method of making petrol from these heavier fractions. This procedure is called **catalytic cracking**. Long-chain hydrocarbon molecules are split (cracked) into short-chain hydrocarbon molecules.

> Catalytic cracking is the breaking down of long-chain hydrocarbon molecules into short-chain molecules for which there is greater demand.

The short-chain hydrocarbons (usually alkanes) obtained from cracking are then used in petrol. In addition, the short-chain hydrocarbons tend to be highly branched. This means that the petrol made from them has a higher octane number.

In oil refineries, the cracking carried out to produce petrol is done by heating the heavier fractions (e.g. fuel oil, diesel oil, kerosene, etc.) in the presence of a catalyst. Hence the term 'cat cracking'. There is quite a variety of cracking reactions. A common one is where an alkane is converted into a branched alkane and an alkene (FIG 21.30). If you split a saturated alkane, at least one of the products must be unsaturated. Think about this and try to explain it!

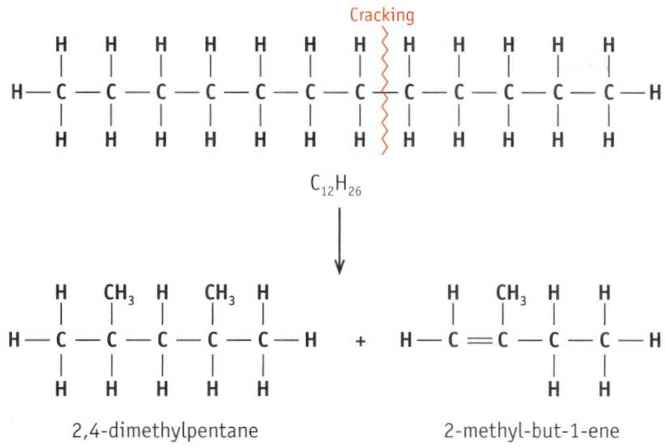

Fig 21.30
A typical cracking reaction.

Altering the conditions of the reaction could cause a different cracking reaction to occur (FIG 21.31).

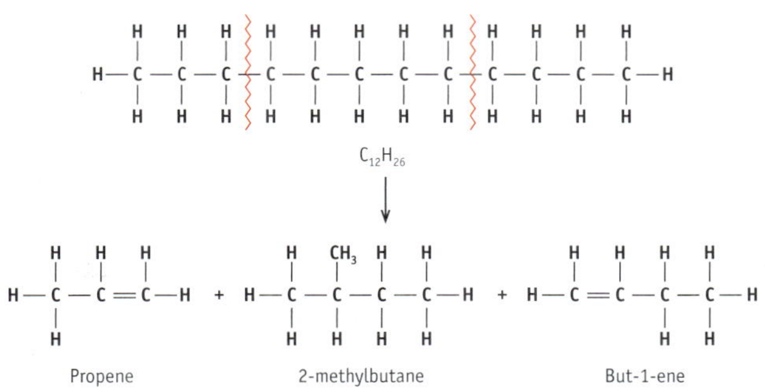

Fig 21.31
A different cracking reaction occurs if the condition of the reaction are changed.

Note that, in both of the above cracking reactions, alkenes are produced. A second benefit of cat cracking is that these alkenes are important feedstock for the petrochemicals industry, i.e. the alkenes are used to manufacture various materials. For example, ethene is used to make poly(ethene) (commonly called polythene) which is used in the manufacture of plastic bags, bowls, food boxes, film for wrapping food, washing-up liquid bottles, buckets, etc. Propene is the chemical used to make poly(propene) which is used in the manufacture of toys, jugs, plastic beakers, chairs, briefcases, etc.

Cat crackers have been used in oil refineries since the 1940s (FIG 21.32). Cat crackers are very flexible and adaptable units. The reaction conditions and catalyst can be varied to give the maximum amount of the desired product, e.g. branched alkanes for petrol or ethene for the petrochemicals industry.

Fig 21.32
A cat cracker in Toronto, Canada.

We will now carry out a small-scale version of an industrial cracking process. This is **not** a mandatory student experiment. It may be carried out by the students or as a teacher demonstration.

EXPERIMENT 21.3

To carry out a cracking reaction of a hydrocarbon.

Introduction

In this experiment, a long-chain hydrocarbon found in liquid paraffin is cracked by passing its vapour over heated porcelain or steel wool. The porcelain and steel wool are catalysts for the reaction. The products formed are analysed by combustion, and by addition of bromine water and acidified potassium permanganate.

One of the substances found in liquid paraffin is the straight-chain alkane $C_{11}H_{24}$. Under the conditions of this reaction, it is likely that the cracking occurs as shown in FIG 21.33.

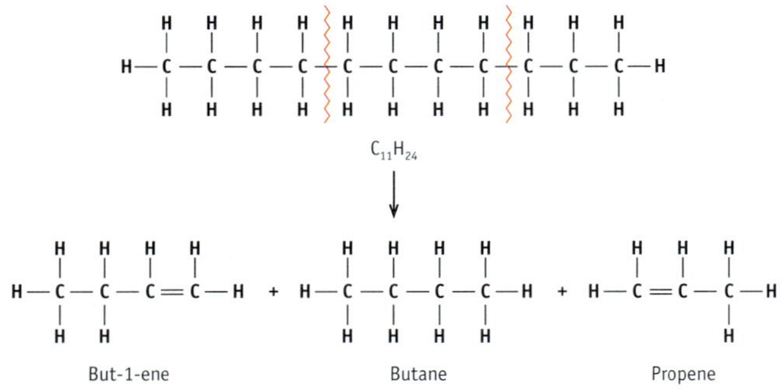

Fig 21.33
The cracking of $C_{11}H_{24}$ found in liquid paraffin.

All three products of the reaction are gases at room temperature. The procedure for carrying out this reaction is very similar to the preparation of ethene.

Apparatus required

Boiling tube, one-hole rubber stopper, delivery tube, trough, test-tube rack, four test-tubes, four corks or rubber stoppers to fit test-tubes, retort stand, spatula, Bunsen burner, wax taper, glass rod.

Chemicals required

Liquid paraffin (**NOT** PARAFFIN OIL), porcelain chips (unglazed if possible) or steel wool , limewater, glass wool, bromine water, acidified potassium permanganate solution.

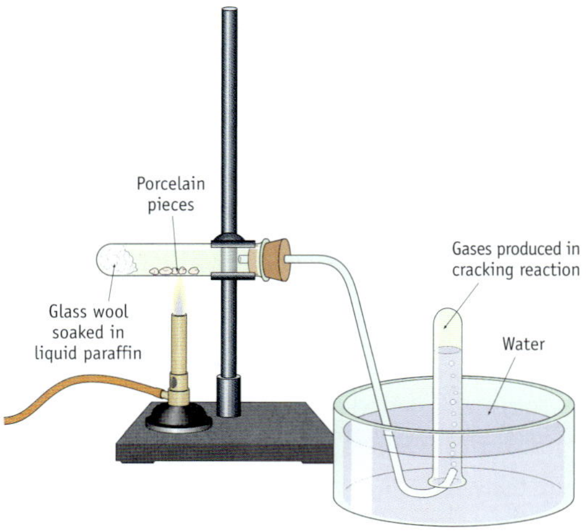

Fig 21.34
The cracking of $C_{11}H_{24}$ found in liquid paraffin.

CAUTION: Beware of suck back (see step 8 below).

Procedure

1. Pour some liquid paraffin into a boiling tube to a depth of about 2 – 3 cm.
2. Add some glass wool to soak up the liquid paraffin. Use a glass rod to push the glass wool down the boiling tube. The glass wool is used to hold the liquid paraffin in place while the cracking reaction is occurring. It is best to put the liquid paraffin in first so that enough glass wool can be added to absorb all the liquid paraffin. Otherwise, some liquid paraffin could just stay on top of the wool and not be absorbed. In addition, the glass wool helps to remove the liquid paraffin from along the inside of the boiling tube. This liquid paraffin could possibly vaporise later on without passing over the catalyst and being cracked.
3. Using a retort stand, clamp the boiling tube in a horizontal position near the mouth of the boiling tube (FIG 21.34).
4. Place a small heap of porcelain pieces about half-way along the boiling tube.
5. Complete the arrangement of the apparatus as shown in FIG 21.34.
6. Light the Bunsen burner, adjust it to give a blue flame and heat the porcelain or steel wool strongly. As the porcelain or steel wool becomes hot, the heat reaches the liquid paraffin at the end of the tube. This causes the liquid paraffin to vaporise and a cloudy vapour is observed inside the tube. From time to time it may be necessary to apply heat to the end of the boiling tube to help vaporisation of the liquid paraffin.

Flammable Gases

7. Allow the bubbles to escape for a short while as these are mainly bubbles of air displaced from the apparatus. Collect four test-tubes of the gas – put a stopper on each tube as it is filled.
8. When all the test-tubes have been filled, to prevent a suck back, loosen the boss head on the retort stand, raise the apparatus up so that the delivery tube is no longer dipping into the water and then turn off the Bunsen burner. Another reason for turning off the Bunsen burner is for safety. The hydrocarbon gases coming out the delivery tube are flammable and form explosive mixtures with air. Under no circumstances should a naked flame be brought near the outlet of the delivery tube.
9. **Combustion**. Remove the stopper from one of the test-tubes and, using a lighted wax taper, apply a light to the mouth of the test-tube. Note that the gas burns with a clear flame. This suggests that the long-chain hydrocarbon has been broken down to a short-chain hydrocarbon. (The fact that flammable gases are produced is evidence that cracking has occurred. A long-chain hydrocarbon would not be collected as a gas. In addition, a long-chain hydrocarbon would burn with a sooty flame due to the high carbon content.) When it has stopped burning, add some limewater to the test-tube, stopper it and shake the test-tube a few times. What do you observe? What can you conclude?

   Assuming that the gas you are burning is pure butane, write a balanced equation for the reaction.
10. **Addition of bromine**. Great care should be taken when handling bromine water. Add about one-tenth of a test-tube of bromine water to a test-tube of the products of the cracking reaction. Replace the stopper and shake the test-tube a few times. What do you observe? Has the cracking reaction produced unsaturated products?

    The first few test-tubes may contain less unsaturated products than the test-tubes filled later. This is because more cracking usually takes place at the higher temperature once the catalyst has heated up.
11. **Addition of acidified $KMnO_4$**. Add about one-tenth of a test-tube of very dilute $KMnO_4$ (acidified) to a test-tube of the gas and shake the test-tube a few times. What do you observe? Has the cracking reaction produced unsaturated products?

(iii) Reforming (Dehydrocyclisation)

Reforming is the third method used to increase the octane number of petrol. **Reforming** involves the **use of catalysts to form ring compounds**. When reforming occurs, straight-chain alkanes are converted to cycloalkanes and cycloalkanes are converted to aromatic compounds (FIG 21.35). Aromatic compounds have high octane numbers. For this reason, petrol contains about 3 – 4% benzene. However, benzene is a carcinogenic compound and its presence in petrol is a cause of concern.

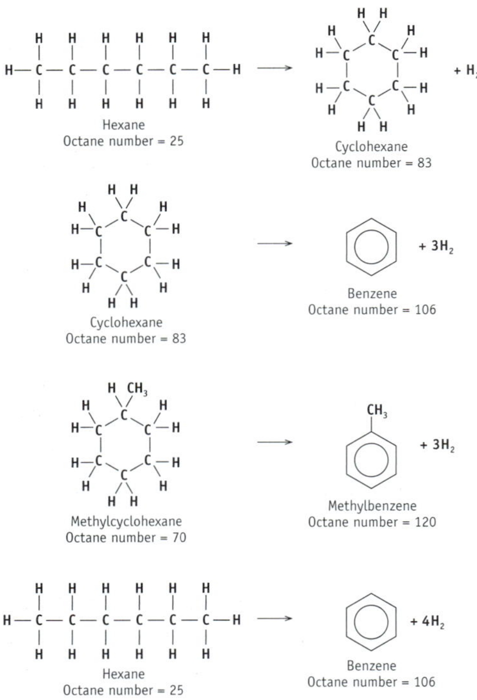

Fig 21.35
Some typical reforming reactions.

Note from FIG 21.35 that, as well as high-octane compounds, hydrogen may also be produced in a reforming reaction. This hydrogen is a valuable product and is piped away to be used for various purposes (SECTION 21.4).

Since hydrogen is being removed and ring compounds formed, this type of reforming reaction is often referred to as **dehydrocyclisation**.

(iv) Adding oxygenates

'Oxygenate' is the name used by petroleum chemists to refer to any fuel that contains oxygen in its molecules. Three oxygen-containing compounds, methanol, ethanol and MTBE, are commonly added to petrol to increase its octane number (FIG 21.36). MTBE stands for *methyl tertiary butyl* ether. The more modern name for this compound is 2-methoxy-2-methylpropane. (You do not need to remember the structure of this compound for your examination.)

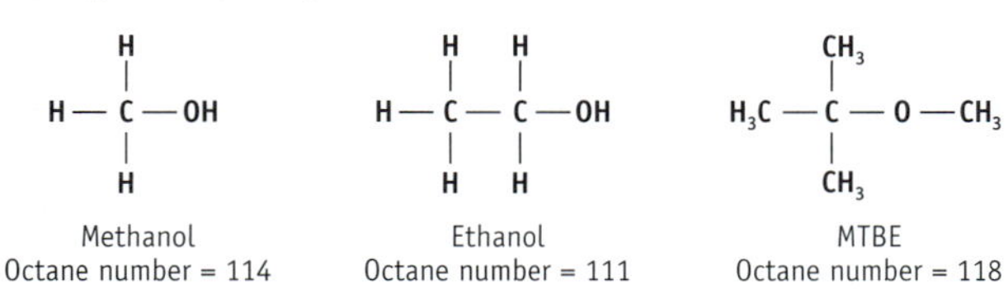

Fig 21.36
These oxygen-containing compounds are commonly added to petrol to increase its octane number

As well as increasing the octane number of petrol, there is another advantage in adding oxygenates to petrol. These oxygenates give rise to very little pollution when they burn, i.e. they are 'cleaner' fuels than hydrocarbons as they give rise to lower levels of carbon monoxide in the exhaust systems of cars. In Brazil, ethanol (obtained from sugar cane) is commonly added to petrol (FIG 21.37).

Fig 21.37
In Brazil, ethanol is added to petrol to increase its octane number. Note the sign on the petrol pump.

From the above discussion, we have seen that petrol is quite a complex mixture. In the oil refinery the job of the petrol blender is to produce a particular type of petrol to meet certain specifications at minimum cost. The composition of petrol even varies from season to season! During the winter, more volatile components (e.g. butane and pentane) are put in the petrol so that the petrol vaporises more easily in the carburettor or fuel injection system. During the summer, less of these volatile hydrocarbons are added to the petrol to prevent loss from the petrol tank by evaporation.

Test Yourself:
Attempt questions 21.9 – 21.10 and W21.8 and W21.10.

21.4 HYDROGEN

So far, we have been discussing hydrocarbon fuels. The only other fuel on your course is hydrogen gas, H_2. There are two main methods for manufacturing hydrogen:

(A) STEAM REFORMING OF NATURAL GAS

This involves reacting methane (natural gas) with steam in the presence of a suitable catalyst.

$$CH_4 + H_2O \longrightarrow 3H_2 + CO$$

This reaction is commonly referred to as the steam reforming of natural gas.

(B) ELECTROLYSIS OF WATER

Hydrogen may also be produced by the electrolysis of water as discussed in CHAPTER 20. The overall equation for the reaction is:

$$H_2O \longrightarrow H_2 + \tfrac{1}{2}O_2$$

This can be an expensive way of producing hydrogen due to the high cost of electricity in many parts of the world.

We have also seen that hydrogen is produced in oil refineries when reforming is carried out.

The most important use of hydrogen is in the manufacture of ammonia. This process is studied in detail in OPTION 1, CHAPTER 25. The essential reaction is the combining of hydrogen and nitrogen in the ratio of one part of nitrogen to three parts of hydrogen.

$$N_2 + 3H_2 \rightleftharpoons 2NH_3$$

Hydrogen is also used for hydrogenating (adding hydrogen to) vegetable oils in the manufacture of margarine, for organic synthesis (i.e. making organic compounds) and as a fuel. At present, only about 10% of hydrogen produced worldwide is used as a fuel. It is predicted by some scientists that, in the future, hydrogen will become a more widely used fuel. This is because burning hydrogen is very 'environmentally friendly' since water is the only product formed ($H_2 + \tfrac{1}{2}O_2 \rightarrow H_2O$). However, since hydrogen forms an explosive mixture with air, certain problems with regard to storage and transportation have to be overcome before it can become widely used.

21.5 THERMOCHEMISTRY

21.5.1 EXOTHERMIC AND ENDOTHERMIC REACTIONS

One of the most important properties of fuels is that they burn in air. We have already seen this combustion reaction when we burned ethene and ethyne. The burning of fuels always produces heat. Any chemical reaction that produces heat is called an **exothermic reaction**. (*Ex* is the Latin word for *out*.) The 'burning' of food inside our bodies supplies us with the energy we need. All combustion reactions are exothermic reactions. You have already met quite a number of exothermic reactions in this course. How many can you remember?

Not all chemical reactions give out heat. In some cases, a drop in temperature is observed when a chemical reaction occurs. In these cases heat is taken in from the surroundings. You will notice the drop in temperature if you hold a beaker of water at room temperature and add some ammonium nitrate or ammonium chloride to it. Heat is taken in from your hand. Such a reaction is called an **endothermic reaction**. An endothermic reaction is one that takes in heat. Another example of an endothermic reaction is the reaction of sherbet with water. Hence, sherbet feels cold in your mouth when you eat it!

The study of the heat changes that accompany chemical reactions is called **thermochemistry** or **chemical thermodynamics**. (The second term has its origin in the study of heat engines where heat and movement gave rise to the name thermodynamics.)

Note: Changes of state also result in heat changes. For example, if you put a volatile liquid like ether on the back of your hand, you will feel the drop in temperature. This is because the heat energy needed to change the liquid to a gas comes from your hand. On the contrary, a burn from steam can be more harmful than a burn from boiling water. This is because the hot steam condenses on your skin and the excess energy of the gaseous molecules is lost (in the form of heat) when they become a liquid. This heat can cause severe burning. Changes of state are not classified as exothermic and endothermic reactions since they are merely physical changes rather than chemical reactions.

21.5.2 HEAT OF REACTION

Chemists use the symbol ΔH ('delta H') to indicate the heat change taking place when a chemical reaction occurs. For example, as discussed already, hydrogen is a fuel which burns in oxygen to form water. We can write a balanced equation as follows:

$$H_{2(g)} + \tfrac{1}{2}O_{2(g)} \longrightarrow H_2O_{(g)} \quad \Delta H = -242 \text{ kJ mol}^{-1}$$

This equation states that when one mole of hydrogen reacts with half a mole of oxygen to form one mole of steam, 242 kilojoules of heat are released. By convention, **a negative value of ΔH indicates an exothermic reaction. A positive value of ΔH indicates an endothermic reaction**. The kJ stands for kilojoules. The SI unit of energy is the joule (J). Since one joule of heat energy is rather small, heat changes when molar quantities react are normally measured in kilojoules (1kJ = 1000 J). The mol^{-1} in the above equation indicates that we are working in molar quantities, i.e. the heat measured relates to 1 mole of H_2 reacting with half a mole of O_2 to form 1 mole of H_2O. The ΔH value above is commonly referred to as the **heat of reaction** and is defined as follows:

> Heat of reaction is the heat change when the numbers of moles of reactants indicated in the balanced equation for the reaction react completely.

If the numbers of moles in the balanced equation are changed, then the heat of reaction also changes. For example, if we burn twice the number of moles of hydrogen, the heat of reaction is doubled:

$$2H_{2(g)} + O_{2(g)} \longrightarrow 2H_2O_{(g)} \quad \Delta H = -484 \text{ kJ mol}^{-1}$$

Note also that the physical states of the reactants and products are always given in such an equation. The reason for this is because changing the physical states of either reactants or products can alter the value of ΔH. Equations like the above are commonly referred to as **thermochemical equations**. Thermochemical equations must be balanced, the value of ΔH must be written in and the physical states of reactants and products must be given. (In books of data containing these thermochemical equations, the temperature at which the given heat of reaction is measured is always stated. This is usually 25 °C.)

HEAT OF COMBUSTION

The term heat of reaction is a general term used for any type of chemical reaction. In some cases, a different term is used to indicate the particular type of chemical reaction being considered. As already stated, a reaction in which a substance is burned is called a **combustion reaction**. Consider the reaction in which carbon is burned in oxygen to form carbon dioxide:

$$C_{(s)} + O_{2(g)} \longrightarrow CO_{2(g)} \quad \Delta H = -393 \text{ kJ mol}^{-1}$$

The heat given out in this reaction (393 kJ mol^{-1}) is called the **heat of combustion** of the reaction. It is defined as follows:

> The heat of combustion of a substance is the heat change when one mole of the substance is completely burned in excess oxygen.

The phrase 'completely burned in excess oxygen' is important in the definition of heat of combustion. This is because some elements have more than one oxide:

$$C_{(s)} + \tfrac{1}{2}O_{2(g)} \longrightarrow CO_{(g)} \quad \Delta H = -111 \text{ kJ mol}^{-1}$$

This value of ΔH is **not** the heat of combustion of carbon since complete combustion of carbon in excess oxygen forms carbon dioxide not carbon monoxide.

Another important phrase in the definition of heat of combustion is 'one mole'. Consider the thermochemical equation to describe what happens when you burn butane from a gas cylinder:

$$2C_4H_{10(g)} + 13O_{2(g)} \rightarrow 8CO_{2(g)} + 10H_2O_{(g)} \quad \Delta H = -5720 \text{ kJ mol}^{-1}$$

Is -5720 kJ mol^{-1} the heat of combustion of butane? It is not since, by definition, the heat of combustion involves one mole of the substance being burned. Therefore, the heat of combustion of butane is $-\frac{5720}{2}$ = -2860 kJ mol^{-1}. This may be represented by the following thermochemical equation:

$$C_4H_{10(g)} + 6\tfrac{1}{2}O_{2(g)} \longrightarrow 4CO_{2(g)} + 5H_2O_{(g)} \quad \Delta H = -2860 \text{ kJ mol}^{-1}$$

Heats of combustion are accurately measured using an instrument called a **bomb calorimeter**. A calorimeter is any container used for the determination of heat changes. (The name comes from an old unit of heat called the calorie; 1 calorie = 4.18 joules.) A bomb calorimeter consists of a small steel container (the bomb) with a screw-on cap (FIG 21.38). The sample whose heat of combustion is to be measured is placed in a crucible inside the bomb. The bomb is placed in a container of water (the calorimeter), oxygen is pumped into the bomb and the sample ignited by electric wires (FIG 21.39). By measuring the rise in temperature of the water and knowing certain information about the ability of water and the material of the calorimeter to absorb heat, it is possible to calculate the heat of combustion of the fuel. (As this type of calculation is not on the syllabus, it will not be discussed further.)

Fig 21.38
The bomb part of a bomb calorimeter.

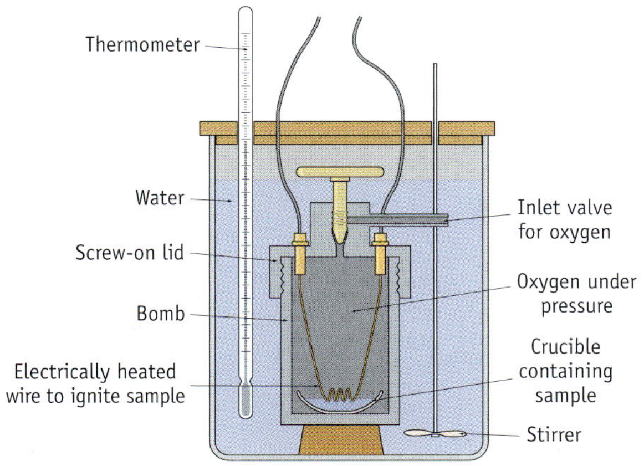

Fig 21.39
A bomb calorimeter is used to accurately measure heats of combustion.

Some examples of heats of combustion of various fuels are given in TABLE 21.3.

Fuel	Formula	Heat of combustion ΔH measured in kJ mol^{-1} at 25 °C	Kilogram calorific value in kJ kg^{-1}
Methane	$CH_4(g)$	-890	55,625
Ethane	$C_2H_6(g)$	-1560	52,000
Propane	$C_3H_8(g)$	-2220	50,454
Butane	$C_4H_{10}(g)$	-2877	49,603
Hexane	$C_6H_{14}(l)$	-4194	48,767
Ethanol	$C_2H_5OH(l)$	-1371	29,804
Carbon	C(s)	-393	32,750
Hydrogen	$H_2(g)$	-286	143,000

Table 21.3
Heats of combustion and kilogram calorific values of some fuels.

In some cases in industry, chemists wish to measure the heat given out by a fuel (or food) in terms of the mass of a fuel rather than the number of moles. For example, if you examine the side of a box of cornflakes, you will find that 100 g of cornflakes provide about 1550 kJ of energy. This value was obtained using a bomb calorimeter (FIG 21.40). Bomb calorimeters are also used to compare the efficiency of various fuels, i.e. to see if the fuel is of good quality in terms of the heat output from a certain mass of fuel. For this purpose, chemists define a term called **kilogram calorific value**.

> The kilogram calorific value of a fuel is the heat energy produced when 1 kg of the fuel is completely burned in oxygen.

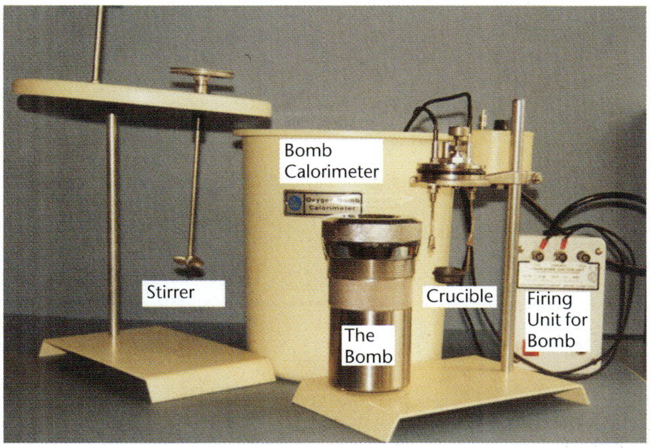

Fig 21.40
Bomb calorimeter and associated apparatus.

Kilogram calorific values can be measured using a bomb calorimeter. They can also be calculated from heats of combustion for fuels which have a definite chemical formula, e.g. methane, butane, etc. However, for substances which have a variable composition (e.g. petrol, coal, peat, etc) they have to be measured using a bomb calorimeter. Some examples of the kilogram calorific values of various fuels are given in TABLE 21.3. The kilogram calorific values of fuels which are not pure substances are not listed in TABLE 21.3, e.g. anthracite (≈ 31,400 kJ kg^{-1}), peat (≈ 11,600 kJ kg^{-1}), petrol (≈ 35,000 kJ kg^{-1}), etc. These are given as approximate values since the actual values can vary due to the quality of the particular fuel.

BOND ENERGY

Let us now consider what determines the value of ΔH for a particular reaction. Suppose we consider the combustion of methane:

$$CH_{4(g)} + 2O_{2(g)} \rightarrow CO_{2(g)} + 2H_2O_{(g)} \quad \Delta H = -890 \text{ kJ mol}^{-1}$$

As already discussed in CHAPTER 5, when a chemical reaction occurs, bonds are broken and new bonds are formed. Energy is required to break a bond and energy is released when a bond is formed. If we examine the left-hand side of the equation (FIG 21.41), we see that four C – H bonds and two O = O bonds must be broken.

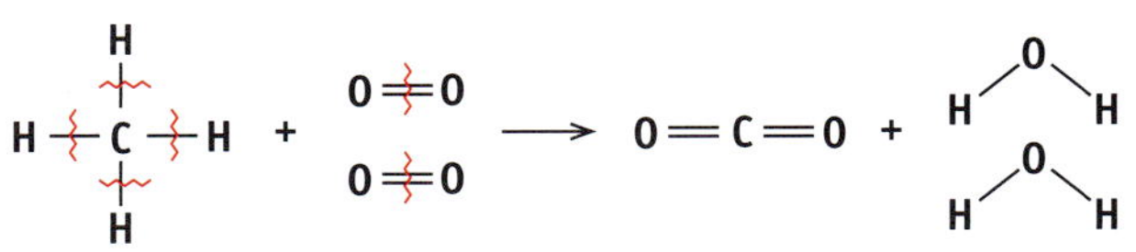

Fig 21.41
Bonds are broken and new bonds are formed when methane burns in oxygen.

The energy involved in breaking bonds is called the **bond energy** and is defined as follows:

> Bond energy is the energy required to break one mole of covalent bonds and to separate the neutral atoms completely from each other.

Bond energy values are measured from studying spectra of molecules and also from values of heats of reaction. They were used by Linus Pauling to draw up a table of electronegativity values (CHAPTER 5). Some average bond energy values are listed in TABLE 21.4. Thus, the value of ΔH for a particular reaction is dependent on the amount of energy that is needed to break the bonds and the amount of energy that is liberated when new bonds are formed.

Bond	Average bond energy value in kJ mol^{-1}
C – H	412
C – C	348
C = C	612
C ≡ C	837
O – H	463
O = O	496
C = O	743

Table 21.4
Average bond energy values for some bonds.

Note: The reason the bond energy values in TABLE 21.4 are **average** values is because it is found that the bond energy value for a particular bond can vary. For example, the energy to break the first C – H bond in methane is not the same as that required to break any of the other C – H bonds.

$CH_{4(g)} \longrightarrow CH_{3(g)} + H_{(g)}$ ΔH = +425 kJ mol^{-1}

$CH_{3(g)} \longrightarrow CH_{2(g)} + H_{(g)}$ ΔH = +470 kJ mol^{-1}

$CH_{2(g)} \longrightarrow CH_{(g)} + H_{(g)}$ ΔH = +416 kJ mol^{-1}

$CH_{(g)} \longrightarrow C_{(g)} + H_{(g)}$ ΔH = +335 kJ mol^{-1}

Therefore, average bond energy =

$$\frac{425 + 470 + 416 + 335}{4} = \frac{1646}{4} = 411.5 \text{ kJ mol}^{-1}$$

HEAT OF NEUTRALISATION

We have already seen (CHAPTER 12) that neutralisation is the reaction between an acid and a base to form a salt and water.

e.g. $HCl + NaOH \longrightarrow NaCl + H_2O$

We have also seen that the essential reaction taking place in this neutralisation reaction is:

$H^+ + OH^- \longrightarrow H_2O$

Neutralisation reactions are nearly always exothermic. When discussing heat changes involved in neutralisation reactions, the term **heat of neutralisation** is defined as follows:

> Heat of neutralisation is the heat change when one mole of H^+ ions from an acid reacts with one mole of OH^- ions from a base.

Heat of neutralisation is measured in kJ mol^{-1}. Consider the heat of neutralisation of a number of strong acids and strong bases when dilute solutions of these acids and bases react together:

$HCl + NaOH \longrightarrow NaCl + H_2O$ ΔH = -57.1 kJ mol^{-1}

$HCl + KOH \longrightarrow KCl + H_2O$ ΔH = -57.2 kJ mol^{-1}

$HNO_3 + NaOH \longrightarrow NaNO_3 + H_2O$ ΔH = -57.3 kJ mol^{-1}

$HNO_3 + KOH \longrightarrow KNO_3 + H_2O$ ΔH = -57.3 kJ mol^{-1}

You will observe that, in every case, the heat of neutralisation is almost the same. The reason for this is because, if we ignore spectator ions, the essential reaction occurring is the same in all cases. For example, if we take a close look at the reaction between nitric acid and potassium hydroxide, we see that the reaction that occurs simply involves a proton and a hydroxide ion coming together to form water.

$$H^+ + \cancel{NO_3^-} + \cancel{K^+} + OH^- \longrightarrow \cancel{K^+} + \cancel{NO_3^-} + H_2O$$

$$\mathbf{H^+ + OH^- \longrightarrow H_2O \quad \Delta H = -57\ kJ\ mol^{-1}}$$

However, consider the reaction between a strong acid and a weak base or between a weak acid and a strong base:

$$HCl + NH_4OH \longrightarrow NH_4Cl + H_2O$$

$$\Delta H = -52.2\ kJ\ mol^{-1}$$

$$CH_3COOH + NaOH \longrightarrow CH_3COONa + H_2O$$

$$\Delta H = -55.2\ kJ\ mol^{-1}$$

When either a weak acid or a weak base is involved, the value of ΔH is numerically less than −57 kJ mol^{-1}. The reason for this is because a weak acid or weak base is not fully dissociated in water. For complete neutralisation to occur, resulting in one mole of water being formed, the weak acid or base must become fully dissociated. This dissociation requires energy, i.e. it is endothermic. Therefore, the heat released will be less than 57 kJ since a certain quantity of heat is required to cause complete dissociation of the acid or base.

Experiment 21.4 Mandatory Student Experiment No. 23

To determine the heat of reaction (heat of neutralisation) of hydrochloric acid with sodium hydroxide.

Introduction

In this experiment, a known number of moles of hydrochloric acid are mixed with an equal number of moles of sodium hydroxide in an insulated container. The rise in temperature of the mixture is then measured and from this the heat of neutralisation is calculated.

Apparatus required

Thermometer (reading to 0.1 °C), two expanded polystyrene cups with lids, two graduated cylinders (100 cm^3), stirrer (plastic-coated stiff wire).

Chemicals required

Hydrochloric acid (1.0 M), sodium hydroxide (1.0 M).

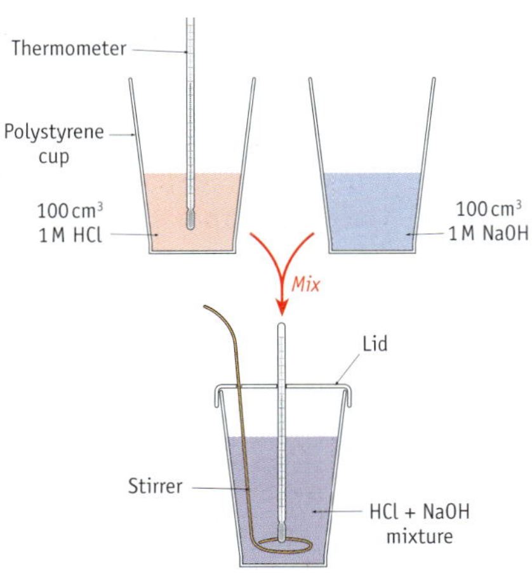

Fig 21.42
Measuring the heat of neutralisation of hydrochloric acid and sodium hydroxide.

Procedure

1. Using a graduated cylinder, place 50 cm^3 of the 1.0 M hydrochloric acid solution into one of the plastic cups (Fig 21.42).
2. Using a second graduated cylinder, place 50 cm^3 of the 1.0 M sodium hydroxide solution into the second plastic cup.
3. Measure the temperature of the hydrochloric acid solution.
4. Wash the thermometer, dry it and measure the temperature of the sodium hydroxide solution.
5. When both solutions are at the same temperature (room temperature), quickly add the base to the acid, stirring well. Take care to avoid any loss of liquid due to splashing.
6. Place a lid on the plastic cup and, with continuous stirring, record the maximum temperature reached.

7. Summarise your results as follows:

Temperature of HCl before mixing	= .... °C
Temperature of NaOH before mixing	= .... °C
Highest temperature reached after mixing	= .... °C
Temperature rise	= .... °C
Number of moles of acid used	= .... °C
Number of moles of base used	= .... °C
Mass of solution	= .... g
	= .... kg

8. Now calculate the heat liberated using the equation:

$$\text{heat liberated} = m \times c \times (t_2 - t_1)$$

m = mass of solution in kg.
c = specific heat capacity of solution in J kg^{-1} K^{-1}
$t_2 - t_1$ = temperature rise

Note:

(a) The **specific heat capacity** of a substance is the amount of heat required to raise the temperature of 1 kg of a substance by 1 K. For example, the specific heat capacity of water is 4200 J kg^{-1} K^{-1}. This means that it takes 4200 joules of heat to raise the temperature of 1 kg of water from, say, 19 °C to 20 °C.

(b) In the calculations, we are assuming that all the heat generated is going to raise the temperature of the solution and that none is absorbed by the plastic container, i.e. the specific heat capacity of the plastic is negligible.

(c) In the calculations, assume that the density of the solution is the same as the density of water, i.e. that 100 cm^3 of the solution have a mass of 100 g or 0.1 kg. Also assume that the specific heat capacity of the solution is 4060 J kg^{-1} K^{-1}.

(d) The most common error in examinations is for students to forget to convert grams to kilograms when substituting into the above equation. It is important to have the mass of the solution in kg since the units of specific heat capacity involve kg.

9. Knowing the heat liberated when this number of moles of acid was neutralised, calculate the amount of heat that would be liberated when one mole of the acid (which , for HCl, is 1 mole of H^+ ions) is neutralised, i.e. the heat of neutralisation. The answer you get may not be the one quoted in data books of heats of neutralisation. What sources of experimental error may be responsible for this?

Calculations

The method of calculating the heat of reaction (heat of neutralisation) is shown in EXAMPLE 21.5.

Example 21.5

A student carried out an experiment to measure the heat of neutralisation of nitric acid by sodium hydroxide in a container made of plastic of negligible specific heat capacity. He used 100 cm^3 of 1.0 M nitric acid and 100 cm^3 of 1.0 M sodium hydroxide. The initial temperature of the solutions was 15.6 °C and the final temperature of the solution was 22.4 °C.

Given that the specific heat capacity of the solution is 4080 J kg^{-1} K^{-1}, calculate the heat of neutralisation. (Assume that the density of the solution is 1 g cm^{-3}.)

SOLUTION

Mass of solution = 200 g = 0.2 kg.

Temperature rise = 22.4 – 15.6 = 6.8 °C

Specific heat capacity = 4080 J kg^{-1} K^{-1}

Heat liberated = mass $\times$ specific heat capacity $\times$ temp rise

= 0.2 $\times$ 4080 $\times$ 6.8

= 5548.8 joules

Number of moles of HNO_3 neutralised = $\frac{\text{volume} \times \text{molarity}}{1000} = \frac{100 \times 1}{1000} = 0.1$

i.e. 0.1 mole of HNO_3 neutralised liberates 5548.8 joules

Therefore, 1 mole of HNO_3 neutralised liberates 5548.8 $\times$ 10 = 55,488 J

= 55.488 kJ

Since for HNO_3, 1 mole of acid gives 1 mole of H^+ ions

$\therefore$ Heat of neutralisation is - 55.488 kJ mol^{-1}

i.e. $HNO_3 + NaOH \rightarrow NaNO_3 + H_2O$ $\quad \Delta H = -55.488$ kJ mol^{-1}

Answer: Heat of Neutralisation = –55.488 kJ mol^{-1}

Note: If you use a diprotic acid like sulfuric acid, care must be taken in the calculations when determining the heat of neutralisation since one mole of acid gives **two** moles of H^+ ions:

e.g. $H_2SO_4 + 2NaOH \rightarrow Na_2SO_4 + 2H_2O$

This is illustrated in EXAMPLE 21.6.

Example 21.6

When 200 cm^3 of a sodium hydroxide solution were added to 200 cm^3 of a 0.4 M solution of sulfuric acid in a plastic container, a neutral solution was produced and the temperature rose by 5.5 °C. The density and specific heat capacity of the neutral solution (assumed equal to that of water) are 1.0 g cm^{-3} and 4.2 kJ kg^{-1} $°C^{-1}$ respectively.

(i) Why was a plastic container used?

(ii) What steps would you have taken to ensure an accurate measurement of the temperature rise?

(iii) Calculate the value of the heat of neutralisation. *[LCH]*

SOLUTION

(i) A plastic container is used because plastic has negligible specific heat capacity (i.e. we can assume the container itself does not absorb any heat).

(ii) A number of steps could be taken to ensure an accurate measurement of the temperature rise: make sure the temperature of both solutions is the same before mixing; mix the solutions quickly; avoid splashing when mixing the solutions; stir the solutions after mixing; make sure you note the **highest** temperature reached; use a cover on the plastic cup to prevent heat loss.

(iii) Mass of solution = 400 g = 0.4 kg.
Temperature rise = 5.5 °C
Specific heat capacity = $4.2\ kJ\ kg^{-1}\ K^{-1} = 4200\ J\ kg^{-1}\ K^{-1}$

(It is not essential to convert the kilojoules to joules. However, specific heat capacity values are usually quoted in $J\ kg^{-1}\ K^{-1}$ rather than in $kJ\ kg^{-1}\ K^{-1}$. Hence, we will work in joules.)

Heat liberated = mass × specific heat capacity × temp rise
= 0.4 × 4200 × 5.5
= 9240 joules

Number of moles of H_2SO_4 neutralised = $\frac{200 \times 0.4}{1000}$ = 0.08 mole

i.e. 0.08 mole of H_2SO_4 neutralised liberates 9240 joules

Therefore, 1 mole of H_2SO_4 neutralised liberates $\frac{9240}{0.08}$ = 115,500 J
= 115.5 kJ

i.e. $H_2SO_4 + 2NaOH \rightarrow Na_2SO_4 + 2H_2O \quad \Delta H = -115.5\ kJ\ mol^{-1}$
Essentially, what is happening is:
$2H^+ + 2OH^- \rightarrow 2H_2O$
Since, by definition, heat of neutralisation involves just one mole of H^+ ions reacting with one mole of OH^- ions to form one mole of H_2O, the heat of neutralisation is half the above value.

i.e. Heat of neutralisation = $-\frac{115.5}{2} = -57.75\ kJ\ mol^{-1}$ or $\Delta H = -57.5\ kJ\ mol^{-1}$

In short, the heat of neutralisation is half the heat of reaction in this case.

Answer: Heat of Neutralisation = $-57.5\ kJ\ mol^{-1}$

Test Yourself:
Attempt questions 21.11 – 21.14 and W21.11 – W21.13.

HEAT OF FORMATION

The final heat of reaction we shall be studying is the heat of formation of a compound. It is defined as follows:

> The heat of formation of a compound is the heat change that takes place when one mole of a compound in its standard state is formed from its elements in their standard states.

The standard state of an element or compound is its normal form at 25 °C and at one atmosphere pressure (101 kPa). For example, the heat of formation of water is $-285.8\ kJ\ mol^{-1}$ and may be represented as follows:

$$H_{2(g)} + \tfrac{1}{2}O_{2(g)} \longrightarrow H_2O_{(l)} \quad \Delta H = -285.8\ kJ\ mol^{-1}$$

The following equation does **not** represent the heat of formation of water:

$$H_{2(g)} + \tfrac{1}{2}O_{2(g)} \longrightarrow H_2O_{(g)} \quad \Delta H = -241.8\ kJ\ mol^{-1}$$

The reason is because the water formed in the above equation is in the gaseous state. At 25°C the normal state of water is a liquid.

Does the following equation represent the heat of formation of water?

$$2H_{2(g)} + O_{2(g)} \longrightarrow 2H_2O_{(l)} \quad \Delta H = -571.6\ kJ\ mol^{-1}$$

The answer is no because this equation shows the heat change that takes place when **two** moles of water are formed from its elements.

The heats of formation of most compounds have negative values of ΔH, i.e. energy is given out when the compound is formed from its elements, TABLE 21.5. We would expect this since compounds are generally more stable than the elements from which they are formed.

	Heat of formation ΔH measured in kJ mol^{-1}
$CH_4(g)$	-74.9
$C_2H_6(g)$	-84.7
$C_2H_5OH(l)$	-278
$CO(g)$	-111
$CO_2(g)$	-393
$NH_3(g)$	-46.2

Table 21.5

Heats of formation of some compounds.

As one would expect from the above definition, the heat of formation of any element in its standard state is zero.

A common error in examinations is for students to write the heat of formation equations incorrectly. Study the following examples carefully and note that in some cases it is necessary to introduce fractions:

$2C_{(s)} + H_{2(g)} \rightarrow C_2H_{2\,(g)}$ ΔH = +233 kJ mol^{-1}

$\frac{1}{2}N_{2(g)} + 1\frac{1}{2}H_{2(g)} \rightarrow NH_{3(g)}$ ΔH = -46 kJ mol^{-1}

$\frac{1}{2}H_{2(g)} + C_{(s)} + \frac{1}{2}N_{2(g)} \rightarrow HCN_{(g)}$ ΔH = +114.5 kJ mol^{-1}

$3C_{(s)} + 1\frac{1}{2}H_{2(g)} + \frac{1}{2}N_{2(g)} \rightarrow C_3H_3N_{(g)}$ ΔH = +194.5 kJ mol^{-1}

21.5.3 HESS'S LAW

Germain Henri Hess was professor of chemistry in the University of St. Petersburg, Russia. He measured the heat changes involved in various reactions. He realised that the quantity of heat produced in going from one substance to another was the same no matter what route was taken. In 1840, he announced his results in what is now called Hess's Law:

Hess's Law states that if a chemical reaction takes place in a number of stages, the sum of the heat changes in the separate stages is equal to the heat change if the reaction is carried out in one stage.

In other words, the amount of heat liberated or absorbed in a chemical reaction is the same whether the reaction takes place in one or several steps (FIG 21.43)

$$A \longrightarrow B \longrightarrow C \longrightarrow D$$

$$\Delta H_{A\rightarrow D} = \Delta H_{A\rightarrow B} + \Delta H_{B\rightarrow C} + \Delta H_{C\rightarrow D}$$

Fig 21.43

Diagrammatic representation of Hess's Law

Consider the two routes for converting carbon and oxygen to carbon dioxide (FIG 21.44).

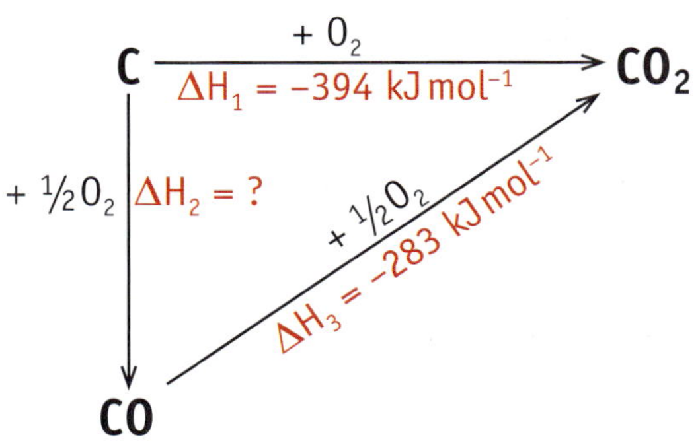

Fig 21.44

Hess's Law may be used to calculate the unknown heat of reaction.

In the first route, carbon is converted directly to carbon dioxide (ΔH_1 = -394 kJ mol^{-1}). In the second route, carbon monoxide is first formed (ΔH_2 = x kJ mol^{-1}) and then the carbon monoxide is further oxidised to carbon dioxide (ΔH_3 = -283 kJ mol^{-1}). According to Hess's Law, the heat change for the single step reaction is the same as the heat change for the two-step reaction.

i.e. $\Delta H_1 = \Delta H_2 + \Delta H_3$

$-394 = x - 283$

$x = -394 + 283 = -111$ kJ mol^{-1}

i.e. $C_{(s)} + \frac{1}{2}O_{2(g)} \longrightarrow CO_{(g)}$ $\Delta H_2 = -111$ kJ mol^{-1}

Thus, Hess's Law may be used to indirectly find the value of the heat of reaction of a chemical reaction that may be difficult to measure directly in the laboratory. For example, it is difficult to measure experimentally the heat of formation of CO since some CO_2 is also formed. However, the value may be calculated very accurately using Hess's Law.

Hess's Law is really just a restatement of the more general Law of Conservation of Energy as applied to heat changes involved in chemical reactions.

The Law of Conservation of Energy states that energy cannot be created or destroyed but can only be converted from one form of energy into another.

The usefulness of Hess's Law is that it means that thermochemical equations can be added, subtracted, multiplied or divided to calculate unknown heats of reaction. For this reason, it is one of the most useful laws in thermochemistry.

There are two types of Hess's Law calculations on this course. (The syllabus specifies that you will only be asked simple calculations!) These types of calculations are illustrated in EXAMPLE 21.7 AND 21.8.

(A) TO CALCULATE THE HEAT OF FORMATION OF A COMPOUND USING OTHER HEATS OF FORMATION AND ONE HEAT OF REACTION

Example 21.7

Calculate the heat of formation of methane from the following data:

(a) $C_{(s)} + O_{2(g)} \longrightarrow CO_{2(g)}$ $\Delta H = -393 \text{ kJ mol}^{-1}$

(b) $H_{2(g)} + \frac{1}{2}O_{2(g)} \longrightarrow H_2O_{(g)}$ $\Delta H = -286 \text{ kJ mol}^{-1}$

(c) $CH_{4(g)} + 2O_{2(g)} \longrightarrow CO_{2(g)} + 2H_2O_{(g)}$ $\Delta H = -879 \text{ kJ mol}^{-1}$ *[LCO]*

SOLUTION

Step 1

Write down the equation you require. We need to write the equation for the heat of formation of methane. Thus, one mole of methane must be on the right-hand side of the equation. The elements from which methane is formed must be on the left-hand side of the equation. The equation must be balanced.

$$C_{(s)} + 2H_{2(g)} \longrightarrow CH_{4(g)} \quad \Delta H = ?$$

We must now build up each part of this equation (i.e. C, $2H_2$ and CH_4) using the given equations (a), (b) and (c) above.

Step 2

Introduce C. [C is in (a) above.]

(a) : $C_{(s)} + O_{2(g)} \longrightarrow CO_{2(g)}$ $\Delta H = -393 \text{ kJ mol}^{-1}$

Step 3

Introduce $2H_2$. We have H_2 in (b) above. Therefore, multiply (b) by 2. Don't forget to also multiply ΔH by 2!

(b) $\times 2$: $2H_{2(g)} + O_{2(g)} \longrightarrow 2H_2O_{(g)}$ $\Delta H = -572 \text{ kJ mol}^{-1}$

Step 4

Get the left-hand side of the required equation by adding (a) to the above equation:

$$\text{(a)}: \quad C_{(s)} + O_{2(g)} \longrightarrow CO_{2(g)} \qquad \Delta H = -393 \text{ kJ mol}^{-1}$$

$$\text{(b)} \times 2: \quad 2H_{2(g)} + O_{2(g)} \longrightarrow 2H_2O_{(g)} \qquad \Delta H = -572 \text{ kJ mol}^{-1}$$

$$\text{(d)} \quad C_{(s)} + 2H_{2(g)} + 2O_{2(g)} \longrightarrow CO_{2(g)} + 2H_2O_{(g)} \qquad \Delta H = -965 \text{ kJ mol}^{-1}$$

Step 5

Introduce CH_4. We have CH_4 on the left-hand side of equation (c). We need it on the right-hand side of the required equation in the box above. Therefore, reverse equation (c) remembering to also reverse the sign of ΔH.

$$\text{(c) reversed:} \quad CO_{2(g)} + 2H_2O_{(g)} \longrightarrow CH_{4(g)} + 2O_{2(g)} \qquad \Delta H = +879 \text{ kJ mol}^{-1}$$

Step 6

Now add (c) reversed to (d) above:

$$\text{(d)}: \quad C_{(s)} + 2H_{2(g)} + 2O_{2(g)} \longrightarrow CO_{2(g)} + 2H_2O_{(g)} \qquad \Delta H = -965 \text{ kJ mol}^{-1}$$

$$\text{(c) reversed:} \quad CO_{2(g)} + 2H_2O_{(g)} \longrightarrow CH_{4(g)} + 2O_{2(g)} \qquad \Delta H = +879 \text{ kJ mol}^{-1}$$

$$C_{(s)} + 2H_{2(g)} + \cancel{2O_{2(g)}} + \cancel{CO_{2(g)}} + \cancel{2H_2O_{(g)}} \rightarrow \cancel{CO_{2(g)}} + \cancel{2H_2O_{(g)}} + CH_{4(g)} + \cancel{2O_{2(g)}} \qquad \Delta H = -86 \text{ kJ mol}^{-1}$$

i.e. $C_{(s)} + 2H_{2(g)} \longrightarrow CH_{4(g)} \quad \Delta H = -86 \text{ kJ mol}^{-1}$

Answer: $\Delta H = -86 \text{ kJ mol}^{-1}$

(B) TO CALCULATE THE HEAT OF REACTION USING HEATS OF FORMATION OF REACTANTS AND PRODUCTS

Example 21.8

Calculate the heat of reaction of the following reaction:

$$2H_2S_{(g)} + SO_{2(g)} \longrightarrow 3S_{(s)} + 2H_2O_{(l)} \qquad \Delta H = ?$$

You are given the following heats of formation:

$$\text{(a)} \quad H_{2(g)} + S_{(s)} \longrightarrow H_2S_{(g)} \qquad \Delta H = -21 \text{ kJ mol}^{-1}$$

$$\text{(b)} \quad S_{(s)} + O_{2(g)} \longrightarrow SO_{2(g)} \qquad \Delta H = -297 \text{ kJ mol}^{-1}$$

$$\text{(c)} \quad H_{2(g)} + \tfrac{1}{2} O_{2(g)} \longrightarrow H_2O_{(l)} \qquad \Delta H = -286 \text{ kJ mol}^{-1}$$

SOLUTION

Step 1

The required equation is:

$$\boxed{2H_2S_{(g)} + SO_{2(g)} \longrightarrow 3S_{(s)} + 2H_2O_{(l)} \qquad \Delta H = ?}$$

Step 2

Introduce $2H_2S$ by reversing (a) and multiplying by 2

(a) reversed: $H_2S_{(g)} \longrightarrow H_{2(g)} + S_{(s)}$ $\Delta H = +21 \text{ kJ mol}^{-1}$
(a) rev × 2: $2H_2S_{(g)} \longrightarrow 2H_{2(g)} + 2S_{(s)}$ $\Delta H = +42 \text{ kJ mol}^{-1}$

Step 3

Introduce SO_2 by reversing (b)

(b) reversed: $SO_{2(g)} \longrightarrow S_{(s)} + O_{2(g)}$ $\Delta H = +297 \text{ kJ mol}^{-1}$

Step 4

Add (a) rev × 2 to (b) reversed:

(a) rev × 2 : $2H_2S_{(g)} \longrightarrow 2H_{2(g)} + 2S_{(s)}$ $\Delta H = +42 \text{ kJ mol}^{-1}$
(b) reversed: $SO_{2(g)} \longrightarrow S_{(s)} + O_{2(g)}$ $\Delta H = +297 \text{ kJ mol}^{-1}$

(d) : $2H_2S_{(g)} + SO_{2(g)} \longrightarrow 2H_{2(g)} + 3S_{(s)} + O_{2(g)}$ $\Delta H = +339 \text{ kJ mol}^{-1}$

Step 5

We already have 3S on the right-hand side of the equation. Therefore, the only remaining item to be introduced is the $2H_2O$. This is done by multiplying (c) by 2 and adding it to (d).

(c) × 2: $2H_{2(g)} + O_{2(g)} \longrightarrow 2H_2O_{(l)}$ $\Delta H = -572 \text{ kJ mol}^{-1}$
(d) : $2H_2S_{(g)} + SO_{2(g)} \longrightarrow 2H_{2(g)} + 3S_{(s)} + O_{2(g)}$ $\Delta H = +339 \text{ kJ mol}^{-1}$

Add: $\cancel{2H_{2(g)}} + \cancel{O_{2(g)}} + 2H_2S_{(g)} + SO_{2(g)} \longrightarrow 2H_2O_{(l)} + \cancel{2H_{2(g)}} + 3S_{(s)} + \cancel{O_{2(g)}}$ $\Delta H = -233 \text{ kJ mol}^{-1}$

i.e. $2H_2S_{(g)} + SO_{2(g)} \longrightarrow 3S_{(s)} + 2H_2O_{(l)}$ $\Delta H = -233 \text{ kJ mol}^{-1}$

Answer: $\Delta H = -233 \text{ kJ mol}^{-1}$

Note: It is important when solving problems using Hess's Law that you approach them in a systematic fashion. The most common error in examinations is for students to approach the problem in a haphazard fashion, ending up 'going around in circles'. You have been warned!

Test Yourself:
Attempt questions 21.15 – 21.17 and W21.14 – W21.23.

CHAPTER SUMMARY

- A homologous series is a series of compounds of uniform chemical type, showing gradations in physical properties, having a general formula for its members. Each member has a similar method of preparation and each member differs from the previous member by a CH_2 unit.
- The structural formula of a compound shows how the atoms of the compound are arranged relative to one another. The full structural formula shows all the atoms and all the bonds in a compound.
- Structural isomers are compounds with the same molecular formula but different structural formulas.
- An aliphatic compound is an organic compound that consists of open chains of carbon atoms and closed-chain compounds (rings) that resemble them in chemical properties.
- Aromatic compounds contain a benzene ring in their molecules.
- The octane number of a fuel is a measure of the tendency of the fuel to resist knocking.
- Catalytic cracking is the breaking down of long-chain hydrocarbon molecules into short-chain molecules for which there is greater demand.
- Heat of reaction is the heat change when the number of moles of reactants indicated in the balanced equation for the reaction react completely.
- The kilogram calorific value of a fuel is the heat energy produced when 1 kg of the fuel is completely burned in oxygen.
- Bond energy is the energy required to break one mole of covalent bonds and to separate the neutral atoms completely from each other.
- The heat of formation of a compound is the heat change that takes place when one mole of a compound is formed from its elements in their standard states.
- Hess's Law states that if a chemical reaction takes place in a number of stages, the total heat change in the separate stages is equal to the heat change if the reaction is carried out in one stage.

EXERCISES

21.1. Draw the full structural formulas and give the systematic names of each of the following compounds.

(a) $CH_3CH(CH_3)CH_3$
(b) $CH_3CH_2CH_2CH(CH_3)CH_3$
(c) $CH_3CH_2CH(CH_3)CH(CH_3)CH_3$
(d) $CH_3CH(CH_3)CH(CH_3)CH_3$

21.2. Draw the structural formula of each of the following compounds

(a) 2-methylbutane
(b) 2,3-dimethylbutane
(c) cyclohexane
(d) 4-ethyl-2,2-dimethylheptane

21.3. Draw the full structural formulas and give the systematic names of each of the following compounds:

(a) $CH_3CH_2CH_2CH{=}CH_2$
(b) $CH_3CH{=}CHCH_2CH_3$
(c) $CH_3CH_2CH_2C(C_2H_5){=}CH_2$ [LCH]
(d) $CH_3CH{=}C(CH_3)CH_3$

21.4. Draw the structural formula of each of the following compounds:

(a) 2,3-dimethylbut-1-ene
(b) 2,4-dimethylpent-2-ene
(c) 1,3,5-trimethylhex-3-ene
(c) cyclohexene

Exercises

21.5. Three isomers of molecular formula C_5H_{12} belong to a homologous series of *saturated aliphatic* hydrocarbons.

(i) Explain the underlined terms.

(ii) Gives the name and the general formula of the homologous series to which the isomers belong.

(iii) Give the systematic (IUPAC) names and the structural formulas of the three isomers. *[LCH]*

21.6. Ethene gas was prepared and collected using the apparatus shown in Fig 21.45.
When a mixture of ethene and hydrogen was passed over a heated catalyst, ethane (contaminated with some unreacted ethene) was produced.

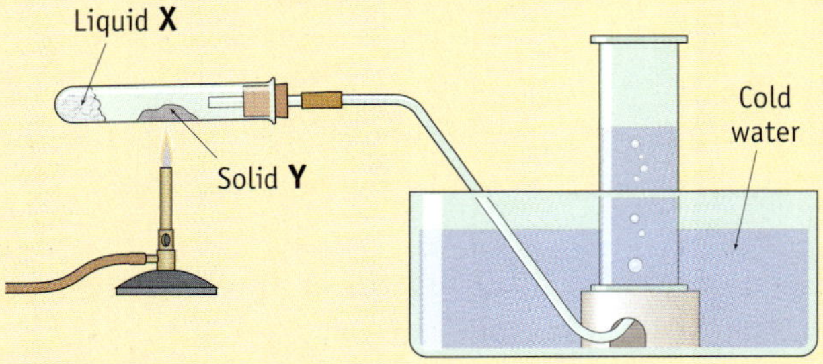

Fig 21.45

(i) Identify liquid X and solid Y. How was liquid X kept at the end of the horizontal tube?

(ii) What test would you have carried out to show that the ethane produced was contaminated with unreacted ethene? *[LCH]*

21.7. Fig 21.46(a) shows the apparatus used by a group of students to prepare ethene in the school laboratory. Fig 21.46(b) shows the apparatus the students used to prepare ethyne (acetylene).

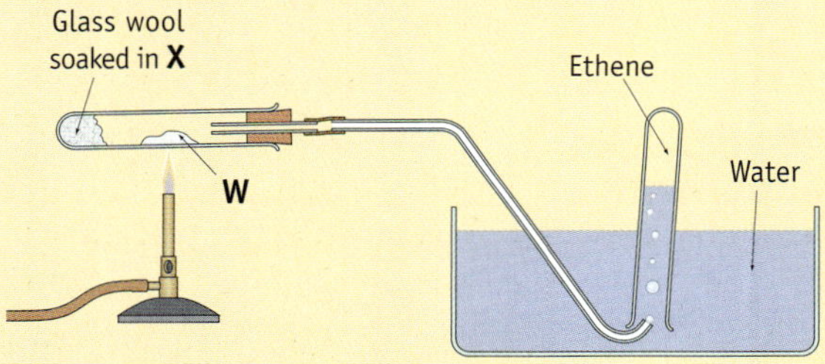

Fig 21.46(a)

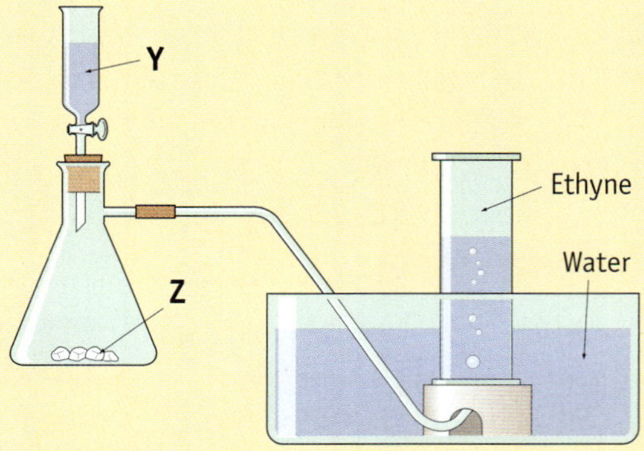

Fig 21.46(b)

(i) Identify W, X, Y, Z and write balanced equations for the reactions involved in the two preparations.

(ii) In preparation (a), what was the purpose of the glass wool?

(iii) In preparation (b), what would the students have observed when Y is dropped onto Z?

(iv) The ethyne obtained in preparation (b) is likely to have contained a number of impurities other than air or water vapour.
How might these impurities have been removed from the ethyne? *[LCH]*

21.8. (a) What is organic chemistry?

(b) Define the term *homologous series*. Give two examples of homologous series.

(c) What are isomers? Write down the names and structural formulas of any two isomers of pentane.

Exercises

(d) Ethene is often referred to as 'industry's most important organic chemical'. What is the reason for this?

(e) What is the colour of bromine water? What do you observe when bromine water is added to a sample of ethene gas and shaken?

(f) If bromine water were not available, what other chemical could you use to test for the presence of an unsaturated compound?

(g) Distinguish between the terms aliphatic and aromatic and give the name and structural formula of one example of each type of compound.

(h) Name the scientist who discovered benzene.

(i) When drawing the structure of benzene, chemists usually use a circle instead of alternating double bonds. Why is the circle preferable to the alternating double bonds?

(j) Draw the structural formulas of methylbenzene (toluene) and ethylbenzene. Write a brief note on the physical properties of methylbenzene (i.e. physical state and solubility).

21.9. With regard to the refining of crude oil, explain the meaning of the terms fractional distillation, LPG, octane number, catalytic cracking and reforming.
Give one use of each of the following: refinery gas, light gasoline, naphtha, kerosene, gas oil and any one of the residue fractions.
Why must one of the products of catalytic cracking always be unsaturated?
What is meant by the term *auto-ignition* in car engines?
Why are oxygenates often added to petrol? Name any one oxygenate commonly used.

21.10. In a simple catalytic cracking experiment using liquid paraffin, the paraffin was vaporised and the vapour was passed over a heated catalyst. This resulted in the production of colourless, flammable gases which were collected in test-tubes over water.

(i) What is catalytic cracking? Name a catalyst that could have been used in the experiment described above.

(ii) State any two benefits of catalytic cracking in the oil refining process.

(iii) Liquid paraffin is a mixture of different compounds. The molecular formula of one of these compounds is $C_{11}H_{24}$. Suppose that, as a result of the cracking experiment, a molecule of this compound gave an alkane molecule with two branches, a branched alkene molecule, and an unbranched alkene molecule. Write down the systematic (IUPAC) name and the structural formula of each of the three molecules obtained from the $C_{11}H_{24}$ molecule. *[LCH]*

21.11. Explain the meaning of the terms **exothermic reaction** and **endothermic reaction** and give one example of each.
Define the term **heat of reaction** and state the units in which it is measured.
Describe how you would accurately measure the heat of combustion of a fuel.
What information about a fuel does its kilogram calorific value give?

21.12. A student mixed 250 cm^3 of 0.5 M HCl with an equal volume of 0.5 M NaOH in a plastic container. The original temperature of both solutions was 14.8 °C and the final temperature was 18.2 °C. Calculate the heat of neutralisation of hydrochloric acid and sodium hydroxide. Assume that the density of the final solution is 1 g cm^{-3} and its specific heat capacity is 4060 J kg^{-1} K^{-1}.

21.13. When 100 cm^3 of a sodium hydroxide solution were added to 100 cm^3 of a 0.5 M solution of sulfuric acid in a plastic container, a neutral solution was produced and the temperature rose by 6.8 °C. The density and specific heat capacity of the neutral solution (assumed equal to that of water) are 1.0 g cm^{-3} and 4.2 kJ kg^{-1} °C^{-1} respectively.
Calculate the value of the heat of neutralisation.

EXERCISES

21.14. Define heat of neutralisation

When 50 cm^3 of a 1.0 M hydrochloric acid solution were added to 50 cm^3 of a potassium hydroxide solution in a plastic container of negligible heat capacity, the acid was completely neutralised and the temperature rose. Calculate the rise in temperature, given that the density and specific heat capacity of the resulting solution (assumed equal to those of water) are 1.0 g cm^{-3} and 4.2 kJ kg^{-1} K^{-1} respectively, and that the heat of neutralisation of hydrochloric acid by potassium hydroxide is $\Delta H = -57.2$ kJ mol^{-1}.

What change, if any, would you expect in the temperature rise if nitric acid were used in place of hydrochloric acid in the above experiment. Explain your answer. *[LCH]*

21.15. Calculate the heat of formation of $C_2H_{6(g)}$ from the following data:

$C_{(s)} + O_{2(g)} \longrightarrow CO_{2(g)}$ $\Delta H = -393.5$ kJ mol^{-1}

$H_{2(g)} + \frac{1}{2}O_{2(g)} \longrightarrow H_2O_{(g)}$ $\Delta H = -285.8$ kJ mol^{-1}

$2C_2H_{6(g)} + 7O_{2(g)} \longrightarrow 4CO_{2(g)} + 6H_2O_{(g)}$ $\Delta H = -3119.6$ kJ mol^{-1}

21.16. Calculate the heat of reaction for the reaction:

$HCOOH_{(l)} \longrightarrow CO_{(g)} + H_2O_{(l)}$

Given the following heats of formation:

$C_{(s)} + \frac{1}{2}O_{2(g)} \longrightarrow CO_{(g)}$ $\Delta H = -111$ kJ mol^{-1}

$H_{2(g)} + \frac{1}{2}O_{2(g)} \longrightarrow H_2O_{(l)}$ $\Delta H = -\ 285.8$ kJ mol^{-1}

$H_{2(g)} + O_{2(g)} + C_{(s)} \longrightarrow HCOOH_{(l)}$ $\Delta H = -\ 424.7$ kJ mol^{-1}

21.17. State Hess's Law.

Calculate the heats of formation of carbon monoxide and nitrogen monoxide from the following data:

$C_{(s)} + O_{2(g)} \longrightarrow CO_{2(g)}$ $\Delta H = -393$ kJ mol^{-1}

$2CO_{(g)} + O_{2(g)} \longrightarrow 2CO_{2(g)}$ $\Delta H = -570$ kJ mol^{-1}

$N_{2(g)} + 2O_{2(g)} \longrightarrow 2NO_{2(g)}$ $\Delta H = 66$ kJ mol^{-1}

$2NO_{(g)} + O_{2(g)} \longrightarrow 2NO_{2(g)}$ $\Delta H = -114$ kJ mol^{-1} *[LCH]*

Some Families of Organic Compounds

22

22.1 INTRODUCTION

In Chapter 21 we learned about families of organic compounds like the alkanes, alkenes and alkynes. In this chapter we will study some other families of organic compounds and learn about the important part that these compounds play in our lives. We will also learn about some organic products found in nature and perform an experiment to extract one of these compounds.

For the purposes of this course, we will divide the families of compounds into two categories. Firstly, we will study some organic compounds in which the atom or group of atoms responsible for the characteristic properties of the compound is attached to a tetrahedral carbon atom. Secondly, we will study some organic compounds in which the group of atoms responsible for the characteristic properties of the compound contains a planar carbon atom.

22.2 TETRAHEDRAL CARBON COMPOUNDS

In addition to the **alkanes**, the other tetrahedral carbon compounds that we will study in this course are the **chloroalkanes** and the **alcohols**.

22.2.1 CHLOROALKANES

In the early days of medicine, surgery was a very painful process. Because of the agony endured by the patient, surgery was only performed if it was considered absolutely essential. With the development of anaesthetics by chemists, surgery became a painless process and progressed at a very rapid rate. One of the early anaesthetics was a compound called chloroform, $CHCl_3$. The modern name for this compound is trichloromethane. Chloroform is an example of a **chloroalkane**. Chloroalkanes are compounds in which **one or more of the hydrogen atoms in an alkane molecule have been replaced by a chlorine atom**. Some examples of chloroalkanes are shown in Fig 22.1. Chloroalkanes have the same basic tetrahedral shape as the corresponding alkanes. Like the alkanes, they are all saturated compounds.

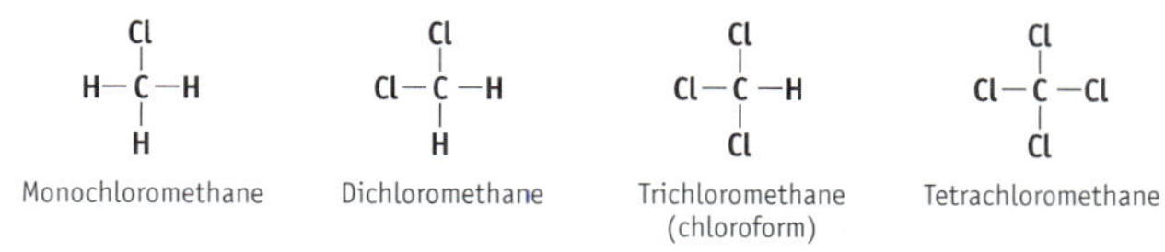

Fig 22.1
Some examples of chloroalkanes.

Chloroalkanes are very rarely found naturally occurring – most of them are synthesised in the laboratory. As can be seen from Table 22.1, most chloroalkanes are liquids at room temperature. Their main use is as solvents for removing oil and grease marks from machinery and in the dry cleaning of clothes. These solvents are very suitable for modern fabrics as they remove the oil and grease and other water insoluble substances from the clothes without harming the fabric. Paint stripper contains the solvent dichloromethane, CH_2Cl_2.

Compound	Name	Boiling point
CH_3Cl	Chloromethane	-23.8 °C
CH_2Cl_2	Dichloromethane	40.2 °C
$CHCl_3$	Trichloromethane	61.2 °C
CCl_4	Tetrachloromethane	76.8 °C
$CH_3CH_2CH_2Cl$	1-chloropropane	46.6 °C
$CH_3CH_2CH_2CH_2Cl$	1-chlorobutane	78.5 °C

Table 22.1
Boiling points of some chloroalkanes.

As expected, since they are organic compounds with little polarity, chloroalkanes are not soluble in water. However, they readily dissolve in (mix completely with) non-polar solvents like cyclohexane. Since many chloroalkanes have low boiling points, this makes them suitable solvents for substances like correcting fluid, e.g. *Tippex*. In the

1980s, it was discovered that some solvents containing chlorine cause damage to the ozone layer. (This will be covered in more detail in CHAPTER 26 of OPTION 1.) For this reason, considerable research is taking place to find alternatives to these compounds.

The method of naming chloroalkanes is illustrated in EXAMPLE 22.1.

Example 22.1

Name the compound $CH_3CH_2CHClCH_3$. *[LCH]*

STEP 1

Draw the full structural formula of the compound. Select the longest continuous chain of carbon atoms containing the carbon atom to which the chlorine atom is attached (FIG 22.2).

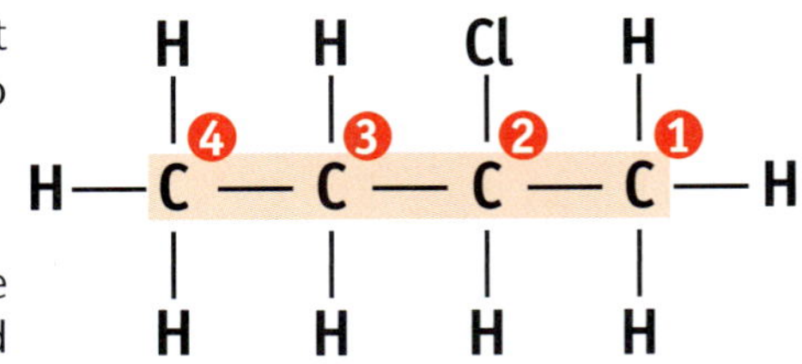

Fig 22.2

STEP 2

The compound is named after the parent alkane. Since there are four carbon atoms in the chain, the name will contain the word **butane**.

STEP 3

Number the carbon chain from the end that gives the lowest possible number to the carbon atom to which the chlorine atom is attached.

STEP 4

Indicate the position of the chlorine atom by a prefix to the name of the parent alkane, i.e. **2-chlorobutane**.

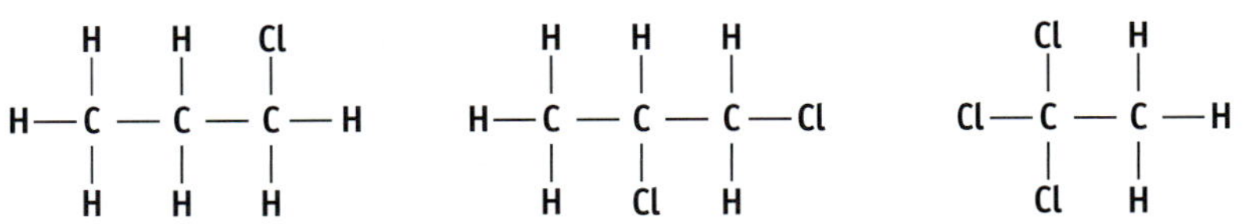

1-chloropropane   1,2-dichloropropane   1,1,1-trichloroethane

Fig 22.3
Examples of chloroalkanes.

Some further examples of chloroalkanes are shown in FIG 22.3. Note that the position of each chlorine atom is indicated by a number before the name. Also, the prefixes di- and tri- are used to indicate two and three chlorine atoms, respectively. (For your examination you must be able to give the names and draw the structures of chloroalkanes containing up to four carbon atoms.)

> **Test Yourself:**
> Attempt questions 22.1 – 22.3 and W22.1 – W22.3.

22.2.2 ALCOHOLS

When most people hear the word 'alcohol' they probably think of the alcohol in alcoholic drinks. The alcohol in alcoholic drinks, ethanol, is just one example of many alcohols. In fact, the alcohols form a homologous series of compounds. Alcohols may be imagined as being derived from alkanes by replacing a H atom by an –OH group. The –OH group is called the **hydroxyl group** and is the functional group of alcohols.

> A functional group is an atom or group of atoms which is responsible for the characteristic properties of a series of organic compounds.

Thus, the functional group of the alkanes is the C – C single bond, that of the alkenes is the C = C double bond and that of the alkynes is the C ≡ C triple bond. To remember that all alcohols contain the –OH functional group, think of the word alc**OH**ol.

The alcohols form a homologous series of general formula $\mathbf{C_nH_{2n+1}OH}$. (It is also possible to have alcohols with two or more –OH groups but these are outside the scope of the syllabus.) The names, formulas and molecular models of the aliphatic alcohols up to C-4 are shown in FIG 22.4. (Although it might appear from the structural formulas that the molecules are flat, remember that the shape around each of the carbon atoms is basically a tetrahedral shape and that the –OH functional group is V-shaped. The use of molecular models will help you to understand the shapes of the molecules.)

Alcohols are named from the corresponding alkanes by changing the **–ane** ending to **–anol**, e.g. methane becomes methanol, ethane becomes ethanol, propane becomes propanol. The method of assigning names to alcohols is shown in EXAMPLE 22.2.

Name	Formula	Structural formula	Molecular model
Methanol (methyl alcohol)	CH_3OH	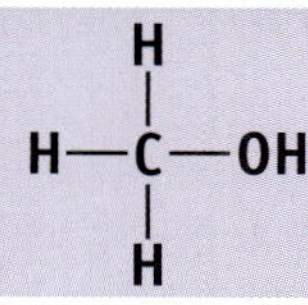	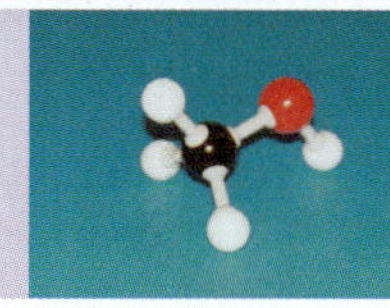
Ethanol (ethyl alcohol)	C_2H_5OH	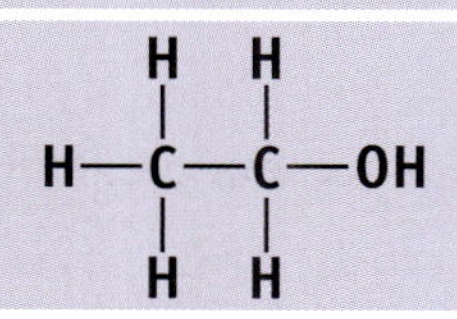	
Propan-1-ol	C_3H_7OH	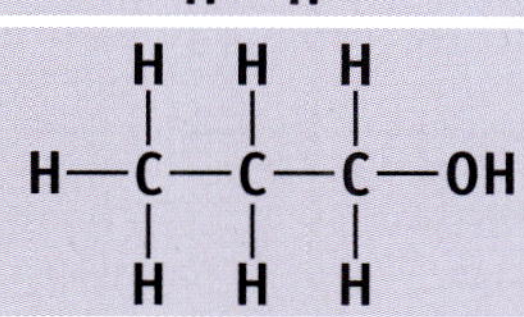	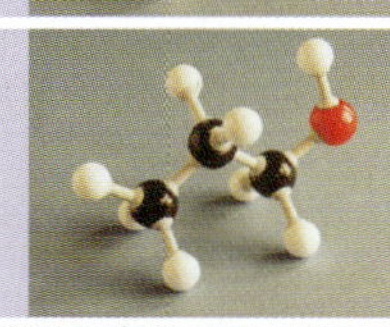
Propan-2-ol (isopropyl alcohol)	C_3H_7OH	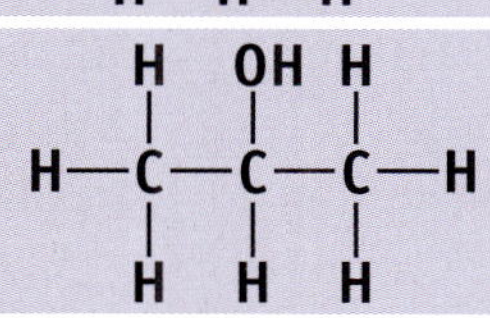	
Butan-1-ol	C_4H_9OH	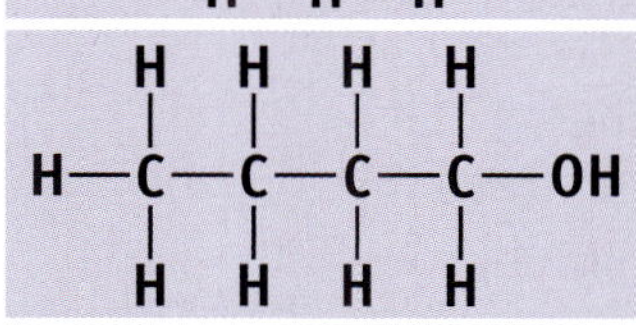	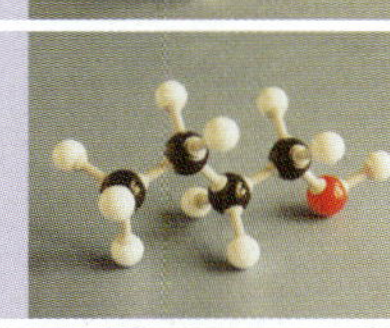
Butan-2-ol	C_4H_9OH	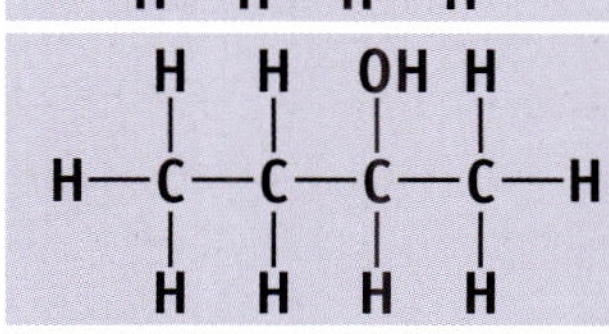	
2-methylpropan-1-ol	C_4H_9OH	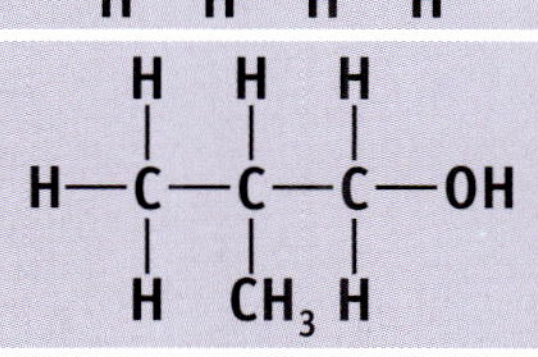	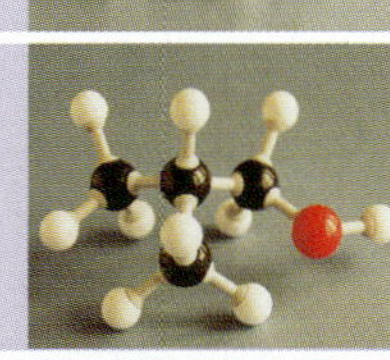
2-methylpropan-2-ol	C_4H_9OH	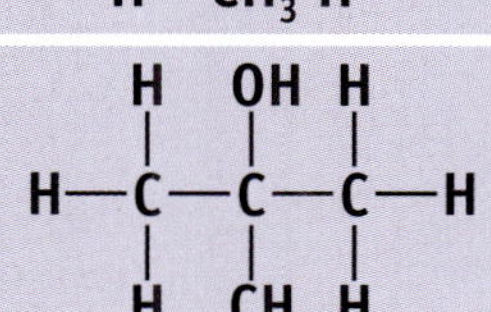	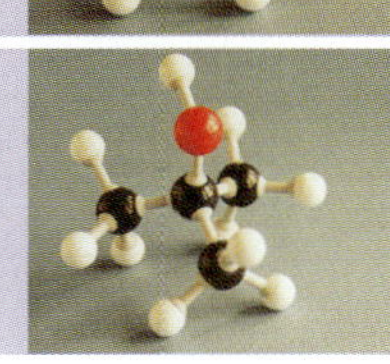

Fig 22.4
The members of the alcohols up to C-4.

Example 22.2

Draw the structural formula and name the compound $CH_3CH_2CH(OH)CH_3$.

STEP 1

Draw the full structural formula of the compound. Select the longest continuous chain of carbon atoms containing the carbon atom to which the –OH group is attached (FIG 22.5)

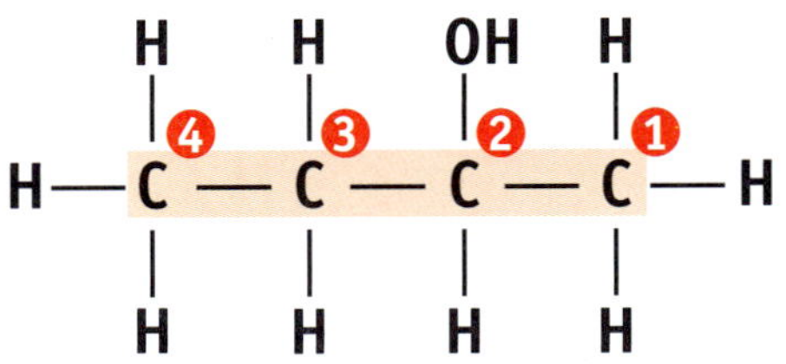

Fig 22.5

STEP 2

The compound is named after the parent alkane by changing the ending –ane to –anol. Since there are four carbon atoms in the chain, the name will contain the word **butanol**.

STEP 3

Number the carbon chain from the end that gives the lowest possible number to the carbon atom to which the –OH group is attached.

STEP 4

Indicate the position of the –OH group by a number, i.e. **butan-2-ol** or **2-butanol.**

Note: If there are any other substituents on the chain, these are also indicated by a number as in 2-methylpropan-2-ol whose structural formula is shown in FIG 22.4.

Alcohols may be classified as primary, secondary or tertiary. A **primary** alcohol is one where the carbon atom joined to the –OH group is attached to only **one** other carbon atom, e.g. ethanol (FIG 22.6). (The letter R is usually used to represent an alkyl group, i.e. methyl group $-CH_3$, ethyl group $-C_2H_5$, etc.) In a **secondary** alcohol, the carbon atom joined to the –OH group is attached to **two** other carbon atoms, e.g. propan-2-ol. In a **tertiary** alcohol, the carbon atom joined to the –OH group is attached to **three** other carbon atoms, e.g. 2-methylpropan-2-ol. Methanol, which contains only one carbon atom, is classified as a primary alcohol.

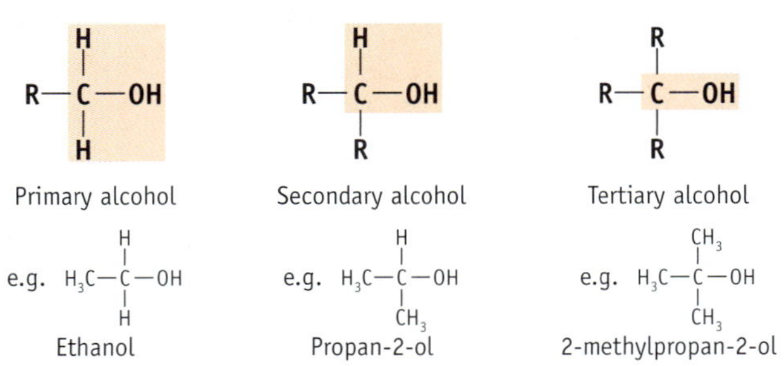

Fig 22.6
Alcohols are classified as primary, secondary or tertiary.

Of all the alcohols that exist, the best known one is ethanol. This is the alcohol found in alcoholic drinks. Ethanol for alcoholic drinks is produced by a process known as fermentation. Fruits like grapes contain the sugar glucose, $C_6H_{12}O_6$. When yeast is added, the sugar forms a solution of ethanol and carbon dioxide according to the following equation:

$$\underset{\text{glucose}}{C_6H_{12}O_6} \xrightarrow{\text{yeast}} \underset{\text{ethanol}}{2C_2H_5OH} + \underset{\text{carbon dioxide}}{2CO_2}$$

Yeast contains an enzyme called zymase which catalyses the above reaction. The source of the sugar for fermentation varies, e.g. grapes (for wine), apples (for cider), malted grain (for beer), etc. Wines contain about 10 – 12% v/v ethanol. Once the ethanol content reaches this level, it kills the yeast and fermentation is stopped. Therefore, it is impossible to produce alcoholic drinks greater than about 12% v/v by fermentation alone. Fermentation is used in the brewing industry to manufacture beer and cider. Beers and cider contain about 7 – 8% v/v ethanol. To produce drinks of higher alcohol concentration, fermented liquids are distilled, e.g. distillation of wine produces brandy. The distilling industry produces spirits like whiskey, gin, vodka and brandy. These contain about 40% v/v ethanol.

All alcohols are toxic to a greater or lesser extent. The first member of the homologous series, methanol, is particularly toxic. It is added to industrial alcohol to prevent people drinking it. This mixture is called methylated spirits. Methanol added in this way is called a denaturing agent. A purple dye is often added as a warning. Drinking methanol leads to blindness and death – hence the phrase 'blind drunk'.

Ethanol is the only one of the alcohols that people drink. It has the effect of relaxing muscles and loosening inhibitions. However, it is a very toxic chemical if taken in large quantities. Many cases of ethanol poisoning are seen in young people in Ireland each year. Excess ethanol can cause unconsciousness and vomiting. This can lead to immediate death as a result of drowning in vomit as well as liver failure in the longer term. Ethanol is a drug that depresses the activity of the central nervous system. This has a number of effects, e.g. it slows our reaction time, makes us less vigilant and impairs our judgement. For this reason, it is very unwise to drink alcohol before driving a vehicle. At present, the legal limit in Ireland for driving a vehicle is 80 mg of ethanol per 100 cm^3 of blood. Ethanol is the only commonly used drug that is sufficiently volatile to pass from the blood to the air in the lungs. Hence, a measurement of the ethanol concentration in the breath gives an indication of the concentration of alcohol in the blood. (In CHAPTER 23 we will discuss how these measurements are carried out.)

Ethanol is now established in our society as an accepted social drug. However, because of the fact that it is an addictive drug, it can lead to heavy drinking which causes liver disease and brain damage. In view of the serious effects of ethanol on physical and mental performance, there is little doubt that if ethanol were a new drug being introduced today, it would be unlikely to be awarded a licence and would probably be listed among the banned drugs!

In addition to its use in alcoholic drinks, ethanol is mainly used as a fuel (CHAPTER 21) and as a solvent. It has a great many uses as a solvent. It is used in substances like perfumes, after-shave lotions, deodorants, hair sprays, polishes, varnishes, dyes, etc. But why is ethanol such a good solvent? In fact, considering that ethane is a gas, why is ethanol a liquid? The answer to both of these questions lies in an understanding of **hydrogen bonding**.

Let us now briefly consider two physical properties of alcohols: boiling point and solubility.

(a) **Boiling point**. It is found that alcohols have higher boiling points than the corresponding alkanes (FIG 22.7). (Recall that the first four members of the alkanes are gases. However, the first four members of the alcohols are liquids.)

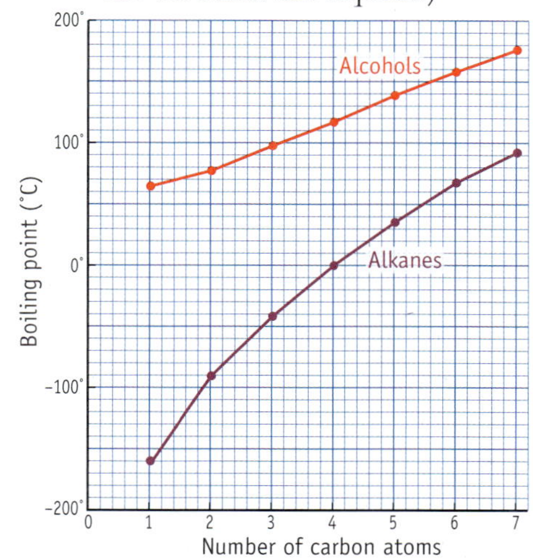

Fig 22.7
Alcohols have higher boiling points than the corresponding alkanes.

The higher boiling points are due to the fact that the highly polar –OH group gives rise to hydrogen bonding between the alcohol molecules (FIG 22.8). Because of this hydrogen bonding, extra energy must be supplied to break these bonds – hence the higher boiling points than the corresponding alkanes. Recall that this effect of hydrogen bonding was also used to explain the higher than expected boiling point of water (CHAPTER 5).

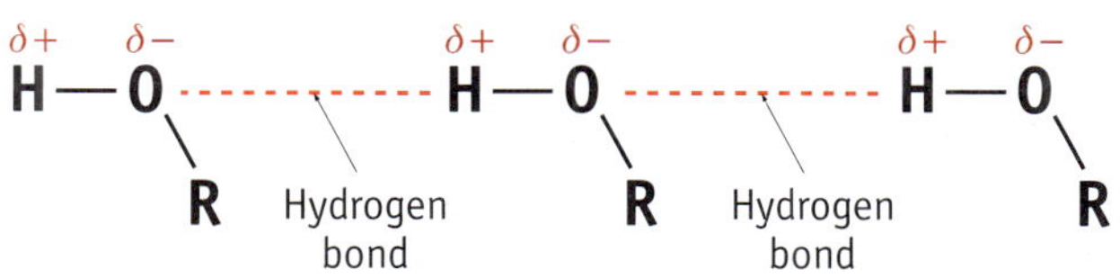

Fig 22.8
The hydrogen bonding between the alcohol molecules is responsible for the boiling points of alcohols being higher than those of the corresponding alkanes.

(b) **Solubility in water**. Methanol, ethanol and propan-1-ol are completely miscible with water. They are said to be infinitely soluble in water, i.e. miscible in all proportions. Verify this for yourself in the laboratory by mixing some ethanol and water in a test-tube. The reason for the solubility in water is because hydrogen bonding occurs between the alcohol molecules and the water molecules (FIG 22.9).

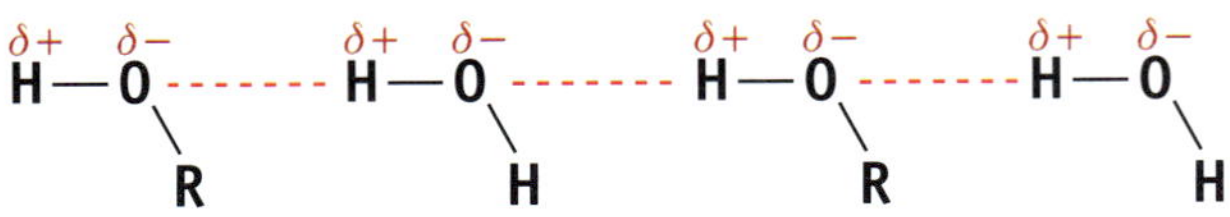

Fig 22.9
Hydrogen bonding between the alcohol molecules and the water molecules causes the alcohol to be completely miscible with water.

In fact, it is impossible to separate ethanol from water by distillation. The two liquids form a mixture with a constant boiling point that distils over as a liquid commonly called **95% ethanol**. (To produce 100% pure ethanol, the water must be removed by a chemical drying agent like calcium oxide, which absorbs water. Pure ethanol is commonly called **absolute alcohol**.) However, the solubility of alcohols decreases with the length of the carbon chain. The tendency of the –OH group (polar) to make the alcohol soluble is counteracted by the insoluble alkyl (non-polar) portion of the molecule (FIG 22.10).

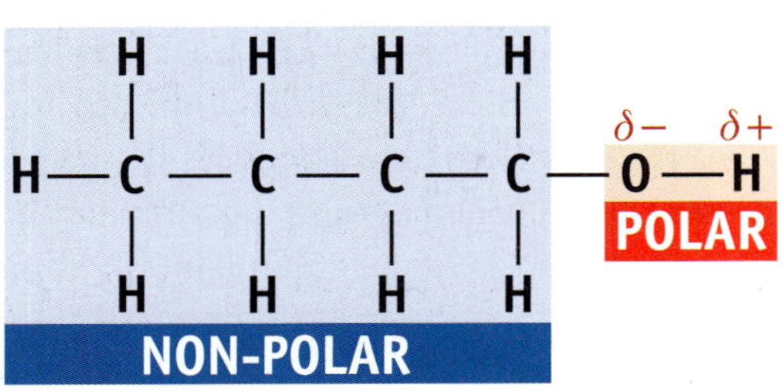

Fig 22.10
Alcohol molecules have a polar and a non-polar part.

The lower members of the series (C_1 – C_3) are completely miscible with water but the solubility decreases rapidly as the length of the carbon chain increases. Thus, as far as solubility is concerned, the higher alcohols behave in a similar fashion to the alkanes. They have poor solubility in water and readily dissolve in solvents like cyclohexane.

We can now understand why ethanol is such an ideal solvent for so many of the products outlined above. In perfumes and after-shave lotions, for example, the ethanol dissolves both the water-insoluble oils (responsible for the aroma) as well as the water-soluble ingredients in the perfume.

As an exercise investigate the solubility of (a) methanol and (b) butan-1-ol in (i) water and (ii) cyclohexane.

> **Test Yourself:**
> Attempt questions 22.4 and W22.4.

22.3 PLANAR CARBON COMPOUNDS

We have already studied some organic compounds containing planar carbon atoms, e.g. the **alkenes** and **alkynes**. We will now study some more families of planar carbon compounds: **aldehydes**, **ketones**, **carboxylic acids** and **esters**. In addition, we will also study some **aromatic compounds**.

22.3.1 ALDEHYDES

Aldehydes form a homologous series of organic compounds containing the –CHO functional group as shown in FIG 22.11. Thus, the general formula for aldehydes is RCHO. The general formula for the homologous series of aliphatic aldehydes is $\mathbf{C_nH_{2n+1}CHO}$.

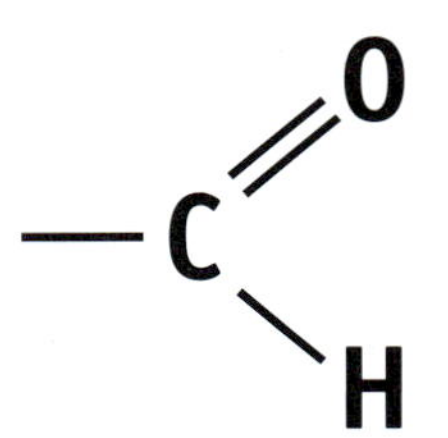

Fig 22.11
The aldehyde functional group

The names and formulas of the aliphatic aldehydes up to C-4 are shown in FIG 22.12. Note that there are two aldehyde isomers of C_3H_7CHO.

Aldehydes are named from the parent alkane by changing the **–ane** ending to **–anal**, e.g. methane becomes methanal, ethane becomes ethanal, propane become propanal. You need only remember the names and structural formulas of aliphatic aldehydes up to C-4 as shown in FIG 22.12. Note that the aldehyde functional group must always occur at the end of the carbon chain. (Think about this!) Thus, the naming of aldehydes is easier

Name	Formula	Structural formula	Molecular model
Methanal (formaldehyde)	**HCHO**	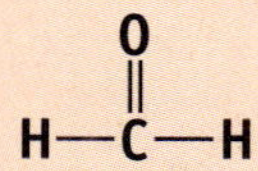	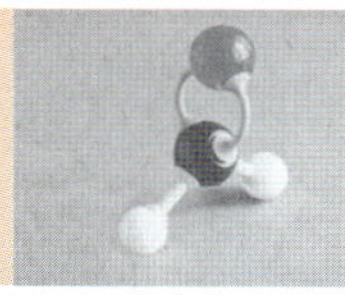
Ethanal (acetaldehyde)	**CH_3CHO**	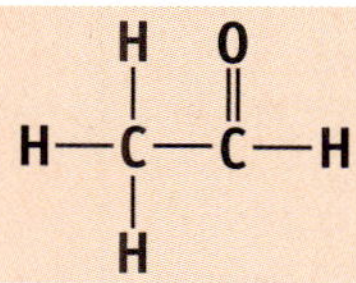	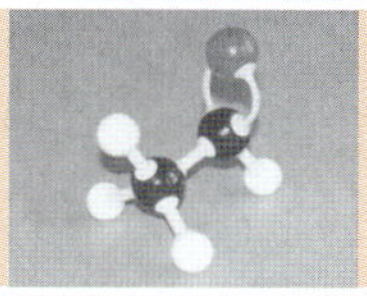
Propanal (propionaldehyde)	**C_2H_5CHO**	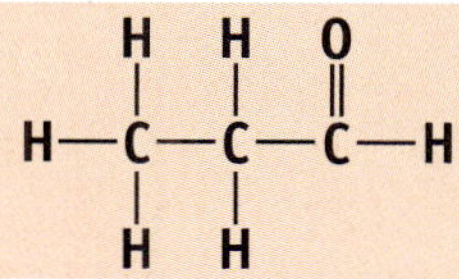	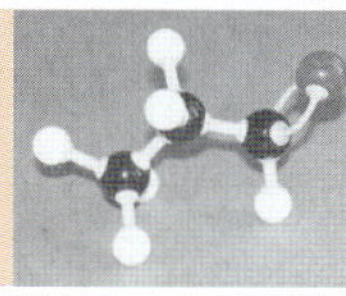
Butanal	**C_3H_7CHO**	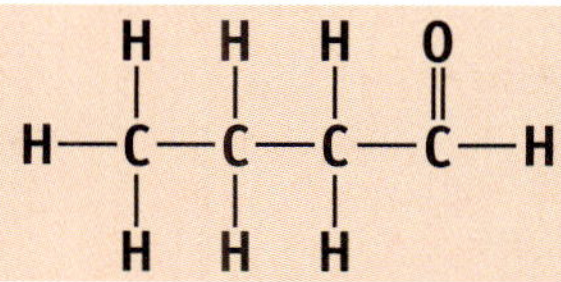	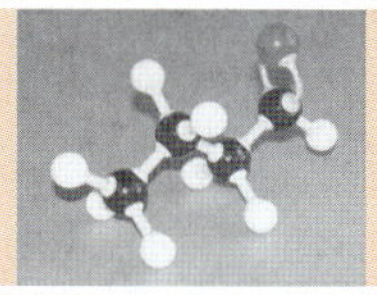
2-methylpropanal	**C_3H_7CHO**	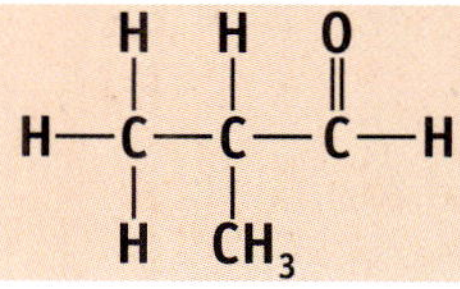	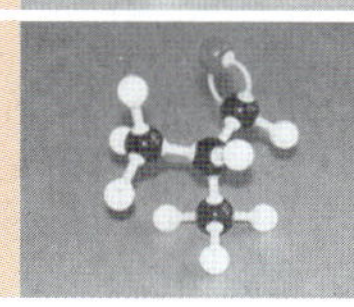

Fig 22.12
The members of the aldehydes up to C-4. Note that C_3H_7CHO has two isomers.

than that of alcohols since there is no need to use a number to indicate the position of the functional group. However, numbers are used to indicate the position of substituents on the main carbon chain, e.g. 2-methylpropanal, FIG 22.12.

The $-\overset{\overset{O}{\|}}{C}-$ group is called the **carbonyl group**. The carbonyl group has a major effect on the physical and chemical properties of the aldehydes. In one respect, the C = O group is similar to the C = C in ethene as the carbonyl group and the atoms to which it is bonded all lie in the one plane with all bond angles of 120°. (This is clearly seen using the molecular model kit in your laboratory.) However, there is one big difference between the C = O group and the C = C group. This difference is that the carbonyl group is strongly polar, $C^{\delta+} = O^{\delta-}$, due to the different electronegativity values of carbon and oxygen (E.N. of C = 2.5, E.N. of O = 3.5).

Let us now consider the influence that the polar carbonyl group has on the physical properties of the aldehydes. It is **not** possible for compounds containing the carbonyl group to form hydrogen bonds with each other since hydrogen bonding only occurs when hydrogen is bonded to an electronegative atom like nitrogen, oxygen or fluorine (CHAPTER 5). In aldehydes, the hydrogen is bonded to carbon. However, there are permanent dipole-dipole interactions between the aldehyde molecules. These dipole-dipole forces were shown in FIG 5.41 (page 65). Therefore, the boiling points of aldehydes are higher than the corresponding alkanes (because of the dipole-dipole forces) but lower than the corresponding alcohols (because of the lack of hydrogen bonding) (FIG 22.13). Note from FIG 22.13 that methanal (formaldehyde) is the only aldehyde that is a gas at room temperature. Note also that ethanal is a very volatile liquid (B.P. = 21 °C) whereas ethanol (B.P. = 78 °C) is far less volatile.

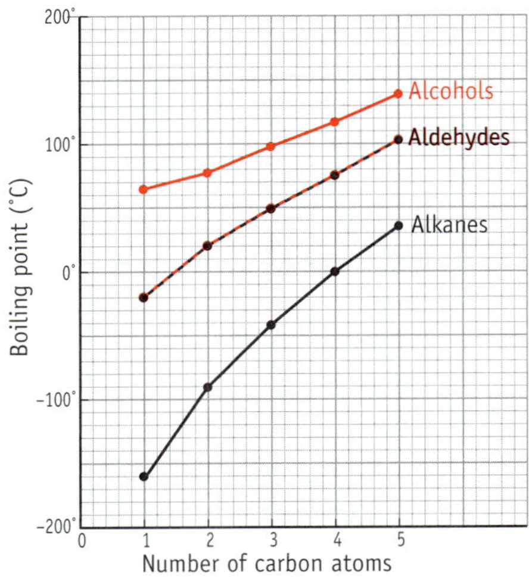

Fig 22.13
The boiling points of aldehydes are higher than the corresponding alkanes because of dipole-dipole forces between the aldehyde molecules. The boiling points of aldehydes are lower than the corresponding alcohols because of the lack of hydrogen bonding between the aldehyde molecules.

A solution of methanal in water, called formalin, is often used for preserving biological specimens and as an embalming fluid. The lower members of the aldehydes are very soluble in water and, like the alcohols, will dissolve both polar and non-polar substances. The high solubility in water is due to the fact that hydrogen bonding takes place between the O atom of the carbonyl group and the H atom of the water molecule (FIG 22.14). However, like the alcohols, the solubility in water decreases with the length of the carbon chain due to the insoluble alkyl portion of the molecule. All aldehydes are quite soluble in organic solvents.

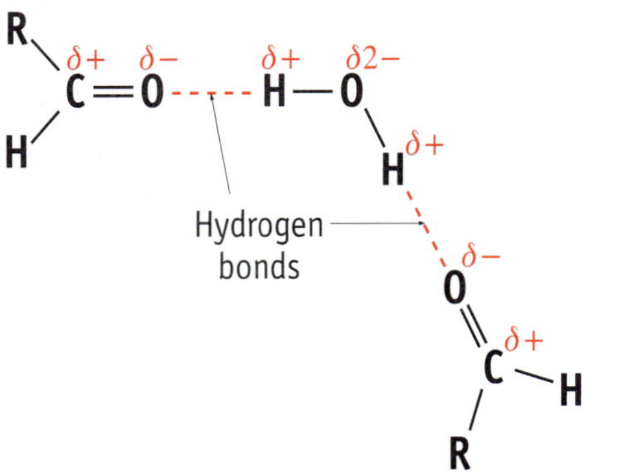

Fig 22.14
Hydrogen bonding causes aldehydes to be very soluble in water.

As an exercise, investigate the solubility of ethanal in (i) water and (ii) cyclohexane.

The only aromatic aldehyde on your course is benzaldehyde (FIG 22.15). You probably have some in your school laboratory. Note the typical smell of almonds. This is because benzaldehyde is found in the oil of almonds.

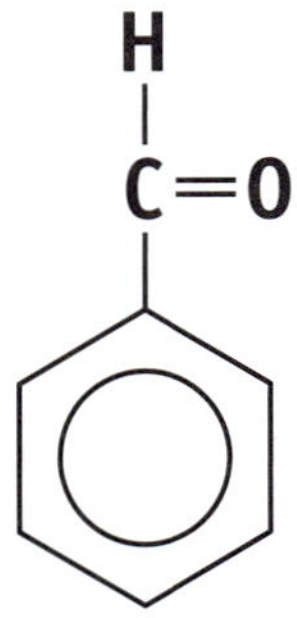

Fig 22.15
Benzaldehyde is found in the oil of almonds.

22.3.2 KETONES

Ketones form a homologous series of organic compounds containing the >C = O functional group attached to two alkyl groups as shown in FIG 22.16. Thus, the general formula for ketones is RCOR'.

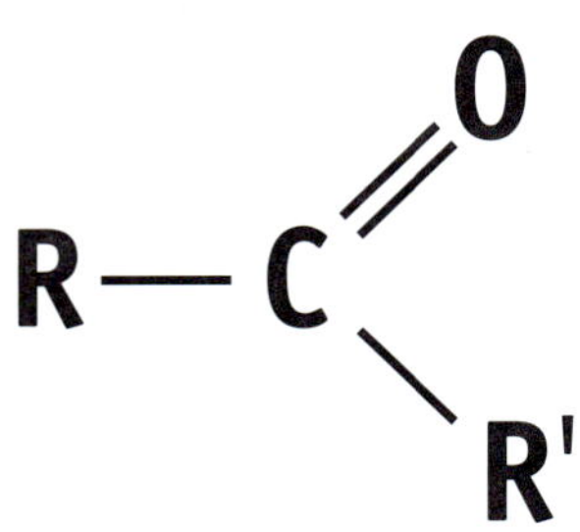

Fig 22.16
The general formula for ketones. R and R' are alkyl groups. These alkyl groups may be the same or different.

If you compare FIG 22.11 and FIG 22.16, you will observe that the only difference in the structural formulas of aldehydes and ketones is the replacement of the hydrogen atom in aldehydes by an alkyl group, R', in ketones. (To remember that the aldehydes are the series containing the H atom, write down the word alde**H**yde.) Since the carbonyl group in ketones has two alkyl groups attached, it cannot be located at the end of the chain. The names and formulas of the first two members of the homologous series of ketones are shown in FIG 22.17. (These are the only two ketones on our course.)

Name	Formula	Structural formula	Molecular model
Propanone (acetone)	CH_3COCH_3	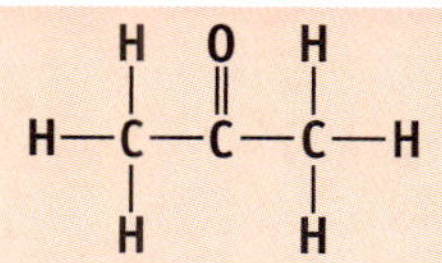	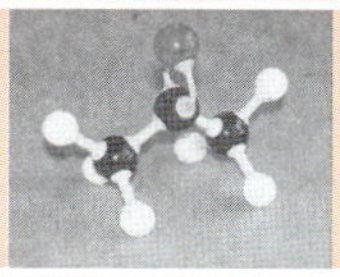
Butanone (methylethyl ketone)	$CH_3COC_2H_5$	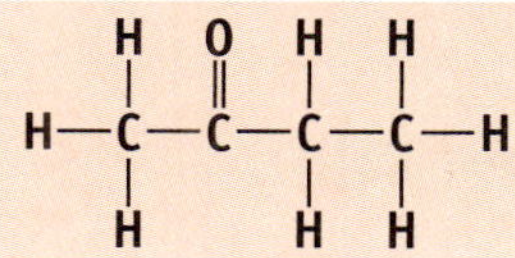	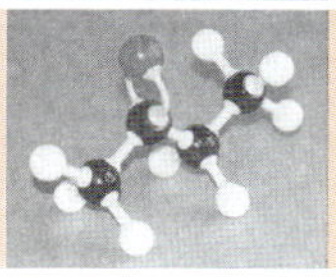

Fig 22.17
The first two members of the homologous series of ketones.

Ketones are named by changing the final **–e** of the parent alkane to **–one**. Thus, propane becomes propanone, butane becomes butanone, etc. The best known ketone is propanone (commonly called acetone) which is widely used as an organic solvent, e.g. nail varnish remover (FIG 22.18).

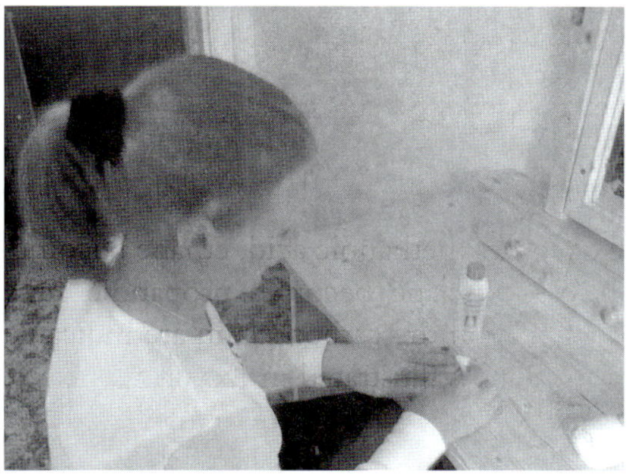

Fig 22.18
Propanone (acetone) is the solvent used in nail varnish

The physical properties of ketones are very similar to those of aldehydes. As in the case of aldehydes, ketones have higher boiling points than the corresponding alkanes, e.g. the boiling point of propane is - 42 °C but that of propanone is 56 °C. Similarly, the boiling point of butane is -0.5 °C but that of butanone is 80 °C. The reason for this is exactly the same as in the case of aldehydes, i.e. the presence of dipole-dipole forces between the molecules of the ketone (FIG 5.40 page 64).

As in the case of aldehydes, the boiling points of ketones are lower than those of the corresponding alcohols due to the lack of hydrogen bonding in ketones, e.g. the boiling point of propanone is 56 °C but that of propan-1-ol is 97 °C. Similarly, the boiling point of butanone is 80 °C but that of butan-1-ol is 117 °C.

The information we learned about the second physical property of aldehydes, solubility, also holds true for ketones. The polarity of the C = O group in ketones causes the lower members of the ketones to be very soluble in water, e.g. propanone and butanone are completely miscible with water. This is due to the hydrogen bonding between the carbonyl group and water molecules as discussed in the case of aldehydes (FIG 22.14). Since propanone and butanone can act as solvents for both polar and non-polar substances, they are widely used as solvents in industry.

As an exercise, investigate the solubility of propanone in (i) water and (ii) cyclohexane.

> **Test Yourself:**
> Attempt questions 22.5 and W22.5.

22.3.3 CARBOXYLIC ACIDS

Carboxylic acids are a family of organic compounds that contain the carboxyl functional group, -COOH (FIG 22.19). Thus, the general formula for carboxylic acids is RCOOH. Note that the carboxyl group contains an –OH group (like the alcohols) and the polar >C = O group (like the aldehydes and ketones). However, the chemical reactions of the carboxylic acids tend to be quite different from either the alcohols or the aldehydes and ketones.

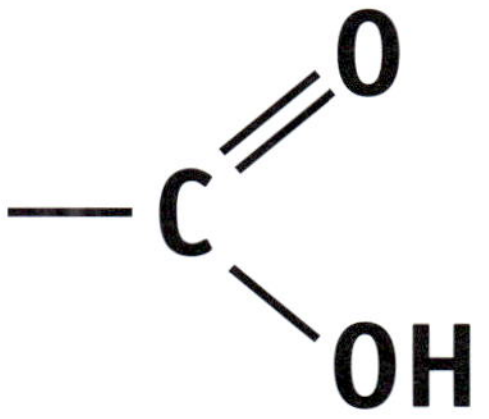

Fig 22.19
The carboxyl group is the functional group in carboxylic acids.

Name	Formula	Structural formula	Molecular model
Methanoic acid (formic acid)	**HCOOH**	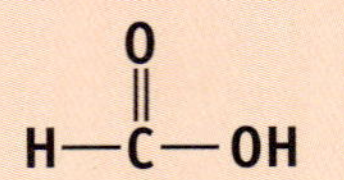	
Ethanoic acid (acetic acid)	**CH_3COOH**	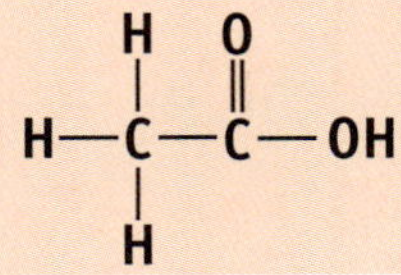	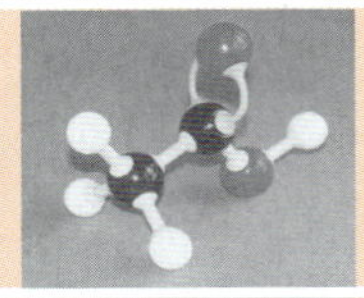
Propanoic acid	**C_2H_5COOH**	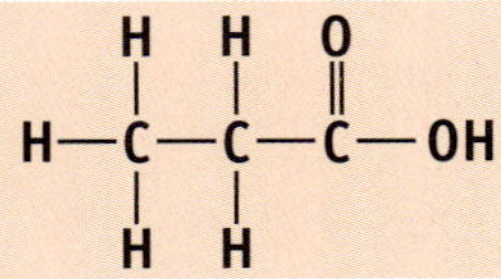	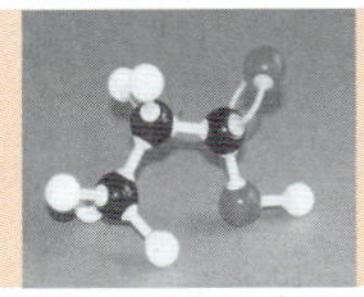
Butanoic acid	**C_3H_7COOH**	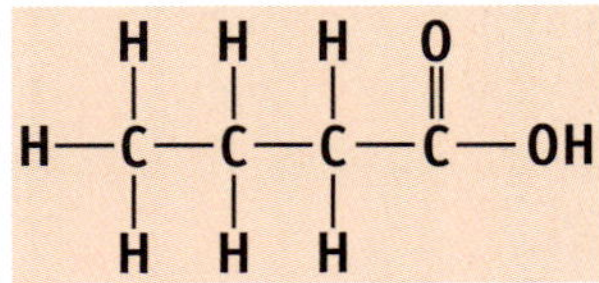	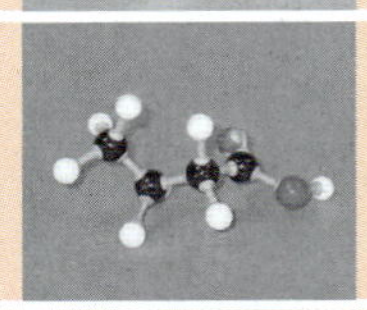
2-methylpropanoic acid	**C_3H_7COOH**	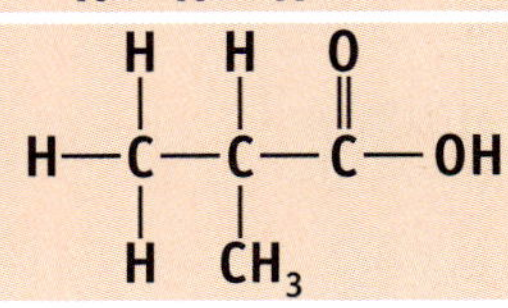	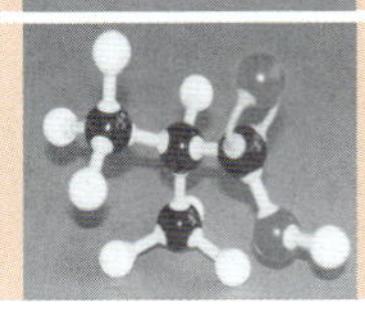

Fig 22.20
Carboxylic acids up to C-4.

The names and formulas of the carboxylic acids up to C-4 are shown in Fig 22.20. Note that there are two carboxylic acid isomers of C_3H_7COOH.

Carboxylic acids are named from the parent alkane by changing the **–e** ending to **–oic acid.** Thus methane becomes methanoic acid, ethane becomes ethanoic acid, propane becomes propanoic acid, etc. Substituents on the carbon chain are indicated by numbers. This is illustrated in Example 22.3.

Example 22.3

What is the systematic (IUPAC) name for $CH_3CCl_2CH_2COOH$? [LCH]

STEP 1

Draw the full structural formula of the compound. Select the longest continuous chain of carbon atoms containing the –COOH group (Fig 22.21).

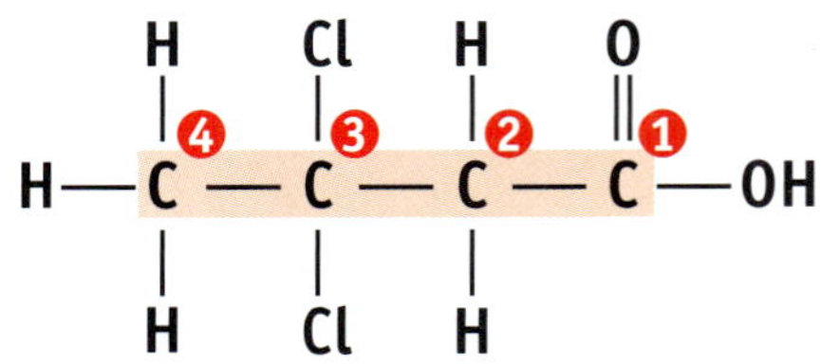

Fig 22.21

STEP 2

The compound is named after the parent alkane by changing the –e ending to –oic acid. Since there are four carbon atoms in the chain, the name will contain the words **butanoic acid**.

STEP 3

Number the carbon chain from the –COOH end. The carbon atom of the carboxyl group is always taken as carbon atom 1.

STEP 4

Indicate the position of substituents by a number, i.e. **3, 3-dichlorobutanoic acid.**

Methanoic acid is found in the stings of nettles and ants (FIG 22.22). (The old name for methanoic acid was formic acid. *Formica* is the Latin word for an ant.) Ethanoic acid is probably the best known member of the carboxylic acid family. (The old name for ethanoic acid was acetic acid. *Acetum* is the Latin word for vinegar.) Vinegar is simply a dilute solution (about 4% v/v) of ethanoic acid. Ethanoic acid is also used to make a chemical called cellulose acetate, which is used in varnishes, lacquers, photographic film for cameras and a type of textile called rayon.

Fig 22.22
Ants inject methanoic acid (formic acid) into their victims when they bite.

Propanoic acid, benzoic acid (FIG 22.23) and some of their salts (e.g. isodium benzoate) are used as food preservatives. (Recall that we recrystallised benzoic acid in EXPERIMENT 5.1 page 54.)

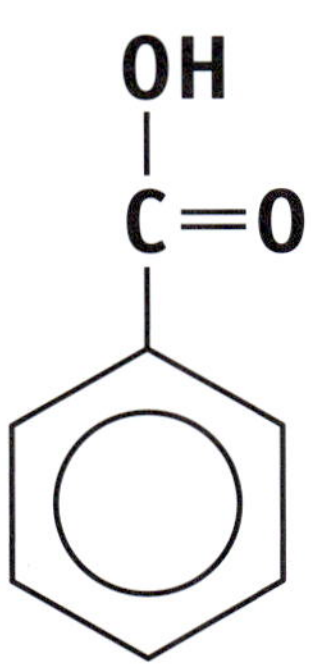

Fig 22.23
Benzoic acid occurs naturally in some plants. This compound and some of its salts are used as food preservatives.

Carboxylic acids have quite characteristic odours. Most people would recognise the sharp smell of vinegar. The longer-chain carboxylic acids have unpleasant odours. For example, the smell of rancid butter is due to the presence of butanoic acid (old name: butyric acid). Human sweat contains a mixture of various carboxylic acids. The odour of unwashed smelly socks is also due to the presence of butanoic acid! The stench of goats is caused by decanoic acid! Our noses are very sensitive to the presence of carboxylic acids and can detect their presence at very low concentrations (10^{-11} mol/L). Dogs are even more sensitive to these smells. Hence, tracker dogs given a scent of a person's clothing, can recognise the smell of the particular mixture of carboxylic acids produced by the sweat glands and follow the scent to track the particular individual (FIG 22.24).

Fig 22.24
A tracker dog recognises the characteristic mixture of carboxylic acids produced by the sweat glands of an individual.

The physical properties of the carboxylic acids are governed by their ability to form hydrogen bonds. Pure ethanoic acid is a liquid at ordinary room temperature. However, if the temperature of the room falls below 17 °C, the ethanoic acid turns into a solid, as the temperature of the room is now below the melting point; the room is not warm enough to melt the pure ethanoic acid. Hence, pure ethanoic acid is often referred to as *glacial acetic acid* as the solid material looks like ice (FIG 22.25).

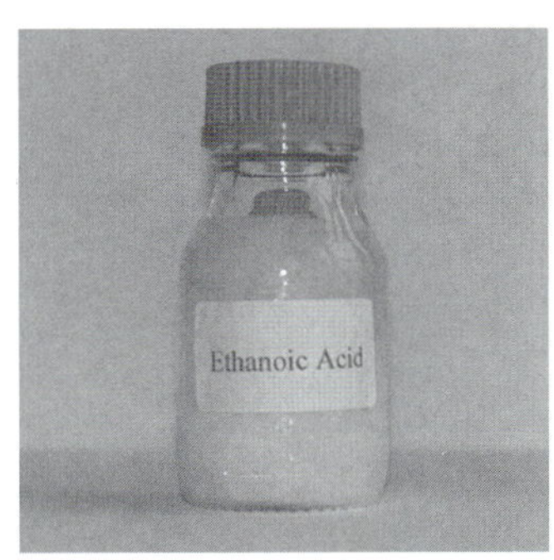

Fig 22.25
Ethanoic acid freezes if the temperature drops to 17 °C or below. Since the ethanoic acid turns into an ice-like solid, it is commonly called glacial ethanoic acid. The high melting point of ethanoic acid is due to the presence of hydrogen bonding.

Why does ethanoic acid have such a high melting point? This is due to the effect of hydrogen bonding. In the pure acid, the molecules group together to form clusters of **dimers** (two molecules) which are held together by two hydrogen bonds (FIG 22.26).

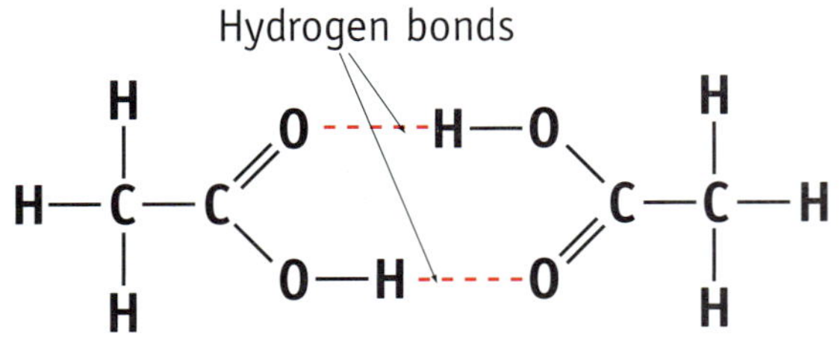

Fig 22.26
Hydrogen bonding gives rise to dimers (groups of two molecules joined together) being formed in pure ethanoic acid.

The effect of this increased amount of hydrogen bonding is clearly seen if we compare the melting point of ethanoic acid with the corresponding members of other organic families, (TABLE 22.2).

NAME	MELTING POINT
Ethane	-183 °C
Ethanal	-124 °C
Ethanol	-117 °C
Ethanoic acid	17 °C

Table 22.2
Melting point of ethanoic acid compared with the corresponding members of other organic families.

Similarly, the boiling point of ethanoic acid (118 °C) is higher than that of ethanol (78 °C). This is due to the greater amount of hydrogen bonding in carboxylic acids compared to alcohols. As expected, methanoic acid, ethanoic acid, propanoic acid and butanoic acid are all liquids at room temperature. All of these compounds are completely soluble in water due to their ability to form hydrogen bonds with water molecules (FIG 22.27).

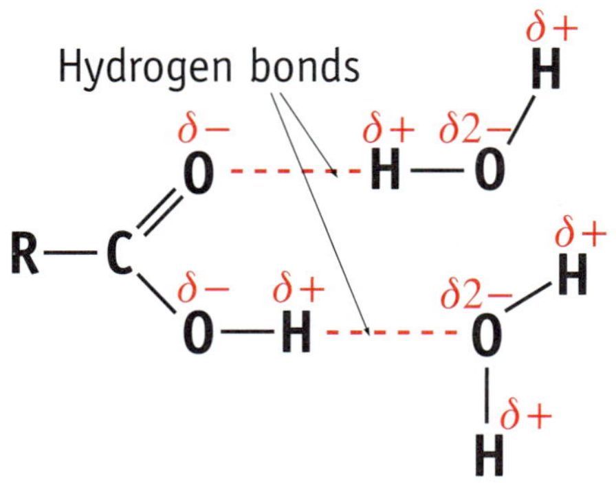

Fig 22.27
The ability of carboxylic acids to form hydrogen bonds with water makes carboxylic acids soluble in water.

As is the case with the other homologous series discussed so far, solubility in water decreases with increasing length of the carbon chain. (In EXPERIMENT 5.1, you may recall that benzoic acid was not very soluble in cold water.) The higher members of the homologous series become more soluble in organic solvents like cyclohexane.

As an exercise, investigate the solubility of ethanoic acid in (i) water and (ii) cyclohexane.

22.3.4 ESTERS

Esters are a family of compounds of general formula RCOOR' (FIG 22.28). Thus, the functional group is the –COO– group. Esters may be considered as being derived from carboxylic acids by replacing the H of the –OH group by an alkyl group.

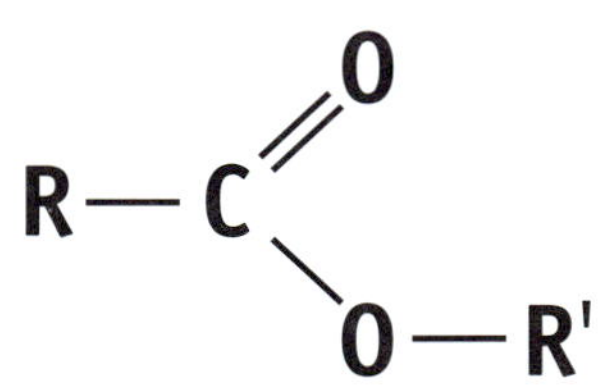

Fig 22.28
The general formula of esters. R and R' may be the same alkyl groups or different alkyl groups.

Esters may be prepared in the laboratory by the reaction between an alcohol and a carboxylic acid. For example, if we react methanol with ethanoic acid, an ester called methyl ethanoate is formed (FIG 22.29).

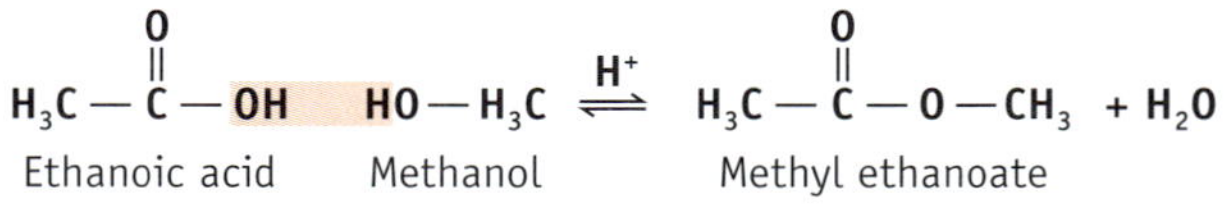

Fig 22.29
This reaction is an example of an esterification reaction. It is catalysed by the presence of H^+ ions.

This reaction is called a **condensation reaction** since it results in the loss of a water molecule. It will be studied in more detail in CHAPTER 23. (Research using radioactive isotopes has shown that the oxygen atom that forms a bridge between the two parts in the ester molecule comes from the alcohol molecule rather than the carboxylic acid molecule.)

The naming of esters is a little more complicated than the naming of the other organic compounds on our course. Study EXAMPLE 22.4 carefully!

Example 22.4

Name the compound $HCOOCH_2CH_3$

Step 1

Draw the full structural formula of the ester.

Step 2

Divide the structural formula into the part derived from the carboxylic acid and the part derived from the alcohol (Fig 22.30).

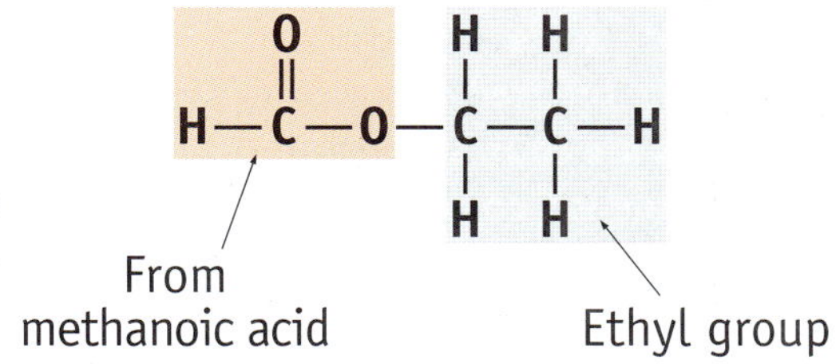

Fig 22.30

Step 3

Write down the name of the alkyl group attached to the bridging oxygen atom. In this case, it is an **ethyl** group.

Step 4

Examine the remaining part of the molecule. Identify the carboxylic acid from which the ester has been derived. In this case it has come from methanoic acid. Change the –oic ending to –oate. This gives us the word **methanoate**.

Step 5

The compound is named by putting the name of the alkyl group before the name obtained in step 4. Thus, the name of the compound is **ethyl methanoate**.

You are only required to know the names and structural formulas of the esters up to C-4. The IUPAC names for these compounds (as well as the traditional names) are given in Fig 22.31.

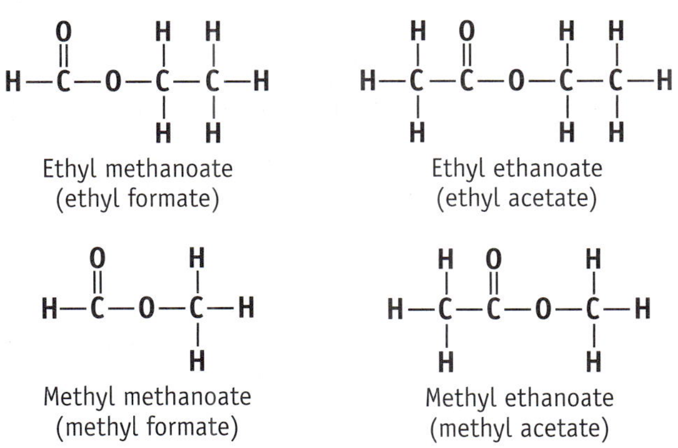

Fig 22.31
The esters up to C-4.

The lower members of the ester family are volatile liquids with pleasant fruity smells. Esters are responsible for many of the scents of flowers and fruits. They have low boiling points (B.P. ethyl ethanoate = 77 °C) because it is not possible for the ester molecules to form hydrogen bonds with each other. (They do not have an –OH group as in the case of carboxylic acids.) However, as they consist of polar molecules they can form hydrogen bonds to a certain extent with water molecules. Therefore, the lower members (up to C-5) are fairly soluble in water. The solubility of ethyl ethanoate in water at 25 °C is 8.5 grams per 100 g of water. As expected, the solubility of the esters decreases as the length of the carbon chain increases. However, they are soluble in organic solvents.

As an exercise, examine the solubility of ethyl ethanoate in (i) water and (ii) cyclohexane.

You may have heard the term **polyester** used to refer to a type of fabric that is used in clothing and in manufacturing items like sails for yachts. Polyesters consist of millions of ester molecules linked together.

Fats and oils are esters that occur in nature (Fig 22.32). They are used by plants and animals for storing energy. If the ester is a solid, it is called a **fat**.

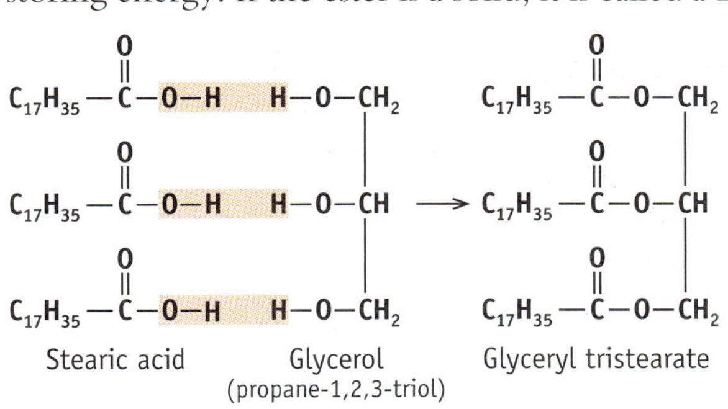

Fig 22.32
Fats are naturally-occurring esters. They are usually formed from glycerol and long-chain acids called fatty acids found in nature. Glyceryl tristearate is one example of a fat.

If it is a liquid, it is called an **oil**. These esters are insoluble in water but soluble in organic solvents. They are usually esters that are made from an alcohol called glycerol and from various long-chain carboxylic acids. These carboxylic acids are commonly referred to as **fatty acids**. (About 50 fatty acids are found in nature.) One example of a fat molecule is a compound called glyceryl stearate. This can be considered as being formed from a particular fatty acid called stearic acid, $C_{17}H_{35}COOH$, found mainly in animal fats and the alcohol glycerol. (Glycerol is the only alcohol with more than one –OH group that you will meet in this course.)

Fats and oils are used extensively in the manufacture of soap. We shall be making soap in the laboratory in CHAPTER 23 where we will discuss this topic in more detail. We will now prepare some esters.

Test Yourself:
Attempt questions 22.6 and W22.6.

EXPERIMENT 22.1

Test-tube preparation of esters

Introduction

Esters made in the laboratory are commonly used in fruit flavourings (pear, cherry, orange, raspberry, banana, etc.) and in the manufacture of perfumes and cosmetics. It is very easy to make esters on a test-tube scale in the laboratory. The syllabus specifies that you should make the ester ethyl ethanoate. In addition, we will make two other esters with characteristic smells.

In this experiment, concentrated sulfuric acid is used as a catalyst.

Apparatus required

Rack of test-tubes, test-tube holder, Pyrex beaker, hot plate or Bunsen burner, spatula, droppers (graduated).

Chemicals required

Ethanol, ethanoic acid (glacial), sulfuric acid (concentrated), pentan-1-ol (amyl alcohol), salicylic acid, methanol.

Procedure

1. Mix 2 cm^3 approx. of ethanol and 1 cm^3 approx. of glacial ethanoic acid in a test-tube.
2. Add three drops of concentrated sulfuric acid to the test-tube.
3. Place the test-tube in a beaker of hot water for about ten minutes. Note the characteristic 'glue-like' smell of the ethyl ethanoate that is formed. (Ethyl ethanoate is actually used as a solvent in many glues. When the ethyl ethanoate evaporates, the solid glue is left behind and binds the materials together. Esters are also used as solvents for varnishes and spray paints.)
4. Repeat the above experiment using pentan-1-ol (also called amyl alcohol) instead of ethanol. A substance called pentyl ethanoate is formed after a few minutes. Note the 'banana-like' smell due to the fact that pentyl ethanoate is one of the substances found in bananas.
5. We will now make a third ester called methyl salicylate. This substance has a very characteristic smell. It is made by mixing a small amount of an acid called salicylic acid (enough to fit on the tip of a spatula) with 2 cm^3 approx. of methanol in a test-tube. Add three drops of concentrated sulfuric acid. Place the test-tube in a beaker of hot water for about ten minutes. The sports people among you will recognise the smell of this ester! What substance bought in a chemist's shop has a similar smell to this compound?

22.3.5 AROMATIC COMPOUNDS

In Chapter 21 we learned that aromatic compounds are compounds that contain a benzene ring structure in their molecules. We saw how all the bonds in the benzene ring were identical and were intermediate between a double bond and a single bond. Hence, it is more correct to draw a circle inside the benzene ring rather than alternating double and single bonds. If we take a closer look at the bonding in benzene, we can see clearly why the circle inside the hexagon was proposed to indicate a molecule of benzene.

Each carbon atom in the benzene ring has four electrons in its outer shell. Of these four electrons, each carbon atom uses three of them to form sigma (σ) bonds. Two sigma bonds are formed between each carbon atom and its two adjacent carbon atoms. The third sigma bond is formed with an atom of hydrogen. The sigma-bonded skeleton is shown in Fig 22.33.

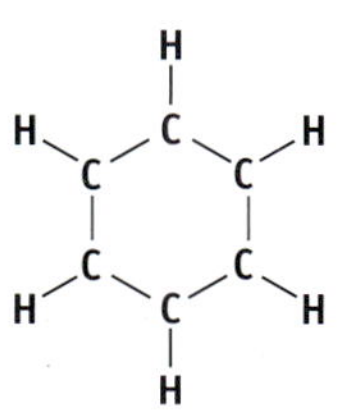

Fig 22.33
The sigma bonds in a molecule of benzene.

The one electron remaining in the outer shell of each carbon atom is in a p orbital. Each p orbital is perpendicular to the plane of the ring as shown in Fig 22.34. Each p orbital overlaps sideways with its two neighbouring orbitals to form two doughnut-shaped electron clouds. One of these electron clouds is above the plane of the ring and the other is below the plane of the ring. These two electron clouds give rise to a pi (π) bond in the molecule of benzene.

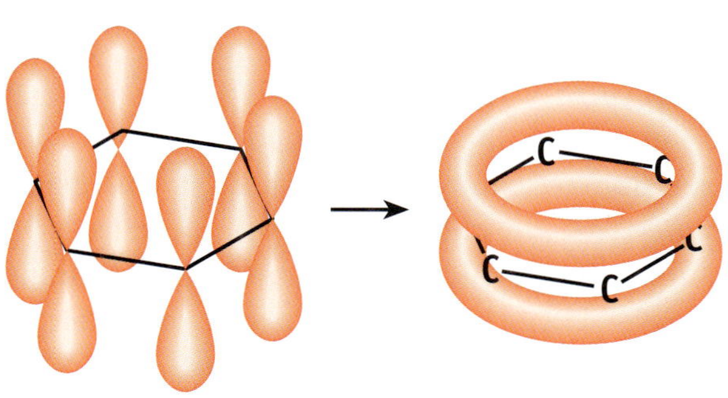

Fig 22.34
Formation of the pi bond in a molecule of benzene. (Hydrogen atoms omitted for clarity.)

The six electrons in the pi bond do not belong to any particular atom. These six electrons are said to be **delocalised**, i.e. not restricted to any particular carbon atom but are free to move around the ring within the pi-bonded system. In this way, the electrons help to bind together not just two carbon atoms but all six carbon atoms. Thus, the delocalised electrons give extra stability to the benzene molecule. The exceptional stability of the molecule is shown by the fact that benzene does not undergo the typical addition reactions that compounds with C = C double bonds (e.g. ethene) undergo.

From the above description, it should be clear that benzene is a planar molecule. It is also a non-polar molecule due to the small electronegativity difference between carbon and hydrogen (E.N. of C = 2.5, E.N. of H = 2.1). For many years, benzene was widely used as an organic solvent. However, due to its carcinogenic nature it is no longer used for this purpose. As mentioned in Chapter 21, methylbenzene (toluene) is the recommended substitute for it and is used as an industrial solvent. Methylbenzene is also used in the manufacture of plastics and explosives and is added to petrol to increase the octane rating.

As an exercise, investigate the solubility of methylbenzene in (i) water and (ii) cyclohexane.

Although benzene is carcinogenic, many aromatic compounds are not dangerous and, in fact, play an important role in our lives. They form the basis of many pharmaceutical compounds: dyes, detergents, insecticides, herbicides, disinfectants, etc. In addition, acid-base indicators like phenolphthalein and methyl orange are aromatic compounds (Fig 18.9, page 258). A number of aromatic compounds are shown in Fig 22.35. You are **not** expected to remember the formulas of any of these compounds but you should be able to remember some of their names to give an indication of the range and scope of aromatic compounds.

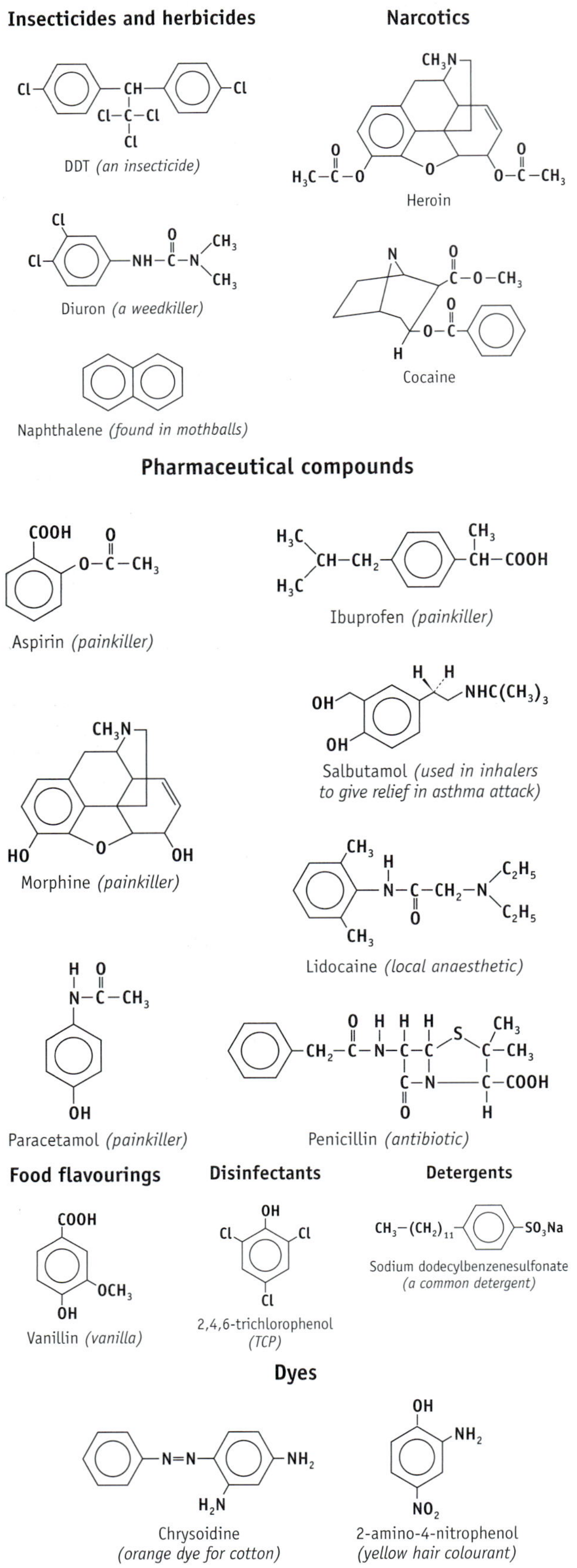

Fig 22.35
The many uses of aromatic compounds.

> **Test Yourself:**
> Attempt question W22.7.

22.4 ORGANIC NATURAL PRODUCTS

Many organic compounds are found in nature. These compounds are commonly referred to as 'natural products'. Examples of natural products are compounds like benzaldehyde (in almonds), caffeine (in tea and coffee), nicotine (in tobacco), opium (in poppies), limonene (in orange peel), quinine (in the bark of cinchona trees), penicillin (in bread mould), etc.

During the nineteenth century, organic chemistry was mainly concerned with the isolation of many of these natural products from plants and human body fluids. Compounds like morphine (1817), strychnine (1818), quinine (1820), caffeine (1821), nicotine (1828), codeine (1832), cocaine (1859), etc. were all isolated because of their particular properties. Many of these natural products are valuable medicines, e.g. quinine is used for the treatment of malaria. Others were isolated because of their poisonous properties, e.g. strychnine was used as a rat poison.

At the time that many of the above substances were discovered, it was not possible to determine the structure of their molecules. With modern analytical techniques, it is now possible to determine the molecular structures and to synthesise these compounds in the laboratory. It is important that many of these natural products can be synthesised in the laboratory. Otherwise, they could be difficult to obtain due to the seasonal nature of plants, weather conditions, etc. In some cases, chemists can improve the medicinal properties of natural products. For example, Felix Hofmann synthesised aspirin in 1898. The structure of this was similar to a naturally-occurring substance but had none of the unpleasant side effects.

In addition to compounds isolated for medicinal purposes, many other compounds were isolated because their fragrant oils could be used for making perfumes. These oils are a mixture of different alcohols, aldehydes, ketones, esters, etc. For example, an alcohol called menthol is one of the major ingredients in oil of peppermint and an alcohol called geraniol is found in rose oil (Fig 22.36).

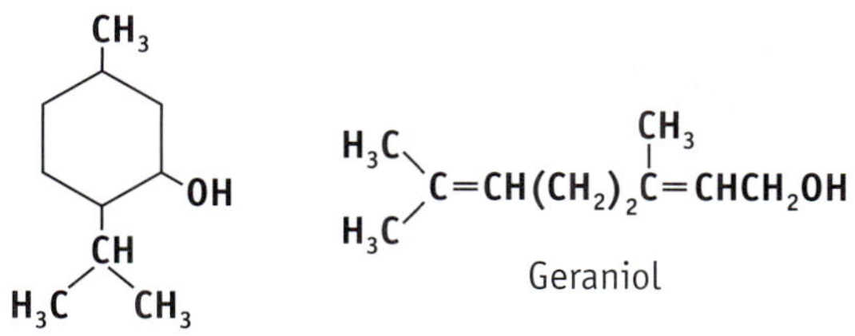

Menthol Geraniol

Fig 22.36
Menthol is found in oil of peppermint and geraniol is found in rose oil.

A technique called **steam distillation** is used to separate the oils from the plants themselves. It is used extensively to extract the oil from plants like roses, lavender, cloves, etc. Steam distillation involves carrying out distillation in a current of steam, i.e. passing steam through the plant material. The main purpose is to avoid too high a temperature during distillation as a high temperature may destroy the plant material. It is found that the mixture of steam and oil distils below the boiling point of water and well below the boiling point of the oil in the plant. The mixture of oil and steam is then condensed by passing it through a Liebig condenser. This mixture is allowed to stand. (The oil usually floats on top of the water.) The two liquids are then separated from each other using a separating funnel.

Sometimes, the distillate comes over as a mixture where the oil droplets are dispersed throughout the water. This is called an **emulsion**. In this case, an organic solvent that dissolves the oil (e.g. cyclohexane) but does not mix with water is added to the emulsion and shaken up with it. The organic solvent is then separated from the water using a separating funnel. The organic solvent is allowed to evaporate leaving the pure oil behind. This process is called **solvent extraction**.

We now use steam distillation to extract clove oil from cloves.

EXPERIMENT 22.2 MANDATORY STUDENT EXPERIMENT NO. 24

To extract clove oil from cloves by steam distillation.

Introduction

In this experiment, steam distillation is used to extract clove oil from cloves. Clove trees are cultivated in places like Zanzibar, Madagascar and Mauritius. Cloves consist of unopened flowers that are picked when their heads develop a pinkish colour. At this period of its growth, the clove is richest in oil. The main constituent of the oil found in cloves is a substance called eugenol (Fig 22.37). You do not need to know the formula of this compound. Oil of cloves is used for flavouring various kinds of food products like meat (spiced beef), sausages, sweets, apple pies, sauces, pickles, etc. Because of its antiseptic properties it is also used in various pharmaceutical preparations and in perfumes, soaps, etc.

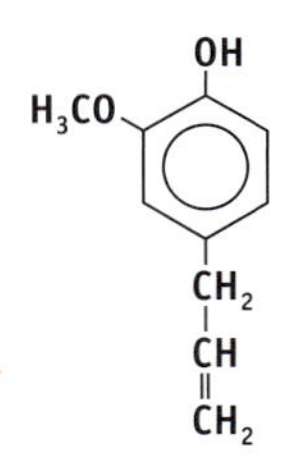

Fig 22.37
Eugenol is the main constituent of oil of cloves.

Pure eugenol can be extracted from the emulsion formed in this experiment using organic solvents like petroleum ether or dichloromethane. However, the extraction process is time consuming and is not required by the syllabus. Hence, details of the extraction process are not given below.

Apparatus required

Steam generator (copper container) fitted with safety tube, steam trap, rubber tubing to fit Liebig condenser, graduated cylinder, glass tubing, plastic tubing, hot plate, three retort stands and clamps, Quickfit set (27BU) – Liebig condenser, stillhead, receiver adaptor, steam inlet tube, 50 cm^3 pear-shaped or round-bottomed flask.

Chemicals required

Whole cloves (ground cloves may have lost a lot of the oil), water.

CAUTION!

1. As steam is being used, exercise great care to avoid scalding.
2. Eugenol can burn the skin. Gloves should be worn when handling the emulsion obtained from this experiment.

Procedure

1. Set up the apparatus for steam distillation as shown in FIG 22.38. The steam trap collects any steam that has condensed to water. If the steam trap were not present, the flask containing the cloves would quickly fill up with water from the condensed steam.
2. Place about 1.5 litres of boiling water into the steam generator.
3. Insert the safety tube and rubber stopper into the steam generator. Ensure that the bottom of the safety tube is well below the level of water in the steam generator.
4. Place about 5 g of cloves in the flask. Add about 10 cm^3 of water to the flask.
5. Turn on the hot plate and pass steam through the flask for about 40 minutes or until a sufficient amount of the clove oil emulsion is carried over during the distillation. The emulsion has a milky colour.
6. From time to time empty the steam trap of water.
7. Note the odour of the emulsion formed. Allow the emulsion to cool and place it in a labelled sample tube.

Note: This experiment could also be carried out using substances like rose petals or lavender.

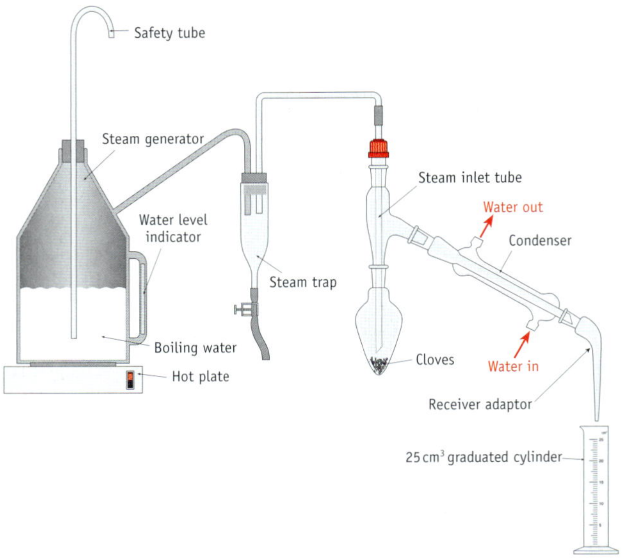

Fig 22.38
Steam distillation is used to extract oil of cloves from cloves.

Test Yourself:
Attempt question 22.7.

CHAPTER SUMMARY

- Chloroalkanes are compounds in which one or more of the hydrogen atoms in an alkane molecule have been replaced by a chlorine atom.
- Chloroalkanes have low boiling points, are insoluble in water but soluble in non-polar solvents.
- A functional group is an atom or group of atoms which is responsible for the characteristic properties of a series of organic compounds.
- The functional group of alcohols is the –OH group. Alcohols may be classified as primary, secondary or tertiary.
- Alcohols are soluble in water and in non-polar solvents. They have higher boiling points than the corresponding alkanes. Alcohols are soluble in water due to hydrogen bonding.
- The functional group of aldehydes is the –CHO group. Aldehydes have higher boiling points than the corresponding alkanes because of dipole-dipole forces but the boiling points are lower than the corresponding alcohols due to the lack of hydrogen bonding between the molecules. Aldehydes are soluble in water as hydrogen bonding takes place between the aldehyde molecules and the water molecules.
- The functional group of ketones is the >C = O group. Ketones have higher boiling points than the corresponding alkanes because of dipole-dipole forces but the boiling points are lower than the corresponding alcohols due to the lack of hydrogen bonding between the molecules. Ketones are soluble in water as hydrogen bonding takes place between the ketone molecules and the water molecules.
- The functional group of carboxylic acids is the -COOH group. Carboxylic acids have higher boiling points than the corresponding alcohols due to the greater amount of hydrogen bonding between the molecules. Carboxylic acids are soluble in water as hydrogen bonding takes place between the carboxylic acid molecules and the water molecules.
- The functional group of esters is the –COO– group. Esters have low boiling points due to the lack of hydrogen bonding between the molecules. The lower members up to C-5 are fairly soluble in water as hydrogen bonding takes place between the ester molecules and the water molecules.
- Fats and oils are esters that occur in nature.
- Aromatic compounds contain a benzene ring in their molecules.
- Natural products are organic compounds found occurring in nature.
- Clove oil may be extracted from cloves by steam distillation.

EXERCISES

22.1. The solvent, 1,1,1-trichloroethane at one time was used as the solvent in correcting fluids like *Tippex*. Draw the structural formula of 1,1,1-trichloroethane. Suggest a reason why its use was discontinued.

22.2. Name the following compounds:

(a) CH_3Cl (b) CH_3CH_2Cl (c) $CH_2ClCHClCH_2Cl$
(d) $CH_3CH_2CHClCH_2Cl$ (e) CCl_3CH_3

22.3. Draw the structural formulas of the following compounds:

(a) Trichloromethane (b) Chloroethane (c) 1-chloropropane
(d) 1,2-dichloroethane (e) 2-chloro-2-methylpropane

22.4. What is meant by the term functional group? Write down the name and formula of the functional group of alcohols.
Draw the structural formula and name the following compound $CH_3CH(OH)CH_3$.
Distinguish between the terms primary alcohol, secondary alcohol and tertiary alcohol. Give the name and structural formula of one example of each type of alcohol.
Explain the following:
(i) Ethane has a boiling point of - 89 °C but ethanol has a boiling point of 78 °C.
(ii) Ethanol is soluble in water in all proportions but ethane is virtually insoluble in water.

22.5. Write down the name and structural formula of the functional group in aldehydes. Why must the aldehyde functional group always occur at the end of a chain?
An aldehyde whose formula is C_3H_7CHO was found to have two aldehyde isomers. Draw the structural formulas of these isomers and give the name of each one.
The carbonyl group in aldehydes and ketones is polar. What is the cause of this polarity?
'The boiling points of aldehydes and ketones are higher than the corresponding alkanes'. Explain.
A student added one third of a test-tube of propanone to (i) one third of a test-tube of water and (ii) one third of a test-tube of cyclohexane. Describe what was observed and explain the observation.

22.6. Write down the functional group of carboxylic acids.
Draw the structural formula of butanoic acid.
Draw the structural formula and name the compound $CH_3CH(CH_3)COOH$.
Name the carboxylic acids found (a) in the stings of ants and nettles and (b) in vinegar.
Give one use for ethanoic acid.
What is the purpose of adding sodium benzoate to some foods?
Why is pure ethanoic acid (acetic acid) often called *glacial* ethanoic acid (acetic acid)? Write a note on the intermolecular bonding in pure ethanoic acid.
Draw the structural formula of ethyl methanoate.

22.7. What is meant by the term 'natural products' in chemistry? Give three examples of natural products.
Give two reasons why chemists are interested in isolating natural products from plants and other sources.
A student extracted clove oil from cloves by the process of steam distillation.
(i) Explain the meaning of the term *steam distillation*. Why are the cloves not simply heated directly to remove the oil?
(ii) Draw a diagram of the apparatus used to carry out the experiment.
(iii) Mention any two safety precautions that should be taken when carrying out the experiment.
(iv) What is the purpose of the steam trap in the experiment?
(v) Name the substance that is the main constituent of clove oil. Give two uses of this substance.
(vi) Describe the appearance of the distillate.
(vii) If the substance mentioned in (v) were to be isolated from the distillate, what type of experimental procedure would be used?

Types of Reactions in Organic Chemistry

23

23.1 INTRODUCTION

So far in our organic chemistry course, we have studied fuels and have taken an overview of the various families of organic compounds. One of the interesting aspects of organic chemistry is that, although there are many organic compounds, the reactions which these compounds undergo are easily classified. In this chapter we will study the various types of chemical reactions that organic compounds undergo. In addition, we will study how chemists synthesise organic compounds in the laboratory. Finally, we will look at the main methods that are used to analyse for the presence of various organic substances.

In the reactions discussed in this chapter, you are not required to know any details about the conditions (temperature, pressure, catalyst or solvent) for the reactions. Hence, these details are left out except in a few cases where it is thought necessary to include them to help your understanding of the reactions.

In this chapter you will be carrying out four experiments – there are exciting times ahead!

23.2 SUBSTITUTION REACTIONS

23.2.1 HALOGENATION OF ALKANES

In our study of alkanes we saw that the alkanes were not very reactive compounds (hence the name paraffins). We learned that they can be burned and are used as fuels. In addition to combustion reactions, alkanes also undergo **substitution reactions**.

> A substitution reaction is a chemical reaction in which an atom or group of atoms in a molecule is replaced by another atom or group of atoms.

An example of a substitution reaction undergone by alkanes is their reaction with halogens in the presence of light. For example, chlorine reacts with methane as shown in Fig 23.1. You will notice that one of the hydrogen atoms in methane has been replaced or substituted by a chlorine atom to form the new compound chloromethane.

$$CH_4 + Cl_2 \rightarrow \underset{\text{chloromethane}}{CH_3Cl} + HCl$$

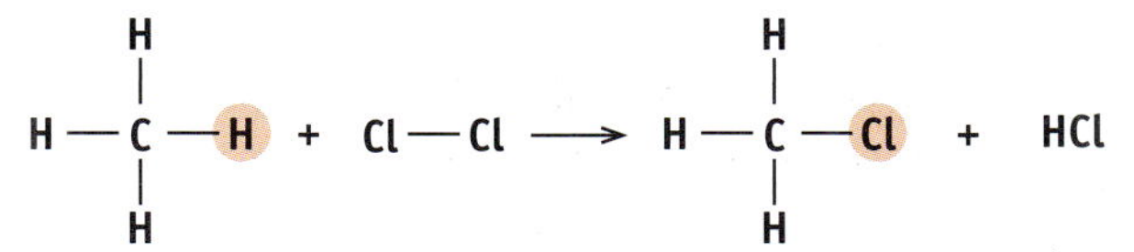

Fig 23.1
A substitution reaction takes place when methane reacts with chlorine. A chlorine atom has replaced a hydrogen atom in a molecule of methane.

If excess chlorine is used, further substitution takes place to form dichloromethane (CH_2Cl_2), trichloromethane ($CHCl_3$) and, eventually, tetrachloromethane (CCl_4). However, in this course we need only be concerned with the first substitution to form monochloromethane. We will now study the **mechanism** of this reaction.

> The mechanism of a reaction is the detailed step-by-step description of how the overall reaction occurs.

In other words, we are interested in how exactly a reaction occurs, i.e. the various steps that lead to the formation of the products. Chemists need to have a knowledge of the mechanisms of reactions in order to establish the best conditions for a particular reaction and hence to maximise the yield of products. For example, knowledge of the mechanism of the reaction between nitrogen and hydrogen to form ammonia (Chapter 25 of Option 1) has enabled chemists to develop new catalysts for this reaction. This has meant that fertilisers can be produced more cheaply and more food can be made available. Similarly, a knowledge of the mechanism of the reaction between carbon monoxide and nitrogen monoxide has enabled chemists to design catalysts for use in the exhaust systems of cars. This has helped to reduce pollution of the atmosphere (CHAPTER 16).

Reaction mechanisms are theories which are put forward in order to explain how a particular reaction occurs. The proposed mechanism must take into account all the observed facts about the reaction. Sometimes reaction mechanisms have to be modified in the light of new facts discovered about a particular reaction.

As already studied when discussing chemical bonding and thermochemistry, chemical reactions involve **bond breaking** and **bond formation**. When bonds are broken and made, there is always a rearrangement of electrons. In order to show this **rearrangement of electrons**, chemists use curly arrows (FIG 23.2). A curly arrow is used to show electron movement from the tail to the head of the arrow. The movement of one electron is shown using a curly arrow with a single head. The movement of two electrons is shown using a double-headed curly arrow. These arrows are often used in explaining the mechanisms of chemical reactions.

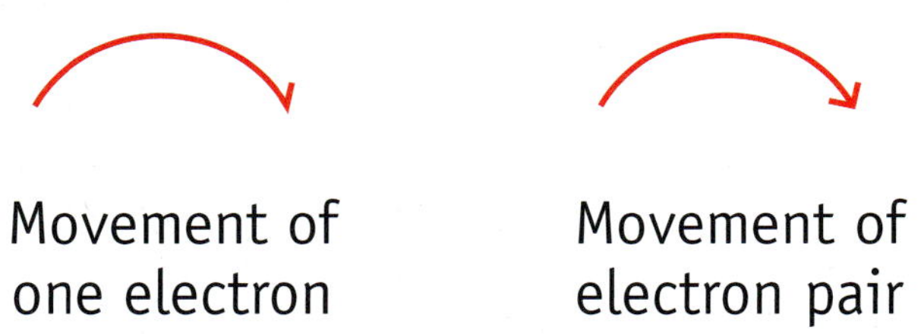

Fig 23.2
Curly arrows are used to indicate electron movement.

We now consider the mechanism of the reaction between methane and chlorine to form chloromethane. This reaction is believed to consist of four steps:

Step 1 Initiation

In this step, a chlorine molecule is broken down into two chlorine atoms in the presence of ultraviolet light (FIG 23.3).

Fig 23.3
The ultraviolet light breaks down the chlorine molecule into two chlorine atoms. The unpaired electron in the chlorine atom is represented by a dot.

Note that the bond breaks in such a way that each atom joined by the bond retains one of the two electrons in the bond. (This type of bond breaking is called **homolytic fission**.)

This reaction is called a **photochemical reaction**, i.e. a reaction that is brought about by light. Evidence for the photochemical nature of the mechanism comes from the fact that the reaction does not occur in the dark at room temperature. However, it takes place rapidly if the mixture is exposed to ultraviolet light. This step is often called the initiation or starting stage of the reaction.

Step 2 Propagation

In this step, a chlorine atom attacks a methane molecule to form hydrogen chloride and a species called a methyl **free radical** (FIG 23.4). A free radical is simply any **atom or group of atoms with an unpaired electron**. The symbol for a free radical shows the unpaired electron as a dot. (A chlorine atom is also a free radical although it is more commonly referred to simply as a 'chlorine atom' rather than as a 'chlorine free radical'.) Free radicals, since they do not have a full octet of electrons in their outer shell, are very reactive and usually only exist for very short periods. The process of ageing is believed to be related to the action of free radicals on the body.

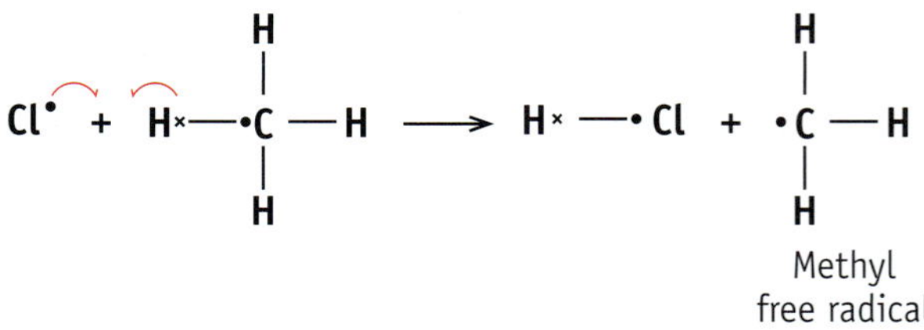

Fig 23.4
A chlorine atom attacks the methane molecule to form hydrogen chloride and a methyl free radical.

Step 3 Propagation

In this step, a methyl free radical attacks a chlorine molecule to form chloromethane and a chlorine atom (FIG 23.5). (Note that the free radical always does the attacking.)

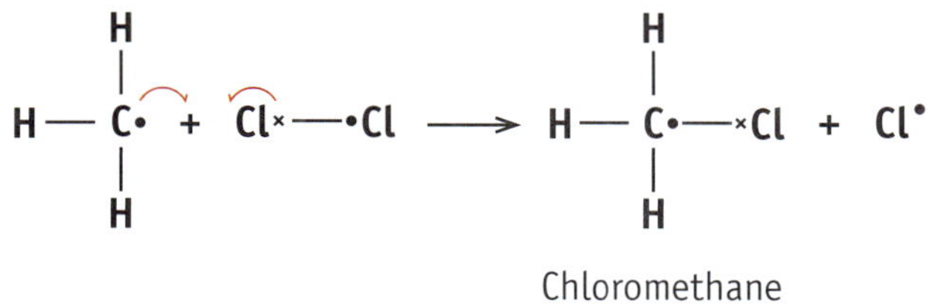

Fig 23.5
A methyl free radical attacks a chlorine molecule to form chloromethane and a chlorine atom.

The chlorine atom produced in step 3 can then react with another molecule of methane as shown in step 2. This sets up a **chain reaction**, i.e. a reaction which continues on and on because a product from one step of the reaction is a reactant for another step of the reaction.

Steps 2 and 3 are referred to as the propagation stage because they keep the chain reaction in operation to generate the chloromethane.

Evidence for this chain reaction comes from a study of the amount of light absorbed by the glass container in which the reaction is taking place. Chemists compared the amount of light falling on the container with the amount of light emerging at the other side. They then determined the number of photons that were absorbed by the reactants. They found that many thousands of molecules of chloromethane are produced for every photon that was absorbed.

Step 4 Termination

A number of reactions take place to stop the chain reaction. All of these involve free radicals or atoms combining with each other (FIG 23.6). Step 4 is called the termination stage because it causes the reaction to come to an end.

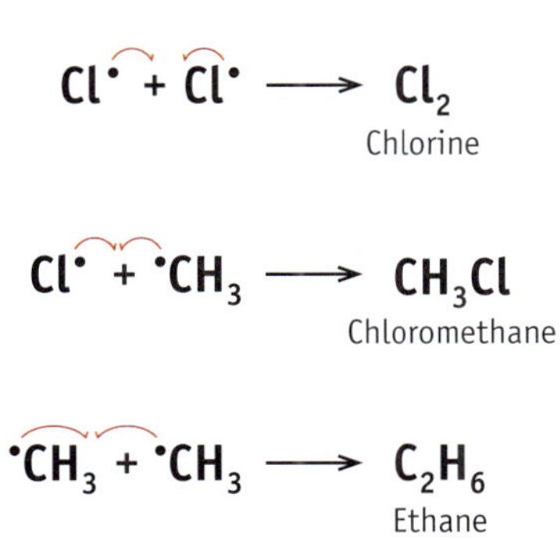

Fig 23.6
The combining together of free radicals and atoms with each other stops the reaction.

The chain reaction can go through hundreds or even thousands of cycles before termination occurs. These reactions are extremely fast and, because of the explosive nature of this reaction, it should never be attempted in schools.

Evidence for the combining of free radicals comes from the fact that small amounts of ethane are found among the products.

A final interesting piece of evidence for the above mechanism comes from the observation that if tetramethyl lead is added to the reaction mixture, there is a marked increase in the rate of reaction. Tetramethyl lead decomposes to form methyl radicals:

$$Pb(CH_3)_4 \rightarrow Pb + 4CH_3^{\bullet}$$

These methyl radicals are then fed into the system increasing the rate of reaction of step 3, followed by step 2 and so on to increase the overall rate of formation of chloromethane.

The same mechanism holds for the chlorination of ethane. (The chlorination of ethane is also listed on the syllabus.) If you are asked about the mechanism for the chlorination of ethane to form chloroethane, remember that the radical formed will be the ethyl radical rather than the methyl radical. Also, in the termination stage butane will be formed rather than ethane. Finally, the lead compound needed to increase the rate of the reaction is tetraethyl lead, $Pb(C_2H_5)_4$, rather than tetramethyl lead.

The chlorination of methane is commonly referred to as a **free radical substitution reaction**.

The halogenation of alkanes can easily be demonstrated in the school laboratory using cyclohexane and bromine water. These two reactants are more convenient to use as they are both liquids compared to methane and chlorine which are both gases. Fill two test-tubes about half-way with cyclohexane. Add sufficient bromine water to give a pale red colour. Protect one test-tube from the light by wrapping it in aluminium foil. Place the other test-tube near a mercury discharge lamp (a good source of UV light) for about 15 minutes (FIG 23.7). Now compare the two test-tubes. What can you conclude from this experiment? What are the names of the products formed?

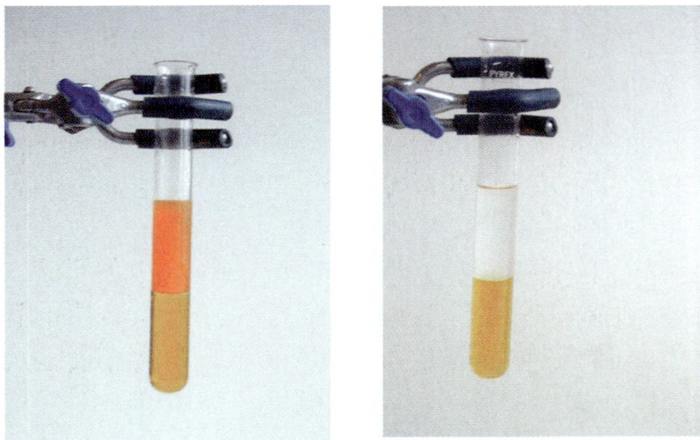

Fig 23.7
Some bromine water is added to cyclohexane in the test-tube on the left. Since the cyclohexane is less dense than water, the upper layer contains cyclohexane and some bromine dissolved in it. The photograph on the right shows what happens when this test-tube is irradiated with uv light. Note that the uv light has brought about a reaction between the cyclohexane and bromine but has no effect on the bromine water.

Almost all organic compounds burn to form carbon dioxide and water. Among the exceptions are the fully halogenated alkanes. These types of compounds are added to fabrics to reduce their tendency to catch fire. They are known as **flame retardants**. Halogenated alkanes like $CBrClF_2$ were used in fire extinguishers for many years since these compounds do not support combustion and are denser than air. However, the use of these compounds in fire extinguishers is being discontinued due to the damage being caused to the ozone layer (CHAPTER 26 of OPTION 1).

> **Test Yourself:**
> Attempt questions 23.1 and W23.1.

23.2.2 ESTERIFICATION

In CHAPTER 22 we saw that esters are formed when an alcohol and a carboxylic acid react together. We called this a **condensation reaction** since water is formed in the reaction. We could also refer to the reaction as a **substitution reaction** since the H of the –OH group in the carboxylic acid molecule has been replaced by an alkyl group (FIG 23.8).

$$R-\overset{\overset{\displaystyle O}{\|}}{C}-OH \quad HO-R' \underset{\text{Hydrolysis}}{\overset{\text{Esterification}}{\rightleftharpoons}} R-\overset{\overset{\displaystyle O}{\|}}{C}-O-R' + H_2O$$

Fig 23.8
The formation of an ester may be regarded as an example of a substitution reaction since the H atom attached to the –OH group of the carboxylic acid has been replaced by an alkyl group.

The formation of an ester is commonly referred to as **esterification**. As may be seen in FIG 23.8, esterification is an equilibrium reaction. The ester formed reacts with water to form the carboxylic acid and alcohol. Thus, a state of equilibrium is set up. The reverse reaction is called **hydrolysis**. It is found that esters are hydrolysed very effectively in the presence of a base like sodium hydroxide or potassium hydroxide (FIG 23.9).

$$R-\overset{\overset{\displaystyle O}{\|}}{C}-O-R' + NaOH \longrightarrow R-\overset{\overset{\displaystyle O}{\|}}{C}-O^- \; Na^+ + ROH$$

$$\text{e.g.}\quad H_3C-\overset{\overset{\displaystyle O}{\|}}{C}-O-C_2H_5 + NaOH \longrightarrow \underset{\text{Sodium ethanoate}}{H_3C-\overset{\overset{\displaystyle O}{\|}}{C}-O^- \; Na^+} + C_2H_5OH$$

Fig 23.9
Hydrolysis of esters is brought about using a base like sodium hydroxide.

Note that the base hydrolysis of esters using sodium hydroxide forms the sodium salt of the carboxylic acid rather than the carboxylic acid itself. This reaction is a very useful one and is used in the manufacture of soap. Base hydrolysis of esters is given a special name and is commonly referred to as **saponification** (Latin *sapona* = soap). This type of reaction is often referred to as the ***base-catalysed hydrolysis of esters***. This is not strictly correct since, as can be seen from FIG 23.9, the sodium hydroxide is actually used up in the reaction.

In the next experiment, we will use the process of saponification to prepare a sample of soap.

EXPERIMENT 23.1 MANDATORY STUDENT EXPERIMENT NO. 25

To prepare a sample of soap.

Introduction

The manufacture of soap is one of the oldest chemical processes. Soap was first manufactured in Roman times and is still the most commonly used substance for personal cleanliness. Soap consists of the salts of long-chain carboxylic acids, e.g. $C_{17}H_{35}COONa$ which is commonly called sodium stearate. (The systematic name for this is a real mouthful – sodium octadecanoate!) Salts of long-chain carboxylic acids are made from fats or oils by the process of saponification. For example, the naturally-occurring ester glyceryl tristearate, found in animal fats, may be hydrolysed to form sodium stearate (soap) and the alcohol glycerol (FIG 23.10). Study this reaction carefully!

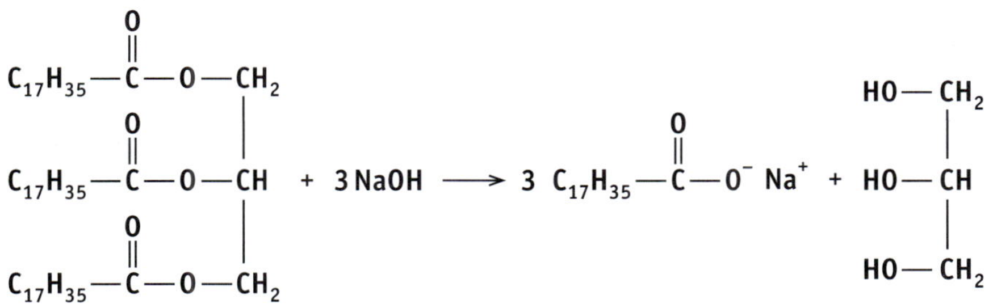

Glyceryl tristearate — Sodium stearate (soap) — Glycerol

Fig 23.10
The naturally occuring ester glyceryl tristearate, found in animal fats, may be hydrolysed to form sodium stearate (soap) and the alcohol glycerol.

Thus, to prepare soap one needs to start off with a substance that contains a naturally-occurring ester, e.g. lard (a fat) or cooking oil (an oil). Oils that are commonly used for making soap are palm oil, olive oil, coconut oil, etc. Therefore, ordinary soap may consist of not only sodium stearate but also many other salts obtained from these various oils. For the sake of convenience, we will assume that the soap we are making consists of just sodium stearate, i.e. that the main reaction occurring is that shown in FIG 23.10.

The glycerol (also called glycerine) obtained in the manufacture of soap is a useful by-product. It is used to manufacture the explosive nitroglycerine.

Both sodium hydroxide and potassium hydroxide are used in the manufacture of soap. Potassium soaps tend to be milder than those made with sodium hydroxide.

If one examines the structure of a molecule of sodium stearate, it is easy to understand how soaps work to remove oils and various salts from our skin. The non-polar part of the molecule (long chain of carbon atoms) dissolves non-polar substances like oil and grease from our skin. The polar part of the molecule ($-COO^- Na^+$) dissolves the various salts produced in our sweat.

Apparatus required

Rubber tubing, conical flask, hot plate, beaker, filter funnel or Buchner funnel, filter paper, spatula, test-tube, retort stand, clamp. Quickfit set (27BU) – Liebig condenser, stillhead, receiver adaptor, 100 cm^3 pear-shaped or round-bottomed flask.

Chemicals required

Lard or sunflower oil, sodium hydroxide pellets, ethanol, saturated sodium chloride solution, anti-bumping chips, water.

CAUTION!

1. Make sure you are wearing eye protection at all times during this experiment.
2. Sodium hydroxide (caustic soda) is very dangerous. Do not allow it to come in contact with your skin or eyes as it causes severe burns.

Procedure

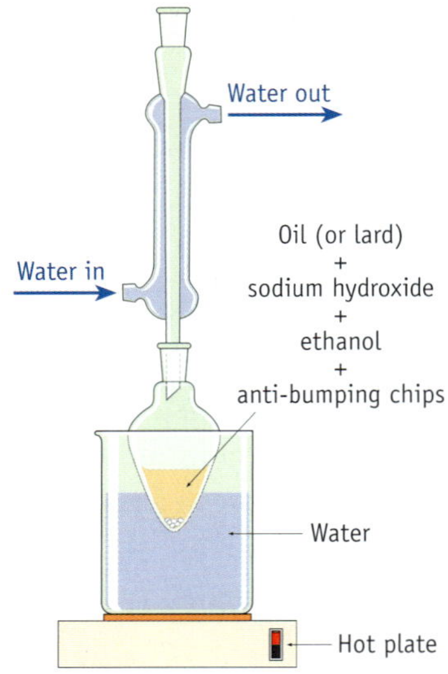

Fig 23.11
Reflux apparatus used in preparation of soap.

1. Place 5 g of lard or 6 cm^3 of sunflower oil in a 100 cm^3 Quickfit flask. Add 3 g of sodium hydroxide and 30 cm^3 of ethanol to the flask. The ethanol acts as a solvent for the oil or lard.
2. Add several anti-bumping chips to the flask. Set up the reflux apparatus as shown in FIG 23.11. Clamp both the Liebig condenser and the flask. **Refluxing** is a laboratory technique in which a liquid is boiled in a container that is attached to a vertical condenser. Vapour from the boiling liquid condenses and flows back into the flask. Thus the liquid is kept at its boiling point without loss of vapour. This action prevents the flask boiling dry and allows sufficient time for organic reactions to occur.
3. Heat the mixture over a water bath for about 30 minutes so that it boils gently. At this stage, saponification should be complete. During the reflux period, the walls of the flask may become coated with the sodium salts of the long-chain acids and with oil or fat that has not been hydrolysed. Therefore, from time to time gently swirl the flask to wash these materials from the walls of the flask back into the reaction mixture.
4. Allow the apparatus to cool and rearrange it for distillation as shown in FIG 23.12. We are now going to distil off most of the ethanol solvent. This will make it easier for us to isolate the soap.
5. Gently heat the flask using the water bath. Continue heating until about 20 cm^3 of ethanol have been collected.
6. Allow the apparatus to cool for a few minutes. Using the clamp to hold the flask, quickly pour the hot reaction mixture into a beaker containing 100 cm^3 of concentrated brine. Brine is a solution of salt (NaCl) in water. Soap does not dissolve in salt solution. Therefore, the soap will be precipitated. The excess sodium hydroxide will dissolve in the salt solution.
7. Filter off the soap using a filter funnel (or a Buchner funnel) and filter paper. Wash the soap with fresh salt solution to remove any sodium hydroxide still present.
8. Using a spatula, take a sample of the soap, add it to about half a test-tube of deionised water and shake the test-tube. What do you observe?
9. Place the soap on filter paper and allow it to dry overnight.

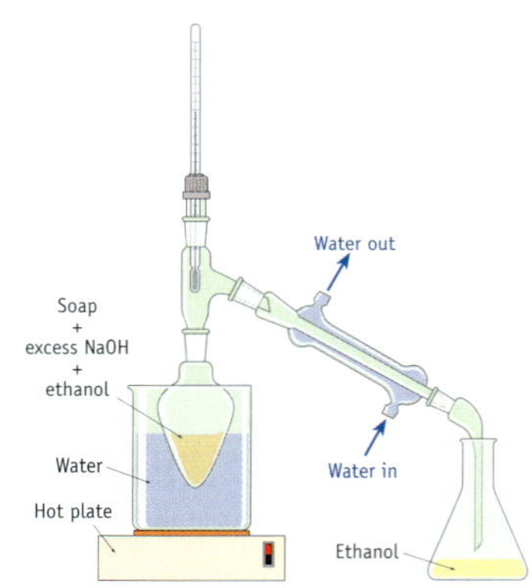

Fig 23.12
The ethanol solvent is removed by distillation.

Note: When soap is made commercially, other ingredients like perfumes and colouring agents are added to make it more attractive.

Test Yourself:
Attempt questions 23.2 and W23.2.

23.3 ADDITION REACTIONS

You may recall that when we prepared ethene (EXPERIMENT 21.1), the addition of bromine to a sample of the ethene caused bromine to add across the C = C double bond to form 1,2-dibromoethane. This is an example of an addition reaction.

> An addition reaction is one in which two substances react together to form a single substance.

However, if double bonds are stronger than single bonds, why do alkenes react so readily with bromine? The answer to this question is found by considering the energy involved in the reaction. When bromine adds across the double bond in ethene, energy is required to break the pi bond. However, energy is released when two single bonds to the bromine atoms are formed. There is a net release of energy to the surroundings (ΔH = -124 kJ mol^{-1}). In other words, the products are more stable than the reactants and this is the driving force for the reaction.

As discussed in CHAPTER 21, ethene is produced on an industrial scale by the process of cracking and is often referred to as 'industry's most important organic chemical'. This is due to the large number of compounds that can be made from it. For example, as well as reacting with bromine, ethene reacts with other chemicals like hydrogen, chlorine, hydrogen chloride and water. These reactions are shown in FIG 23.13.

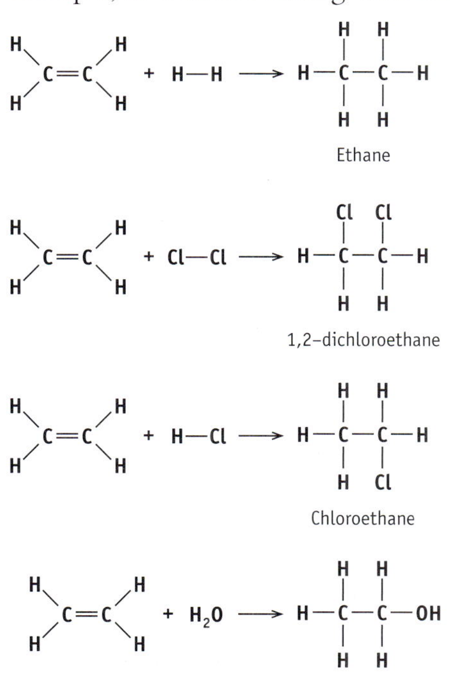

Fig 23.13
Some examples of addition reactions.

The compound 1,2-dichloroethane, formed by the reaction between ethene and chlorine, is used to manufacture chloroethene (commonly called vinyl chloride). This is then used to make polyvinyl chloride (PVC), a widely used plastic. The systematic name for PVC is poly(chloroethene). It will be studied in more detail in CHAPTER 27 of OPTION 2.

As discussed in the previous chapter, ethanol, formed by the reaction between ethene and water, is widely used as an industrial solvent.

Hydrogenation of C = C double bonds is used in the manufacture of margarine. Margarine was invented as an alternative to butter and was used widely by those who could not afford to buy butter. However, in recent times the use of margarine has become widespread. Edible oils like sunflower oil, palm oil and corn oil are often referred to as **polyunsaturated**, i.e. they have lots of C = C double bonds. This is because they are esters of long-chain unsaturated carboxylic acids (SECTION 23.2.2). These unsaturated compounds are thought to be less damaging to our health than the saturated fats found in dairy products like butter and cream. The problem is that the edible oils are not much use for spreading on bread! However, chemists discovered that by adding hydrogen to some of the C = C double bonds in these oils (in the presence of a suitable catalyst), the oils could be changed into soft solids with low melting points (FIG 23.14). In other

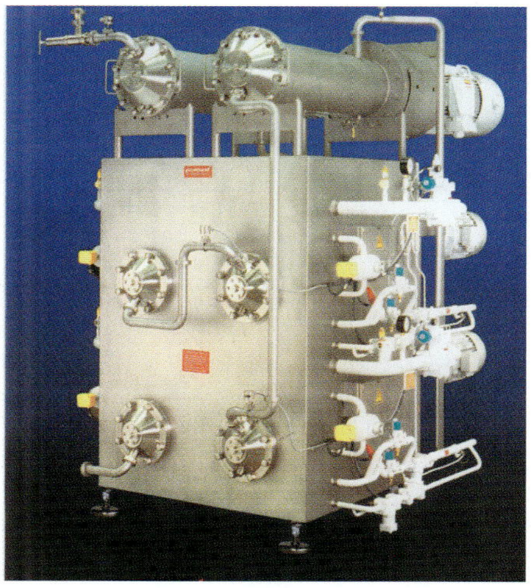

Fig 23.14
This photograph shows equipment used in the hydrogenation of some of the C = C double bonds in vegetable oils to manufacture margarine.

words, the oils can be converted into fats (margarine) with low melting points. Thus, margarine has better spreading properties than butter (due to the lower melting point) and contains less of the saturated fats that many people want to avoid. By controlling the degree of hydrogenation, the margarine can be made as hard or as soft as needed.

We now consider the mechanism of the reaction between ethene and bromine.

The reaction between ethene and bromine can occur in the dark. This suggests that the mechanism is different than that between methane and chlorine (free radical substitution). It is believed that the mechanism involves four steps.

Step 1

The C = C double bond in ethene is a region of high concentration of negative charge. The Br_2 molecule, as it approaches the C = C double bond, becomes polarised (Fig 23.15).

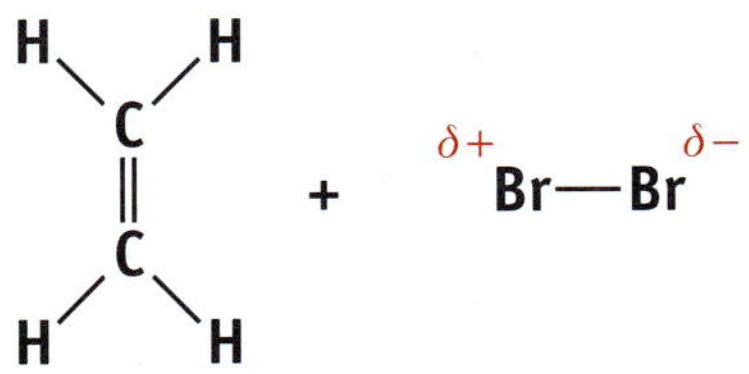

Fig 23.15
The high concentration of negative charge in the C = C double bond, causes the Br - Br bond to become polarised.

Step 2

The induced polarisation becomes so great, the Br_2 molecule splits into Br^+ and Br^- species (Fig 23.16). This breaking of the bond so that the two electrons end up on one atom is called **heterolytic fission** because two different species are formed.

Fig 23.16
Heterolytic fission occurs.

Step 3

The Br^+ species, in order to gain the two electrons it needs to give it a stable octet of electrons in its outer shell, attacks the C_2H_4 molecule. (The C_2H_4 molecule has the required two electrons in its pi bond.) The Br^+ ion forms a covalent bond with one of the carbon atoms. The other carbon atom is left with a positive charge since it has lost one of its outer electrons. This positively charged carbon atom is called a **carbonium ion** (Fig 23.17).

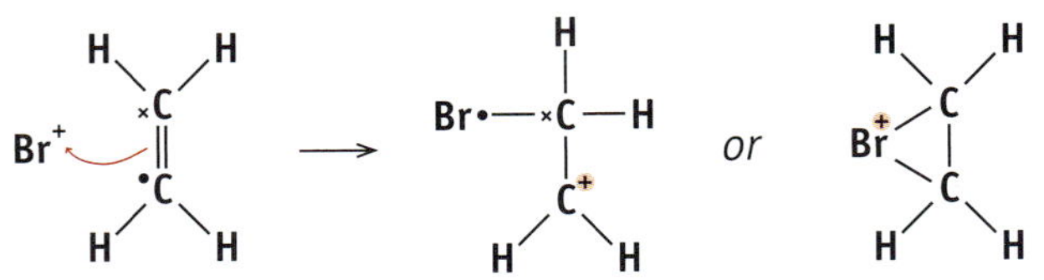

Fig 23.17
Carbonium ion formation.

There is evidence to suggest that the carbonium ion has the cyclic structure shown in Fig 23.17. This structure is called a **cyclic bromonium ion**. The relatively large size of the bromine atom allows the formation of this three-membered ring structure.

Step 4

The carbonium ion is then attacked by the Br^- ion. This results in the formation of 1,2-dibromoethane (Fig 23.18).

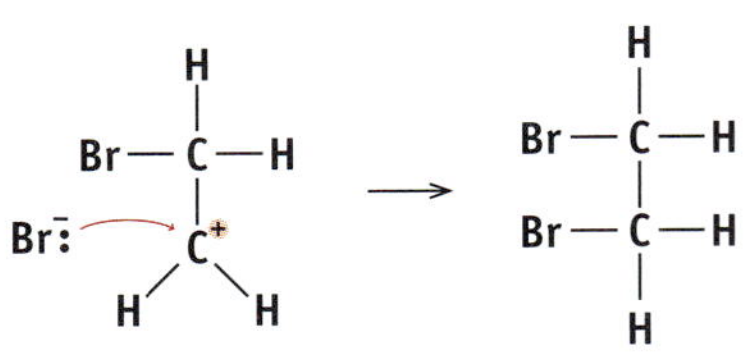

Fig 23.18
Attack on carbonium ion by bromide ion.

The mechanism of this reaction is often referred to as **ionic addition** since ions add across the C = C double bond. Evidence for this mechanism comes from the observation of what happens when ethene reacts with bromine in water in the presence of sodium chloride solution. As before, 1,2-dibromoethane is formed but two other compounds are also formed. One of these compounds is 1-bromo-2-chloroethane (Fig 23.19). This is formed when the carbonium ion is attacked by the Cl^- ion. Hence, this is evidence for the existence of the carbonium ion.

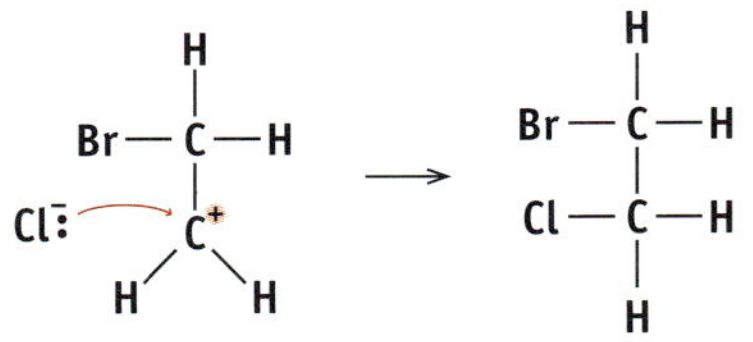

1-bromo-2-chloroethane

Fig 23.19
When the bromination reaction is carried out in the presence of chloride ions, 1-bromo-2-chloroethane is formed. This is evidence for the existence of the carbonium ion.

The other compound formed is called 2-bromoethanol. This is formed when the carbonium ion is attacked by the water molecule (Fig 23.20). Again, this is evidence for the presence of the carbonium ion.

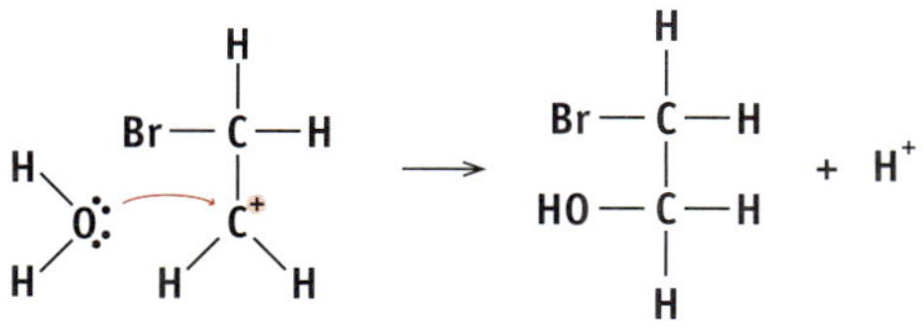
2-bromoethanol

Fig 23.20
When ethene reacts with bromine in water, 2-bromoethanol is formed. This is evidence for the existence of the carbonium ion.

The mechanism for the addition of bromine to ethene also holds for the addition of Cl_2 and HCl to ethene. In the case of Cl_2, the Cl^+ species attacks the C = C double bond first. Then, the Cl^- ion attacks the carbonium ion to form 1,2-dichloroethane (Fig 23.21(a)).

In the case of HCl, the H^+ ion attacks the C = C double bond first. The Cl^- ion then attacks the carbonium ion to form chloroethane (Fig 23.21(b)).

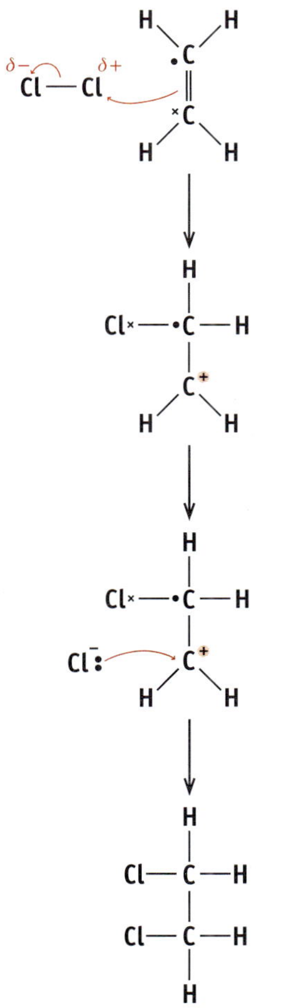
1,2-dichloroethane

Fig 23.21(a)
The mechanism of the addition of chlorine to ethene.

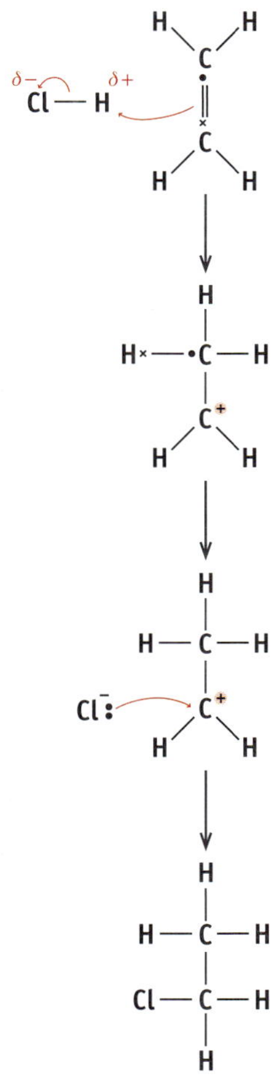
1-chloroethane

Fig 23.21(b)
The mechanism of the addition of HCl to ethene.

A cyclic intermediate is not formed in the case of the addition of Cl_2 or HCl to ethene. This is because the Cl atom and the H atom are too small to form a ring compound.

Test Yourself:
Attempt questions 23.3 and W23.3.

23.3.1 POLYMERISATION REACTIONS

The addition of bromine to ethene is an example of a simple addition reaction. However, it is also possible for molecules containing double bonds to undergo addition reactions among themselves. For example, it is possible for molecules of ethene to link with each other to form a long chain of carbon atoms, Fig 23.22.

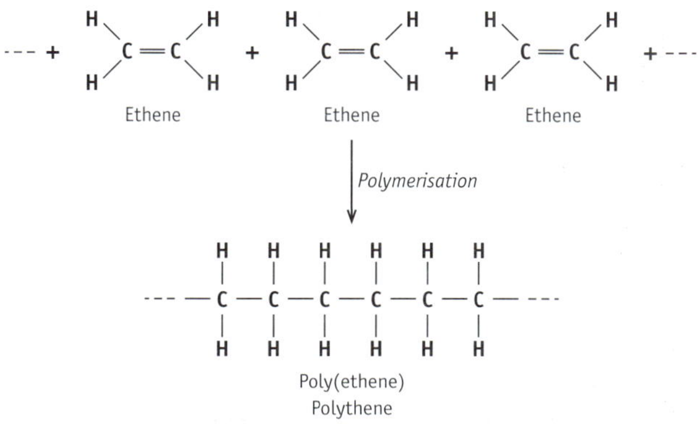

Fig 23.22
Ethene molecules join together to form poly(ethene).

The compound formed is called poly(ethene), or more simply, polythene. Polythene is an example of a **polymer**.

Polymers are long chain molecules made by joining together many small molecules.

Polymers consist of a repeating structure and can have a molecular mass of several thousand. The small molecules from which the polymer is made are called **monomers**. Thus, ethene is the monomer used in the manufacture of the polymer polythene.

monomer + monomer + monomer + ... → polymer

(The word *polymer* comes from the Greek language, *poly* = many, *meros* = parts.) Polymers are commonly referred to as 'plastics'. As mentioned in Chapter 21, polythene is a very useful material and is used in the manufacture of plastic bags, bowls, lunch boxes, food-wrapping film, washing-up liquid bottles, buckets, etc. We will be studying polythene in more detail in CHAPTER 27 of OPTION 2. Another example of a very useful polymer is poly(propene). This is made by the polymerisation of propene (FIG 23.23).

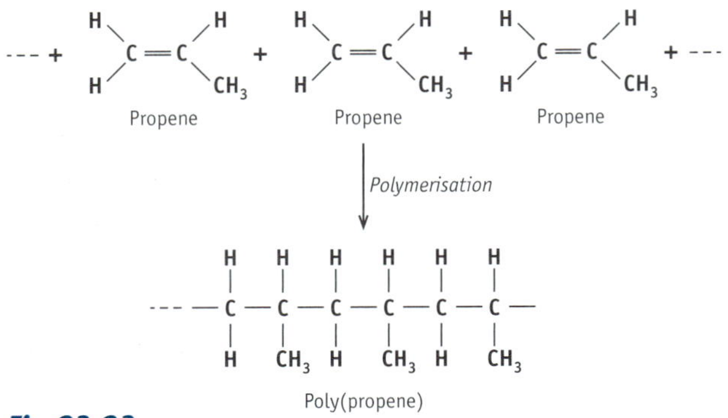

Fig 23.23
Propene molecules join together to form poly(propene).

Poly(propene) is used in the manufacture of toys, jugs, plastic beakers, chairs, briefcases, etc. Poly(propene) can be made into a fibre that is used in the manufacture of ropes. It is better than rope made from natural fibre as it does not rot and can be produced in many different colours.

Polymers like polythene and poly(propene) are categorised as **addition polymers** because their manufacture involves addition reactions. Usually, only one type of monomer molecule is involved in their manufacture. Addition polymers are the most important synthetic polymers. (The other type of polymer is called a condensation polymer but we will not be studying condensation polymers in this course.)

Crude oil is the raw material for the manufacture of polymers. Approximately 4% of the world production of crude oil is turned into plastics. As discussed in CHAPTER 21, alkenes like ethene and propene, produced from crude oil, are the raw materials used in the petrochemical industry for the manufacture of addition polymers. The petrochemical industry, that part of the chemical industry based on crude oil, is largely dependent on the alkenes to manufacture a wider series of compounds. Examples of compounds manufactured by the petrochemical industry are plastics, detergents, medicines, solvents, fibres, fabrics, cosmetics, insecticides, herbicides, paints, etc.

23.4 ELIMINATION REACTIONS

In EXPERIMENT 21.1 we saw that ethene could be produced from ethanol. We learned that the reaction involved the removal of a molecule of water from a molecule of ethanol by passing the ethanol vapour over aluminium oxide (FIG 23.24).

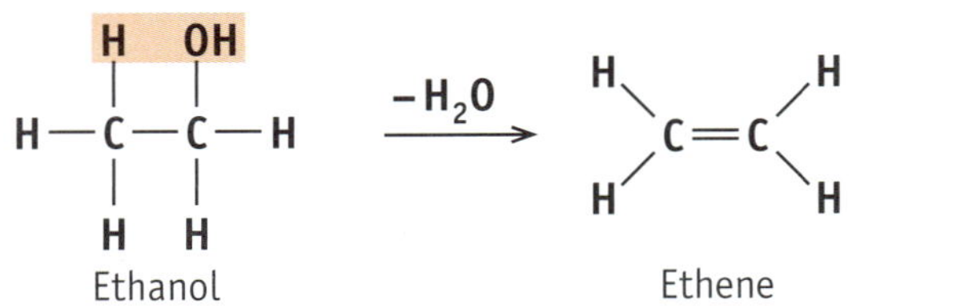

Fig 23.24

This type of reaction is an example of an elimination reaction.

An elimination reaction is one in which a small molecule is removed from a larger molecule to leave a double bond in the larger molecule.

Thus, the removal of water from an alcohol to form an alkene is an elimination reaction. The formation of propene from propan-2-ol is another example of an elimination reaction (FIG 23.25). Since water is removed, this particular type of elimination reaction is also referred to as a **dehydration reaction**.

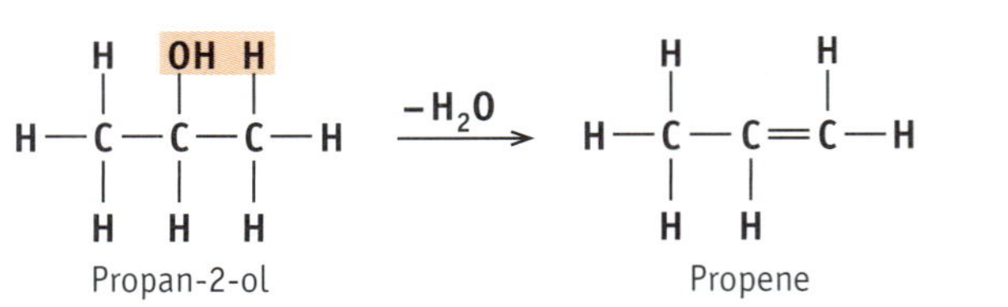

Fig 23.25

There are many other chemical reactions that are classified as elimination reactions. However, the removal of water from an alcohol to form an alkene is the only type of elimination reaction on our course.

23.5 REDOX REACTIONS

When a primary alcohol reacts with an oxidising agent, the primary alcohol is converted to an aldehyde. Suitable oxidising agents are acidified sodium dichromate, $Na_2Cr_2O_7$, [also called sodium dichromate(VI)] or acidified potassium permanganate, $KMnO_4$. If the oxidising agent is in excess, the aldehyde is further oxidised to a carboxylic acid (FIG 23.26).

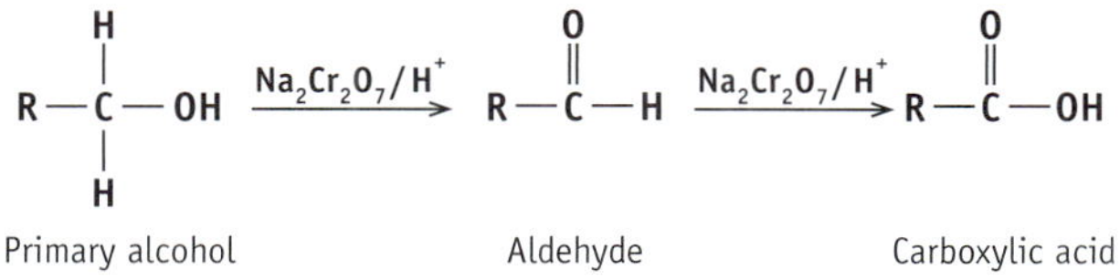

Fig 23.26
Primary alcohols are oxidised to aldehydes which are further oxidised to carboxylic acids.

It may easily be seen that the conversion of the primary alcohol to the aldehyde is an oxidation reaction since two hydrogen atoms have been removed from the primary alcohol. (Recall from CHAPTER 14 that reduction was explained in terms of the addition of hydrogen. Thus, oxidation is the removal of hydrogen, i.e. dehydrogenation.). In fact, the word aldehyde comes from ***al***cohol ***dehyd***rogenation.

The conversion of an aldehyde to a carboxylic acid is clearly an oxidation reaction since an extra oxygen atom has been added to the aldehyde.

If a secondary alcohol reacts with the above oxidising agents, a ketone is formed (FIG 23.27).

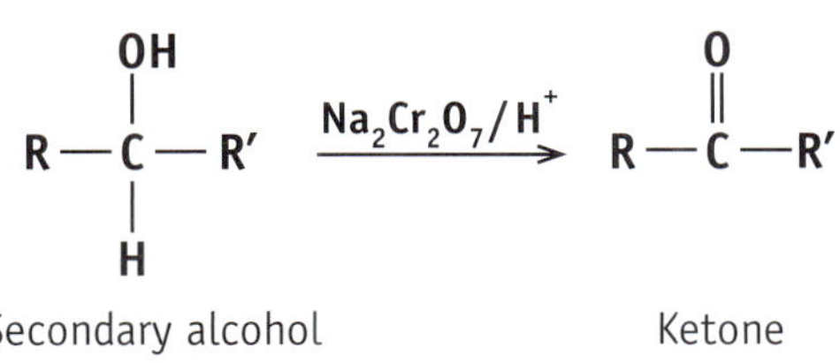

Fig 23.27
Seconday alcohols are oxidised to ketones.

The oxidation of ethanol to ethanoic acid causes problems for wine producers. A bottle of wine, that is left open to the air, slowly changes into a dilute solution of ethanoic acid (vinegar).

In order to get rid of ethanol in the body, it is oxidised to ethanal. This takes place in the liver, which can oxidise about 8 g of ethanol an hour. Excessive quantities of alcohol can damage the liver and give rise to a condition known as cirrhosis of the liver. The ethanal is eventually oxidised to carbon dioxide and water.

As mentioned in CHAPTER 21, ethanol obtained from sugar cane is commonly added to petrol in Brazil. However, in Ireland its use as a fuel is limited by the high cost of ethanol. The only place you are likely to see ethanol being burned in Ireland as a fuel is in a methylated spirits burner.

Almost all organic compounds burn to form carbon dioxide and water. The combustion of ethanol may be represented by the following equation:

$$C_2H_5OH + 3O_2 \rightarrow 2CO_2 + 3H_2O$$

In CHAPTER 22 we learned that the legal limit in Ireland for driving a vehicle is 80 mg of ethanol per 100 cm^3 of blood. Alcohol is a drug that is sufficiently volatile to pass from the blood to the air in the lungs. When the Gardaí stop a motorist who is suspected of drink driving, a measurement of the concentration of ethanol in the breath gives a good indication of the amount of alcohol in the blood. The breathalyser is a very successful method of detecting the presence of alcohol in the breath of a motorist (FIG 23.28). The orange crystals of potassium dichromate, $K_2Cr_2O_7$, in the breathalyser turn green when they oxidise ethanol to ethanal and ethanoic acid. The extent to which the crystals turn green is a measure of the concentration of alcohol in the blood.

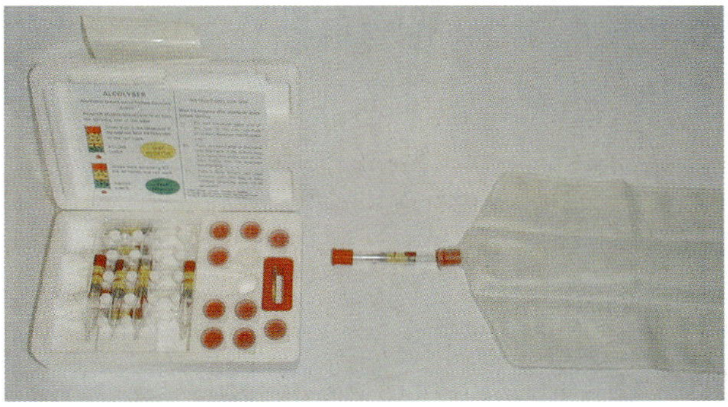

Fig 23.28
The breathalyser is used to detect the presence of alcohol in the breath of a motorist.

In recent years, another instrument called an alcometer has been developed to detect the presence of alcohol in the breath. This contains a type of electrical cell in which ethanol is oxidised to ethanoic acid at one of the electrodes. The greater the concentration of alcohol, the higher the voltage reading on the cell. If the readings from the roadside test indicate the presence of a high level of alcohol, the motorist is taken to a Garda station for a blood sample. A more accurate determination of the concentration of alcohol in the blood is then carried out in a laboratory (PAGE 390).

EXPERIMENT 23.2 MANDATORY STUDENT EXPERIMENT NO. 26

To prepare ethanal and examine its properties.

Part 1: Preparation of ethanal

Introduction

In this experiment, we will prepare a sample of ethanal by oxidising ethanol. The oxidising agent we will use is sodium dichromate, $Na_2Cr_2O_7$. Sodium dichromate is also called sodium dichromate(VI). Crystals of sodium dichromate have the formula $Na_2Cr_2O_7.2H_2O$.

The balanced equation for the reaction is:

$3C_2H_5OH + Cr_2O_7^{2-} + 8H^+ \rightarrow 3CH_3CHO + 2Cr^{3+} + 7H_2O$

You do not have to remember this equation. All you need remember are the following two half equations:

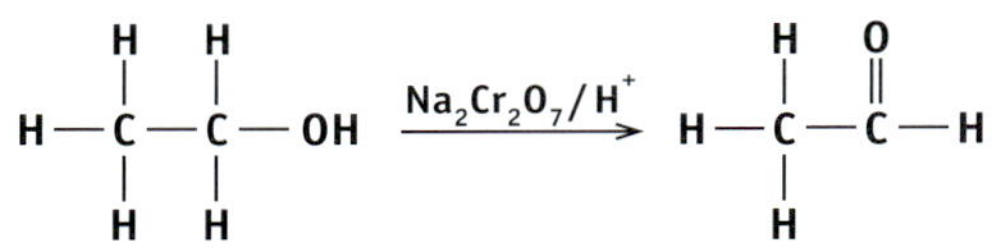

Ethanol Ethanal

$Cr^{+6} + 3e^- \longrightarrow Cr^{+3}$

(*or* $Cr^{VI} + 3e^- \longrightarrow Cr^{III}$)

Orange Green

Chromium in an oxidation state of +6 has an orange colour. Chromium in an oxidation state of +3 has a green colour. Hence, we would expect to see a colour change from orange to green as the reaction takes place.

By using excess alcohol rather than excess sodium dichromate and by distilling off the ethanal as it is formed, further oxidation of the ethanal to ethanoic acid is kept to a minimum.

Apparatus required

Conical flask, beaker to accommodate conical flask, rubber tubing to fit Liebig condenser, hot plate or Bunsen burner, graduated cylinder (0 – 10 cm^3) or graduated plastic droppers, small beaker (50 – 100 cm^3), stirring rod, two retort stands and clamps, Quickfit 27BU – Liebig condenser, dropping funnel, receiver adaptor, still head, pear-shaped flask (50 cm^3).

Chemicals required

Sodium dichromate, ethanol, concentrated sulfuric acid, anti-bumping chips, ice, deionised water.

CAUTION!

Great care should be taken when adding concentrated sulfuric acid to water. The reaction is highly exothermic.

Procedure

1. Weigh out 5 g of $Na_2Cr_2O_7.2H_2O$ in a small beaker. (Note that the crystals contain water of crystallisation.)
2. Add 5 cm^3 of deionised water to the beaker. Using a stirring rod, stir the solution to dissolve the sodium dichromate.
3. Add 4 cm^3 of ethanol to the sodium dichromate solution in the beaker.

4. Pour 6 cm^3 of water into a pear-shaped flask of 50 cm^3 capacity. (A pear-shaped flask is particularly useful when dealing with small quantities of liquid. If a round-bottomed flask were used, the liquid would be spread over a larger surface area. This could cause decomposition and charring to occur whereas the pear-shaped flask keeps the liquid in a more compact volume.)
5. Slowly add 2 cm^3 of concentrated sulfuric acid to the water in the flask. The concentrated sulfuric acid is added to the water (rather than the other way around) as sulfuric acid is more dense than water. Since the reaction is highly exothermic, as the acid sinks down through the water, the heat generated spreads throughout the solution. Therefore, gently swirl the flask and cool the flask using running tap-water. (If water were added to concentrated sulfuric acid, it would float on the surface of the acid and the heat generated would cause a dangerous hot acid spray to be generated.)
6. Add some anti-bumping chips to the flask.
7. Set up the apparatus as shown in Fig 23.29. Since ethanal has a very low boiling point of 21 °C, the conical flask is surrounded by ice. Ensure that the T-piece (still head) connecting the flask to the condenser is upright. Otherwise, when the tap of the dropping funnel is opened, some of the liquid from the dropping funnel may pass down the condenser instead of into the flask.
8. Using a beaker of boiling water, carefully heat the dilute acid in the flask to about 70 °C. The beaker of boiling water may be held in a beaker tongs or placed on a hot plate. A Bunsen burner can also be used to heat the dilute acid but this must be done **very gently**. The beaker of boiling water or Bunsen burner should then be removed once the sulfuric acid has reached the required temperature. If the heat is not removed, the reaction could become quite violent when the ethanol and sodium dichromate mixture is added to the hot sulfuric acid. The reason for this is because the reaction of ethanol with acidified sodium dichromate is exothermic.

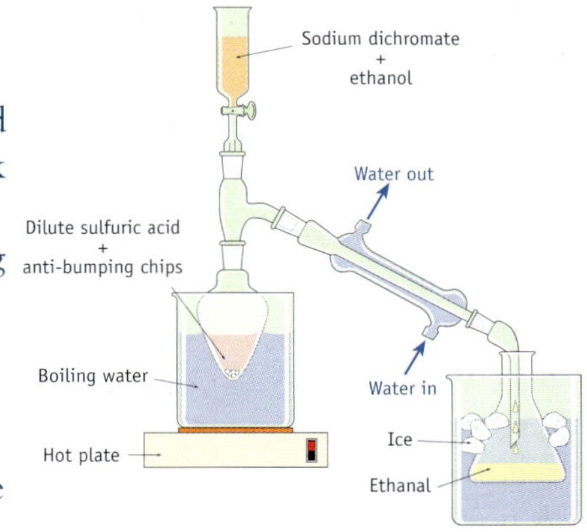

Fig 23.29
The laboratory preparation of ethanal.

9. Add the mixture of alcohol and dichromate from the dropping funnel at such a rate that the liquid is maintained at the boiling point. In other words, keep the drops of liquid falling from the dropping funnel into the pear-shaped flask so that the liquid in the pear-shaped flask is kept boiling. Since this reaction is exothermic, heating should not be necessary once the reaction starts. Don't forget to close the tap of the dropping funnel when the addition is finished!
10. If the reaction appears to stop, resume heating as above.
11. A colourless liquid is collected in the conical flask. Continue heating until about 10 cm^3 of the distillate has been collected. At this stage, practically all of the ethanal has been distilled over. The distillate contains mainly ethanal with some impurities like ethanol, ethanoic acid and water. The water could be removed by adding a substance like anhydrous sodium sulfate which absorbs water. The sodium sulfate is then filtered off to give dry ethanal. Note the smell of the product and compare it with a sample of pure ethanal. Place the sample of ethanal in a labelled sample tube.
    If time permits, the sample of ethanal from each group could be collected. These samples could be distilled over a water bath. The fraction boiling between 20 °C and 23 °C could be collected in a receiver surrounded by ice. This will give a very pure sample of ethanal.

Note: Aldehydes and ketones can be reduced back to alcohols in the presence of hydrogen and a nickel catalyst. Thus, ethanal is reduced to ethanol and propanone is reduced to propan-2-ol (FIG 23.30).

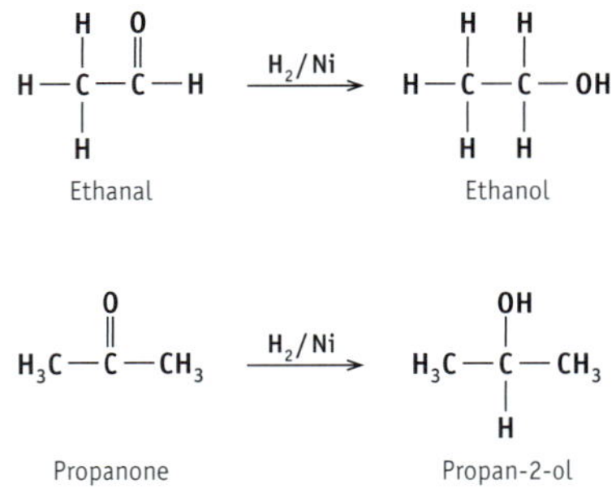

Fig 23.30
The reduction of aldehydes and ketones.

Part 2: To study the reactions of ethanal with (a) acidified potassium permanganate solution, (b) Fehling's solution and (c) ammoniacal silver nitrate.

Introduction

We will now examine the properties of ethanal. The sample of ethanal used in the second part of this experiment can be the purified sample of ethanal prepared above or a sample of ethanal from the chemical store.

We will study how ethanal reacts with the three compounds listed above. In these tests, we need only use small quantities of the chemicals. We will repeat each of the tests with propanone (acetone) also in order to show that ketones are not easily oxidised. In these tests we will be using two chemicals that we have not met before. A brief note on each of these now follows:

Fehling's solution (Fehling's reagent)

This is a mixture, in equal amounts, of Fehling's A and Fehling's B. It must always be freshly made up before use.

Fehling's A is a solution of copper sulfate in water (blue colour). Fehling's B is a solution of a chemical called potassium sodium tartrate (Rochelle salt) and sodium hydroxide in water. This solution is colourless. When equal quantities of Fehling's A and B are mixed, a royal blue colour is obtained. This royal blue solution is Fehling's solution.

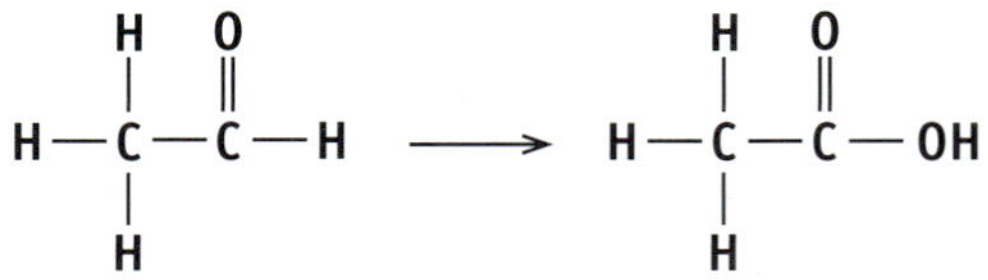

$$\underset{\text{Blue}}{Cu^{2+}} + e^- \longrightarrow \underset{\text{Red}}{Cu^{+}}$$

Fehling's solution is an oxidising agent and one of its uses is to distinguish between aldehydes and ketones. When Fehling's solution is warmed with an aldehyde, a brick red precipitate of copper (I) oxide, Cu_2O, is obtained. You are not required to know the fully balanced equation for this reaction but you must know the two half reactions shown above.

Thus, aldehydes are said to be good reducing agents since they reduce the Cu^{2+} ions to Cu^+ ions. Since ketones are difficult to oxidise, no colour change is observed when propanone is heated with Fehling's solution.

Another test that is used to distinguish between aldehydes and ketones is the silver mirror test.

Ammoniacal silver nitrate. (Silver Mirror Test)

Aldehydes reduce a solution of silver ions in dilute ammonia to silver metal. A solution of silver oxide, Ag_2O, dissolves in ammonia solution to form a solution called ammoniacal silver nitrate. This is also commonly referred to as **Tollen's reagent.** If ammoniacal silver nitrate reacts with some ethanal in a test-tube and is warmed, a silver mirror is deposited on the inside of the test-tube.
The half reactions are:

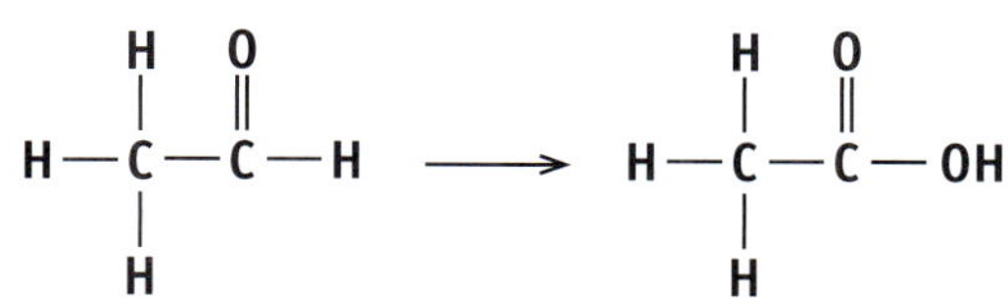

Ethanal → Ethanoic acid

$$Ag^+ + e^- \longrightarrow Ag\downarrow$$

The silver mirror test gives a negative result with ketones since ketones are difficult to oxidise.

Apparatus required

Bunsen burner or hot plate, rack of test-tubes, test-tube holder, wash bottle, Pyrex beaker, droppers.

Chemicals required

Ethanal, propanone, potassium permanganate, dilute sulfuric acid, Fehling's A and B, silver nitrate (0.1 M), sodium hydroxide solution (3 M), dilute ammonia (3 M).

Procedure

(a) Reaction of ethanal with acidified potassium permanganate

1. Acidify about one-tenth of a test-tube of very dilute potassium permanganate solution (pale pink) with about the same quantity of dilute sulfuric acid.
2. Add about five drops of ethanal to the test-tube of acidified potassium permanganate.
3. Using a test-tube holder, stand the test-tube in a beaker of hot water for a few minutes.
4. Note the colour change and explain your observation. Write down the two half reactions taking place.
5. Repeat the experiment with propanone. Explain your observation.

(b) Reaction of ethanal with Fehling's solution

1. Make up some Fehling's solution by mixing about one-tenth of a test-tube of Fehling's A with the same quantity of Fehling's B. The mixture should be a clear royal blue colour.
2. Add about five drops of ethanal to the test-tube containing the Fehling's solution.
3. Using a test-tube holder, stand the test-tube in a beaker of hot water for a few minutes.

4. Write down what you observe and explain your observation. Also write down the two half reactions taking place.
5. Repeat the experiment with propanone. Explain your observation.

(c) Reaction of ethanal with ammoniacal silver nitrate (Silver Mirror Test)

The formation of a good silver mirror can be quite tricky.

1. To 1 cm^3 of a solution of silver nitrate, add 1 drop of dilute sodium hydroxide solution. A brown precipitate of silver oxide is formed.
2. Add dilute ammonia solution, drop by drop, until the brown precipitate just redissolves. You have now prepared Tollen's reagent. Tollen's reagent should always be freshly made up. If stored, it is likely that explosive products will be formed.
3. Wash out another test-tube thoroughly with detergent, rinse with ordinary water and finally with propanone (acetone) to remove any droplets of water. Allow the test-tube to dry.
4. Pour the Tollen's reagent prepared in 2 above into the clean test-tube.
5. Add about three drops of ethanal and, using a test-tube holder, stand the test-tube in a beaker of boiling water for a few minutes.
6. Write down what you observe and explain your observation. Also write down the two half reactions taking place.
7. Repeat the experiment with propanone. Are ketones easily oxidised?

Note: It is only possible to oxidise ketones if very strong oxidising agents like hot concentrated nitric acid are used. This has the effect of breaking up the molecule into at least two molecules of carboxylic acid.

Test Yourself:
Attempt questions 23.4 and W23.4 – W23.5

EXPERIMENT 23.3 MANDATORY STUDENT EXPERIMENT NO. 27 (I)

To prepare ethanoic acid and examine its properties.

Part 1: Preparation of ethanoic acid

Introduction

We have already seen that primary alcohols may be oxidised to aldehydes. However, if the oxidising agent is in excess, a carboxylic acid may be formed. We use this fact to prepare ethanoic acid.

In this experiment, ethanol is oxidised to ethanoic acid using a mixture of sodium dichromate and dilute sulfuric acid. When we prepared ethanal by adding a mixture of ethanol and sodium dichromate solution to hot sulfuric acid, oxidation was stopped at the aldehyde stage by distilling off the volatile ethanal before further oxidation could take place. However, in this experiment, the use of excess oxidising agent and the refluxing of the mixture ensures that further oxidation to ethanoic acid occurs.

The reaction is carried out by slowly adding an aqueous solution of ethanol to a solution of sodium dichromate in dilute sulfuric acid. The overall reaction may be represented by the equation:

$$3C_2H_5OH + 2Cr_2O_7^{2-} + 16H^+ \rightarrow 3CH_3COOH + 4Cr^{3+} + 11H_2O$$

You are not expected to remember the above equation. All you need remember are the following two half equations:

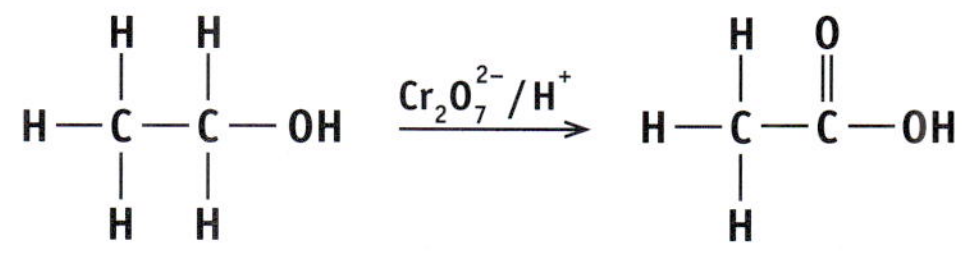

$$Cr^{+6} + 3e^- \longrightarrow Cr^{+3}$$

$$(\text{or } Cr^{VI} + 3e^- \longrightarrow Cr^{III})$$

Orange → Green

Apparatus required

Graduated dropper, small glass funnel, graduated cylinder (0 – 10 cm^3), conical flask, two retort stands and clamps, rubber tubing to fit Liebig condenser, hot plate, Bunsen burner, tripod, wire gauze, Pyrex beaker, stirring rod, Quickfit 27BU – Liebig condenser, dropping funnel, receiver adaptor, still head, pear-shaped flask (50 cm^3).

Chemicals required

Sodium dichromate, ethanol, concentrated sulfuric acid, anti-bumping chips, blue litmus paper, deionised water.

CAUTION!

Concentrated sulfuric acid should be handled with great care.

Procedure

1. Using a graduated dropper (or a glass funnel and a small graduated cylinder), carefully measure out 3.5 cm^3 of concentrated sulfuric acid. Slowly add this sulfuric acid in small quantities to 6 cm^3 of deionised water in a 50 cm^3 pear-shaped flask, cooling the flask under the tap and gently swirling the contents.
2. Add 5 g of sodium dichromate to the diluted acid in the flask. (If you use a funnel made from a piece of paper, it will help to prevent the solid sticking to the inside of the neck of the flask). Also add a few anti-bumping granules to the flask and set up the apparatus as show in Fig 23.31. When setting up the apparatus, ensure that both the flask and the condenser are clamped.
3. Place 1.5 cm^3 of ethanol and 5 cm^3 of water in the dropping funnel. Immerse the flask in a beaker of cold water and add the mixture drop by drop to the flask. Wait for the reaction to subside before adding the next few drops. Since the reaction is very exothermic, the beaker of cold water helps to control the temperature and keep the reaction from becoming too vigorous. Adding the mixture of water and alcohol drop by drop down the cold condenser into the flask helps to cool the mixture of alcohol and water. This helps to control the reaction between the alcohol and acidified sodium dichromate solution. If the alcohol and water were directly added to the flask without the condenser being in place, the reaction would become so vigorous that the contents of the flask would spurt out of the flask.

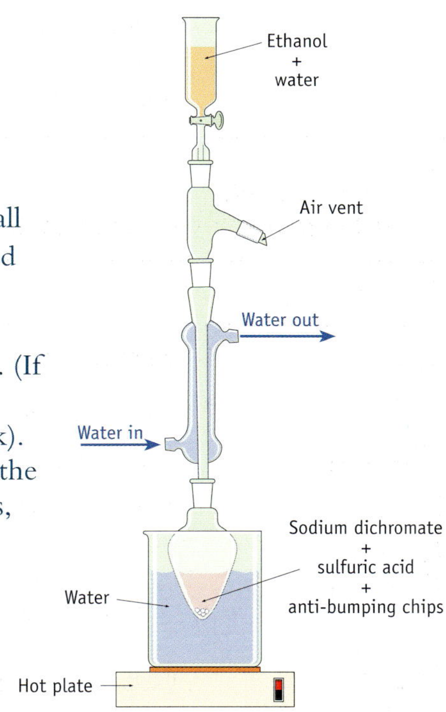

Fig 23.31
The preparation of ethanoic acid involves the use of this reflux apparatus.

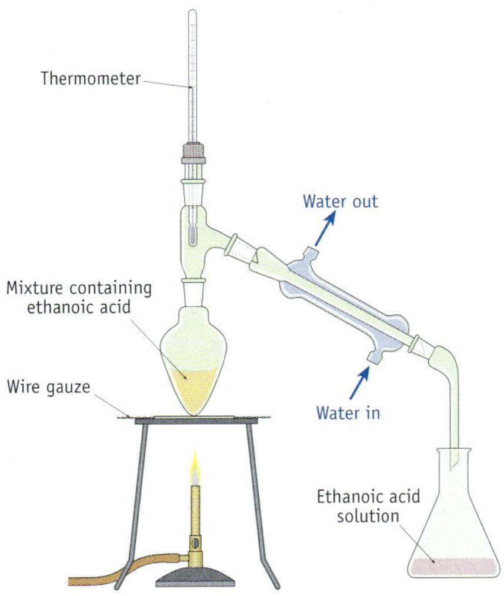

Fig 23.32
The ethanoic acid is separated from the mixture by distillation

4. Having added all of the mixture of water and ethanol, bring the water in the beaker to the boil and reflux the contents of the flask for about 20 minutes. Note the colour change from orange to green as the reaction proceeds.
5. Remove the hot water bath, allow the apparatus to cool for a few minutes and arrange the flask for distillation using the apparatus shown in FIG 23.32. Note that the flask is heated gently over a wire gauze. If the reaction becomes too vigorous, remove the Bunsen burner. (The mixture is heated using a Bunsen burner rather than a water bath as ethanoic acid has a boiling point of 118 °C.)
6. Continue heating until about 5 cm^3 of aqueous ethanoic acid has collected in the receiver. Note the characteristic smell of the ethanoic acid. You could also test for the presence of the acid using a piece of damp blue litmus paper.
Place the sample of ethanoic acid in a labelled sample tube. A little ethanal may be present as an impurity. However, it is unlikely that much will be present due to the severe conditions, i.e. refluxing for 20 minutes in the presence of excess oxidising agent. Almost all of the ethanal formed will be oxidised to ethanoic acid under these conditions. Other possible impurities are water, ethanol and ethyl ethanoate.
The properties of ethanoic acid will be studied in part 2 of EXPERIMENT 23.3 (PAGE 380).

Note the difference between the procedures followed in the preparation of ethanal and the preparation of ethanoic acid. In the preparation of ethanal, the sodium dichromate solution was in the dropping funnel. As the drops of sodium dichromate and alcohol reacted with the hot sulfuric acid solution in the flask, the ethanal formed was distilled off immediately. This was done to prevent it being oxidised to ethanoic acid. However, in the preparation of ethanoic acid, the acidified sodium dichromate solution is placed in the flask. The reason for this is to ensure complete oxidation of the ethanol to ethanoic acid.

Note: Carboxylic acids can be reduced to aldehydes and to alcohols in the presence of hydrogen and a nickel catalyst (FIG 23.33).

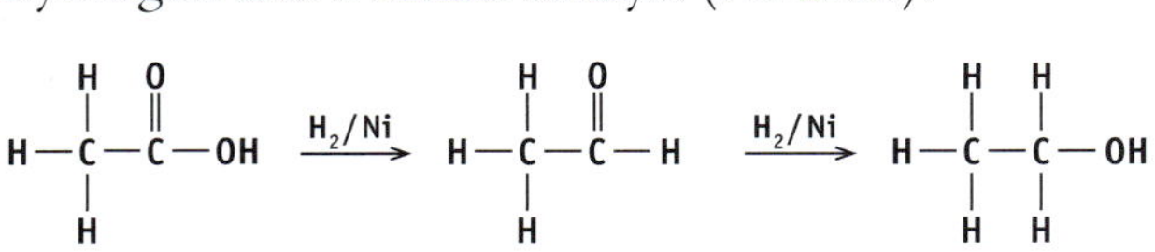

Ethanoic acid (acetic acid) — Ethanal (acetaldehyde) — Ethanol

Fig 23.33

Test Yourself:
Attempt questions 23.5 and W23.6.

23.6 REACTIONS AS ACIDS

In this section we will take a brief look at two classes of organic compounds that can act as acids, i.e. alcohols and carboxylic acids.

23.6.1 REACTION OF ALCOHOLS WITH SODIUM

If a piece of sodium metal is added to ethanol, it is observed that a gas is given off. (The reaction is similar to that of sodium with water but less vigorous.) When the gas is analysed, it is found to be hydrogen. The sodium reacts with ethanol to form a salt called sodium ethoxide and hydrogen gas.

$$2Na + 2C_2H_5OH \rightarrow 2C_2H_5ONa + H_2$$

sodium + ethanol → sodium ethoxide + hydrogen

If we examine the above equation carefully, we see that the H atom of the –OH group in ethanol has been replaced by a sodium atom. In other words, the ethanol is behaving as an acid since it is losing a H^+ ion. The hydrogen gas is formed when the H^+ ions from the ethanol react with the sodium atoms:

$$2Na + 2H^+ \rightarrow 2Na^+ + H_2$$

The polarity of the O-H bond ($-O^{\delta-}- H^{\delta+}$) makes it easier for the H atom to break off as a H^+ ion. The fact that the reaction of sodium with ethanol is far less vigorous than its reaction with water suggests that ethanol has a lower tendency than water to donate H^+ ions, i.e. ethanol is a weaker acid than water. In fact, ethanol is about one hundred times weaker in acidity than water. As a general rule, alcohols are much weaker acids than water. They are so weak that they do not react with bases like sodium hydroxide and sodium carbonate. They only act as acids in the presence of very reactive metals like sodium which have a great tendency to lose an electron.

23.6.2 ACIDIC NATURE OF THE CARBOXYLIC ACID GROUP

Why do carboxylic acids have the ability to act as acids, i.e. what makes it possible for the hydrogen atom of the carboxyl group to break off as a proton? There are two reasons for this:

(a) Inductive Effect.

If we examine the carboxyl functional group, we see that the H atom is attached to an oxygen atom in a C – O – H chain (Fig 23.34). Since the carbon atom of the carbonyl group is slightly positive, this carbon atom tends to attract electrons from the oxygen atom of the -OH group. This 'electron pulling' effect along a chain of atoms is called an **inductive effect**. Thus, the carbonyl group has the effect of drawing electrons away from the –OH group and this facilitates the ionisation of the H atom in the O – H bond. Therefore, carboxylic acids lose the proton of the COOH group quite readily and hence show acidic properties.

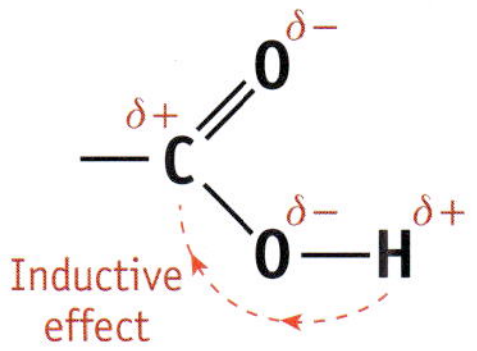

Fig 23.34
The inductive effect helps the proton to be lost more easily.

(b) Stability of carboxylate ion

When the carboxyl group loses a proton, it forms a negative ion called the carboxylate ion (Fig 23.35).

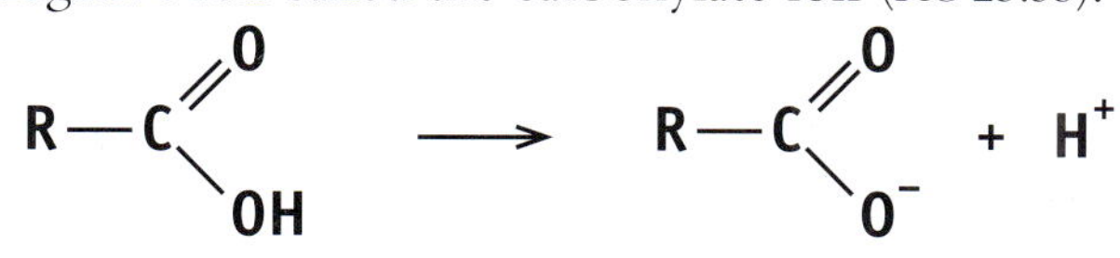

Fig 23.35

Chemists have found (using X-ray diffraction) that the two carbon-oxygen bonds in the carboxylate ion are of equal length. This suggests that, instead of having a C – O single bond and a C = O double bond, the actual structure is a mixture of these two possible structures shown in Fig 23.36.

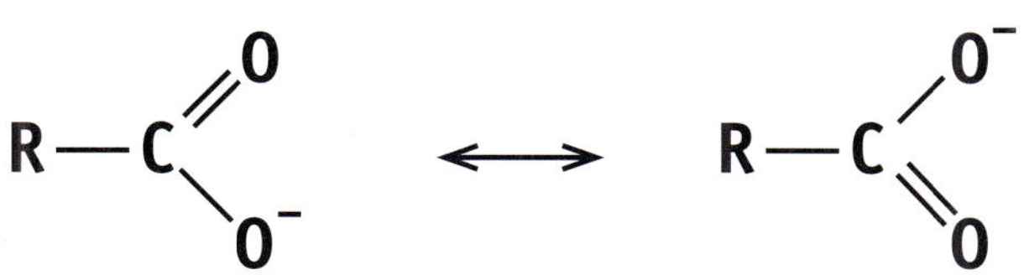

Fig 23.36

In other words, the negative charge is not localised on one oxygen atom but is spread over the three atoms (Fig 23.37). As in the case of benzene, we say that the new structure is a **resonance hybrid** of the two possible structures and this gives it extra stability.

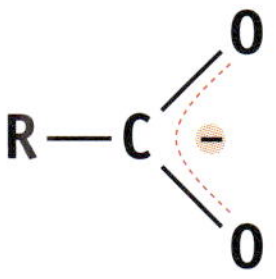

Fig 23.37
The carboxylate ion is formed when a carboxylic acid molecule loses a proton.

The fact that the carboxylate ion is so stable is the driving force for the carboxylic acid to lose a proton. Thus, carboxylic acids lose protons readily.

EXPERIMENT 23.3 MANDATORY STUDENT EXPERIMENT NO. 27 (II)

Part 2: To study the reactions of ethanoic acid with (a) sodium hydroxide, (b) sodium carbonate and (c) magnesium metal.

Introduction

In part 1 of EXPERIMENT 23.3 we prepared ethanoic acid. In this part of the experiment, we will look at three reactions of ethanoic acid. Ethanoic acid is a far weaker acid than the common inorganic acids like sulfuric acid, hydrochloric acid and nitric acid. However, it is strong enough to show some typical properties of acids. Examples of these properties of acids are:

(a) Ethanoic acid reacts with sodium hydroxide to form the salt sodium ethanoate and water:

$$CH_3COOH + NaOH \rightarrow \underset{\text{sodium ethanoate}}{CH_3COONa} + H_2O$$

(b) Ethanoic acid reacts with sodium carbonate liberating carbon dioxide and forming the salt sodium ethanoate and water.

$$2CH_3COOH + Na_2CO_3 \rightarrow \underset{\text{sodium ethanoate}}{2CH_3COONa} + CO_2 + H_2O$$

(c) Ethanoic acid reacts with magnesium metal to form the salt magnesium ethanoate and hydrogen gas is liberated.

$$2CH_3COOH + Mg \rightarrow \underset{\text{magnesium ethanoate}}{(CH_3COO)_2Mg} + H_2$$

Apparatus required

Test-tubes, test-tube holder, spatula, sand paper, droppers.

Chemicals required

Ethanoic acid solution (1.0 M), magnesium ribbon, sodium hydroxide solution (2 M), sodium carbonate powder, phenolphthalein solution.

Procedure

1. Fill a test-tube to about one-quarter of its capacity with dilute ethanoic acid. Add about three drops of phenolphthalein solution. Using a dropper, add sodium hydroxide solution drop by drop to the ethanoic acid until you observe a colour change. Write down what you observe. Explain your observation with the aid of a balanced chemical equation.
2. Place some sodium carbonate powder (about one spatula load) in a test-tube. Add about one-quarter of a test-tube of ethanoic acid solution to the sodium carbonate powder. Write down what you observe. Place a lighted taper in the mouth of the test-tube. What do you observe? Explain your observations with the aid of a balanced chemical equation.
3. Using a piece of sand paper, clean a small length of magnesium ribbon (about 4 cm). Fill a test-tube to about one-quarter of its capacity with dilute ethanoic acid. Add the magnesium ribbon to the ethanoic acid. Write down what you observe. Place a lighted taper near the mouth of the test-tube. What do you observe? Explain your observation with the aid of a balanced chemical equation.

Conclusions

What conclusions can you draw about the properties of ethanoic acid from the above tests?

23.7 ORGANIC SYNTHESIS

There are about seven million organic compounds known to chemists and more are being made or discovered each year. Many of these organic compounds are made from simpler materials. This procedure is called **organic synthesis**.

> Organic synthesis is the process of making organic compounds from simpler starting materials.

Chemists who specialise in organic synthesis have to plan the steps and then design experiments for making new and useful substances. There are usually a number of steps involved in the synthesis of a new compound. The new molecule being made is often referred to as the **target molecule**. Each step involves preparing a new compound from a previous compound. For example, the synthesis of ethanoic acid from ethene may be carried out as shown in FIG 23.38.

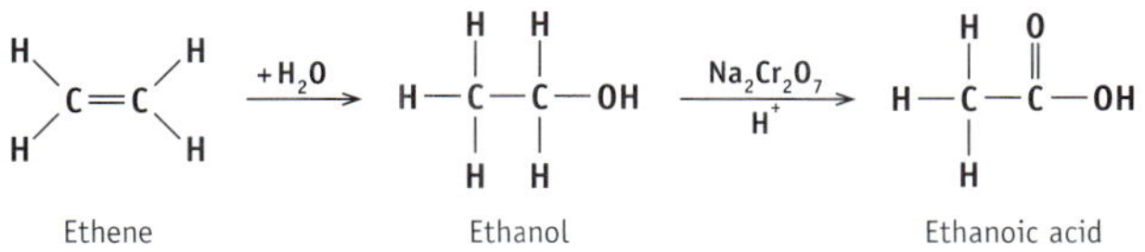

Fig 23.38
Synthesis of ethanoic acid from ethene.

There are many examples of useful products of organic synthesis, e.g. you probably have mild painkillers like aspirin and paracetamol (FIG 23.39) in your home. You may also have heard of other painkillers like ibuprofen and morphine, antibiotics like penicillin, weedkiller like paraquat and insecticides like DDT and malathion. (You are not required to know the structures of any of these compounds but you must know two examples of useful products of organic synthesis.)

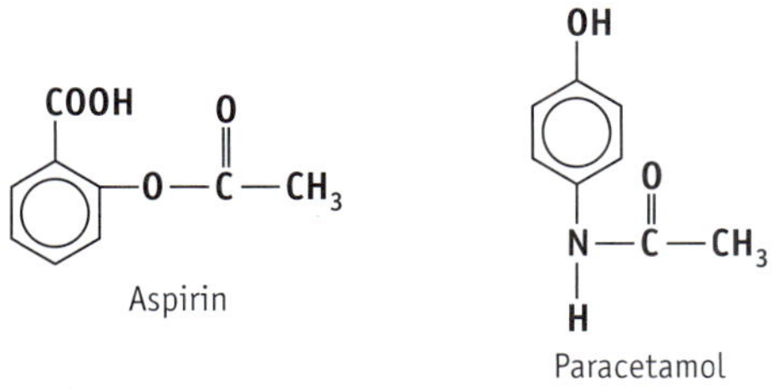

Fig 23.39
Aspirin and paracetamol are two very useful compounds synthesised by organic chemists. Aspirin was first synthesised in 1898 by a German chemist called Felix Hofmann of the Bayer company.

The majority of chemicals synthesised by the Irish chemical industry are pharmaceuticals. Most pharmaceuticals have fairly complicated structures. The synthesis of these compounds usually involves quite a number of steps and the formation of intermediate compounds before the final compound is obtained. Pharmaceuticals are usually named using short, trivial names rather than by their chemical names, e.g. 2-ethanoyloxybenzoic acid or acetylsalicyclic acid is simply called aspirin. In the pharmaceutical industry, new organic compounds are being continually designed and synthesised in order to provide possible new drugs to combat various illnesses.

Organic synthesis involves **bond breaking** and **bond making.** We have already discussed in this chapter how bonds are broken (homolytic and heterolytic fission) and have studied the energy changes involved in bond breaking and bond making in CHAPTER 21. For your examination, you need to be able to work out reaction schemes of up to three conversions, i.e. you need to know the various reactions we have met in the course and how these reactions can be used to convert one compound into another. This is not as difficult as you might think! To make it easy for you, all of the main reactions we have met so far in organic chemistry are summarised in FIG 23.40. Study this diagram very carefully!

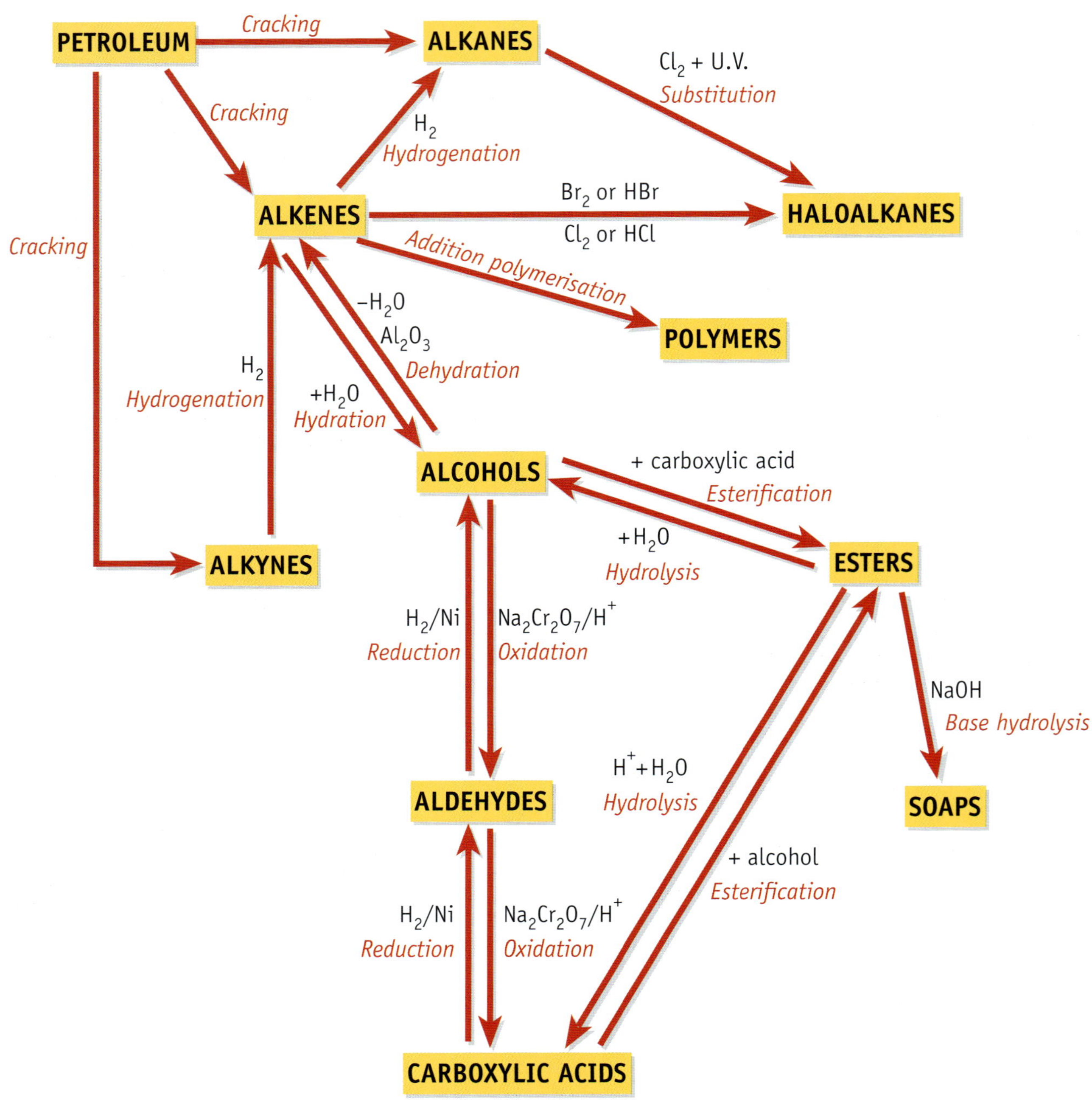

Fig 23.40
A summary of the main reactions in organic chemistry that are on our course.

Example 23.1

Name the reagents X and Y that are used to bring about the following reactions:

$$H_2C = CH_2 \xrightarrow{X} CH_3CH_3 \xrightarrow{Y} CH_3CH_2Cl$$

ANSWER

In the first step, we are converting an alkene to an alkane. Referring to FIG 23.40, we see that hydrogen is the reagent required, i.e. X = H_2.

In the second step, we are converting an alkane to a haloalkane (or more specifically a chloroalkane). Referring to FIG 23.40, this is carried out using chlorine in the presence of ultraviolet light, i.e. Y = Cl_2.

Example 23.2

Identify the reagents A and B that are used to carry out the following reactions:

$$CH_3\text{-}\overset{\overset{\large O}{||}}{C}\text{-}H \xrightarrow{A} CH_3CH_2OH \xrightarrow{B} H_2C = CH_2$$

ANSWER

In the first step, an aldehyde is being reduced to an alcohol. Referring to FIG 23.40, we see that this reduction reaction is brought about by hydrogen, i.e. A = H_2.

In the second step, an alcohol is being dehydrated to form an alkene. Referring to FIG 23.40, we see that this is carried out using aluminium oxide, i.e. B = Al_2O_3.

Example 23.3

A widely used plastic called PVC is manufactured by reacting ethene with substance X to produce 1,2-dichloroethane. This substance is then subjected to thermal cracking which removes HCl to form compound Y. Compound Y is then polymerised to form PVC.

$$H_2C = CH_2 \xrightarrow{X} CH_2Cl - CH_2Cl \xrightarrow{-HCl} Y \xrightarrow{\text{polymerisation}} PVC$$

(i) Name the compound commonly referred to as PVC.

(ii) Give the name and formula of substance X. What name is normally given to the type of reaction between X and ethene?

(iii) Draw the structural formula and give the name of compound Y.

(iv) Show the repeating unit in PVC.

ANSWER

(i) PVC stands for polyvinyl chloride. This is the traditional name. The systematic name for PVC is poly(chloroethene).

(ii) X is chlorine, Cl_2. This is an example of an addition reaction.

(iii) Y is called chloroethene. The structural formula is shown in FIG 23.41.

(iv) The formation of PVC from chloroethene is shown in FIG 23.42. Note that the repeating unit is indicated on the diagram.

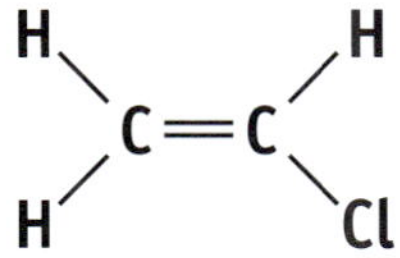

Fig 23.41
Chloroethene

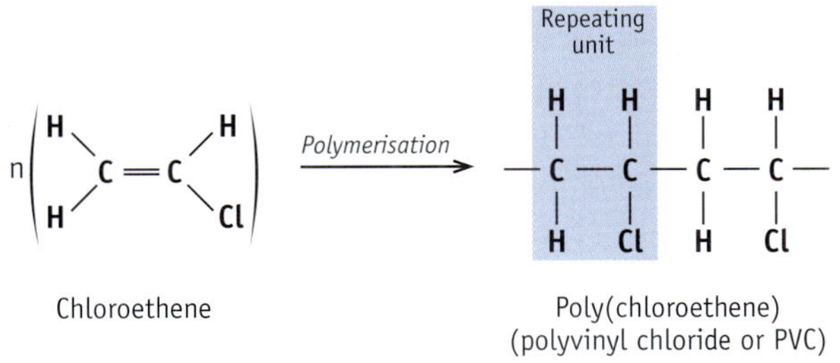

Fig 23.42
Chloroethene is polymerised to form poly(chloroethene) [PVC]. Many molecules of the monomer join together to form the polymer.

Note: EXAMPLE 23.3 should be studied carefully as this organic synthesis is specifically mentioned on the syllabus.

> **Test Yourself:**
> Attempt questions 23.6 – 23.7 and W23.7.

23.8 CHROMATOGRAPHY AND INSTRUMENATION IN ORGANIC CHEMISTRY

23.8.1 CHROMATOGRAPHY

The word chromatography comes from two Greek words meaning 'colour writing'. Chromatography is a very important separation technique in chemistry. It was invented by a Russian botanist called Mikhail Tswett in 1903. He used the technique of chromatography to separate the colour pigments in plants from each other. Chromatography is now widely used in chemistry not only for separating the different substances in a mixture, but also for identifying and measuring the amounts of these substances present in the mixture. Although originally chromatography was used to separate coloured substances, the technique has now developed beyond just coloured substances and is commonly used for a wide variety of different chemical compounds. However, we will restrict our practical work in chromatography to examples involving the separation of coloured substances.

The technique of chromatography is used in a number of different ways in the laboratory. We will now examine a number of these different ways and you will perform an experiment in your school laboratory using one of these techniques. All of the various techniques in chromatography depend on the same basic principle, i.e. a mobile phase carries the components of a mixture at different rates through a stationary phase. In other words, something like water or ethanol (mobile phase) could carry a dissolved mixture through a material like paper or aluminium oxide (stationary phase). The mobile phase flows over the stationary phase and the components of the mixture are separated out on the stationary phase. The separation is caused by the fact that components which tend to be held by the stationary phase move more slowly than those which mix well with the mobile phase.

> Chromatography is a separation technique in which a mobile phase carrying a mixture moves in contact with a selectively adsorbent stationary phase.

Let us now consider some of the various techniques of chromatography by conducting the following experiment:

EXPERIMENT 23.4 MANDATORY STUDENT EXPERIMENT NO. 28

To separate a mixture of coloured substances (e.g. inks in felt-tipped pens, dyes in food colouring, mixture of indicators, etc.) using paper chromatography or thin-layer chromatography or column chromatography.

(i) Paper Chromatography

In this technique, a small amount of the mixture to be separated is placed on a strip of adsorbent paper, e.g. filter paper. (It is also possible to buy special chromatography paper.) The end of the paper is placed in a suitable solvent. The solvent soaks up through the paper and dissolves the sample of material being analysed. Each of the components of the mixture dissolves to a different extent in the solvent. As the solvent continues to rise up through the paper, components that are not very soluble will come out of solution and appear as a spot on the filter paper. Components that are very soluble in the solvent will remain in solution and will take longer to separate out as a spot on the filter paper. Thus, the original sample will be separated out into its various components at different intervals along the paper. (If necessary, it is possible to get a precise analysis of each component by cutting the paper into strips and extracting each compound with the solvent before subjecting it to further tests.)

The technique of paper chromatography is illustrated in EXPERIMENT 23.4A.

EXPERIMENT 23.4a

To separate the components of the ink in felt-tipped pens using paper chromatography.

Introduction

In this experiment, we will examine the composition of inks of various colours. Some inks are made by mixing together two or more different colours. We will attempt to separate these colours by making use of the fact that each colour will travel a different distance up along a length of chromatography paper that is dipping into a suitable solvent.

Apparatus required

Chromatography paper (width 5 cm), scissors, gas jars or beakers, pencil, ruler, glass rod, clip (to attach paper to rod) and water-soluble felt-tipped pens of different colours, e.g. black, green, brown and blue.

Chemicals required

Water, methanol. Other solvents that can be tried out are ethanol, butan-1-ol, propanone, etc.

Procedure

1. Cut a suitable length of chromatography paper to fit the gas jar or beaker. Try to avoid handling the paper a lot, as the oils from your hands can affect the results.
2. Using a pencil, draw a faint line about 2 cm from the bottom of the paper.
3. Place a small dot of each of the two or three different coloured inks side by side on the pencil line. Make sure they are well spaced from each other. Using the pencil, write the colour of the ink near the dot. (Why is it advisable to use a pencil rather than a pen?)
   Note: Brown and black inks usually give good separation of colours.
4. Pour water into the gas jar to a depth of about 1 cm. (If the felt-tipped pens are marked 'water insoluble', you will have to use organic solvents.) Place the length of chromatography paper in the gas jar as shown in FIG 23.43. Allow time for the solvent to rise through the paper until a good separation of colours is observed.
5. Repeat the experiment with a different solvent or mixture of solvents, e.g. a 50 : 50 mixture of methanol and water. In many cases, the nature of the solvent can improve the separation of the different colours. Which inks are mixtures of other inks?

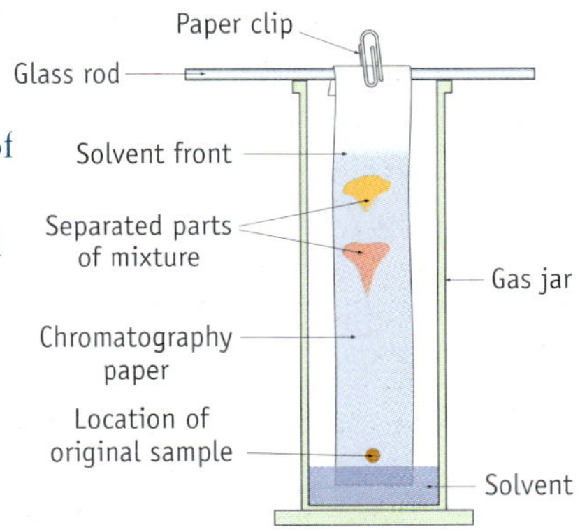

Fig 23.43
Paper chromatography may be used to separate the different colours in ink.

Instead of using a length of chromatography paper to demonstrate paper chromatography, you could also use a filter paper with ink spots of various colours (FIG 23.44). A hole is placed in the centre of the filter paper to hold a wick (also made of filter paper) which allows the solvent to move through the filter paper.

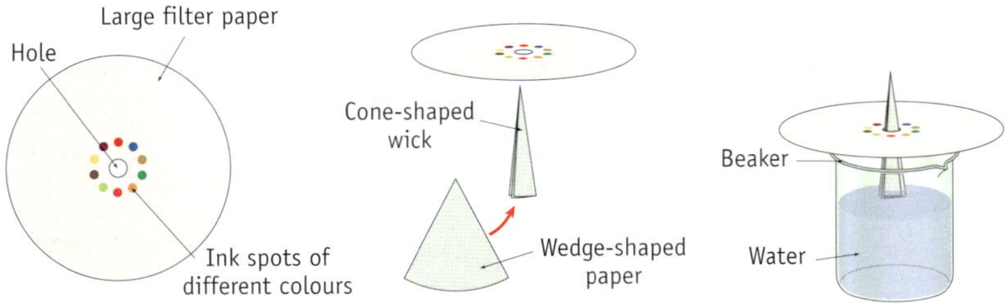

Fig 23.44
A large filter paper may also be used instead of a strip of chromatography paper to demonstrate paper chromatography.

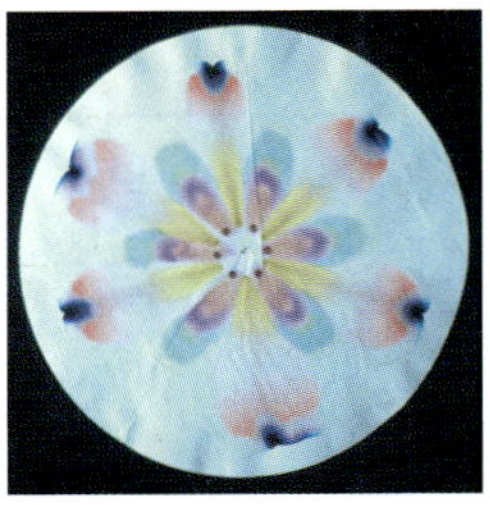

Fig 23.45
Chromatogram showing mixture of dyes in various inks.

A photograph of a chromatogram obtained by this method is shown in FIG 23.45.

(ii) Thin-layer chromatography

Thin-layer chromatography (commonly referred to simply as TLC) is a type of chromatography where the stationary phase is a thin layer of aluminium oxide or silica gel supported on a glass plate. (Silica gel is simply a compound of silicon and oxygen whose basic formula is SiO_2.) The aluminium oxide or silica gel is made into a slurry with water and is spread thinly and evenly over a glass or plastic plate and allowed to dry.

However, it is more convenient to buy ready-made TLC plates with aluminium foil instead of glass. It is these plates which will be used in the experiment outlined below.

In thin-layer chromatography, a spot of the mixture being analysed is placed near the bottom of the plate. When the spot has dried, the TLC plate is placed standing in a suitable solvent in a beaker. As in the case of paper chromatography, the solvent rises up the plate. The different substances in the mixture are then separated out as the solvent moves over the mixture. The TLC plate is removed from the beaker when the solvent approaches the top of the plate. The plate is then allowed to dry. If an organic solvent is being used, a lid is normally put on the beaker to prevent evaporation of the solvent.

TLC is a more efficient separation technique than paper chromatography. The reason for this is because the particles in the stationary phase are smaller in TLC than in paper chromatography and this gives better separation of the components of the mixture.

Thin-layer chromatography is often used in the pharmaceutical industry to determine the purity of drugs. If the drug that has been manufactured is impure, it will separate out into a number of components. Thin-layer chromatography is also used in forensic science to separate the colours in dyes extracted from fibres and hence to see if a fibre found at the scene of a crime and a fibre found on a suspect contain the same dye. In addition, thin-layer chromatography is used in the analysis of drugs like cannabis (FIG 23.46).

Fig 23.46
This photograph shows thin-layer chromatography being used in the Forensic Science Laboratory, Dublin to analyse cannabis.

In many cases, the components being separated may not be coloured. In these cases, the plate is sprayed with certain chemicals to make the spots visible or the plate may be viewed under UV light which makes some types of spots visible.

We now perform an experiment to separate out the components of food dyes using thin-layer chromatography.

EXPERIMENT 23.4b

To separate the components of food dyes using thin-layer chromatography.

Introduction

In this experiment, we will analyse some commercial food dyes to investigate if particular dyes are mixtures of a number of dyes. We will use thin-layer chromatography to carry out this separation process (FIG 23.47). Instead of making up TLC plates ourselves, we will use a commercially available roll of TLC plating (with aluminium backing).

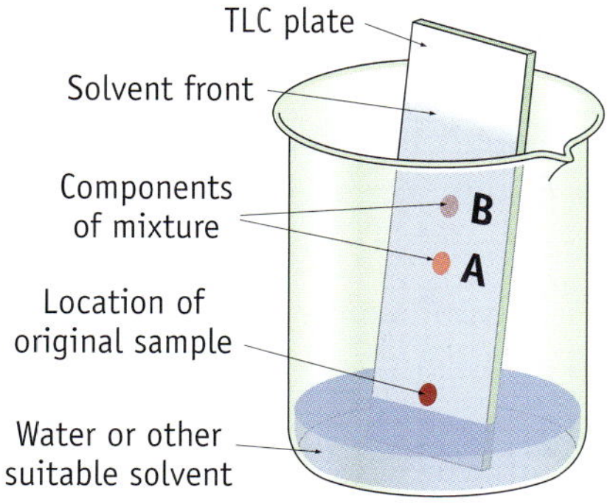

Fig 23.47
Thin-layer chromatography may be used to separate the components of food dyes. We will carry out this experiment on a smaller scale using a sample tube instead of a beaker.

Apparatus required

Roll of TLC plating, scissors, pencil, ruler, capillary tubes, sample tubes (5 cm × 2.5 cm).

Chemicals required

Methanol, commercial food dyes of various colours (green, yellow, blue, red, etc.).

Procedure

1. Cut a strip of TLC plating (about 5 cm × 2 cm) from the roll.
2. Using a pencil, draw a faint line about 1 cm from the bottom of the plate.
3. With the aid of a fine capillary tube, place a small spot of food dye on the pencil line. Using the pencil, write the colour of the ink near the dot.
4. Pour methanol into a sample tube to a depth of about 0.5 cm. Place the TLC plate in the sample tube. Allow time for the solvent to rise up through the plate until a separation of colours is observed.
5. Repeat the experiment with a different coloured dye and a fresh TLC plate and compare your results. Investigate if any of the dyes are mixtures of the other dyes.

Thin-layer chromatography can also be used to separate the ink in felt-tipped pens, the mixture of indicators in universal indicator solution, the dyes in flowers, etc.

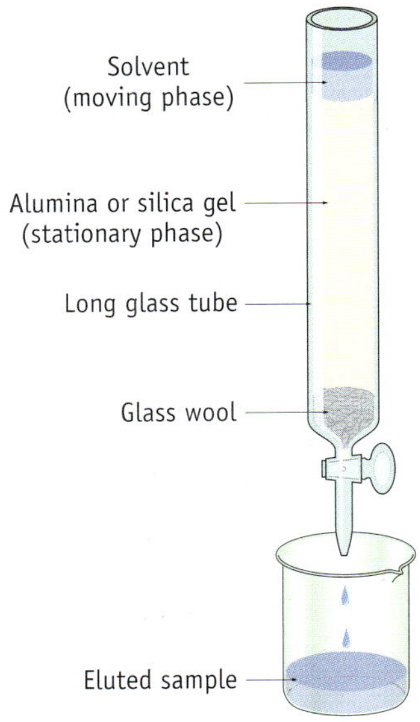

Fig 23.48
In column chromatography the stationary phase consists of solid adsorbent materials like alumina or silica gel packed into a long glass tube.

(iii) Column chromatography

Column chromatography is another type of chromatography in which the stationary phase is placed in a long glass tube commonly referred to as a column. The stationary phase usually consists of a solid adsorbent material like alumina or silica gel packed into a long glass tube (FIG 23.48). A piece of glass wool or a special type of glass disc is placed at the bottom of the column to prevent the material packed in the column from running out.

The stationary phase is made into a slurry using the appropriate solvent, e.g. if you were separating organic compounds from a mixture, solvents like ethanol, propanone, etc. could be used. If you were separating water-soluble compounds, the solvent used would be water. The stationary phase is kept saturated with the solvent to prevent the material in the column from drying out. If this happens, cracking of the material in the column could occur giving rise to channels that allow the mobile phase to flow through the column without any separation of the components of the mixture occurring.

The mixture to be separated is dissolved in a solvent and added to the top of the column. The solution is allowed to pass through the column. More solvent is added to keep the level of solvent above the stationary phase as shown in FIG 23.48. If the components of the mixture are coloured, each coloured substance will be adsorbed to a different extent along the column. If the column is very long, a series of coloured bands will appear along the column. This series of bands is called a **column chromatogram**. By adding more solvent, each of the coloured substances can be dissolved out of the column and collected separately in beakers.

The procedure of passing a solvent through the column is called **elution**. The general term for the solvent is the **eluent**. Thus, chemists say that the coloured substances have been eluted out of the column.

In EXPERIMENT 23.4C, we will use column chromatography to separate the mixture of dyes in food dyes. We will use a short column and the different colours in the food dye will be collected separately.

EXPERIMENT 23.4c

To separate the components of food dyes using column chromatography.

Introduction

In this experiment, we will analyse some commercial food dyes to investigate if particular dyes are mixtures of a number of dyes. We will use column chromatography to carry out this separation process. You will make your own column using a plastic disposable pipette, some cotton wool and silica gel. (FIG 23.49)

Apparatus required

Plastic disposable pipette (2.5 cm^3 graduated), cotton wool, clothes peg, retort stand, narrow length of wire, sample tubes (5 cm × 2.5 cm), scissors.

Chemicals required

Methanol, commercial food dyes of various colours (green works best), silica gel (particle size 0.13 – 0.25 mm, 60 – 120 mesh).

Procedure

1. Using a scissors, cut off the tip of the bulb and all of the lower narrow section of a plastic pipette.
2. Using a length of stiff wire, put a small wad of loosely fitting cotton wool at the bottom of the plastic pipette.
3. Fill the plastic pipette with silica gel up to the base of the bulb.
4. With the aid of a clothes peg or test-tube holder, stand the plastic pipette over a sample tube. Using a dropper, add about 3 cm^3 of methanol to the column and allow it to drain through it.
5. Using a dropper, add three drops of the green food dye to the washed column and allow it to drain through.
6. As the food dye drains through the column, continue adding methanol to ensure that a small reservoir of methanol remains above the top of the column.
7. As the separation continues, different coloured dyes will flow from the column. Change the sample tubes to collect each coloured substance individually.
   Instead of food dyes, you could also try to separate other mixtures like the ink in felt-tipped pens, the mixture of indicators in universal indicator solution, the dyes in flowers, etc.

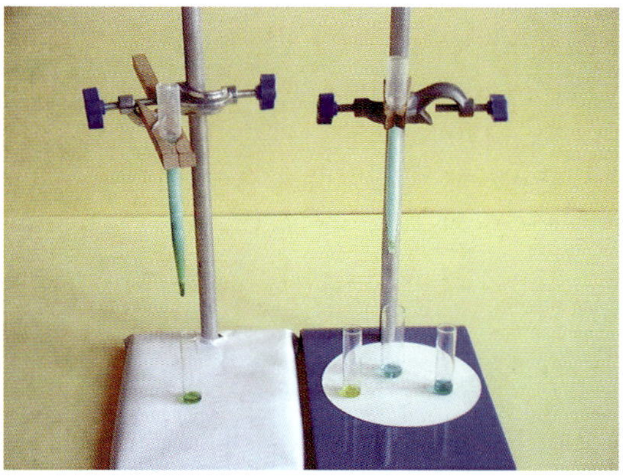

Fig 23.49
Apparatus for extracting components of green food dye. The yellow and blue colours have been eluted out of the column.

In industrial laboratories, professional chemists use more sophisticated separation techniques and 'high tech' apparatus than we use in the school laboratory. We now consider two of these chromatography techniques.

(I) Gas chromatography (GC)

Gas chromatography is so called because it uses a gas as the mobile phase. The stationary phase is a high boiling point liquid (e.g. a long-chain alkane) spread on solid particles (alumina or silica gel) that are packed into a long coiled tube called the column. This coiled tube, of small diameter and several metres in length, is kept inside a temperature-controlled oven.

The sample to be analysed is injected into the hot column using a syringe (Fig 23.50).

Fig 23.50
Apparatus used in gas chromatography. This chemist is using gas chromatography to determine the concentration of ethanol in alcoholic drinks.

The sample is vaporised and is carried through the tube using an unreactive gas like nitrogen, helium, or argon. The gas is commonly called a **carrier gas** as it carries the mixture through the hot column (Fig 23.51). Some components of the mixture are less soluble in the solvent than others. Therefore, the various components in the mixture are separated out as they flow through the column. As each component of the mixture leaves the column, it passes into a detector that records a signal. This detector plots a chart called a gas chromatogram on which each component of the mixture shows up as a peak.

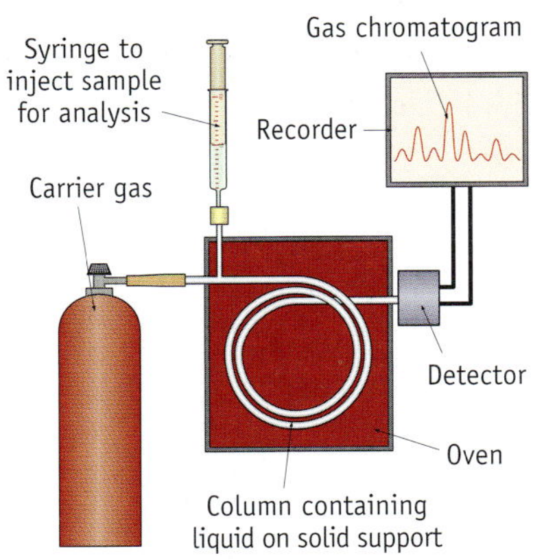

Fig 23.51
Schematic representation of gas chromatography.

Gas chromatography is commonly used to measure the level of alcohol in blood or urine samples, to analyse drug-test samples from athletes, etc.

One of the most powerful analytical tools available to a chemist is the combined use of gas chromatography and mass spectrometry. (The details of the mass spectrometer have already been studied in CHAPTER 4.) In this technique, commonly called **GC-MS**, the separated components that come from the gas chromatography apparatus are passed into a mass spectrometer. This enables each individual compound to be analysed separately and identified. Comparing the mass spectrum obtained with a large number of mass spectra stored in the mass spectra library database helps in this analysis. This powerful technique is commonly used in the Forensic Science Laboratory, Dublin, to investigate abuse of drugs.

In addition, GC-MS is used to monitor the performance of a particular pharmaceutical in the body. Volunteers are given a dose of the drug and then blood samples are taken at regular intervals and subjected to GC-MS analysis. This enables the researcher to follow how the drug is being metabolised inside the body.

GC-MS is also used to detect gases from waste dumps and organic pollutants in water. The technique is so sensitive that it can detect concentrations of these organic pollutants as low as 10^{-9} g/L. Small types of mass spectrometer have been used on board spaceships. When the Viking spacecraft landed on Mars, the atmosphere and soil were analysed using GC-MS equipment on board the spacecraft.

(II) HIGH-PERFORMANCE LIQUID CHROMATOGRAPHY (HPLC)

High-performance liquid chromatography (HPLC) developed from gas chromatography. In HPLC the substance being analysed is added to a liquid solvent which is passed over a solid. It was discovered that better separation of the components of the mixture occurs if the particles in the stationary phase are very small. However, it was also found that if very small particles were used in the column, the liquid passed very slowly through the column. Therefore, a pump is used to force the liquid through the column. The liquid is drawn from a large bottle of the solvent which is called the solvent reservoir.

Chromatography in which a pump is used to force a liquid through a tightly packed column is called **high-performance liquid chromatography**. (Since high pressures are used, HPLC is also used as an abbreviation for high-pressure liquid chromatography.) A schematic diagram of a HPLC instrument is shown in FIG 23.52.

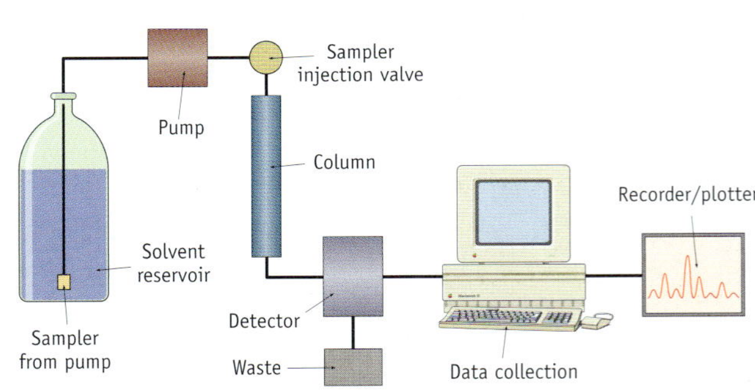

Fig 23.52
Schematic representation of HPLC instrument.

Since separation is more efficient in HPLC, the column need not be as long as in GC. In general, the length of the column in HPLC is in the region of only 10 – 30 cm. It is usually made of stainless steel to withstand the high pressure. Another advantage of HPLC over GC is that HPLC is carried out at a lower temperature than GC and, therefore, can be used to analyse compounds that would decompose at the high temperature needed for GC.

HPLC is commonly used in the analysis of foods, e.g. to test for the presence of growth promoters in meat, additives in food, vitamins in food, concentration of caffeine in Coca Cola, etc (Fig 23.53). Mass spectrometry may also be used in conjunction with HPLC to aid identification.

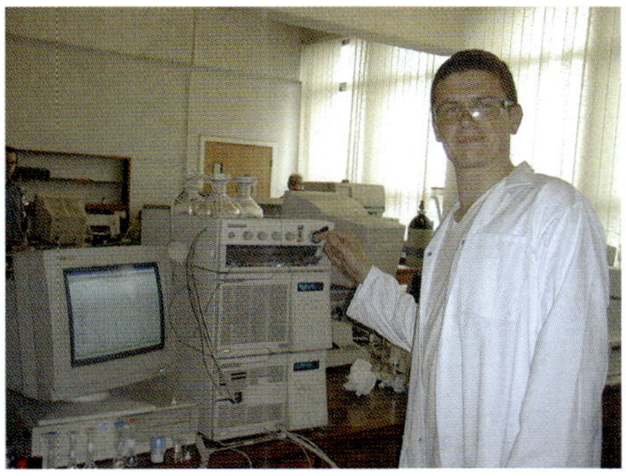

Fig 23.53
This HPLC instrument is being used by the research chemist to assist in the separation of various pharmaceutical compounds

We have seen above that GC and HPLC are very useful analytical techniques used by chemists. We have also seen that mass spectrometry is another very powerful tool used on its own or in conjunction with GC and HPLC.

23.8.2 INFRA-RED AND ULTRAVIOLET SPECTROMETRY

Let us now consider briefly two other types of instruments commonly used by chemists: (a) infra-red spectrometers and (b) ultraviolet spectrometers.

(A) Infra-red spectrometry (IR)

Infra-red radiation is radiation of wavelength longer than visible light (10^{-3} – 10^{-6} m). This radiation is invisible and is given off by hot objects, e.g. an electric light bulb or even our own bodies. Special infra-red cameras are often used to find people buried in collapsed buildings.

It is found that organic compounds absorb infra-red radiation. The infra-red radiation is aborbed by the vibrations of the bonds of the molecules. For example, a C = O bond will absorb infra-red radiation at a particular frequency and an O – H bond will absorb infra-red radiation at a different frequency. Thus, every organic compound has a unique IR spectrum that can serve as a 'fingerprint' for it. The infra-red spectrum of the plastic polystyrene is shown in Fig 23.54. No other compound has the same infra-red spectrum as polystyrene. Databases containing thousands of IR spectra of organic compounds are available to help in the identification of compounds. Chemists can identify the presence of particular functional groups by studying the IR spectrum of a compound.

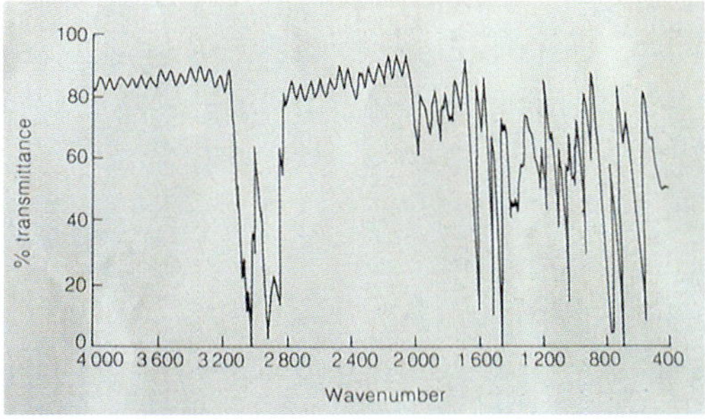

Fig 23.54
Infra-red spectrum of polystyrene.

The IR spectrum of a compound is obtained using an infra-red spectrometer similar to that shown in Fig 23.55. Infra-red spectrometry is commonly used to identify plastics, illegal drugs, etc. It is also used to back up evidence for prosecutions arising from breathalyser tests in motoring offences.

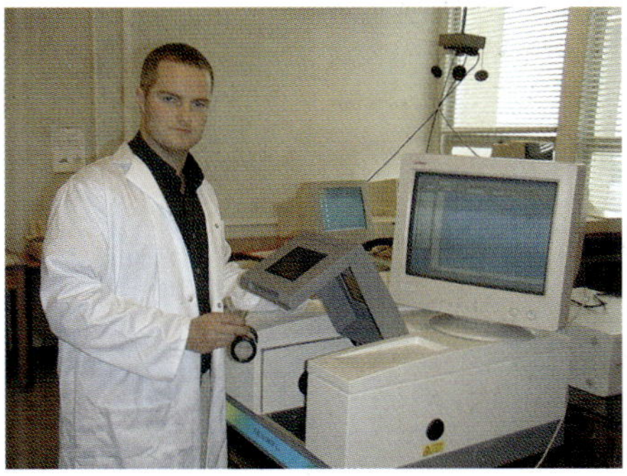

Fig 23.55
This infra-red spectrometer is being used to obtain the IR spectrum of an organic compound. The IR spectrum will be used to help identify the compound.

(B) Ultraviolet spectrometry (UV)

Ultraviolet radiation has a wavelength ($\approx 10^{-8}$ m) that is shorter than the wavelength of visible light. The main source of ultraviolet radiation is the sun (Chapter 26 of Option 1). You may have observed the effects of UV radiation in discos when clothes glow when exposed to UV radiation from a mercury lamp. Many modern washing powders contain chemicals that absorb UV light and give off visible light. These chemicals are called 'optical whiteners'. The fact that many compounds absorb UV light is used in the analysis of these compounds.

UV spectrometry is used to detect the presence of certain functional groups in molecules. It may also be used to measure the concentrations of certain organic compounds in solution, e.g. plant pigments, drugs, etc. In biochemistry laboratories, it may be used to measure the amount of a drug metabolite reaching a certain part of the body.

A modern UV spectrometer is shown in Fig 23.56. This instrument measures the absorption of UV radiation by the solution over a range of wavelengths.

In general, depending on the compound, maximum absorption tends to occur at a particular wavelength (Fig 23.57). This helps to identify the compound and also to measure the concentration of the compound in solution.

Fig 23.56
This UV spectrometer is being used to help identify an organic compound.

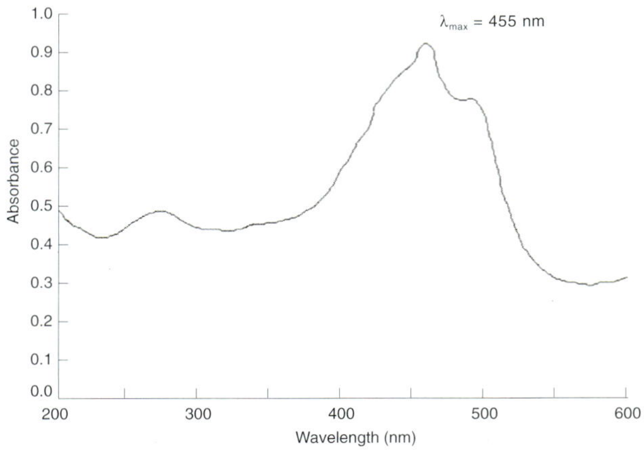

Fig 23.57
This an example of a UV spectrum of β-carotene found in carrots.

In practice, UV spectrometry is commonly used in conjunction with HPLC to detect the presence of many compounds and measure their concentrations.

Test Yourself:
Attempt questions 23.8 and W23.8.

CHAPTER SUMMARY

- A substitution reaction is a chemical reaction in which an atom or group of atoms in a molecule is replaced by another atom or group of atoms.
- The mechanism of a reaction is the detailed step-by-step description of how the overall reaction occurs.
- The reaction between methane and chlorine involves initiation, propagation and termination.
- Saponification is the base-catalysed hydrolysis of esters.
- An addition reaction is one in which two substances react together to form a single substance.
- The mechanism of the reaction between methane and bromine (ionic addition) involves heterolytic fission of Br_2, followed by carbonium ion formation and attack by the Br^- ion.

CHAPTER SUMMARY

- Polymers are long-chain molecules made by joining together many small molecules.
- An elimination reaction is one in which a small molecule is removed from a larger molecule to leave a double bond in the larger molecule.
- Ethanol is oxidised to ethanal using acidified sodium dichromate. If the oxidising agent is in excess, ethanoic acid is formed.
- Alcohols react with sodium to form a sodium salt and hydrogen gas.
- The acidic nature of the carboxylic acid group is due to an inductive effect and to the stability of the carboxylate ion.
- Organic synthesis is the process of making organic compounds from simpler starting materials.
- Chromatography is a separation technique in which a mobile phase carrying a mixture moves in contact with a selectively adsorbent stationary phase.
- There are different types of chromatography – paper, thin-layer, column, gas and high-performance liquid chromatography.
- Chemists commonly use infra-red spectrometers and ultraviolet spectrometers.
- Infra-red spectrometry is used to identify compounds. Every compound has a unique IR spectrum.
- Ultraviolet spectrometry is used to detect the presence of certain functional groups and to measure the concentration of certain organic compounds in solution.

EXERCISES

23.1. Explain the terms (i) substitution reaction, (ii) mechanism of a reaction and (iii) free radical. Give one example where knowledge of the mechanism of a reaction is of benefit to mankind.
Outline the mechanism of the reaction between methane and chlorine to form chloromethane.
What evidence is there to indicate that (i) a chain reaction takes place during the course of this reaction and (ii) free radicals combine with each other during the reaction?
What would you expect to happen if tetraethyl lead were added to the reaction mixture? Explain your reasoning.

23.2. A student prepared a sample of soap by refluxing a mixture of lard, sodium hydroxide and ethanol for about 30 minutes. The apparatus was then arranged for distillation. When distillation was completed, the mixture was quickly transferred to a concentrated solution of brine. The soap was filtered off, washed and dried.

(i) Draw a labelled diagram of the reflux apparatus used in this experiment. Why is it necessary to reflux the reactants?
(ii) What is the common name for sodium hydroxide?
(iii) What is the function of the ethanol?
(iv) One of the products of the reaction is also an alcohol. Name this product and write down its structural formula.
(v) What is brine? What is the purpose of adding the mixture to brine?
(vi) The soap is separated from the brine by filtration. What substances, apart from soap, would you expect to find in the filtrate?
(vii) What colour was the soap obtained in this experiment?
(viii) Describe what happens when this soap is added to hard water.

23.3. Explain the meaning of the term *addition reaction*. Using the structural formulas of reactants and products, show the reaction of ethene with (i) chlorine and (ii) water. Name the product formed in each case and give one use for each product.
Name one common substance which is manufactured by the hydrogenation of C = C double bonds.
Describe the mechanism of the reaction between ethene and bromine. What evidence exists to support the mechanism you have described.

23.4. A student prepared ethanal by oxidising ethanol with sodium dichromate in the presence of dilute sulfuric acid.
(i) Draw a labelled diagram of the apparatus you would use to prepare ethanal.
(ii) Why is it important to remove the ethanal as soon as it is formed?
(iii) What precautions would you take when mixing concentrated sulfuric acid with water? Why are these precautions necessary?
(iv) What colour is the dichromate/ethanol solution? What would you notice as this solution is added to the hot sulfuric acid solution? Explain your answer.
(v) Why was the heating stopped before starting to add the mixture of sodium dichromate and ethanol to the hot sulfuric acid solution?
(vi) What is the purpose of adding anhydrous sodium sulfate to the crude ethanal formed?
(vii) What would you observe if ethanal is warmed gently with (a) acidified potassium permanganate solution, (b) Fehling's solution and (c) ammoniacal silver nitrate solution. In each case, write the two half reactions that are occurring.

23.5. Ethanoic acid was prepared in the laboratory by slowly adding a mixture of ethanol and water to an acidified solution of sodium dichromate. The final mixture was then refluxed for a time. At the end of the reflux period, the ethanoic acid was isolated by distillation.
(i) Why is the mixture of ethanol and water *slowly* added to the acidified solution of sodium dichromate?
(ii) Name the acid used to acidify the sodium dichromate solution.
(iii) The addition of the ethanol–water mixture to the acidified solution of sodium dichromate could become very violent. What is the reason for this? What precaution is normally taken to ensure that the reaction does not go out of control?
(iv) What is observed as the ethanol reacts with the acidified sodium dichromate solution?
(v) Why is it necessary to reflux the contents of the flask for some time before the distillation is carried out?
(vi) Describe and explain any colour changes that take place during the refluxing.
(vii) What precaution is taken to ensure that boiling takes place smoothly during the reflux period?
(viii) Draw a labelled diagram of the distillation apparatus.
(ix) Why is it unlikely that much ethanal will be present in the distillate?
(x) Give one important difference between the procedures followed in the preparations of ethanal and ethanoic acid. What is the reason for this difference?

23.6. (i) Name the reagent needed to bring about the following conversion:

$$CH_2 = CH_2 \xrightarrow{X} CH_3CH_2OH$$

(ii) Write down the name and structural formula of the product P in the following reaction:

$$CH_2 = CH_2 \xrightarrow{HCl} P$$

(iii) Give the name and structural formula of the starting material S in the following reaction:

$$S \xrightarrow{Cr_2O_7^{2-} / H^+} CH_3CH_2CHO$$

(iv) Identify the reagents A and B needed to bring about the following reactions:

$$CH_2 = CH_2 \xrightarrow{A} CH_3CH_3 \xrightarrow{B} CH_3CH_2Cl$$

(v) Name the reagents E and F needed to bring about the following reactions:

$$CH_3\text{-}\overset{\overset{\displaystyle O}{\|}}{C}\text{-}CH_3 \xrightarrow{E} CH_3\overset{\overset{\displaystyle OH}{|}}{C}HCH_3 \xrightarrow{F} H_2C = CH\text{-}CH_3$$

Give the names of each of the three compounds shown above.

23.7. Show how you could bring about the following conversions in no more than three steps. For each step, give the name of the reagent or reagents used and the structural formulas of all compounds.

(i) $CH_2 = CH_2 \rightarrow CH_3COOH$

(ii) $CH_3COOH \rightarrow CH_2 = CH_2$

(iii) $CH_3CH = CH_2 \rightarrow CH_3COCH_3$

(iv) $CH_3CHO \rightarrow CH_3COOCH_2CH_3$

(v) $CH_3CH(OH)CH_3 \rightarrow CH_3CHClCH_2Cl$

23.8. Explain what is meant by the term *chromatography*.
Describe how paper chromatography or thin-layer chromatography or column chromatography may be used to separate the components of a mixture of inks.
Give one use of thin-layer chromatography.
What does the abbreviation HPLC stand for? With the aid of a diagram, outline the basic principle of a HPLC instrument. Give two uses of this instrument.
Mass spectrometry is often used in conjunction with HPLC. What is the reason for this?
Write a brief note on ultraviolet spectrometry. Give one use for this analytical technique.

24

Stoichiometry II

24.1 INTRODUCTION

We have already studied the fundamentals of stoichiometry in CHAPTER 11. In this chapter we will study how chemists use stoichiometry to work out what reactant is present in excess and what reactant is present in limiting amount in a chemical reaction. We will also study how chemists calculate the percentage yield of a chemical reaction.

24.2 THE LIMITING REACTANT

When carrying out chemical reactions, it is quite common for chemists to use an excess of one or more of the reactants. This often helps to increase the yield of products. For example, you may have carried out an experiment in Junior Certificate science where you reacted together iron and sulfur to form iron sulfide. The equation for this reaction is:

$Fe + S \rightarrow FeS$

This equation tells us that one atom of iron reacts with one atom of sulfur. Alternatively, you could say that the iron and the sulfur react in the ratio of 1:1. But consider what happens if we react four atoms of iron with six atoms of sulfur (FIG 24.1). Clearly, two atoms of sulfur will be left over at the end of the reaction. Since the sulfur is left over in the reaction, we say that the sulfur is present in **excess**. Therefore, as the sulfur is in excess, the amount of product formed depends on the amount of iron present. The iron will continue to react with the sulfur until all of the iron atoms have taken part in the reaction.

When the iron is used up, the reaction stops. Thus, the iron limits or determines the amount of product formed. The iron is called the **limiting reactant**.

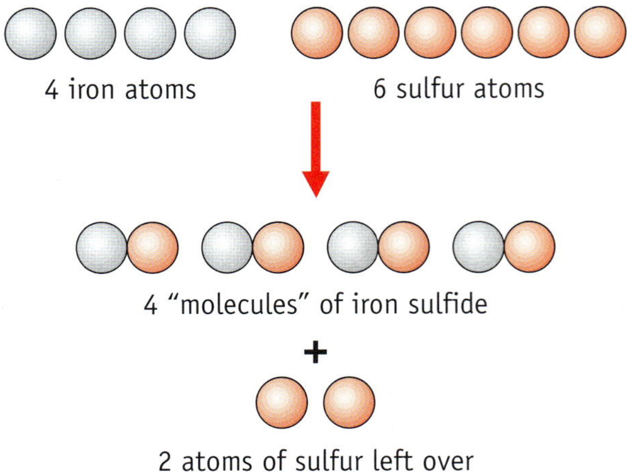

Fig 24.1
Since iron and sulfur react in the ratio 1:1, two atoms of sulfur are left over when four atoms of iron react with six atoms of sulfur. We say that sulfur is in excess and that iron is the limiting reactant. In other words, iron determines the amount of iron sulfide formed.

If you are given the amounts of two or more reactants, it will be necessary to find out which one is the limiting reactant before you can calculate the amount of product formed.

From the experiments you have carried out in this course, you are aware that amounts of reactants are often measured in grams. Therefore, you will have to convert grams to moles in order to find out which of the reactants is the limiting reactant.

Example 24.1

A student weighed out 10 grams of iron and 10 grams of sulfur. After mixing them, she placed them in a boiling tube and heated it to bring about the following reaction:

$Fe + S \rightarrow FeS$

Calculate the maximum amount of iron sulfide that could be formed in this reaction.
(S = 32, Fe = 56)

SOLUTION
First calculate the number of moles of each reactant.

$$\text{number of moles of Fe} = \frac{\text{mass}}{\text{rel atomic mass}} = \frac{10}{56} = 0.18$$

$$\text{number of moles of S} = \frac{\text{mass}}{\text{rel atomic mass}} = \frac{10}{32} = 0.31$$

From the balanced equation, one mole of iron reacts with one mole of sulfur. Therefore, 0.18 moles of iron will require 0.18 mole of sulfur for the reaction to go to completion. However, since the student has weighed out 0.31 mole of sulfur, clearly the sulfur is in excess. Therefore, the iron is present in limiting amount.

$Fe + S \rightarrow FeS$

Since the sulfur is in excess, you can forget about it when carrying out the calculations.

Fe → FeS
1 mole → 1 mole
0.18 mole → 0.18 mole
= 0.18 × 88 g (Rel molecular mass FeS = 56 + 32 = 88)
= 15.84 g

Answer: The maximum amount of iron sulfide that could be formed is 15.84 grams.

24.3 CALCULATING THE PERCENTAGE YIELD

The mass of a product formed in a chemical reaction is called the **yield**. The yield of a product that is calculated from the chemical equation is called the **theoretical yield**. The calculation of the theoretical yield assumes that the reaction is 100% efficient, i.e. all of the starting material is converted to the desired product. However, there are many reactions in chemistry which do not go to completion, i.e. there is not a 100% conversion of reactants to products. The yield that is obtained in practice is called the **actual yield.**

In order to check if a process is economically viable, it is important for chemists to know the percentage of reactants converted to products. The concept of **percentage yield** is used in order to express the percentage of reactants converted to products. It is defined as follows:

$$\text{Percentage yield} = \frac{\text{Actual yield of product}}{\text{Theoretical yield of product}} \times 100$$

Note: Actual yield and theoretical yield need not necessarily be given in units of mass. Yields may also be expressed in units of moles, litres, etc. However, it is important when calculating percentage yield that the units of actual yield and theoretical yield are the same.

Example 24.2

Ethene, C_2H_4, can be made from ethanol, C_2H_5OH, according to the following equation:

$C_2H_5OH \rightarrow C_2H_4 + H_2O$

Starting with 20 g of ethanol, a student prepared 8 g of ethene. Calculate the percentage yield of ethene.

Solution

Relative molecular mass $C_2H_5OH = 2(12) + 5(1) + 16 + 1 = 46$
Relative molecular mass $C_2H_4 = 2(12) + 4(1) = 28$

$$\text{Number of moles of ethanol} = \frac{\text{Mass}}{\text{Rel molecular mass}} = \frac{20}{46} = 0.435$$

$C_2H_5OH \rightarrow C_2H_4 + H_2O$
1 mole $\rightarrow$ 1 mole
0.435 moles $\rightarrow$ 0.435 moles
$= 0.435 \times 28$
$= 12.18$ g

i.e. the theoretical yield of ethene is 12.18 g.
Actual yield = 8 g of ethene.

Therefore,

$$\text{Percentage yield} = \frac{\text{Actual yield of product}}{\text{Theoretical yield of product}} \times 100$$

$$= \frac{8}{12.18} \times 100$$

$$= 65.68\%$$

Answer: Percentage yield of ethene is 65.68%.

Example 24.3

A group of students prepared ethene by passing ethanol vapour over heated aluminium oxide. The equation for the reaction is:

$C_2H_5OH \rightarrow C_2H_4 + H_2O$

If the students used 2.3 cm^3 of ethanol (density = 0.8 g/cm^3) in the above experiment and obtained a 15% yield of ethene, what volume of ethene (measured at room temperature and pressure) would have been collected? (Molar volume at room temperature and pressure = 24.0 L) *[LCH]*

Solution

First calculate the mass of ethanol used.

$$\text{Density} = \frac{\text{Mass}}{\text{Volume}}$$

$\Rightarrow$ Mass = Density $\times$ Volume $= 0.8 \times 2.3$
$= 1.84$ g ethanol

Relative molecular mass $C_2H_5OH = 2(12) + 5(1) + 16 + 1 = 46$

Number of moles of ethanol $= \dfrac{\text{Mass}}{\text{Rel molecular mass}} = \dfrac{1.84}{46} = 0.04$

$C_2H_5OH \rightarrow C_2H_4 + H_2O$

1 mole $\rightarrow$ 1 mole

0.04 moles $\rightarrow$ 0.04 moles

$= 24 \times 0.04$ (Molar Volume = 24 L)

$= 0.96$ L

i.e. the theoretical volume of ethene which could be formed is 0.96 L.

15% yield obtained $\Rightarrow$ 15% of 0.96 L obtained

$= \dfrac{15}{100} \times 0.96$

$= 0.144$ L

$= 144\ cm^3$

Answer: 144 cm^3 of ethene would have been collected at room temperature and pressure.

Example 24.4

A group of students prepared a sample of ethanal (acetaldehyde), CH_3CHO, by reacting ethanol, C_2H_5OH, with sodium dichromate, $Na_2Cr_2O_7.2H_2O$ (FIG. 24.2). The students used 14.9 g of sodium dichromate and 11.5 cm^3 of ethanol (density = 0.80 g/cm^3). It was found that 1.6 g of ethanal, was formed. The equation for the reaction is:

$3C_2H_5OH + Cr_2O_7^{2-} + 8H^+ \rightarrow 3CH_3CHO + 2Cr^{3+} + 7H_2O$

(i) From the information given above, show which of the two substances, ethanol or sodium dichromate is in excess.

(ii) Calculate the percentage yield of ethanal. *[LCH]*

SOLUTION

(i) Note that in this problem we are given a volume of ethanol rather than its mass. Its mass is calculated using the value of the density given.

Density $= \dfrac{\text{Mass}}{\text{Volume}}$

Therefore, Mass = Density $\times$ Volume $= 0.8 \times 11.5$

$= 9.2$ g ethanol

Fig 24.2
These students are preparing ethanal.

Now calculate the numbers of moles of ethanol and sodium dichromate.

Relative molecular mass $C_2H_5OH = 2(12) + 5(1) + 16 + 1 = 46$

Number of moles of ethanol $= \dfrac{\text{Mass}}{\text{Rel molecular mass}} = \dfrac{9.2}{46} = 0.2$

Relative molecular mass $Na_2Cr_2O_7.2H_2O = 2(23) + 2(52) + 7(16) + 2(18) = 298$

Number of moles of sodium dichromate $= \frac{\text{Mass}}{\text{Rel molecular mass}} = \frac{14.9}{298} = 0.05$

$3C_2H_5OH + Cr_2O_7^{2-} + 8H^+ \rightarrow 3CH_3CHO + 2Cr^{3+} + 7H_2O$

3 moles   1 mole

0.15 moles   0.05 moles

From the balanced equation, one mole of dichromate reacts with three moles of ethanol. Therefore, 0.05 moles of dichromate would react with $3 \times 0.05 = 0.15$ moles of ethanol. However, we have **more** than 0.15 moles of ethanol present. Our calculation above shows that we have 0.2 moles of ethanol present, i.e. **the ethanol is present in excess.** In other words, the sodium dichromate is present in limiting amount.

(ii) Relative molecular mass $CH_3CHO = 12 + 3(1) + 12 + 1 + 16 = 44$

Since the sodium dichromate is the limiting reactant, we calculate the percentage yield of ethanal from the amount of sodium dichromate present.

$3C_2H_5OH + Cr_2O_7^{2-} + 8H^+ \rightarrow 3CH_3CHO + 2Cr^{3+} + 7H_2O$

1 mole $\rightarrow$ 3 moles

0.05 moles $\rightarrow$ 0.05×3

$= 0.15$ moles

$= 0.15 \times 44$ g

$= 6.6$ g

i.e. the theoretical mass of ethanal which could be formed is 6.6 g

In the question, we are told that the actual amount of ethanal formed was 1.6 g

Therefore,

$$\text{Percentage yield} = \frac{\text{Actual yield of product}}{\text{Theoretical yield of product}} \times 100 = \frac{1.6}{6.6} \times 100 = 24.24\%$$

Answer: (i) The ethanol is in excess.

(ii) Percentage yield is 24.24%

Example 24.5

A group of students (FIG 24.3) prepared ethanoic acid by the reaction, in aqueous solution, between 9.2 cm^3 of ethanol (density 0.8 g/cm^3) and 35.76 g of sodium dichromate(VI), $Na_2Cr_2O_7.2H_2O$, acidified with excess sulfuric acid. The reaction mixture was boiled in a suitable apparatus for about 30 minutes, after which the ethanoic acid produced was recovered by distillation. After further purification the yield of ethanoic acid was found to be 47.4%. The equation for the reaction is:

$3C_2H_5OH + 2Cr_2O_7^{2-} + 16H^+ \rightarrow 4Cr^{3+} + 3CH_3COOH + 11H_2O$

(i) Show clearly that there was an excess of sodium dichromate(VI) over ethanol in the experiment carried out by the students.

(ii) Calculate the volume of ethanoic acid (density 1.05 g/cm^3) obtained as a result of the experiment. *[LCH]*

SOLUTION

(i) Relative molecular mass $C_2H_5OH = 2(12) + 5(1) + 16 + 1 = 46$

Relative molecular mass $Na_2Cr_2O_7.2H_2O = 2(23) + 2(52) + 7(16) + 2(18) = 298$

Mass of ethanol = Density $\times$ Volume $= 0.8 \times 9.2$

$= 7.36$ g

Number of moles of ethanol $= \frac{\text{Mass}}{\text{Rel molecular mass}} = \frac{7.36}{46} = 0.16$

Number of moles of sodium dichromate = $\frac{\text{Mass}}{\text{Rel molecular mass}} = \frac{35.76}{298} = 0.12$

$3C_2H_5OH + 2Cr_2O_7^{2-} + 16H^+ \rightarrow 4Cr^{3+} + 3CH_3COOH + 11H_2O$

3 moles 2 moles

i.e. 1 mole of ethanol reacts with $\frac{2}{3}$ mole of dichromate

Therefore, 0.16 mole of ethanol reacts with $\frac{2}{3} \times 0.16 = 0.107$ mole of dichromate

i.e. 0.16 mole of ethanol would react with 0.107 mole of dichromate. However, there is 0.12 mole of dichromate present. Therefore, the dichromate is present in excess, i.e. the ethanol is the limiting reactant.

(ii) Relative molecular mass $CH_3COOH = 12 + 3(1) + 12 + 2(16) + 1 = 60$

The ethanol is the limiting reactant, so the percentage yield of ethanoic acid is calculated on the amount of ethanol present.

$3C_2H_5OH + 2Cr_2O_7^{2-} + 16H^+ \rightarrow 4Cr^{3+} + 3CH_3COOH + 11H_2O$

3 moles ⟶ 3 moles

$\Rightarrow$ 0.16 moles ⟶ 0.16 moles

$= 0.16 \times 60$

$= 9.6$ g (theoretical mass of ethanoic acid)

Density = $\frac{\text{Mass}}{\text{Volume}}$

$\Rightarrow$ Theoretical volume of ethanoic acid = $\frac{\text{Mass}}{\text{Density}} = \frac{9.6}{1.05} = 9.14 \text{ cm}^3$

Since we only get a 47.4% yield of ethanoic acid, then only 47.4% of the 9.14 cm^3 will be obtained.

47.4% yield $= \frac{47.4}{100} \times 9.14 \text{ cm}^3$

$= 4.33 \text{ cm}^3$

Answer: (i) The sodium dichromate is in excess.

(ii) Volume of ethanoic acid obtained is 4.33 cm^3.

Fig 24.3
These students are preparing ethanoic acid

Note that the calculations for determining the limiting reactant in the preparation of ethanoic acid (EXAMPLE 24.5) are slightly more difficult than the calculations for the preparation of ethanal (EXAMPLE 24.4). The reason for this is that the balanced equation for the preparation of ethanal has a ratio of ethanol : dichromate of 3 : 1. However, in the preparation of ethanoic acid, this ratio is 3 : 2 and hence one of these numbers has to be brought to 1 in the calculations as shown in EXAMPLE 24.5.

Always remember that in the preparation of ethanoic acid the dichromate must be in excess to ensure complete oxidation of the ethanol to ethanoic acid (EXPERIMENT 23.3). On the contrary, in the preparation of ethanal, the dichromate is present in limiting amount to prevent further oxidation to ethanoic acid (EXPERIMENT 23.2). Therefore, in the case of these two experiments, you know the answer before you begin the calculations but you must prove it using the experimental data given.

Test Yourself:
Attempt questions 24.1 – 24.6 and W24.1 – W24.5

CHAPTER SUMMARY

- In a chemical reaction, the reactant that determines the amount of product formed is called the limiting reactant.
- Apart from the limiting reactant, the other reactants are said to be in excess.
- Theoretical yield is the yield of a product that is calculated from the chemical equation.
- Actual yield is the yield that is obtained in practice.
- Percentage yield $= \frac{\text{Actual yield of product}}{\text{Theoretical yield of product}} \times 100$

EXERCISES

24.1. Sucrose, $C_6H_{12}O_6$, a sugar found in grapes, ferments to form ethanol in wine. The reaction which takes place is:
$C_6H_{12}O_6 \rightarrow 2C_2H_5OH + 2CO_2$
When 20 g of sucrose ferments, it is found that 7.5 g of ethanol is formed. Calculate the percentage yield of ethanol.

24.2. A sweet smelling substance called ethyl ethanoate, $CH_3COOC_2H_5$, may be obtained by reacting ethanoic acid, CH_3COOH, with ethanol, C_2H_5OH, according to the following equation:
$CH_3COOH + C_2H_5OH \rightarrow CH_3COOC_2H_5 + H_2O$
It is found that 20 g of ethyl ethanoate are obtained from 21 g of ethanol. Calculate the percentage yield of the reaction to the nearest whole number.

24.3. A group of students prepared ethanal (acetaldehyde) by the reaction between acidified sodium dichromate(VI) solution and ethanol. The masses used were 17.88 g of sodium dichromate(VI), $Na_2Cr_2O_7.2H_2O$, and 11.04 g of ethanol. The ethanal was removed from the reaction mixture as soon as it was formed and was collected in a flask surrounded by ice-water. After further purification, the yield of ethanal was found to be 2.97 g. The equation for the experiment is:
$3C_2H_5OH + Cr_2O_7^{2-} + 8H^+ \rightarrow 3CH_3CHO + 2Cr^{3+} + 7H_2O$
(i) Show clearly that the ethanol is in excess.
(ii) Calculate the percentage yield of ethanal. *[LCH]*

24.4. A student prepared a sample of ethanoic acid, CH_3COOH, by the reaction in aqueous solution between 6.9 cm^3 of ethanol, C_2H_5OH, (density 0.80 g/cm^3) and 29.8 g of sodium dichromate(VI), (formula $Na_2Cr_2O_7.2H_2O$), acidified with excess sulfuric acid. The reaction may be represented by the following equation:
$3C_2H_5OH + 2Cr_2O_7^{2-} + 16H^+ \rightarrow 4Cr^{3+} + 3CH_3COOH + 11H_2O$
In the experiment, 5.4 cm^3 of ethanoic acid were collected.
(i) Show that there is more than enough sodium dichromate to oxidise all the ethanol.
(ii) Calculate the percentage yield of ethanoic acid.
(Density of ethanoic acid = 1.06 g/cm^3) *[LCH]*

24.5. Iron may be manufactured according to the following reaction:
$2Fe_2O_3 + 3C \rightarrow 3CO_2 + 4Fe$
Calculate the maximum mass of iron (to the nearest gram) that can be obtained from 500 g of iron(III) oxide and 50 g of carbon.

24.6. Ethanal and ethanoic acid were prepared in the school laboratory, in two separate experiments, by the oxidation of ethanol using an acidified solution of sodium dichromate(VI), $Na_2Cr_2O_7.2H_2O$. The two liquids were then purified by distillation. The quantities of reactants and the yields of products are given in the following table.

Mass of sodium dichromate(VI)	Volume of ethanol	Name of product	Yield of product
8.94 g	6.9 cm^3	Ethanal	1.62 g
8.94 g	2.3 cm^3	Ethanoic acid	1.73 g

The equations for the reactions involved in the above preparations are:

$$3C_2H_5OH + Cr_2O_7^{2-} + 8H^+ \rightarrow 3CH_3CHO + 2Cr^{3+} + 7H_2O$$
$$3C_2H_5OH + 2Cr_2O_7^{2-} + 16H^+ \rightarrow 3CH_3COOH + 4Cr^{3+} + 11H_2O$$

Calculate the percentage yields of (a) ethanal and (b) ethanoic acid. (The density of ethanol is 0.8 g/cm^3.) *[LCH]*

ANSWERS TO NUMERICAL PROBLEMS – TEXTBOOK

CHAPTER 4 THE PERIODIC TABLE

4.1 (l) 69.8

4.5 (iii) 63.6

CHAPTER 9 THE MOLE CONCEPT

9.1 32 g

9.2 (a) 2 g (b) 28 g (c) 32 g (d) 71 g (e) 38 g (f) 160 g

9.3 (a) 98 (b) 44 (c) 128 (d) 58.5 (e) 267 (f) 64 (g) 123.5 (h) 106 (i) 40 (j) 62 (k) 286 (l) 310 (m) 278 (n) 342

9.4 (a) 3 (b) 2.5 (c) 0.125 (d) 4 (e) 0.15

9.5 (a) 6 (b) 0.025 (c) 5 (d) 2 (e) 0.025 (f) 0.1

9.6 (a) 3.24×10^{23} (b) 7.44×10^{23} (c) 3.72×10^{24} (d) 1.2×10^{26} (e) 3×10^{19}

9.7 (a) 1.2×10^{24} (b) 9.33×10^{23} (c) 2.4×10^{24} (d) 3.6×10^{22} (e) 6×10^{21}

9.8 (a) 8.33 (b) 1.33×10^{3} (c) 4.33×10^{-12} (d) 1.63×10^{4} (e) 2.17×10^{-18}

CHAPTER 10 PROPERTIES OF GASES

10.1 628.92 cm^3

10.2 1.8×10^5 Pa

10.3 4.55 L

10.4 9819 L

10.6 (a) 0.76 g/L (b) 1.25 g/L (c) 3.17 g/L (d) 0.71 g/L (e) 1.43 g/L (f)2.14g/L

10.7 112

10.8 (d) 100.63

10.10 77.4 L

10.11 106

10.12 120

CHAPTER 11 STOICHIOMETRY

11.1 (a) Mg = 60%, O = 40%
(b) Al = 52.94%, O = 47.06%
(c) Mg = 20%, S = 26.67%, O = 53.33%
(d) Ca = 40%, C = 12%, O = 48%
(e) Na = 39.32%, Cl = 60.68%
(f) N = 21.21%, H = 6.06%, S = 24.24%, O = 48.48%
(g) C = 26.67%, H = 2.22%, O = 71.11%
(h) C = 40%, H = 6.67%, O = 53.33%
(i) Pb = 62.54%, N = 8.46%, O = 29%
(j) K = 39%, C = 12%, H = 1%, O = 48%

11.2 urea (46.67%)

11.3 CH, CH_2O, CH_2O

11.4 N_2H_4CO

11.5 (a) NO (b) NO_2 (c) CH_4 (d) CH_2 (e) $FeCl_3$ (f) CH (g) SO_3 (h) SO_2 (i) K_2CrO_4 (j) $K_2Cr_2O_7$ (k) $CuFeS_2$

11.6 $Na_2CO_3.10H_2O$

11.7 CaO

11.8 Al_2O_3

11.9 $MgSO_4.7H_2O$

11.10 64

11.11 58.04 g

11.12 250 g

11.13 12.78 g

11.14 0.56 L

11.15 6 g

11.16 6.92 L

CHAPTER 13 VOLUMETRIC ANALYSIS ACID-BASE

13.1 11.38% w/w

13.2 (a) 750 ppm (b) 60 ppm (c) 170 ppm (d) 3 ppm

13.3 (a) 40 g/L (b) 112 g/L (c) 12.25 g/L (d) 1.06 g/L (e) 0.315 g/L

13.4. (a) 0.025 M (b) 0.49 M (c) 3.01 M (d) 5 M (e) 0.13 M

13.5 (a) 2.5×10^{-4} (b) 0.045 (c) 4.35×10^{-3} (d) 5×10^{-4} (e) 4.5×10^{-3}

13.6 2.1 g

13.7 4 g

13.8 12.5 cm^3

13.9 555.56 cm^3

13.11 (a) 0.46 M (b) 18.4 g/L

13.12 80.4%

13.13 (a) 0.87 M (b) 52.2 g/L (c) 5.22%

13.14 (a) 0.847 M (b) 50.82 g/L (c) 5.08%

13.15 (a) 62.99% (b) x = 10

13.16 (a) 25.07% (b) x = 2

CHAPTER 15 VOLUMETRIC ANALYSIS OXIDATION-REDUCTION

15.1 (a) 0.0328 M (b) 5.18 g/L

15.2 (a) 0.0938 M (b) 26.08 g/L

15.3 (a) 0.0225 M (b) 3.56 g/L

15.4 (d) x = 7

15.5 x = 6

15.6 (ii) 0.08 moles/litre (iii) x = 6

15.7 (iii) (a) 0.174 g (b) 0.064 g (c) 51.48%

15.8 (iv) (a) 0.128 g (b) 0.047 g (c) 53.33%

15.9 (a) 0.1348 M (b) 33.43 g/L

15.10 (v) (a) 0.1442 M (b) 35.76 g/L

15.11 (vii) (a) 0.986 M (b) 73.46 g/L (c) 7.35%

15.12 (v) (a) 0.499 M (b) 37.18 g/L (c) 3.72%

CHAPTER 16 RATES OF REACTIONS

16.3 (ii) 210 secs

(iii) $\frac{3}{4}$ or 0.75 or 75%

16.5 (ii) 56 cm^3 (0.056 L)

(iv) Any value between 9.5 - 11.0 cm^3/min acceptable.

CHAPTER 17 EQUILIBRIUM

17.6 4.12

17.7 0.0156

17.8 (ii) 16 (iv) 4

CHAPTER 18 pH AND INDICATORS

18.2 (a) 7 (b) 2.40 (c) 7.58 (d) 2.28 (e) 10.06

18.3 (a) 2.24×10^{-6} moles/L (b) 7.08×10^{-12} moles/L
(c) 0.08 moles/L (d) 1×10^{-7} moles/L
(e) 2.24×10^{-10} moles/L.

18.4 (a) 1 (b) 1.70 (c) 0.40 (d) 0.52 (e) 0.32

18.5 (a) 13.98 (b) 14.70 (c) 12.70
(d) 10.63 (e) 12.30

18.6 (ii) 2.87

18.7 2.28

18.8 1.34×10^{-3} moles/L

18.9 $[H^+] = 4.27 \times 10^{-3}$, pH = 3.02

18.10 1.89

18.11 11.88

18.13 (i) 25 cm^3

18.17 (i) 27 cm^3

CHAPTER 19 ENVIRONMENTAL CHEMISTRY - WATER

19.3 625 p.p.m.

19.4 (vi) Total hardness = 190 p.p.m.
Temporary hardness = 130 p.p.m.

19.6 Suspended solids = 2080 p.p.m.
Dissolved solids = 3400 p.p.m.

19.7 14.08 p.p.m.

19.9 11.58 p.p.m.

19.10 7.84 p.p.m., 2.56 p.p.m.,
BOD = 105.6 p.p.m.

CHAPTER 21 FUELS

21.12 ΔH = -55.216 kJ mol^{-1}

21.13 -57.12 kJ mol^{-1}

21.14 6.81 °C

21.15 ΔH = -84.6 kJ mol^{-1}

21.16 ΔH = +27.9 kJ mol^{-1}

21.17 (i) carbon monoxide: ΔH = -108 kJ mol^{-1}
(ii) nitrogen monoxide: ΔH = 90 kJ mol^{-1}

CHAPTER 24 STOICHIOMETRY II

24.1 74.11%

24.2 49%

24.3 (ii) 37.5%

24.4 (ii) 79.5%

24.5 311 g

24.6 (a) 40.91% (b) 72.08%

A

B

C